Design Institute for Emergency Relief Systems (DIERS) 2015

Topical Conference at the 2015 AIChE Spring Meeting and 11th Global Congress on Process Safety

Austin, Texas, USA
26-30 April 2015

ISBN: 978-1-5108-0689-4

Printed from e-media with permission by:

Curran Associates, Inc.
57 Morehouse Lane
Red Hook, NY 12571

CURRAN ASSOCIATES INC.
proceedings
.com

Some format issues inherent in the e-media version may also appear in this print version.

Printed by Curran Associates, Inc. (2015)

For permission requests, please contact AIChE
at the address below.

AIChE
3 Park Avenue
New York, NY 10016-5991

Phone: (203) 702-7660
Fax: (203) 775-5177

www.aiche.org

Additional copies of this publication are available from:

Curran Associates, Inc.
57 Morehouse Lane
Red Hook, NY 12571 USA
Phone: 845-758-0400
Fax: 845-758-2634
Email: curran@proceedings.com
Web: www.proceedings.com

TABLE OF CONTENTS

409505 Chemical Engineering: Is This the Golden Age?

Abstract:

Schlanger will offer his insights on the current opportunities in the chemical process industries in his talk entitled "Chemical Engineering: Is This the Golden Age?" on Monday morning, April 27. At the conclusion of his lecture, Schlanger will receive the AGILE Award in recognition of his leadership at CEVA Group, a $9 billion logistics supplier to industry worldwide; LyondellBasell, a major player in the plastics, chemicals, and refining industries; Arco Chemical; and at the other companies where he has held leadership positions.

Overpressure Protection of a Pressure Vessel by System Design through the Application of ASME VIII UG-140 in lieu of a Relief device by an Appropriate Choice of MAWP and/or by Safety Instrumented System

(2015 AIChE Spring Meeting and 11th Global Congress on Process Safety, April 26-29, 2015, Austin, TX)

Dilip K. Das, P.E., Fellow of AIChE, Principal Engineer, Bayer CropScience LP, Kansas City, MO.

Abstract

This technical paper addresses the new provision of UG-140 of ASME VIII Division 1[1] and how it impacts on the traditional overpressure protection of a pressure vessel. It offers the basics of a Safety Instrumented System (SIS), and explains the circumstances under which the user can replace the traditional relief devices to protect a pressure vessel from overpressure either by an appropriate choice of Maximum Allowable Working Pressure (MAWP) so that the unmitigated pressures from all credible scenarios are less than MAWP of the vessel or through a safety instrumented system or a combination of safety instrumented system and traditional relief devices. Overpressure protection by system design in lieu of a relief device is deceptively simple and deceptively economical.

The paper also explains the dependence of ASME B31.3[2] (process piping code) on ASME VIII Division 1 (unfired pressure vessel code), and explains how the concatenation effect of the latter on the former implicitly mandates High Integrity Protection System (HIPS) -subset of the Safety Instrumentation System- to replace the traditional relief devices for the protection of a piping system.

Overpressure protection of a pressure vessel using ASME VIII Division 1, UG-140, has a place in industrial process safety. A safety instrumented system is worth considering when traditional relief devices are inoperable due to the hostile nature of the fluid or the required relief device is impractical. When traditional relief devices are replaced by a safety instrumented system, it may be the optimum choice if viewed through an optimization study with the traditional protection of a pressure vessel by relief devices as the base case, if the base case is a viable option. The cost of initial engineering, hardware and associated software cost, and life cycle cost of the safety instrumented system should be compared with the base case to justify the option of the safety instrumented system.

Historical Perspectives

Traditional last line of protection of a pressure vessel or piping from overpressure beyond code allowable limit involves the ***literal*** use of a pressure safety device or devices. There have been cases where the process environment is so hostile, corrosive, or energetic that a relief device is either inoperable or impractical to install. To provide an alternative method of overpressure protection, Code Case 2211 was initiated in 1996, and three years later, Code Case 2211-1 was approved as an alternative method of overpressure protection by system design in lieu of a relief device in ASME V111, Division 1 & 2.

The same principle is now used in ASME VIII, DIVISION 1 paragraph ***UG-140: OVERPRESSURE PROTECTION BY SYSTEM DESIGN.*** UG-140 paragraph (b)(5) refers to WRC-498 (2005)[3] which is based on Code Case 2211-1. The ANSI/API 521 is now open to the use of SIS[4].

Decision to Go for Overpressure Protection by System Design (OPBSD)

The decision to go for OPBSD involves consideration of some basic criteria:

- Physical condition of equipment and nature of fluid to examine the dependability of a relief device. [RV inlet requires freeze protection or clean purge, required relief device too large, etc.]
- Mandate of Jurisdictional Authority: Does the State, insurer or other authorities allow OPBSD?
- Willingness of the user to assume full responsibility of OPBSD and fabricate equipment accordingly with UG-140 stamp.
- The Responsibility-clause of UG-140 is a heavy burden on the part of the user, and must never be underestimated. "It is not sufficient to show that an instrumented system provides reliability equivalent to that of a conventional overpressure protection device."

Understanding the Difference between Conventional Protection and Protection

By System Design

- The conventional protection of a pressure vessel from overpressure requires relief devices.
- Protection by system design avoids relief devices or uses a combination of relief devices and safety instrumented system to protect a pressure vessel from overpressure.
- Overpressure protection by system design is *deceptively simple and deceptively economical*. It (UG-140) opens the door for smart engineering, and it has a place in industry.
- The major motivation behind UG-140 is the inherent safety.

- But if one uses UG-140 in applications where Basic Process Control System (BPCS) with a relief device would work, one may be wasting company funds not only in initial investment, but also in life-cycle cost.

Necessary Actions after Deciding to Go for OPBSD

- Formation of a multidisciplinary team with experience in, FMECA (Failure Mode, Effect & Criticality Analysis), "What-if" or other equivalent methodology, and SIS (Safety Instrumented Systems)
- Collection of necessary documents
- Performing contingency analysis or identification of credible overpressure scenarios
- Application of HAZOP, FMECA, What-if, LOPA, HAZAN, or other equivalent methodology, and SIS
- Documentation of tabulated results of maximum operating pressure and coincident operating temperature from all credible scenarios
- Acceptance of responsibility by signing the analysis (all members)

The two flowcharts shown below outline the logic flow diagram in making the right decision. [The alpha numeric expressions inside the various flow chart symbols refer to the paragraph or subparagraph in the ASME VIII Division 1 code, UG 140]. **The flow charts are for a guideline only, and are not intended to cover all possible options. The reader should understand that an optimum design of an overpressure protection may be a hybrid of SIS and relief device(s), in particular when existing equipment with existing SIS is being adapted for new scenarios. The reader must use his/her engineering judgment to override the flow charts included in this article. The reader should remember that the use of SIS in lieu of a relief device must meet the UG 140 (b) 5 and should have SIL -3 or better. The user may not obey this recommendation, and in so doing, assumes full responsibility and the consequence of failure, which is a heavy responsibility that ASME transfers to the user. One can use a SIS with SIL <3 to reduce the frequency of a scenario, but a relief device must be sized to cover the relief load of that scenario. If overpressure protection due the same scenario is avoided by instrumentation then such instrumentation should conform to SIL-3. Similarly, when the bypass control valve around a safety relief valve is used to lower the diameter of header network system and flare load, the control valve and the associated instruments should conform to SIL-3. The reader must be aware that the design temperature of the vessel wall during a fire relief condition must be considered if overpressure protection through SIS alone is sought; this is not required outside UG 140 when a pressure vessel under a fire scenario is protected by a relief device thereby allowing the user to use thinner vessel. The reader must understand that for the same design conditions and material thickness, calculations done by ASME VIII Division 2 will result in higher MAWP than the resulting calculations done by ASME VIII Division 1. However, the engineering cost is more in the Division2 method. One of the** main doctrines of DIERS (Design Institute for Emergency Relief Systems) is that the emergency relief system design is not just a design issue: it is an optimization issue.

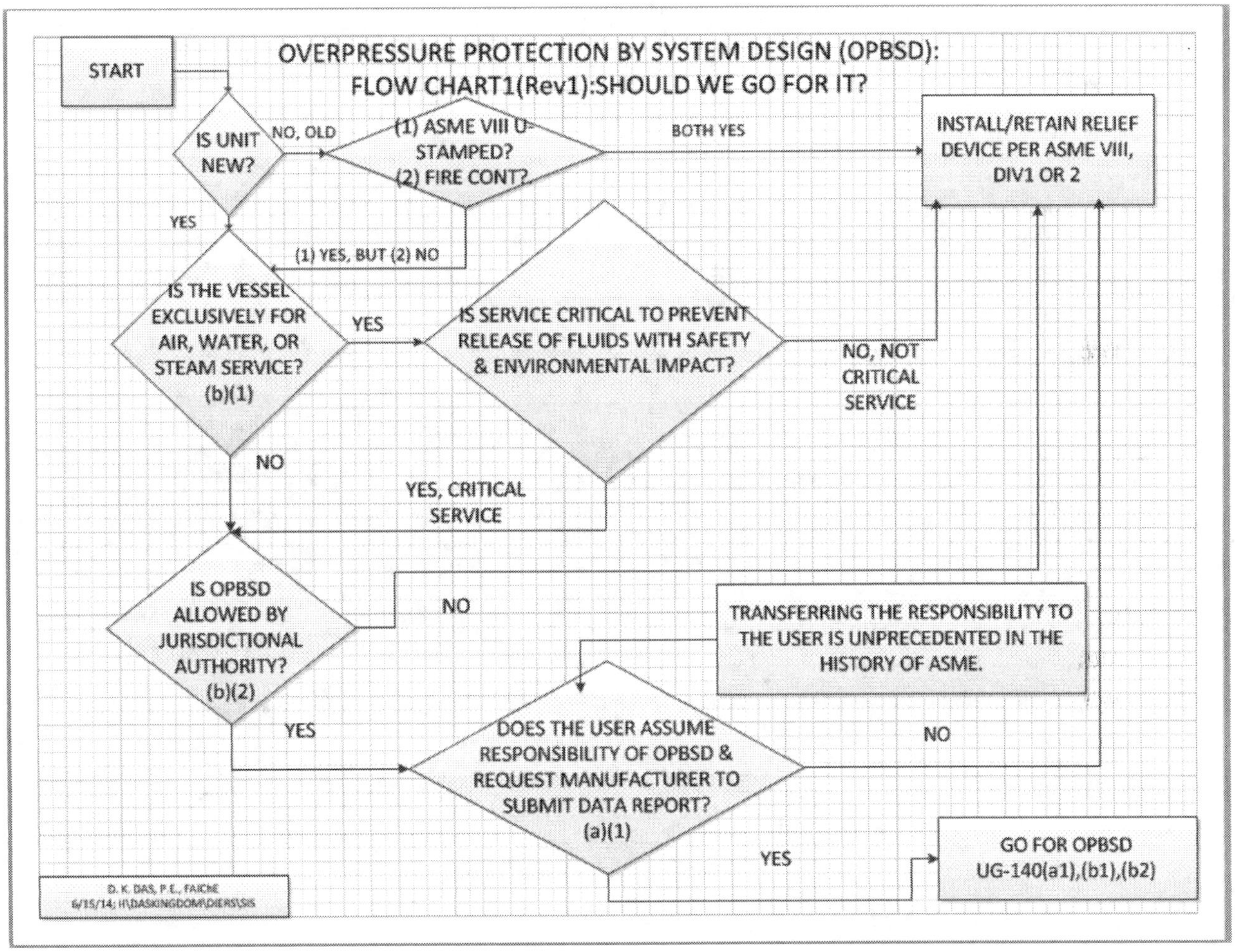

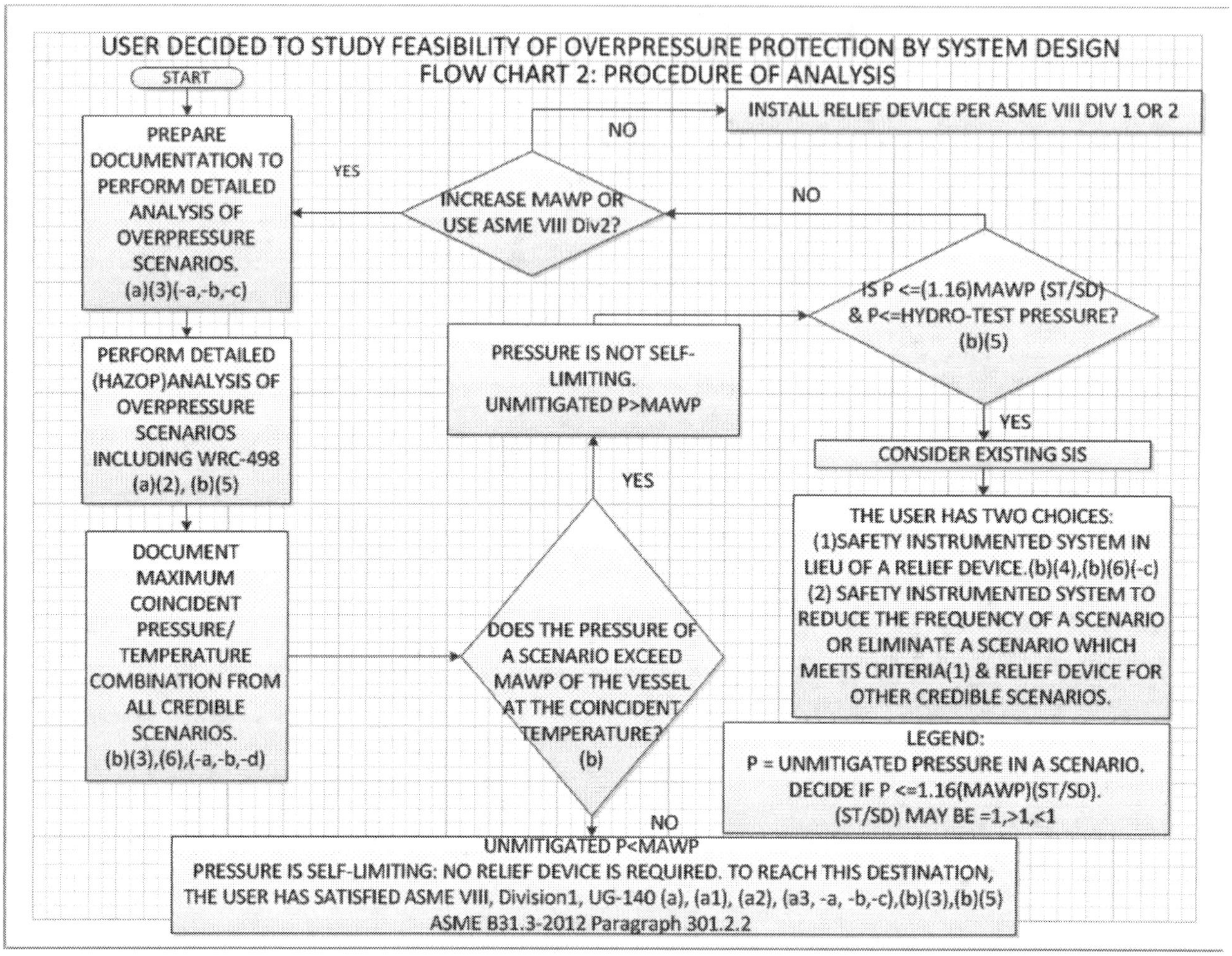

USER DECIDED TO STUDY FEASIBILITY OF OVERPRESSURE PROTECTION BY SYSTEM DESIGN
FLOW CHART 2: PROCEDURE OF ANALYSIS
START
PREPARE DOCUMENTATION TO PERFORM DETAILED ANALYSIS OF OVERPRESSURE SCENARIOS. (a)(3)(-a,-b,-c)
PERFORM DETAILED (HAZOP)ANALYSIS OF OVERPRESSURE SCENARIOS INCLUDING WRC-498 (a)(2), (b)(5)
DOCUMENT MAXIMUM COINCIDENT PRESSURE/ TEMPERATURE COMBINATION FROM ALL CREDIBLE SCENARIOS. (b)(3),(6),(-a,-b,-d)
DOES THE PRESSURE OF A SCENARIO EXCEED MAWP OF THE VESSEL AT THE COINCIDENT TEMPERATURE? (b)
YES
NO
PRESSURE IS NOT SELF-LIMITING. UNMITIGATED P>MAWP
IS P <=(1.16)MAWP (ST/SD) & P<=HYDRO-TEST PRESSURE? (b)(5)
NO
YES
INCREASE MAWP OR USE ASME VIII Div2?
YES
NO
INSTALL RELIEF DEVICE PER ASME VIII DIV 1 OR 2
CONSIDER EXISTING SIS
THE USER HAS TWO CHOICES: (1)SAFETY INSTRUMENTED SYSTEM IN LIEU OF A RELIEF DEVICE.(b)(4),(b)(6)(-c) (2) SAFETY INSTRUMENTED SYSTEM TO REDUCE THE FREQUENCY OF A SCENARIO OR ELIMINATE A SCENARIO WHICH MEETS CRITERIA(1) & RELIEF DEVICE FOR OTHER CREDIBLE SCENARIOS.
LEGEND: P = UNMITIGATED PRESSURE IN A SCENARIO. DECIDE IF P <=1.16(MAWP)(ST/SD). (ST/SD) MAY BE =1,>1,<1
UNMITIGATED P<MAWP
PRESSURE IS SELF-LIMITING: NO RELIEF DEVICE IS REQUIRED. TO REACH THIS DESTINATION, THE USER HAS SATISFIED ASME VIII, Division1, UG-140 (a), (a1), (a2), (a3, -a, -b,-c),(b)(3),(b)(5)
ASME B31.3-2012 Paragraph 301.2.2

Overpressure Protection of Piping System

- **A common misconception is that piping system is outside ASME VIII, and therefore, none of the rules of ASME VIII applies.**
- Piping system design is governed by ASME B31.3 which is a sibling of ASME VIII, Division 1
- Two paragraphs of ASME B31.3 refers to pressure relief devices:
  - Para. 301.2.2(a)
  - Para. 322.6.3 Pressure Relieving Devices

  "Pressure relieving devices...shall be in accordance with the Boiler Pressure Vessel (BPV) Code, Section VIII, Division 1. UG-125 (c), UG-126, UG-127, and UG-132 through UG-136 excluding UG-135(e) and UG-136(e). The terms ***design pressure*** **and** ***piping system*** shall be **substituted for** ***Maximum Allowable Working Pressure*** **and** ***vessel*** respectively."
- Since the piping code B31.3 refers to ASME VIII, Division 1 for guidance, UG-140 and its rules apply to piping system also through the effect of concatenation.

Safety Instrumented System

- A Safety Instrumented System (SIS) is an independent instrumented system with higher availability on demands compared with Basic Process Control System (BPCS) which is the standard instrumented system for a typical process control.
- The standards governing the SIS are developed jointly by the International Electrotechnical Committee (IEC 61511) and International Society of Automation (ISA) 84.00.01-2004 Parts (1-3)[5].
- The only difference between the above two standards is that the American standard has added a grandfather clause for existing SISs which transfer the responsibility for the SISs to the users[6].

The Components of a SIS loop

The components of a SIS loop are:

- Sensor
  - Devices or combination of devices that measure the process condition such as pressure, temperature, etc. Examples are transmitters, transducers, process switches, position switches and measuring probe. It measures and sends the measured information to logic solver.

- Logic Solver
    - Devices or combination of devices that perform logic function(s). Examples are electrical logic system, electronic logic system, and logic system for programmable and electronic system.

- Final Control Element (Actuator)
    - Devices or combination of devices that executes the physical action necessary to bring the condition to a safe state. Examples are valves, switch gear, motors including their auxiliary elements, for example, a solenoid valve.

Redundancy in SIS design

The need for redundancy, in particular diverse redundancy, to provide high availability in demand is evident in design of the human brain. Humans did not invent this: They simply discovered it. About sixty five years ago, a scientist named Karl Lashley experimented on the capability of rats to find food through a maze as a function of the total available brain cells. He removed a portion, gradually up to 90%, of the brain cells from the rats and subjected the rats to the same experiment. To his surprise, he observed that the rats could get to the food source with ninety percent of their brain cells removed. He concluded, wrongly, that most of the brain cells were unnecessary for the purpose. The true reason was discovered later: **Many different forms of same brain function neurons are redundant, and are embedded all over the brain.** This diverse redundancy is behind the high availability of brain function on demand in the event of brain damage. It is the same redundancy that improves the availability on demand of a SIS in the event of a failure of a component part.

Common mode failure of a component in SIS

There is no free lunch: There is a price to pay when one wants high availability on demand. The components of a SIS performing the same function should have difference in their inherent design to avoid the common mode failure and, thereby, improve the availability on demand of the system. For example, if two temperature sensors are used, one could be a thermocouple and the other could a RTD, rather than both being either thermocouples or RTDs.

The Required SIL Rating of a SIS to Substitute a Relief Device

- The High Integrity Pressure Protection Systems (HIPPS)[7] serving as the only line of defense as called for UG-140 must have Safety Integrity Level **(SIL) rating of 3 or *better***, as provided IEC 61508. Why?
    - Because, a relief valve that fails to operate at the set pressure may nevertheless operate at higher pressure, whereas the HIPPS is more likely to fail completely.
    - The failure-to-open-on-demand uncertainty, higher response time, dependence on process dynamics, requirement of multiple actions, coupled with the difference in the failure modes, results in the majority of user-setting of an SIL-3 target or better for the HIPPS to substitute the function of a relief device.

Conclusion

- Like the constitutional freedom of the USA, the freedom implicit in the clause of "User-Responsibility" in UG-140 of ASME VIII, Division 1 has brought the opportunity of being smart as well as the opportunity of being stupid.
- One of the main doctrines of the DIERS technology is that ERS (Emergency Relief System) design is NOT just a design issue: it is an optimization issue.
- The same code (ASME VIII Division 1) offers you the freedom in choosing whatever path you desire, but you assume the responsibility.
- The UG-140 of ASME VIII, Division 1 has a place in industry. Its wise use will bring virtuous business value in a corporation. Emergency relief system design is not just a design issue: it is an optimization issue.

Acknowledgement:

1) Mr. Cecil Cai, Bayer Technology Services, Baytown, TX, Personal Communication, June – July 2014, February 2015
2) Mr. Fisher, Harold, Chair, DIERS (CCPS), Personal Communications, July 2014
3) Mr. David Gaydos & Mr. Gene Gorski, ioMosaic Corporation, personal communication, July, 2014
4) Dr. Kumpinsky, Enio, Ashland Chemicals, personal communications, August 2014
5) Dr. Summers, A.E., SIS-TECH Solutions, Personal Communications, September 2014; January, 2015
6) Mr. Howell, Peter, Mark V, Inc., Editorial Review, February - March 2015

References:

1) ASME Boiler and Pressure Vessel Code VIII Division 1, 2013, The American Society of Mechanical Engineers, The Park Ave, New York, NY
2) ASME B31.3, 2012, ASME Code for Pressure Piping, The American Society of Mechanical Engineers, The Park Ave, New York, NY
3) Sims, J.R, and W. G. Yeich, Guidance on the Application of Code Case 2211-Overpressure Protection by System Design, WRC Bulletin 498, Welding Research Council, Inc., P.O. Box 1942, NY
4) ANSI/API Standard 521, January 2007, Annex E
5) International Electrotechnical Commission standards 61511(ISA ANSI/ISA 84.00.01-2004) parts 1, 2, 3, Research Triangle Park, NC (2004)
6) Summers, A.E., "Can you safely "grandfather" your SIS?", Chemical Processing, August 2005
7) International Association of Oil & Gas Production, Report 443, High Integrity Protection System, February 2015.

Further Readings:

1) Smith, Dustin, "Safety Instrumented Systems in Lieu of Pressure Relief Valves", DIERS Spring Meeting, 2014, Houston, TX.
2) Summers, A.E., "Safety Controls, Alarms, and Interlocks as IPLs", Process Safety Progress (Vol. 33, No. 2), June, 2014.
3) Summers, A.E., "Working Under Pressure", SIS TECH, AIChE 8th Global Congress on Process Safety, Houston, TX, 2012
4) Summers, A.E., Consider HIPS for Reactive Processes, Chemical Processing, March 2004, p 33
5) Summers, A. E., "Consider An Instrumented System For Overpressure Protection", *Chemical Engineering Progress*, (pp 65-68), November 2000.
[A zealous engineer aspiring to learn SIS is advised to read the scholarly, lucid, and angelic thoughts on the subject in the publications of Dr. Summers.]
6) Medoff, Michael & Rainer Faller, "Functional Safety –An IEC 61508 SIL 3", *exida.com L.L.C*, Sellersville, PA, 2010
7) LePree, J., Today's SIS: Integrated Yet Independent, Chemical Engineering, August 2009
8) Melhem, Georges, Maximize the Use of Your Existing Flare Structures, an ioMosaic White Paper, 2006.
9) Mitchell, K. J., Understanding the difference between continuous Mode and Demand Mode Safety Instrumented Function, Kenexis Consulting Corporation, 2005
10) Williams, J. Patrick, "Reliability for Safety Instrumented Systems", KBC Advanced Technologies, CEP, September 2004
11) Urbanik, S. A., Evaluating Relief Valve Reliability When Extending the Test and Maintenance Interval, Process Safety Progress, September 2004

12) “Guidelines for Process Equipment Reliability Data With Data Tables”, Center for Chemical Process Safety, New York, NY, 1989
13) Lawley, H. G and Trevor Kletz, High Pressure Trip Systems For Vessel Protection, Chemical Engineering, May 12, 1975.

Understanding Gas Blowby Scenario Calculations

Nancy Faulk
Siemens Energy, Inc.
4615 Southwest Fwy, Suite 900
Houston, TX 77027
nancy.faulk@siemens.com

Abdul Aldeeb
Siemens Energy, Inc.
4615 Southwest Fwy, Suite 900
Houston, TX 77027
abdul.aldeeb@siemens.com

Prepared for Presentation at
American Institute of Chemical Engineers
2015 Spring Meeting
11th Global Congress on Process Safety
Austin, Texas
April 27-29, 2015

UNPUBLISHED

Understanding Gas Blowby Scenario Calculations

Nancy Faulk
Siemens Energy, Inc.
4615 Southwest Fwy, Suite 900
Houston, TX 77027
nancy.faulk@siemens.com

Abdul Aldeeb
Siemens Energy, Inc.
abdul.aldeeb@siemens.com

Keywords: Gas blowby, control valve failure, vapor breakthrough

Abstract

Gas blowby overpressure scenarios are routinely identified when analyzing vessels with liquid-level inlet control valves. Blowby can occur when this type of inlet control valve fails open, emptying the upstream vessel liquid inventory into the downstream vessel of interest, with subsequent vapor or two-phase flow into the downstream vessel. While it is easy to determine if this scenario leads to a credible overpressure scenario, based on the operating pressure in the upstream vessel compared to the downstream vessel MAWP, the ensuing calculations are often not well understood, even by engineers regularly involved in pressure relief analysis. This paper provides guidance on determining the relief phase (liquid, vapor or two-phase), calculating the required relief rate, and taking credits where appropriate.

1. Introduction

This paper is intended to describe the engineering basis for the gas blowby (also known as vapor breakthrough) overpressure scenario. This complex variation of inlet control valve failure may occur in cases of a pressure vessel operating at a high pressure, whose liquid bottoms are on level control, discharging into a lower-pressure system, as seen in Figure 1. The inlet control valve failure scenario is an applicable overpressure contingency when the upstream normal operating pressure exceeds the downstream systems' maximum allowable working pressure (MAWP). In addition, gas blowby must be considered. Blowby can occur when this type of inlet control valve fails open, emptying the upstream vessel liquid inventory into the downstream vessel of interest, with subsequent vapor or two-phase flow into the downstream vessel. When this occurs, the relief devices on the low-pressure system should be designed to handle the vapor flow through the liquid control valve [1]. This paper provides guidance on determining the relief phase (liquid, vapor, or two-phase), calculating the required relief rate, and taking credits where appropriate to calculate a realistic required relief rates.

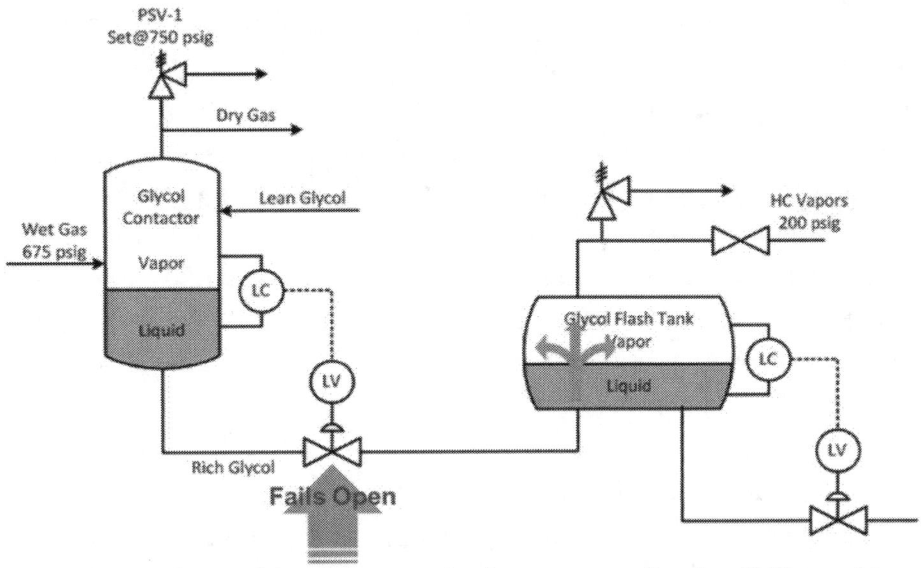

Figure 1. Example gas blowby scenario due to control valve failing wide-open

2. Baseline Assumptions

2.1 Process Model Definition

The process system is defined based on the typical arrangement of an upstream separator with a downstream separator operating at a lower pressure. A control valve connects the liquid outlet of the upstream separator to the downstream separator inlet. System is presented in Figure 1.

Following assumptions are made to further define the system in question for the purpose of this paper's modeling:

- There are no other liquid inlets to the vessels. Vapor inlets are possible but are not included in the methodology below.
- There are no pumps in the line connecting the separators, or, if there are, that they have no impact on the maximum flow between the separators.
- There are no further operations to be considered on the downstream equipment after determination of relief rate and properties. For example, if the downstream system is a column, after the relief rate and relief stream are determined, simulation of the system with the abnormal flow may be required.

2.2 Development of Relief Conditions and Pressurization Rates

This methodology is intended to model what is fundamentally a dynamic process using steady-state assumptions. In order to arrive at a more accurate answer, many details have been fine-tuned. However, there are some cases where simplification is required to avoid significant additional computation, as following:

- It is assumed that the quantity of vapor in the upstream system is large enough that the depressuring of this vapor inventory does not occur on the same time scale as the liquid transfer and overfill determination.
- It is assumed that there is sufficient flashing in the liquid entering the downstream separator at the normal operating pressure and at relief pressure that overpressure of the downstream system happens quickly enough to consider the downstream system at relief pressure when calculating the failure open rate of the control valve and the liquid volume transfer (higher pressure leads to more liquid, as lighter components are kept in the liquid phase, resulting in both a lower liquid density and a more conservative volume and level height). This allows the control valve flow rates to be calculated using only a single downstream pressure (i.e., relief pressure). If little to no flashing occurs, this assumption should be revisited.
- When flashing the upstream liquid to the downstream relief pressure, there may be differences in composition of the remaining liquid portion of the stream, resulting in different densities between the liquid entering the downstream vessel at relief pressure and the normal downstream liquid density. Over time it is expected that the liquid density at the downstream outlet would approach the density of the downstream inlet liquid at relief pressure. However, as it is more conservative to use the larger density at normal operating pressure (i.e., it results in a greater difference and therefore a larger maximum downstream volume), the default assumption should be to use the normal density. If the volume of the vessel is small, this assumption may be revisited.

3. Required Data

3.1 Upstream Vessel

The upstream vessel and its contents must be defined so that the liquid volume and composition in the upstream vessel are known. This requires knowledge of:

- Vessel orientation – vertical, horizontal or spherical
- Vessel dimensions
- Type of head – elliptical, conical, round, or flat
- Liquid level; if unknown, a conservative estimate may be made based on vessel orientation, heads, and dimensions
- Elevation of outlet nozzle connected to piping which flows to the control valve – measured relative to the bottom tangent line if a vertical vessel, and relative to the bottom of the vessel if horizontal
- Additional liquid volume; only used if this liquid volume (in boot or piping) is relevant to the volume draining into the downstream vessel.

3.2 Upstream Liquid and Vapor Streams

It should be possible to define composition of the vapor outlet stream and liquid outlet stream independently, or to define the combined feed stream(s) to the vessel and perform a T-P flash to separate into vapor and liquid. Additional information needed includes normal operating temperature and pressure, normal liquid flowrate out of the vessel (which at steady-state equals the liquid flowrate into the downstream vessel).

3.3 Downstream Vessel

The downstream vessel and its contents must be defined similarly to the upstream vessel, in order to determine the available volume for overfilling and to determine the relief flow phase regime:

- Vessel orientation – vertical, horizontal or spherical
- Vessel dimensions
- Type of head – elliptical, conical, round, or flat
- Liquid level; if unknown, a conservative estimate may be made based on vessel orientation, heads, and dimensions
- Elevation of inlet nozzle connected to piping which flows from the control valve – measured relative to the bottom tangent line if a vertical vessel, and relative to the bottom of the vessel if horizontal
- Pressure relief device set pressure
- Indication of whether a high level alarm is present, independent of the potential failure mechanisms. If there is intent to take credit for operator intervention in the event of overfilling, the time for operator intervention is also needed.

3.4 Downstream Liquid and Vapor Streams

It should be possible to define composition of the vapor outlet stream and liquid outlet stream independently, or to define the combined feed stream(s) to the vessel and perform a T-P flash to separate into vapor and liquid. Additionally, the normal operating temperature and pressure, and normal liquid and vapor flowrate out of the vessel are required.

3.5 Control Valve and Bypass Information

The control valve size, manufacturer and model specifications are required in order to obtain the appropriate valve coefficients. Either the control valve manufacturer's or the ISA sizing equations can be used [2]. The bypass valve calculations can be performed by defining the valve as a piping element, if the size and type of valve are known; alternatively, the bypass valve coefficients can be obtained from the manufacturer.

4. Calculations

After all data necessary for calculations has been determined, the original vessel inventories and new liquid levels can be determined. The new liquid level is used as the basis for determining if the relief stream is all vapor, all liquid, or two-phase..

4.1 Upstream Vessel Calculations

4.1.1 Liquid Inventories

Calculate the volume of liquid present in the upstream vessel at the normal liquid level; this represents the total liquid inventory. Next, determine the volume of liquid in the upstream vessel, at the elevation of the outlet nozzle, which represents the liquid inventory that will remain in the upstream vessel. The upstream carryover volume is the difference between these two inventories, plus any additional liquid volume as noted in Section 3.1 above.

4.1.2 Continuing Liquid Inflow

If the system was at steady state prior to the upset, then the liquid inflow to the upstream vessel should be equal to the normal liquid outflow.

4.1.3 Initial Liquid Flow through Control Valve

At the time the control valve fails wide open and the flow across the control valve is considered to be all liquid. Initial liquid flow rate through the wide open control valve and/or bypass (during de-inventorying) can be calculated assuming maximum C_V and other valve coefficients in the appropriate liquid equation.

4.1.4 De-Inventorying of Upstream Vessel

Determine the fill rate from the failure open of the control valve and the de-inventorying of the upstream liquid. Since there is still liquid flow to the upstream vessel, this is based on the incremental flow out of the upstream vessel.

The time to deplete the liquid volume from the upstream vessel can be calculated and this should be based on the upstream carryover volume, not the total liquid inventory. Note that there is still liquid flow into the upstream vessel after it drains, which continues to flow into the downstream vessel.

4.2 Downstream Vessel Calculations and Relief Flow Regime

4.2.1 Liquid Volumes

The volume of liquid initially in the downstream vessel and volume of liquid in the downstream vessel up to the inlet nozzle should calculated using the normal liquid level. The total volume of the downstream vessel is calculated assuming liquid full. Add the additional volume available for retention capacity listed given in the downstream vessel definition (Section 3.3 above). If there is a high level alarm on the downstream vessel, calculate the volume of liquid at the high level alarm set point.

Calculate the volume of liquid that the carryover volume will occupy at the downstream relief pressure. This includes the effects of flashing, which can affect the composition and therefore the density. Finally, calculate the new downstream volume, after de-inventorying the upstream vessel into the downstream vessel, by adding the carryover volume to the normal liquid volume.

4.2.2 New Liquid Level: Lower than Inlet Nozzle

Figure 2 depicts depleting the upstream system liquid inventory, based on normal liquid level, which does not result in overfilling the downstream system. If the downstream system new liquid level is below the downstream system inlet nozzle, all vapor release should be evaluated initially.

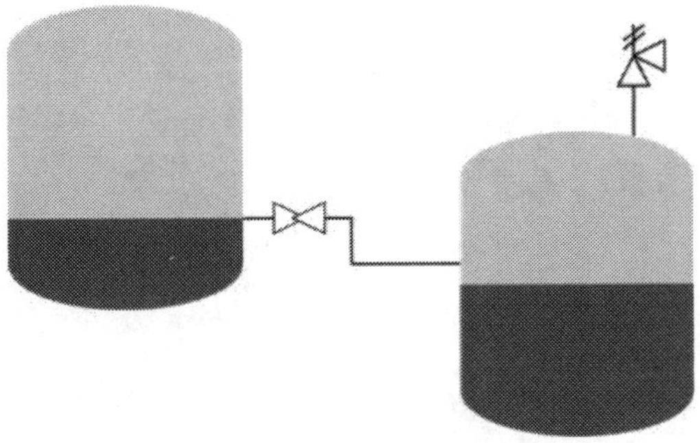

Figure 2. New liquid level (downstream vessel) is below inlet nozzle

4.2.3 New Liquid Level: Above Inlet Nozzle; Vessel Not Overfilled

Figure 3 depicts depleting the upstream system liquid inventory, based on normal liquid level, which does not result in overfilling the downstream system. If the downstream system new liquid level is above the downstream system inlet nozzle, vapor-liquid disengagement analysis is required.

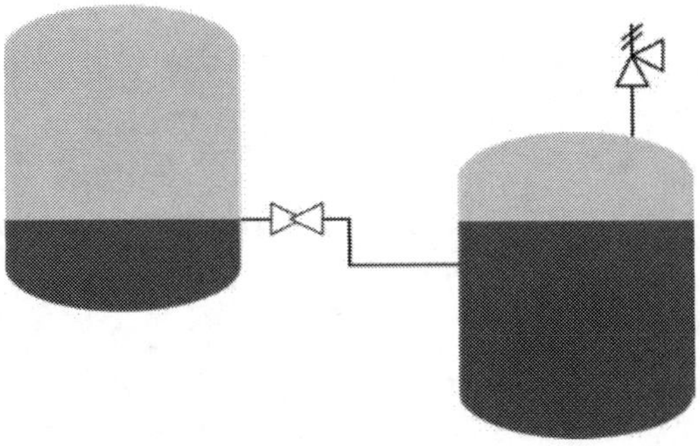

Figure 3. New liquid level (downstream vessel) is above inlet nozzle

4.2.4 New Liquid Level: Downstream Vessel Overfilled

Figure 4 depicts depleting the upstream system liquid inventory, based on normal liquid level, resulting in overfilling the downstream system. In such a case, liquid displacement analysis based on the equivalent volumetric inlet flow rate should be performed to determine the liquid relief rate. If overfilling is possible, calculate time to fill the downstream vessel up completely from the elevation of the high liquid level alarm to the vessel top, accounting for downstream vessel liquid outflow; this time can be used in case credit for operator intervention can be applied.

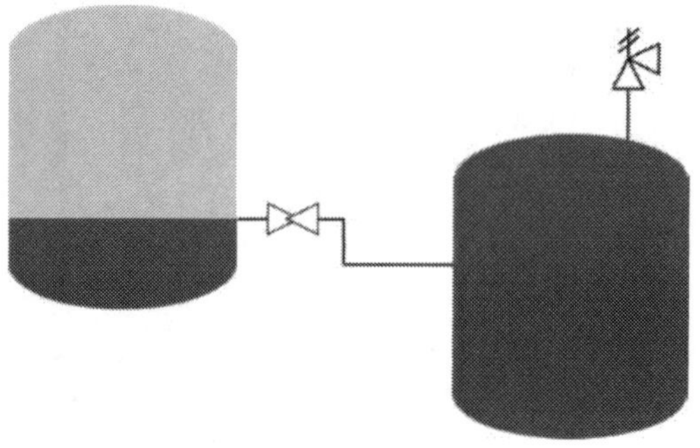

Figure 4. Downstream vessel overfills

4.3 Vapor Relief

The vapor flow rate through wide open control valve and/or bypass is calculated using the maximum C_V and 100% flow for both valves and the appropriate vapor equation and using the upstream vapor composition, pressure, temperature and density. This is the required mass relief rate. The relief stream properties are the vapor properties at relief pressure of the vapor entering the downstream vessel. Applying credit for the overhead vapor outlet flow path or for continued liquid flow across the control valve is discussed in section 4.5 and 4.6 of the paper, respectively.

4.4 Liquid Relief

Note that in this scenario, the vessel overfills and the volume of vapor entering the system transfers momentum to the liquid sitting in the top of the vessel. A volumetric flow rate of liquid equal to the volumetric inflow of vapor is taken at the relief stream.

Calculate the vapor flow rate through wide open control valve and/or bypass, using the maximum C_V and 100% flow for both valves and the appropriate vapor equation. Use the upstream vapor composition, pressure, temperature, and density.

Perform liquid displacement analysis; convert the vapor flowrate (W_{Vapor}) into a liquid displacement flowrate ($W_{Liq\ disp}$), based on the liquid density of the upstream liquid and the vapor density of the upstream vapor:

$$W_{Liq\ disp} = W_{Vapor}\left(\frac{\rho_{Liq\ ,relief}}{\rho_{Vapor\ ,relief}}\right) \quad \text{[Eq. 1]}$$

If the flow across the control valve is being modeled as two phase (e.g., if credit for continued liquid flow across the control valve is being taken) then the vapor density should be based on the vapor portion of the combined stream after being flashed adiabatically (PH flash) to the relief pressure.

The required mass relief rate is $W_{Liq\ disp}$. The relief stream properties are the liquid properties at relief pressure of the liquid resulting from taking the upstream liquid, flashing adiabatically to the relief pressure, and separating to vapor and liquid streams. Note that the relief rate calculated in this way almost invariably results in inadequate relief capacity. If taking credit for the overhead vapor outlet flow path, go to step 4.5. If taking credit for continued liquid flow across the control valve, go to step 4.6.

4.5 Vapor Outflow Credit

Determine if credit can be taken for the normal actual volumetric flow rate out of the normal vapor outlet on the downstream equipment system. This credit can only be taken provided it can be demonstrated that the outlet vapor flow path will remain available during the upset condition. If this credit is taken for liquid displacement or two phase relief, then satisfying the above requirement should also consider the impact of liquid on downstream equipment (e.g. liquid entering compressor suction) and should consider if the overhead vapor outlet line can sustain the weight load of the liquid flow (as required in API Std 521 [1]).

If this credit can be taken, subtract the normal volumetric outlet flow rate from the volumetric flow rate calculated. Any more complex analysis would require a hydraulic analysis.

4.6 Continued Liquid Inflow Credit

This credit is based on the continued liquid inflow into the upstream vessel; this flow will naturally progress through the upstream vessel liquid outlet and flow across the control valve during the gas blow by event. The credit can significantly reduce the required relief rate as it reduces the area through which the vapor can travel across the control valve. It is important to note that some practioners do not recommend taking this credit, as the calculation is not conservative and the behavior of the fluid at the entrance of the control valve is not well known. The potential for pockets of vapor or an annular volume of vapor to pass across the control valve at certain points in time could result in peak relief rates approaching or equal to the all-vapor case. However, if facility guidelines allow credit to be taken for continued liquid flow through the control valve, proceed with the steps below.

Use the upstream vessel normal liquid outlet flow rate to calculate the control valve C_V occupied by the liquid phase. Use the appropriate liquid equation, e.g. ISA, and the normal upstream liquid T, P and composition. Manually vary the C_V until the calculated rate matches the normal upstream liquid outlet flow rate. Use this value as the normal C_V. Re-calculate the maximum vapor flow rate across the control valve, taking credit for the continued liquid flow by using the reduced C_V, equal to the maximum C_V minus the normal C_V. Use the appropriate vapor equation and the normal upstream liquid T, P and composition.

Once the credit for continued liquid flow has been taken, and the mass rates of the vapor and liquid flow rates across the control valve are known, justification for taking the credit should be provided by determining the flow regime. The justification calculation has no impact on the relative mass flow of vapor or liquid across the control valve. Note that if the two phase fluid has been assumed to be a pseudo-single phase homogeneous fluid, then no further steps need to be taken; the justification is the assumption itself. Otherwise, determine the two-phase flow regime based on pipe orientation, using upstream flowrates and cross-sectional area of the piping. If the control valve is horizontally oriented, use GPSA Figure 17-15; for vertically oriented valves, use GPSA Figure 17-16 [3]. Justification for taking this credit may be made for the homogeneous, bubbly, stratified, wave or non-boiling flow regimes. No credit should be taken for mist flow, annular mist flow, annular flow, churn flow, or slug flow.

4.7 Evaluation of Onset of Two-Phase Relief

For the potential two-phase scenario, liquid vapor disengagement analysis must be performed, based on guidance provided by DIERS Project Manual [4].

Calculate the vapor flow rate, $W_{CV,V}$, through wide open control valve and/or bypass, using the appropriate vapor equation with the maximum Cv and 100% flow for both valves. Use the upstream vapor composition, pressure, temperature and density.

Characterize the fluid behavior as homogenous (foaming tendency), bubbly (liquid viscosity ≥100 cP), or churn-turbulent (liquid viscosity <100 cP). Next, calculate the vessel average void fraction $\bar{\alpha}_{VES}$:

$$\bar{\alpha}_{VES} = \frac{V_{tot} - V_{liq}}{V_{tot}} \quad \text{[Eq. 2]}$$

Where V_{tot} is the total downstream vessel volume and V_{liq} is the liquid level after the upstream vessel has de-inventoried. Select the value for C_0 to be used:

Table 1. C_0 based on Void Fraction

Model	Best Estimate	Max $\bar{\alpha}_{VES}$	Conservative	Max $\bar{\alpha}_{VES}$
Churn Turbulent	1.5	0.667	1	~1
Bubbly	1.2	0.833	1.01	~1
Homogeneous	1.01	~1	1.01	~1

4.7.1 Bubbly and Churn-Turbulent Fluid Behavior

Calculate the superficial vapor velocity, $j_{g\infty}$, at the liquid surface using the vapor mass flow rate F equal to $W_{CV,V}$::

$$j_{g\infty} = \frac{F}{\rho_g A_{CR}} \quad \text{[Eq. 3]}$$

Where ρ_g is the gas density and A_{CR} is the cross-sectional area of the downstream vessel. Next, calculate the characteristic bubble rise velocity, U_∞:

$$U_\infty = K \frac{\{\sigma g(\rho_f - \rho_g)\}^{1/4}}{\sqrt{\rho_f}} \quad \text{[Eq. 4]}$$

Where K = 1.53 for churn-turbulent and 1.18 for bubbly. The constant g is the gravitational acceleration constant, σ is the interfacial tension, and ρ_f is the liquid density. Note also that in USC units an additional conversion factor g_c needs to be considered in the numerator. If the units for surface tension are in dynes/cm, then to convert to lb_m/s^2, multiply by 2.204623×10^{-3}. This conversion includes the g_c term.

Now calculate the dimensionless superficial vapor velocity based on the vapor mass flow rate, Ψ_F:

$$\Psi_F = \frac{j_{g\infty}}{U_\infty} \quad \text{[Eq. 5]}$$

Depending on the model used, calculate the average void fraction of the swelled fluid (Forrest method), for churn turbulent (Equation 6) or bubbly (Equation 7, not explicit for α):

$$\bar{\alpha} = \frac{\Psi_F}{2+C_0\Psi_F} \quad \text{[Eq. 6]}$$

$$\Psi_F = \frac{\bar{\alpha}(1-\bar{\alpha})^2}{(1-\bar{\alpha}^3)(1-C_0\bar{\alpha})} \quad \text{[Eq. 7]}$$

Determine if vapor or two-phase flow will result using the following criteria. For $\bar{\alpha} \leq \bar{\alpha}_{VES}$ or $\Psi_F < \Psi$; all vapor flow results; use the vapor flow rate as the required relief rate, and no further calculations are required. In the event that $\bar{\alpha} > \bar{\alpha}_{VES}$ or $\Psi_F \geq \Psi$, two phase flow results. If two phase is the result, reset the value of the average void fraction of the fluid to the vessel average void fraction $\bar{\alpha} = \bar{\alpha}_{VES}$. This also consequently means that $\Psi_F = \Psi$.

Calculate the void fraction at the liquid surface, for churn turbulent (Equation 8) or bubbly (Equation 9):

$$\alpha_m = \frac{2\bar{\alpha}}{1+C_0\bar{\alpha}} \quad \text{[Eq. 8]}$$

$$\alpha_m = \bar{\alpha} \quad \text{[Eq. 9]}$$

Calculate the mass fraction at the liquid surface:

$$X_m = \frac{\alpha_m \rho_g}{\alpha_m \rho_g + (1-\alpha_m)\rho_f} \quad \text{[Eq. 10]}$$

Assume a starting value for the mass fraction of the fluid entering the relief device X_0 by setting it equal to the mass fraction at the swelled liquid surface, X_m. Create a new relief stream with new relief fluid properties based on the combination of X_0 mass percent vapor and (1- X_0) mass percent liquid. This can be done using the mass flow rates (e.g., taking the mass flow rate of vapor and mass flow rate of liquid and mixing them together in a mixed stream operation in PS PPM). Note that these are combined under non-equilibrium conditions. However, given the limitations of the available tools, for the purposes of calculating the mass flux, the conditions may be reconsidered at equilibrium conditions.

Calculate the estimated mass flux of the relief device using an appropriate two-phase sizing method. This can be omega, 2pt or 3pt HEM, or rigorous 2-phase sizing. The mass flux can be simplified by using the relief pressure and the choke pressure (i.e. ignoring the effects of inlet and outlet piping). If the API orifice area is used for the valve area, the mass flux should be calculated with the coefficient of discharge (and any other sizing coefficients) factored in.

Some iteration may be required; consult the DIERS Project Manual for further information [4].

4.7.2 Homogeneous Fluid Behavior

If the fluid is homogeneous, the fluid is assumed to be uniform at all points. Therefore, the relief fluid phase will be two-phase, with a vapor quality based on the void fraction. The equivalent mass fraction can be calculated using a similar approach as in Equation 10 above. No further calculations are required.

5. Conclusion

Gas blowby overpressure scenarios are routinely identified when analyzing vessels with liquid-level inlet control valves. Blowby can occur when this type of inlet control valve fails open, emptying the upstream vessel liquid inventory into the downstream vessel of interest, with subsequent vapor or two-phase flow into the downstream vessel. The relief flow regime depends on the liquid level after de-inventorying. The ensuing calculations are often not well understood, and can be quite complex in the event of two-phase flow. This paper provides guidance on determining the relief phase (liquid, vapor or two-phase), calculating the required relief rate, and taking credits where appropriate.

6. References

[1] American Petroleum Institute, *API Standard 521, Pressure-relieving and Depressuring Systems, 6th* Edition, 2014.

[2] International Society of Automation, *ISA Standard ISA-75.01, Industrial Process-Control Valves – Part 2-1: Flow Capacity – Sizing Equations for Fluid Flow Under installed Conditions,* 2012.

[3] Gas Processors Suppliers Association, *Engineering Data Book*, 12th Edition, 2004.

[4] Fisher, H.G., Forrest, H.S., Grossel, S.S., Huff, J.E., Muller, A.R., Noronha, J.A., Shaw, D.A., and Tilley, B.J., *Emergency Relief System Design Using DIERS Technology: The Design Institute for Emergency Relief Systems (DIERS) Project Manual,* Wiley, New York, 1993.

Disclaimer:

Can I Use My Cooling Water Header As A Relief Device?

Rahul Raman
Provenance Consulting
301 W. 6th Street Suite 200 Borger Texas 79007
Rahul.raman@provenanceconsulting.com

Prepared for Presentation at
American Institute of Chemical Engineers
2015 Spring Meeting
11th Global Congress on Process Safety
Austin, Texas
April 27-29, 2015

Can I Use My Cooling Water Header As A Relief Device?

Rahul Raman
Provenance Consulting
301 W. 6th Street Suite 200 Borger Texas 79007
Rahul.raman@provenanceconsulting.com

Keywords: Cooling water header, DIERS, Emergency Relief System, OSHA 1910.119, ASME Section VIII Div.1, Relief Device,

Abstract

Process safety information (PSI) needs to be evergreen and must contain the relief system design and design basis per OSHA 1910.119. Cooling water is used as a cooling medium in heat exchangers, often on the low-pressure side. During analysis of overpressure scenarios on the low-pressure side, there is often times debate about the quantity of relief credit that can be taken for flow out of the protected system and into the cooling water network. Current industry practice offers guidance that is generic and might be applied incorrectly without appropriate relief systems knowledge. This paper provides a flowchart-based approach and methodology that can be used for qualifying and quantifying relief credit through the cooling water side of heat exchangers.

1. Introduction

According to current industry practice, a heat exchanger tube rupture scenario is considered when the design pressure or maximum possible system pressure of the high-pressure side exceeds the corrected hydrotest pressure of the low-pressure side. Typically, an adequately-sized pressure relief device, PRD, is installed to relieve the fluid from the high-pressure side. In some older facilities, overpressure protection credit was given to the low-pressure side's hydraulic circuit (e.g., a cooling water header) as long as the tube rupture relief load did not accelerate the normal flow in the low-pressure side's hydraulic circuit. In other words, if the volumetric relief load into a cooling water header was less than the volumetric cooling water rate, the cooling water header was considered adequate and utilized as the sole means of overpressure protection. In cases where external fire was the only source of overpressure for the cooling water side of an exchanger, isolation valves upstream and downstream of the exchanger were considered acceptable without administrative controls as long as the exchanger was drained during isolation. Normal cooling water flow is often considered adequate overpressure protection without rigorous analysis or consideration of potential effects on other interconnected exchangers. Many operating companies use the above conditions to mitigate external fire scenario on the cooling water side. Review of previous literature on the subject provides a velocity criterion. This criterion was based on heuristics and cannot be validated. The guidelines provided here are based on interpretation of the ASME Code and API documents [1] [2]. When in doubt, a conservative

Note: Do not add page numbers. Do not refer to page numbers when referencing different portions of the paper

design approach and strong engineering judgment must be followed. Additionally, all local and federal regulations need to be followed in the evaluation of the emergency relief system.

2. Flowchart-Based Relief Credit Approach

It is important for engineers to use good engineering judgment, Recognized and Generally Accepted Good Engineering Practice (RAGAGEP), and a logical approach for Emergency Relief System (ERS) design. To aid the design process, an easy-to-follow flowchart is presented. This flowchart is applicable for protecting the cooling water side of heat exchangers only. This section details the flowchart-based approach and explains the components, subroutines, and logic for emergency relief system design.

2.1 Types of Relief Capacity Credit

2.1.1 Full Capacity Credit (ASME Section VIII Div.1 UG-127 (d))

The most common type of relief device is a pressure relief valve, followed by a rupture disc, or a rupture disc in combination with a pressure relief valve. It is uncommon to model piping systems as a relief device, but ASME code [1] allows piping systems to protect pressure vessels if the following criteria are satisfied: the piping system must follow the installation guidelines of UG-135 [1], the calculated relief capacity must be multiplied by 0.9, and the relief path must either directly or indirectly discharge to the atmosphere. See Figure 1 for a schematic of calculating the relieving capacity, where P_1 is the maximum allowable accumulation pressure, P_2 is the exit pressure, and $\dot{m}$ is the maximum flow rate for the given pressure drop. The calculated $\dot{m}$ is multiplied by 0.9 to be conservative in lieu of uncertainties, i.e. acceptable error tolerances with the calculation methodology and procedure [1].

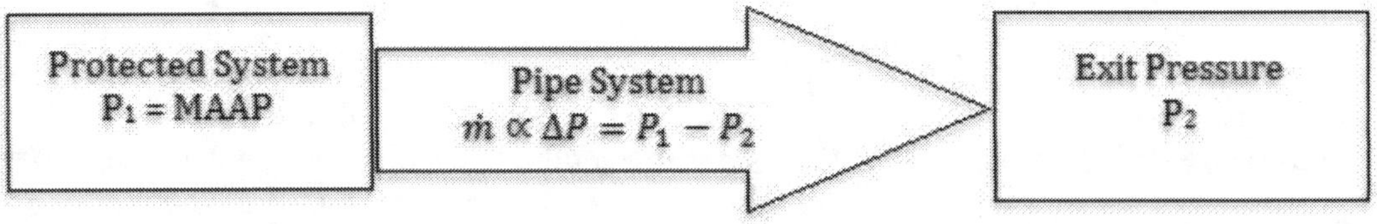

Figure 1: Schematic of relieving requirement determination

The cooling water network in a plant acts like a pulmonary system with main headers branching to sub-headers and area headers. All of the individual branches come back to form cooling water return sub-headers, which discharge the water to a cooling tower, which is open to atmosphere. The cooling water outlet of a heat exchanger can be considered as a relief path as it indirectly discharges to the atmosphere. However, the exit pressure, P_2, should be the stagnation pressure of the sub-header or the pressure at which the outlet branch connection reaches the return header and not atmospheric pressure. The exit pressure of the header equals P_2 as this will limit flow as the header needs to be maintained at a constant pressure.

2.1.2 Partial Relief Credit (API 521 Section 4.4.8.2)

Partial relief credit is subjective and is an area that often requires experience and sound engineering judgment. Partial credit is typically associated with automatic controls such as control valves that may not act favorably during an overpressure event [2]. Hence, an experience-based partial relief credit calculation is recommended. A case in point would be a temperature controller on the outlet of the low-pressure, cooling water side of heat exchanger that maintains the temperature of the high-pressure, hot process side. Firstly, the engineer must consult an expert or other authoritative personnel who can explain the mode of operation of the control valve. Secondly, if process historical data were available, then percentage opening of the valve would be beneficial in estimating the partial credit capacity more accurately. A two-year historical average may be considered as a good representation of the valve performance during an overpressure event. Thirdly, operator response during an overpressure event must be discussed and must be documented as part of the calculation. Finally, the most conservative percentage opening should be considered for the calculation. For example, if operators respond by closing the control valve during an overpressure event, then zero credit must be taken through the intended relief path. If the operator response is favorable and opens the control valve in the full open position, then the position at which the control valve was last positioned before the overpressure event should be considered for the partial credit and should be documented in the design basis.
In order to perform two-phase flow calculations through a control valve, a review of the available methodology provided by Dr. Diener et al. is recommended [3]. The calculated partial relief credit can be subtracted from the total required capacity, where the remainder must flow through another relief device. The relieving temperature and pressure can be used in lieu of normal operating conditions that may be higher than the former. For example, the stagnation temperature and pressure of an external fire scenario may be used.

2.1.3 Assumptions for the External Fire Case

Conventional relief analysis for the external fire scenario requires that the vessel remain isolated to simplify the analysis. API 521 section 4.4.13.2.4.1 allows the user to take credit for any open flow paths as part of the analysis [2]. It does not give additional guidance for the flow path, but in the case of the cooling water side of a heat exchanger, the cooling water network can be considered a normal flow path. Due to the potential for boiling due to external fire, two-phase flow into the cooling water network is considered. Firstly, the total vapor content in the two-phase flow equals the conventional vapor generation calculations based on the wetted surface area of the exchanger. The maximum relieving capacity, $\dot{m}$ is calculated by an iterative calculation by increasing the liquid flow rate. Secondly, the quality, x, of the flow can go as low as 0.0001. In such cases where the quality is extremely low, the fluid may be treated as saturated water. Finally, it is important that calculations and assumptions are part of the relief system documentation.

2.2 Flowchart Elements

There are flowchart elements that need to be highlighted to understand the logic better, namely the subroutine element. Subroutines are predetermined tasks that require a separate flowchart or are a set of complex processes, which reduce the complexity of the overall flowchart. They act as the point of information entry or exit, start or end of a process, and can be introduced in multiple, distinct locations in the overall flowchart.

2.2.1 Subroutine Element

The subroutines defined below are normal process workflows that can be performed by an engineer by understanding and using industry practices such as API 521 [2], API 520 part I [4] & II [5], and ASME Section VIII Div. 1 [1]. The following subroutine elements are defined:

1. Size Relief Device: The engineer considers the system to be unprotected or considers that existing protection is insufficient, requiring an adequately-sized relief device. This element acts as the end of the flowchart and the relief system documentation needs to be updated with the recommendations and changes proposed by the engineer.
2. Partial Credit: The engineer calculates relief capacity credit of the cooling water network refer to section 2.1.2 [2].
3. Full Credit: The engineer calculates maximum relief capacity of the cooling water network based on ASME Section VIII Div. 1 UG-127 (d) (2) [1].
4. Appendix M 5.8 (ASME Section VIII Div.1 Non-mandatory appendix) [1]: There are four criteria that are required for a relief path to have full area stop valves without administrative controls:
    4.1. External fire is the only applicable scenario.
    4.2. A documented standard operating procedure is required if the pressure vessel is to be isolated from its intended service.
    4.3. All pressure sources are isolated and depressurized if vessel is isolated from service.
    4.4. The pressure vessel must be drained and be kept free of liquids if vessel is isolated from service.

In order to implement Appendix M 5.8 in a facility, a model approach is suggested that can be executed at site or enterprise level. As the cooling water sides of most exchangers are similar, a common procedure or a site-wide procedure can be drafted and approved. This procedure will need to be a part of Process Safety Information (PSI) and becomes an integrated part of the relief system design and documentation [6]. Any changes to such a procedure affect the document lifecycle and relief system documentation of the cooling water side of heat exchangers.

2.2.2 Documentation

Operating facilities today have ready access to PSI that is supported by a data repository where such information can be accessed, edited, archived, and submitted using IT infrastructure. It is important that the relief system design basis be documented and that it follows a document life cycle to keep it evergreen. It is imperative that the engineer documents all the assumptions and calculations in a comprehensive and coherent manner. This is the last step in the flowchart and marks the completion of the evaluation.

2.2.3 Detailed Description of Flowchart

The flowchart starts with an examination of the current protected system to see what type of protection is present. See Figure 2 for the flowchart. This requires access to the relief system documentation database and review of existing calculations, if a relief device is present to protect the system. The reviewer must make sure that documents are in accordance with RAGAGEP. If it is adequate, no further evaluation is necessary. However, if the relief device is inadequate then we inspect the relief path for the following types of components.

- Check Valve: It is often stressed that check valves should not be installed in the relief path, so it is imperative that the engineer size an alternative relief device if a check valve exists in the intended piping system relief path.
- Full Credit: Appendix M 5.8 [1] allows the user to take full credit for the relief path if external fire is the only applicable scenario. Typically in operating facilities, administrative controls are in place for protecting the cooling water side for the thermal expansion case. Furthermore, if there are no other overpressure scenarios, such as heat exchanger tube failure, and external fire is the only applicable scenario, then appendix M 5.8 can be invoked—see 2.2.1 for additional details. If no components are present between the vessel nozzle and the relief header then this can also be considered for full credit.
- Automatic control: Firstly, if there is no relief device (e.g., a relief valve) installed to protect the system, then it is important to check if the intended relief path has automatic controls, such as, control valves, pneumatic valves, and on/off valves. If an automatic control is present then the flow path ends in the size relief device subroutine. Secondly, if a relief device is present and is not adequate, a partial credit can be evaluated for the relieving capacity based on API guidelines 4.4.8.2, see section 2.1.2 for details on partial relief credit.

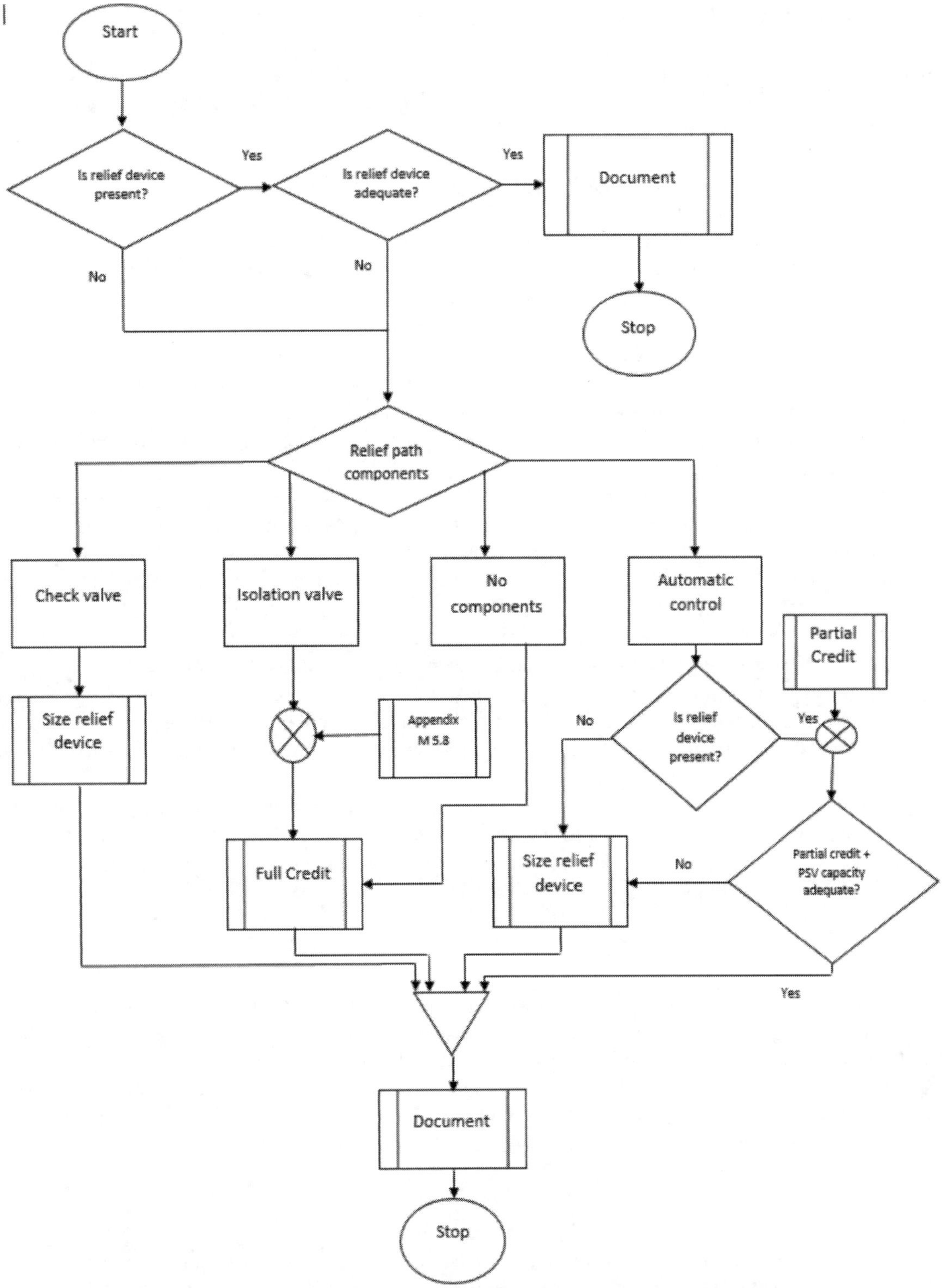

Figure 2. Flowchart-Based Capacity Credit

3. Review of Previous Work and Network Headers

Previous work on using cooling water flow networks as a relief path was based on a velocity criterion. This paper seeks to highlight the challenges of this velocity-based approach and shows some of the pitfalls associated with it.

3.1 Review of Previous Work (Velocity-Based Approach)

A cooling water header is typically maintained at a consistent pressure, supplying a relatively constant flow rate to the heat exchangers in a process unit. Based on previous literature survey, a velocity criterion was developed to determine the adequacy of the cooling water network when used as a means of overpressure protection. In other words, an overpressure event must not result in cooling water accelerating in the outlet of the heat exchanger [7]. This basis was developed in order to simplify the analysis and for evaluating the pressure drop through the pipe system. This velocity-based approach may be too conservative and impractical for the following reasons:

1) Favorable response or unfavorable response by operator intervention is neglected in such evaluations.
2) Previous methods prescribed a constant velocity in the cooling water leaving the heat exchanger that is arbitrarily selected. Standard chemical engineering handbooks give a typical range of velocity values that are used to size lines, for instance, 5 to 20 ft/s. This is subject to different analysis and may be open to interpretation. Therefore, a specific data point in the range cannot be used.
    - Velocity of a two-phase and vapor mixtures determined at a static point may not be accurate as the velocity changes as pressure attenuates in pipe.
3) There is some minimal amount of fluid acceleration that occurs during an overpressure event. The acceleration and its incremental increase will not overpressure the vessel beyond its maximum allowable working pressure. It needs to be blocked in to have a pressure excursion beyond the MAWP.
4) The capacity of the piping system is determined by the restrictions such as pipe and pipe fittings. If contents are at saturated conditions, then they are going to accelerate as pressure decreases downstream of the vessel. The capacity is dependent on the mass flux that the pipe can handle.

3.2 Network Headers

The branch connections to and from the return and supply header form a complex interconnected system of exchangers. Any pressure excursion in a given exchanger may overpressure other interconnected equipment [8]. Rigorous evaluation may be required and this evaluation becomes more complex when considering additional up and downstream networks. Computational tools are required to do such an analysis, which is out of the scope of this article, but the reader is cautioned and should consider such network effects.

4. Conclusion

This paper gives a flowchart based approach for capacity credit through the cooling water side of a heat exchanger only. This provides a contemporary design and calculation methodology that allows relief system engineers to calculate accurate overpressure protection credit for the low pressure hydraulic circuit. The reader can use it as a design aid to quantify the credit and must document all the assumption and calculations in an accessible database.

5. References

[1] ASME Boiler and Pressure Vessel Code Section VIII Division 1, Rules for Construction of Pressure Vessels, American Society of Mechanical Engineers, New York, NY, 2011.

[2] API 521, Pressure-relieving and Depressuring Systems, Sixth edition, American Petroleum Institute, Washington, DC, 2014.

[3] Dr.-Ing. Ralf Diener, Dr.-Ing. Jörg Kiesbauer, and Dr.-Ing Jürgen Schmidt, Improved Control Valve Sizing for Multiphase Flow, Available at http://www.samson.de/pdf_en/w01750en.pdf, Accessed on February 8, 2015.

[4] API 520, Sizing, Selection, and Installation of Pressure-relieving Devices, Part I – Sizing and Selection, Ninth edition, American Petroleum Institute, Washington, DC, 2014.

[5] API 521, Sizing, Selection, and Installation of Pressure-Relieving Devices in Refineries, Part II – Installation, Fifth edition, American Petroleum Institute, Washington, DC, 2003.

[6] Process Safety Management of Highly Hazardous Chemicals, Occupational Safety & Health Administration, Available at

https://www.osha.gov/pls/oshaweb/owadisp.show_document?p_table=STANDARDS&p_id=9760 Accessed on February 8, 2015.

[7] W. Y. Wong, Capacity Credit Calculation for Exchanger Tube Rupture, Hydrocarbon Processing, December 1992.

[8] W. Y. Wong, PRV Sizing for Exchanger Tube Rupture, Hydrocarbon Processing, February 1992.

Influence of Overpressure in Pressure Vacuum Safety Valves on Emission Reduction and Explosion Risk Minimization of Atmospheric Storage Tanks

Presenter/Dr.-Ing. Davide Moncalvo
Dr.-Ing. Michael Davies
Braunschweiger Flammenfilter GmbH (PROTEGO Germany)
Industriestrasse 11, 38122 Braunschweig, Germany
Davide.moncalvo@protego.com

Matthew Barfield, E.I.T.
Chris Mason
PROTEGO USA Inc.
497 Jessen Lane, Charleston, SC, 29492
Matthew.Barfield@protego.com

Prepared for Presentation at
American Institute of Chemical Engineers
2015 Spring Meeting
11th Global Congress on Process Safety
Austin, Texas
April 27-29, 2015

UNPUBLISHED

Influence of Overpressure in Pressure Vacuum Safety Valves on Emission Reduction and Explosion Risk Minimization of Atmospheric Storage Tanks

Presenter/Dr.-Ing. Davide Moncalvo
Dr.-Ing. Michael Davies
Braunschweiger Flammenfilter GmbH (PROTEGO Germany)
Industriestrasse 11, 38122 Braunschweig, Germany
Davide.moncalvo@protego.com

Matthew Barfield, E.I.T.
Chris Mason
PROTEGO USA Inc.
497 Jessen Lane, Charleston, SC, 29492
Matthew.Barfield@protego.com

Keywords: emissions, relief valve, leakage, loss of products, overpressure

Abstract

Containment of emissions from atmospheric or low-pressure tanks is, for economic reasons, a welcome feature by plant operators in the choice of Pressure Vacuum Relief Valves (PVRV), beside the necessary requirement of explosion risk minimization. Best engineering practice is to select a valve with high seat tightness close to set pressure and a minimum overpressure.

This paper compares two methods for estimating product losses when no measurement is available. Regarding the estimated product losses, different venting technologies are compared with overpressures ranging from 10 %, as usually required for safety relief valves (SRV) on pressurized tanks, to 100 %, which is the common technology for PVRV on the market. The experimental tests confirm that gaseous emissions are minimized by using the 10% overpressure technology without compromising the safety of the tank.

The results presented in this and previous papers by PROTEGO® were a cornerstone for the improvement of the blowdown requirement in the API 2000 standard.

1. Introduction

In today's economically competitive environment it is important to utilize a venting technology on storage tanks, like Pressure Vacuum Relief Valves (PVRV), which provides the maximum degree of safety from explosion of flammable liquids for the protected item and personnel coupled with minimum product losses. In API 620 [1] pressure relief valves are required to open at latest when the pressure at the top of the protected tank reaches the nominal pressure rating of the tank and they must discharge enough capacity to avoid exceeding the maximum allowable accumulated pressure (MAAP) of the tank. The difference between the set pressure, when the valve begins to open, and the relieving pressure is called overpressure. By choosing a valve with little overpressure the set pressure can be specified to be as close as possible to – without exceeding - the nominal pressure rating for the tank in conformity to API 620. Most of the available PVRV on the market require overpressures of 80% to 100% of the set pressure to reach full relieving capacity [2], which is far above the 10 % requirement for high-pressure safety valves [3,4]. This results in very low set pressure levels and supposedly increased emission losses due to increased operating cycles. At the moment there is almost no study to assess the impact of overpressure on emissions from low-pressure tanks.

This paper summarizes some recent studies done at PROTEGO® on the impact of overpressure on product loss. Some of the results described here were already discussed by Bosse and Davies [2] and others have been presented at conferences without papers. Although the results in this paper are not new, the approach given here is more scientific and rigorous. The paper begins with a comparison between the calculation methodology in the API Bulletin 2516 [5] and the one within VDI 3479 [6] in terms of leakage prediction in tanks of several sizes and nominal pressures. The second part of the paper focuses on the impact on leakage by the specification of set pressure and overpressure of the relief device.

2. Evaporation losses from low pressure tanks

The conservation of mass imposes that the change in mass for a tank is the difference between the mass flow rates via pumping and the loss of product, as given in Eq. 1.

$$\dot{m}_{tank} = \dot{m}_{pump-in} - \dot{m}_{pump-out} - \dot{m}_{losses} \quad \text{[Eq.1]}$$

When more liquid is pumped in the tank than the amount which is removed, than product is added into the tank: the rising liquid surface presses up the gases and vapors in the gaseous phase until the tank pressure is so high that the relief valve opens. At that moment the tank is losing product to the ambient

In the following chapters the evaporation rates as according to API Bulletin 2516 [5] and VDI 3479 [6] are described and commented. In order to avoid any future misunderstanding, the standard referred here deal with tanks breathing to or from ambient air.

3. Product losses as per API Bulletin 2516

The API Bulletin 2516 applies to tanks whose pressure is between 2.5 and 15 psig (172.4 and 1034.2 mbar-g) and whose vacuum is between 1 and 2 oz./in² gauge (4.3 and 8.6 mbar-g) .

This bulletin defines breathing losses as the losses of product from a tank due to thermal expansion of the vapors or due to an increase of vapor pressure caused by heating the contained liquid. The minimum tank pressure to prevent breathing losses is given in Eq. 2.

$$P_2 = 1.1\,[P_{atm} + P_1 - p_1] - [P_{atm} - p_2] \quad \text{for} \quad P_{atm} + P_1 \geq p_1 \; \text{[Eq. 2]}$$

In Eq. 2, P_2 and P_1 are respectively the gauge pressures when the pressure and the vacuum valve open and p_1 and p_2 are the vapor pressures at the liquid surface temperature of 90°F (32.2°C) and 100°F (37.7°C). In order to prevent breathing losses, the absolute vacuum vent pressure must be higher than the saturation pressure at 90°F. In other words, some air from the last inbreathing should be present in the vapor space. No correlation is provided to estimate the breathing losses.

Indeed, at higher tank pressures breathing losses are minimized but evaporation is facilitated and therefore the so called boiling losses must be accounted for. Boiling losses are present in a tank when the tank pressure equals or exceeds the saturation pressure at the liquid surface. It is here expected that the saturation pressure at 90°F is higher than the absolute vacuum vent pressure. No correlation is given for the boiling losses as well.

Indeed, the standard proposes a correlation for the working losses as percentage of the pumped-in load F_v, which is given in Eq. 3. The working losses are the product losses through the pressure relief valve when the tank pressure exceeds the set pressure of the valve. Additionally, to the symbols in Eq. 2, p_v is the vapor pressure at the liquid surface. Eq. 3 is valid under the assumptions that the vapor pressure and temperature remain unchanged during filling.

$$F_v = \frac{p_v}{p_u} \cdot \frac{P_{atm}+P_1-p_v}{P_{atm}+P_2-p_v} \quad \text{with} \quad p_u = 33.3\,psi \qquad \text{[Eq. 3]}$$

No formula is given either for the leakage losses from lax flanged connections between the tank and the valves as they differ from one tank to another randomly.

4. Product losses as per VDI 3479

The calculation method in VDI 3479 is valid for fixed-roof tanks with an internal available floating deck. The annual mass losses L_{tank} in Eq. 4 are the leakages, which cannot be prevented by using the floating deck. In this equation η_{SD} is the efficiency of the floating deck (in German: *SchwimmDecke*), η_{VD} (in German: *VakuumDruckventile*) is the efficiency of the pressure and vacuum valves, f is the coating factor, L_A and L_B are respectively the breathing and filling annual leakages. Some examples of coating factors f are given in Table 1 for different shades of gray.

$$L_{tank} = (1-\eta_{SD}) \cdot [(1-\eta_{VD}) \cdot f \cdot L_A + L_B] \quad \text{[Eq. 4]}$$

The breathing and filling leakages could be measured. In case no measurement is available VDI 3479 gives a conservative estimate via Eq. 5. According to this equation the annual breathing losses can be calculated as the weighted sum of the daily mass losses during summer and winter days, L_{As} and L_{Aw}. For simplicity a year is broken into d_s summer days and d_w winter days.

Table 1. Some examples of coating factors for shades of gray taken from VDI 3479

Types of gray	Coating factors
White	1.0
Silver	1.1
Light gray	1.3

Mouse gray	1.6

The mean daily breathing mass losses can be estimated as the product of the dimensionless saturation level f_A, the saturation concentration density c and the volumetric flow rate V_n, which is the daily change of the gas volume in the tank. The saturation level f_A is the ratio between the actual concentration and the one at saturation.

$$L_A = L_{As} \cdot d_s + L_{Aw} \cdot d_w = f_{As} \cdot c_s \cdot \dot{V}_{ns} \cdot d_s + f_{Aw} \cdot c_w \cdot \dot{V}_{nw} \cdot d_w \quad \text{[Eq. 5]}$$

The efficiency of the pressure and vacuum valves is the percentage amount of the losses L_A, which does not leave the tank through the valves, L_{AVD}. Based on Eq. 5 the efficiency of the valves can be calculated as shown in Eq. 6. In this equation V_{nsVD} and V_{nwVD} are the daily volumetric rates through the relief valves due to vapor expansion or contraction.

$$\eta_{VD} = 1 - \frac{L_{AVD}}{L_A} = 1 - \frac{f_{As} \cdot c_s \cdot \dot{V}_{nsVD} \cdot d_s + f_{Aw} \cdot c_w \cdot \dot{V}_{nwVD} \cdot d_w}{f_{As} \cdot c_s \cdot \dot{V}_{ns} \cdot d_s + f_{Aw} \cdot c_w \cdot \dot{V}_{nw} \cdot d_w} \quad \text{[Eq.6]}$$

The filling losses in Eq. 7 are defined in a similar way as Eq. 5 for the breathing losses with f_B and c_B being respectively the saturation ratio at a reference value of 0.85 and the saturation concentration density during filling and Q being the annual filled-in volume.

$$L_B = f_B \cdot c_B \cdot Q \quad \text{[Eq.7]}$$

4. Comparison between the two methodologies

For the comparison between the two methodologies the authors refer to the experimental data in the DGMK Forschungsbericht 225 [7] for a middle European climate. In Fig. 1 the vapor losses as a function of the vapor pressure calculated using the methodologies in API RP 2516 and VDI 3479 are compared for different tank volumes. For this study the tank no. *X* has a volume of *X* times 1000 m³ (35 315 ft³), so the tank sizes range from one to ten thousand cubic meter. The curves show that the methodology in API RP 2516 estimates always higher vapor losses than VDI 3479 for any combination of tank nominal pressure rating and size. However, both methodologies agree that the vapor losses for a tank of given size increase with the vapor pressure, i. e. the working pressure in the tank.

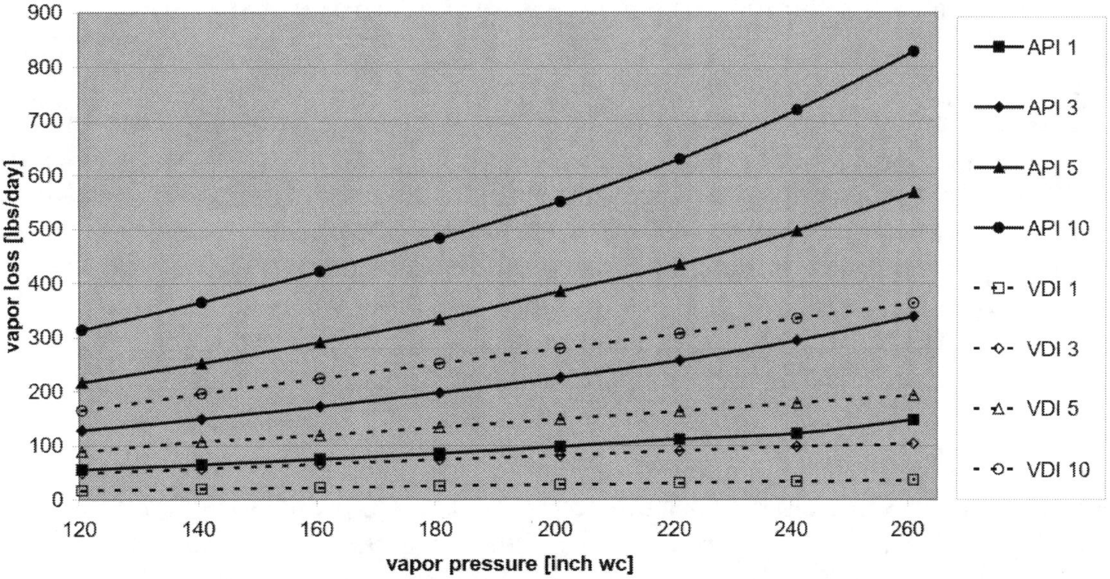

Fig. 1 Vapor losses in function of vapor pressure for various tank volumes

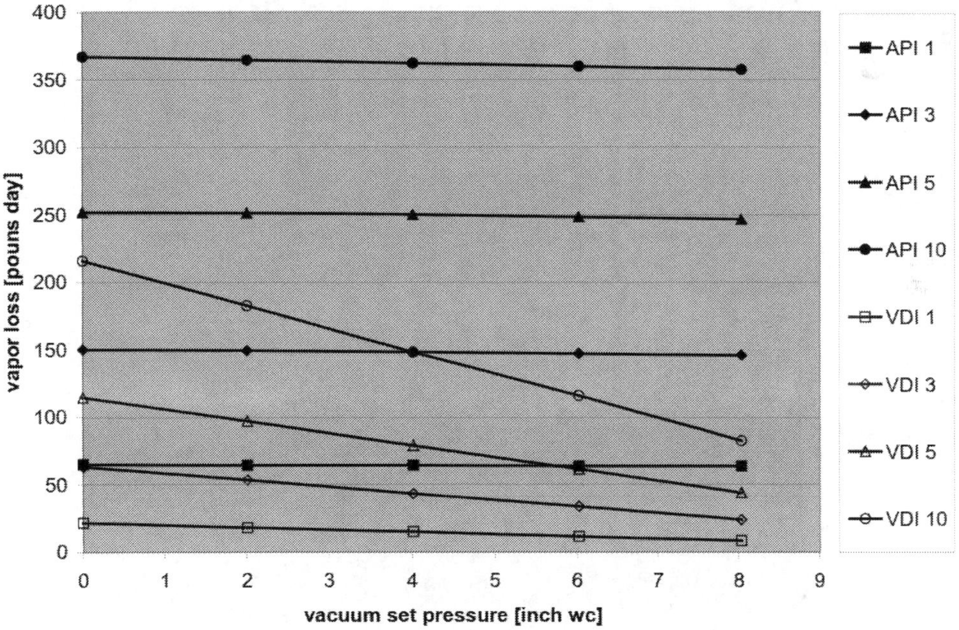

Fig. 2 Vapor losses in function of vapor vacuum for various tank volumes

In Fig. 2 the API 2516 standard does not show any significant change in the vapor losses by increasing the vacuum, as instead VDI 3479 does, since the latter includes the pump-out saving in the breathing losses.

From Fig. 1 and 2 it can be evinced that the emissions calculated using the API Bulletin 2516 are roughly two times those of the VDI 3479 standard.

5. Set pressure and overpressure effect on emission minimization

Independently from the methodology used to assess the pressure losses, for a tank of given nominal pressure rating and size, pressure relief valves with a smaller overpressure and therefore with higher set pressures are expected to deliver less leakage. In other words, a valve like those of PROTEGO® that require 10 % overpressure for full lift, a technology many readers already know from high-pressure safety valves, are expected to be subject to less leakage than valves that require higher overpressures. In order to validate this expectation, PROTEGO® has developed a portable test rig to measure small leakages from pressure relief valves up to DN 200 (8 inch) in adherence to the API Bulletin 2521 [8]. In agreement with that standard leakage tests should be conducted at a pressure equal to 75 % of the set pressure.

In Fig. 3 two tests are shown; for both tests the valves were supposed to be fully open at 1 oz./in² (4.3 mbar) gauge. The first test was performed at a set pressure of 0.5 oz./in² (2.15 mbar) gauge and a pressure at leakage of 0.375 oz./in² gauge; this set pressure has been chosen since many PVRVs require an overpressure of 100 % to reach full opening. The second test was performed at the set pressure of 0.9 oz./in² (3.88 mbar) gauge and at a leakage pressure of 0.675 oz./in² (2.98 mbar) gauge. Fig. 3 clearly shows that competitor valves, which require a much higher overpressure, fail to remain tight at low tank pressures. In economic terms high leakages translate into huge product losses and high pollution taxes to the EPA or similar agencies worldwide.

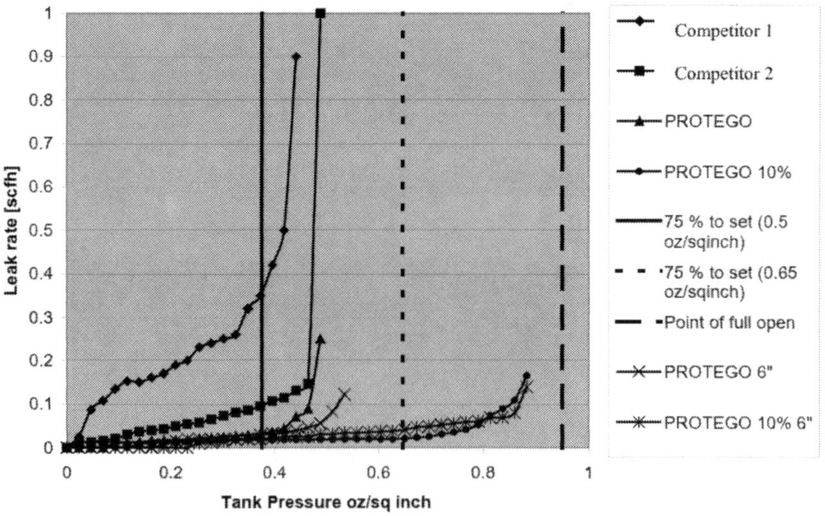

Fig. 3 Leakages in function of tank pressure for several

5. Improvement of API 2000 standard on leakage minimization

On the basis of PROTEGO® results there has been an improvement of the API 2000 standard. In the Fifth Edition [9] the leakage tests have to be conducted at an inlet pressure equal to 75% of the set pressure, since at that pressure the seat in the relief valve is assumed to be tight. A further point was that conservation vents, e. g. spring and weight loaded valves, must have a 0% blowdown, thus meaning that these valves should reclose at the set pressure. This requirement is not only too stringent; it is physically unfulfillable since the lift force on the disk when the valve is reclosing can be neutralized only with pressures below the set pressure, e.g. the blowdown.

According to the Seventh Edition [10] Annex C, seat leakage and the amount of blowdown are dependent on the relief-valve design. Seat leakage is assumed to start between 75 % and 90 % of the set pressure.

Nevertheless, there are currently many installed direct venting systems designed using API 2000 Fifth Edition [9] presenting high product losses in potentially explosive atmospheres. The simplest and most immediate recommendation would be to verify previous sizing using the Seventh Edition. Moreover, consideration should be given to selecting valves with minimal required overpressure against other devices with lower required set pressures. It would also be recommendable to monitor the protected tank and assess the product losses over a large period of time. Valves with low set pressures not only tend to leak more frequently but also to cycle more; for a plant owner it means a double expense both in terms of product loss and spare part purchase. These future costs may overcome the savings from the purchase of a cheaper device.

Furthermore, some PVRV manufacturers support the use of 50% overpressure. However, this requirement should be pondered seriously, especially when the reseating pressure may fall below the set pressure of the blanketing gas regulator. In this case the PVRV cannot reseat and continuous heavy losses due to bleeding of the blanketing gas may result.

6. Conclusions

This paper shows the leakage assessment methods in API Bulletin 2516 and in VDI 3479. At the moment there is no experimental validation which methodology delivers more accurate results. Some additional work is necessary in the field.

To fulfill the demands of stringent environmental legislations and effectively minimize the loss of product, plant owners should start considering the application of technologies with minimum overpressure over others that rely on far lower set pressures. A 10% overpressure, which many users know from high-pressure safety valves, should become a standard also for low-pressure relief valves, especially in virtue of the seat leakage test requirements in the newest API 2000.

An additional design advantage of the low overpressures is the higher flexibility in designing the set pressure of the conservation vent. The authors expect that a set pressure of the conservation vents which can be even 10% to 20% below the set pressure of the emergency relief valves should guarantee a chatter-free vent opening and avoid conflict with the valves.

7. References

[1] API 620 Third Edition, Design and Construction of Large, Welded, Low-Pressure StorageTanks, 2002

[2] P. Bosse and M. Davies, Venting technologies for reducing vapour losses, Hydr. Eng., 2008.

[3] API 520 Part I Ninth Edition, Sizing, Selection, nd Installation of Pressure-relieving Devices, 2014.

[4] ISO 4126-1, Safety devices for protection against excessive pressure - Part 1: Safety valves, 2013.

[5] API Bulletin 2516, Evaporation loss from low-pressure tanks, 1962.

[6] VDI 3479, Guideline Emission Control Marketing Installation Tank Farms, 2002

[7] DGMK Forschungsbericht 225 "Kohlenwasserstoff – Emissionen aus Festdachtanks – Vergleich von Berechnungsformeln unter besonderer Berücksichtigung der VDI Richtlinie 3479, 1985.

[8] API Bulletin 2521, Use of pressure-vacuum vent valves for atmospheric loss, 1993.

[9] API 2000 Fifth Edition, Venting Atmospheric and Low pressure Storage Tanks, 1998

[10] API 2000 Seventh Edition, Venting Atmospheric and Low pressure Storage Tanks, 2014.

11TH GLOBAL CONGRESS ON PROCESS SAFETY

A Comprehensive Guide to Accurately Size Pressure and Vacuum Relief Devices for Atmospheric and Low Pressure Storage Tanks

Steve Streblow
Contek Solutions LLC
6221 Chapel Hill Blvd., Suite 300
sstreblow@contekllc.com

Ushma Patel
Contek Solutions LLC
6221 Chapel Hill Blvd., Suite 300
upatel@contekllc.com

Prepared for Presentation at
American Institute of Chemical Engineers
2015 Spring Meeting
11th Global Congress on Process Safety
Austin, Texas
April 27-29, 2015

UNPUBLISHED

A Comprehensive Guide to Accurately Size Pressure and Vacuum Relief Devices for Atmospheric and Low Pressure Storage Tanks

Steve Streblow
Contek Solutions LLC
6221 Chapel Hill Blvd., Suite 300
sstreblow@conteklIc.com

Ushma Patel
Contek Solutions LLC
6221 Chapel Hill Blvd., Suite 300
upatel@conteklIc.com

Keywords: PVRV; API 2000; OSHA; API 12 F; Emergency Pressure Relief, Tank, Vacuum,

Abstract

Increased environmental concerns and regulation have made proper Pressure and Vacuum Relief Valve (PVRV) design a priority. PVRVs protect tanks against rupture due to increases in internal pressure or implosion due to vacuum. They also minimize emissions, loss of product, equipment corrosion due to emissions, and decrease maintenance costs.

The use of Pressure and Vacuum Relief Valves is mandated by almost all of the recognized regulatory institutions involved in tank safety, such as API, OSHA, NFPA, etc. API 12F, Section 6, further stipulates the use of Emergency Pressure Relief Vents, in addition to PVRV's when storage tanks containing flammable liquids are exposed to fire, unless the tanks in question are installed in remote locations. The venting system must be designed in accordance to API 2000 which supplies the rigorous formulas to calculate both normal venting and emergency (fire case) venting. Annex B and C of API 12 F provide "short cut" tables with values for thermal venting and emergency venting for different tank sizes, respectively.

Most API 12F tanks come furnished with a standard size thief hatch, usually 8 inches. It is often mistakenly assumed that this will be adequate to handle the emergency venting requirement. This paper presents a guide to ensuring the tank venting system is designed to handle the maximum relieving requirements in keeping with the most stringent stipulations of the codes. It discusses topics such as: common mistakes made when using API 12 F, inconsistencies between API 2000 and API 12F, use of a dynamic method to calculate the worst case venting scenarios, and the effective reduction in flow thru a standard 8" opening when a thief hatch is installed.

1. Introduction

One of the key components of a tank venting system is the Pressure and Vacuum Relief Valve (PVRV). Pressure and Vacuum Relief Valves protect tanks against rupture due to increases in internal pressure or implosion due to vacuum, thus avoiding permanent damage to the equipment. Fixed roof tanks are susceptible to rupture from many scenarios including whenever

Note: Do not add page numbers. Do not refer to page numbers when referencing different portions of the paper

liquid is pumped into the tank or due to changes in the vapor pressure of the contents, caused by significant thermal changes. During these instances, the PVRV protects the tanks from over-pressure by depressurizing the tank. This procedure is known as out-breathing. Alternately, imploding of a tank may occur for various reasons, including the pumping out procedure or due to thermal changes. This is due to the fact that as the liquid level in the tank decreases, the pressure in the vapor space also decreases creating a vacuum. The vacuum can only be broken by allowing either atmospheric air, an inert gas such as Nitrogen, or blanket gas to enter the tank. This is known as inbreathing. PVRVs typically vent to atmosphere but can also have provisions to allow the vapors to be piped away for recovery or destruction.

The PVRV is typically installed on a flanged nozzle at the top of the fixed roof of an atmospheric or low pressure storage tank. Besides protecting the equipment from permanent damage, PVRVs also serve to minimize emissions to the environment, as well as minimize loss of product due to evaporation, by controlling tank venting. When combined with a well-designed vapor recovery system, the loss of product can be cut to almost zero, thus leading to significant money savings, especially for tank farms. These devices also help reduce corrosion of the surrounding equipment due to less fugitive emissions. Therefore, the cost of overall plant maintenance is also decreased.

The use of PVRVs is mandated by API, OSHA and NFPA. API 12F, Section 6, further stipulates the use of additional Emergency Pressure Relief Vents, for flammable liquid storage tanks that could potentially be exposed to fire, unless installed in remote locations. The venting system must be designed per API 2000 which supplies the rigorous formulas to calculate normal and emergency venting requirements. Annex B and C of API 12 F provide "short cut" tables for normal and emergency venting, based on tank sizes.

On smaller tanks, the PVRV valve alone may be adequate to provide sufficient flow capacity for emergency venting. Larger tanks however, may require additional emergency pressure relieving devices in the form of larger or additional vent valves or additional thief hatches. Most tanks already come furnished with a standard size thief hatch, usually 8 inches. However, it is often a mistake to assume that it will be adequate to handle the emergency venting requirement, especially for larger size tanks.

This paper presents a viable approach to ensuring that the tank venting system is adequately designed to handle the maximum relieving requirements in keeping with the most stringent stipulations of the various codes such as API 12F and API 2000. It discusses in detail several topics such as: case studies of common mistakes made when using API 12 F, inconsistencies between API 2000 and API 12F, use of a quick inexpensive dynamic method to calculate the worst case tank venting scenarios, the effective reduction in flow thru a standard 8" opening when a thief hatch is installed, and the further reduction in capacity by the installation of a low emission model.

In the course of our consulting work, Contek has come across storage tanks that were not properly engineered on several projects. Many companies simply order an API 12F tank, with standard openings, and then if environmental requirements dictate, they install a thief hatch on the 8" opening. As mentioned above, this may not always be the correct approach. After reviewing this paper, if the reader has any doubts that the tanks at their company are not properly engineered, we encourage them to contact a qualified company to study the tank design and provide the best path forward.

2. Emergency Venting Capacities of standard API 12F Tanks

API 12F tanks come in standard capacities, dimensions, and design pressures. As seen in Table 1, below, the design pressure starts decreasing once the capacity exceeds 400 barrels. The maximum allowable internal pressure during emergency venting (column 4) starts decreasing once the capacity exceeds 150 barrels. The overpressure protection driving force is a function of the difference between the design pressure and the allowable pressure during emergency venting (see column 5). This driving force starts decreasing once the capacity exceeds 150 barrels. Considering the fluid flowing out of the tank to be incompressible, (this is a good assumption for the small percentage of absolute pressure drop), and applying Bernoulli's mechanical energy balance equation to the problem, results in the driving forces being proportional to the square root of the difference between the design pressure and the maximum allowable internal pressure during emergency venting. Scaling the driving force shows that the smallest tank has 2.83 times the driving force of the largest tank. Also note that at a given capacity, the larger the diameter, the lower the design pressure, the allowable internal pressure and the pressure difference. Pressures less than that for the smallest tank are in red text.

Table 1 - Tank Design Pressure, Maximum Allowable Emergency Pressure and Overpressure Protection Driving Force For Standard API 12F Tanks

Nominal Capacity Barrels	Diameter x Height Ft, In	Design Pressure oz/in^2	Maximum Allowable Pressure During Emergency Venting oz/in^2	Emergency Pressure - Design Pressure oz/in^2	Scaled Driving Force
90	7,11 x 10,0	16	48	32	2.83
100	9,6 x 8,0	16	48	32	2.83
150	9,6 x 12,0	16	48	32	2.83
200	12,0 x 10,0	16	24	8	1.41
210	10,0 x 15,0	16	48	32	2.83
250	11,0 x 15,0	16	24	8	1.41
300	12,0 x 15,0	16	24	8	1.41
400	12,0 x 20,0	16	24	8	1.41
500	12,0 x 25,0	16	24	8	1.41
500	15,6 x 16,0	8	12	4	1.00
750	15,6 x 16,0	8	12	4	1.00
(1)	(2)	(3)	(4)	(5)	(6)

Note : Columns 1 thru 4 are taken from Table C.1 of API 12F, Twelfth Edition, October 2008, Specification for Shop Welded Tanks for Storage of Production Liquids.

This analysis illustrates that merely using the same size opening, number of openings, etc. on a larger tank may not be as safe as it is on a smaller size tank. Let's continue with the analysis.

2.1 Venting Capacity of the Standard 8" Opening for Various Tank Sizes:

From a fundamental standpoint, for a pool fire case scenario, the more surface area exposed to the fire, the greater the required venting rate. However, as was seen in Table 1, the driving force for overpressure relief decreases as the tank capacity increases. Will there be a point where the 8" opening to the atmosphere will not be adequate? It turns out that once the tank capacity exceeds 150 barrels, a single 8" opening is no longer adequate to handle the required emergency venting. As shown in Column 7 of Table 2, below, the standard 8" opening only provides 31% of the total required venting capacity for a 750 barrel tank. The scaled driving force concept from Table 1 agrees well with the scaled flow rates. Tanks for which the 8" opening is inadequate are shown in red text in Table 2.

Table 2 - Tank Emergency Venting Requirements Versus Venting Capacity of the Standard 8" Opening

Nominal Capacity Barrels	Diameter x Height Ft, In	Required Venting Without Drainage SCFH	8" Opening Venting Capacity SCFH	8" Opening Scaled Flowrate	Table 1 Scaled Driving Force	8" Opening Percentage Of Required Venting
90	7,11 x 10,0	239,000	471,000	2.79	2.83	197
100	9,6 x 8,0	233,000	471,000	2.79	2.83	202
150	9,6 x 12,0	292,000	471,000	2.79	2.83	161
200	12,0 x 10,0	301,000	239,135	1.41	1.41	79
210	10,0 x 15,0	342,000	471,000	2.79	2.83	138
250	11,0 x 15,0	361,000	239,135	1.41	1.41	66
300	12,0 x 15,0	378,700	239,135	1.41	1.41	63
400	12,0 x 20,0	446,350	239,135	1.41	1.41	54
500	12,0 x 25,0	507,000	239,135	1.41	1.41	47
500	15,6 x 16,0	455,200	169,094	1.00	1.00	37
750	15,6 x 16,0	543,600	169,094	1.00	1.00	31
(1)	(2)	(3)	(4)	(5)	(6)	(7)

Note : Columns 1 thru 3 are taken from Table C.1 and Column 4 is taken Table C.2 of API 12F, Twelfth Edition, October 2008, Specification for Shop Welded Tanks for Storage of Production Liquids.

What will happen to the relieving capacity if one installs a thief hatch, as is often done, on the 8" opening? Let's continue the analysis to find out.

2.2 Venting Capacity of the Standard 8" Opening for Various Tank Sizes upon Installation of a Thief Hatch:

An Enardo 660 thief hatch was chosen as the basis for this analysis, since it is commonly used in the industry where low emissions are required. As shown in Table 3, below, none of the tanks have adequate overpressure protection once a thief hatch is installed on the standard 8" opening. The percentage of overpressure relief capacity provided ranges from approximately 13% for the smallest tank to 2% for the largest tank. An overpressure protection system that is undersized by one to two orders of magnitude is what is commonly found installed. The scaled driving force concept from Table 1 agrees well with thescaled flowrates. Tanks where the Enardo 660 is inadequate are shown in red text in Table 3.

Table 3 - Tank Emergency Venting Requirements Versus Thief Hatch Venting Capacities

Nominal Capacity Barrels	Diameter x Height Ft, In	Required Venting Without Drainage SCFH	Enardo 660 Venting Capacity SCFH	Enardo 660 Scaled Flowrate	Table 1 Scaled Driving Force	Enardo 660 Percentage Of Required Venting
90	7,11 x 10,0	239,000	32,198	2.44	2.83	13.47
100	9,6 x 8,0	233,000	32,198	2.44	2.83	13.82
150	9,6 x 12,0	292,000	32,198	2.44	2.83	11.03
200	12,0 x 10,0	301,000	16,670	1.26	1.41	5.54
210	10,0 x 15,0	342,000	32,198	2.44	2.83	9.41
250	11,0 x 15,0	361,000	16,670	1.26	1.41	4.62
300	12,0 x 15,0	378,700	16,670	1.26	1.41	4.40
400	12,0 x 20,0	446,350	16,670	1.26	1.41	3.73
500	12,0 x 25,0	507,000	16,670	1.26	1.41	3.29
500	15,6 x 16,0	455,200	13,203	1.00	1.00	2.90
750	15,6 x 16,0	543,600	13,203	1.00	1.00	2.43
(1)	(2)	(3)	(4)	(5)	(6)	(7)

Note : Columns 1 thru 3 are taken from Table C.1 of API 12F, Twelfth Edition, October 2008, Specification for Shop Welded Tanks for Storage of Production Liquids.

The values in Column 4 were determined by using EnardoSoftCalcII, version 1.2.5.7.

3. Case Studies

The following section discusses the typical findings on some of the projects Contek has worked on involving tank venting systems.

3.1 Project No. 1 - API 2000 versus API 12F:

Project No. 1 is an upstream oil production facility with three tanks: a 500 bbl capacity Gun Barrel Tank, a 400 bbl Condensate Tank, and a 400 bbl Water Tank. Each of the tanks was equipped with an Enardo 660 Pressure / Vacuum relief device set at 4 oz/in^2.

Based on the results of the study, detailed in Table 4 below, the following conclusions were drawn:

1. With a fixed tank, the API 2000 fire case flow rate varies based on the tank contents. The water tank has a lower API 2000 flow rate than the tank that contains hydrocarbon.

2. Comparing the generic API 12F tables with the more specific API 2000 calculations, the conclusions are:
   a) The gunbarrel tank has an actual fire flow rate, based on API 2000, about 2% higher than API 12F.
   b) The condensate tank has an actual fire flow rate about 2% higher than API 12F.
   c) The water tank has an actual fire flow rate of about 37% of API 12F.

3. Using the API 12F method and not using the API 2000 method means that the hydrocarbon containing tanks may be under protected, while the water tanks may be overprotected. There may be an opportunity for cost savings by not placing the same devices on all tanks.

4. The Enardo 660 does not provide nearly enough capacity for overpressure protection. Depending on the tank, it provides between 8 and 27 percent of what is needed.

5. The Enardo 660 does not provide enough capacity for vacuum protection. Depending on the tank it provides between 63 and 76 percent of what is needed.

The recommendations to make the system safe were:

1. For pressure, increase the vent header piping nominal diameter from 2" to 3"

2. For vacuum, add Enardo 2000s to the tanks. 8" size for gun barrel and condensate and 4" for water.

Table 4 - Project 1

	Tank 1	Tank 2	Tank 3
Service	**Gun barrel**	**Condensate**	**Water**
Capacity [Barrels]	500	400	
Pressure/Vacuum Relief Device	Enardo 660		
Pressure Setting [oz/in^2]	4		
Maximum Filling Rate [GPM]	50	61.67	287.5
Maximum Draining Rate [GPM]	60	30	30
Latitude Factor	0.32		
Insulation Factor	1		
C Factor	6.5		
Environmental Factor	1		
Initial Fire Case Flowrate Per API 2000 [SCFH]	517,238	455,868	164,926
Emergency Venting Per API 12F [SCFH]	507,000	446,350	446,350
API 2000 Fire Flow/API 12F Fire Flow	1.02	1.02	0.37
Device Pressure Rate [SCFH]	45,000	45,000	45,000
Percent capacity provided for fire case	8.70	9.87	27.28
Required Vacuum Flowrate [SCFH]	5,672	4,681	4,681
Device Vacuum Flowrate [SCFH]	3,586	3,586	3,586
Percent capacity provided for vacuum	63.22	76.61	76.61
Solution for pressure	Change vent header to atmosphere from 2" to 3"	Change vent header to atmosphere from 2" to 3"	Change vent header to atmosphere from 2" to 3"
Solution for vacuum	Add Enardo 2008 to tank	Add Enardo 2008 to tank	Add Enardo 2004 to tank

3.2 Project No. 2 - API 2000 versus API 12F:

Project No 2 is an upstream oil production facility with two tanks: a 400 bbl capacity Oil Tank, and a 400 bbl Water Tank. Each of the tanks was equipped with an Enardo 660 Pressure / Vacuum relief device set at 8 oz/in^2. Based on the results of the study, detailed in Table 5 below, the following conclusions were drawn:

1. During normal operation, there is a significant amount of liquid flashing from the light oil, resulting in vapors needing to exit the tank. This is, however, not taken into account in the API 12F emergency venting calculation method, which results in the API 12F relief flow rate being only about 10% of actual required relief for the oil and 76% of actual required relief for the water. This means that if just the API 12F method is used to determine the required vent rate it could result in a tank being under protected during normal operation.

2. Performing final fire case provides insight into how hot the tank is going to get. As the fire case proceeds, the oil tank temperature increases as the more volatile components are preferentially boiled off. Since the water is much closer to a pure component the temperature does not change significantly as the fire progresses.

3. For both the oil and water tanks, the flow rate at the start of the fire is highest.

4. With a fixed tank, the API 2000 fire case flow rate varies based on the contents. The water tank has a lower API 2000 flow rate than the tank that contains hydrocarbon.

5. Comparing the generic API 12F tables with the more specific API 2000 calculations the conclusions are:
   a) The oil tank has an actual fire case relief flow rate that is about 39% higher than that calculated using API 12F.
   b) The water tank has an actual fire relief flow rate that is about 38% of that calculated using API 12F.

6. Using the API 12F method and not using the API 2000 method means that the hydrocarbon containing tanks may be underprotected, while the water tanks may be overprotected. There may be an opportunity for cost savings by not placing the same devices on all tanks.

7. The Enardo 660 does not provide nearly enough capacity for overpressure protection. Depending on the tank it provides between 5 and 18 percent of what is needed.

8. The Enardo 660 does provide enough capacity for vacuum protection.

The recommendations to make the system safe were:

1. For pressure, add an Enardo 2000 to each tank, 12" for oil and 8" for water

Table 5 - Project 2

	Tank 1	Tank 2
Service	**Oil**	**Water**
Capacity [Barrels]	400	
Pressure/Vacuum Relief Device	Enardo 660	
Pressure Setting [oz/in^2]	8	
Maximum Filling Rate [GPM]	12.79	92
Maximum Draining Rate [GPM]	210	210
Latitude Factor	0.25	
Insulation Factor	1	
C Factor	5	
Environmental Factor	1	
Normal Flashing Rate [SCFH] API 2000	5,300	717
Normal out-breathing resulting from a maximum inflow of liquid into the tank [SCFH] API 2000	206	1476
Normal out-breathing resulting from a maximum inflow of liquid into the tank [SCFH] API 12F	220	1577
Thermal Outbreathing [SCFH] API 12F	400	400
Thermal Outbreathing [SCFH] API 2000	392	392
Total Outbreathing [SCFH] API 2000	5,898	2,585
Total Outbreathing [SCFH] API 12F	620	1,977
% API 12F Outbreathing/ API 2000 Outbreathing	10.5	76.5
Initial Fire Case Temperature [Degrees F]	89	199
Final Fire Case Temperature [Degrees F]	249	199
Initial Fire Case Flowrate Per API 2000 [SCFH]	618,676	167,993
Final Fire Case Flowrate Per API 2000 [SCFH]	513,635	161,076
Emergency Venting Per API 12F [SCFH]	446,350	446,350
API 2000 Fire Flow/API 12F Fire Flow	1.39	0.38
Device Pressure Rate [SCFH]	31,348	31,348
Percent capacity provided for fire case	5.07	18.66
Required Vacuum Flowrate [SCFH]	2,200	2,200
Device Vacuum Flowrate [SCFH]	3,586	3,586
Percent capacity provided for vacuum	163.00	163.00
Solution for pressure	A 12" Enardo 2000 or a 10" Enardo 950	A 8" Enardo 2000 or a 8" Enardo 950

3.3 Project No. 3 - API 2000 versus API 12F:

Project No 3 is an upstream oil production facility with two tanks: a 400 bbl capacity Oil Tank, and a 400 bbl Water Tank. Each of the tanks was equipped with an Enardo 660 Pressure / Vacuum relief device set at 16 oz/in^2. Based on the results of the study, detailed in Table 6 below, the following conclusions were drawn:

1. Comparing the generic API 12F tables with the more specific API 2000 calculations the conclusions are: a) the oil tank has an actual fire flow rate of only about 28% of API 12F.

2. Using the API 12F method and not using the API 2000 method means that the water tanks may be overprotected. There may be an opportunity for cost savings by not placing the same devices on all tanks.

3. For this particular water tank, there was not a fire case scenario. However, there were 2 additional upset scenarios that needed to be considered.

4. The Enardo 660 does not provide nearly enough capacity for overpressure protection. Depending on the tank it provides between 3 and 13 percent of what is needed.

5. The Enardo 660 does not provide enough capacity for vacuum protection. Depending on the tank, it supplies only about 71 to 79 percent of the requirement.

The recommendations to make the system safe were:

1. For pressure, add an Enardo 2008 to each tank and increase the balance line size between tanks from 3" to 4"

2. For vacuum, add Enardo 950 – 3" to each tank

Table 6 - Project 3

	Tank 1	Tank 2
Service	**Oil**	**Water**
Capacity [Barrels]	400	
Pressure/Vacuum Relief Device	Enardo 660	
Pressure Setting [oz/in^2]	16	
Maximum Filling Rate [GPM]	7	188
Maximum Draining Rate [GPM]	7	73
Latitude Factor	0.32	
Insulation Factor	1	
C Factor	6.5	
Environmental Factor	1	
Normal Flashing Rate [SCFH] API 2000	-	0
Normal out-breathing resulting from a maximum inflow of liquid into the tank [SCFH] API 2000	55	1504
Thermal Outbreathing [SCFH] API 2000	502	502
Total Outbreathing [SCFH] API 2000	557	2,006
Initial Fire Case Temperature [Degrees F]	115	
Initial Fire Case Flowrate Per API 2000 [SCFH]	125,620	
Emergency Venting Per API 12F [SCFH]	446,350	446,350
API 2000 Fire Flow/API 12F Fire Flow	0.28	-
Additional pressure case - upstream control valve failure [SCFH]		531,174
Additional pressure case - well blowdown [SCFH]		458,333
Device Pressure Rate [SCFH]	16,670	16,670
Percent capacity provided for maximum case	13.27	3.14
Required Vacuum Flowrate [SCFH]	4,493	5,025
Device Vacuum Flowrate [SCFH]	3,586	3,586
Percent capacity provided for vacuum	79.81	71.36
Solution for pressure	Enardo 2008 & Increase line size between tanks from 3" to 4"	Enardo 2008 & Increase line size between tanks from 3" to 4"
Solution for vacuum	Enardo 950 3"	Enardo 950 3"

3.4 Project No. 4 - API 2000 versus API 12F:

Project No 4 is an upstream oil production facility with three tanks: a 300 bbl capacity Gun Barrel Tank, a 300 bbl Condensate Tank, and a 300 bbl Water Tank. The Gun Barrel Tank was equipped with a Jayco JT-8 and the other two tanks were equipped with an Enardo 660 Pressure /

Vacuum relief device set at 8 oz/in^2. Based on the results of the study, detailed in Table 7 below, the following conclusions were drawn:

1. Comparing the generic API 12F tables with the more specific API 2000 calculations the conclusions are:
   a) The gunbarrel tank has an actual fire case relief flow rate that is 108% of that calculated using API 12F.
   b) The oil tank has an actual fire case relief flow rate of only 93% of that calculated using API 12F.
   c) The water tank has a fire case relief flow rate of only 32% of that calculated using API 12F.

2. Using the API 12F method and not using the API 2000 method means that the gunbarrel tank may be underprotected, while the oil and water tanks may be overprotected. There may be an opportunity for cost savings by not placing the same devices on all tanks.

3. The Jayco JT-8 does not provide nearly enough capacity for overpressure protection. It only provides 27% of what is needed for the gunbarrel tank.

4. The Enardo 660 does not provide nearly enough capacity for overpressure protection. Depending on the tank, it provides between 8 and 25 percent of what is needed.

5. The Jayco JT-8 does not provide enough capacity for vacuum protection. For the gunbarrel tank, it supplies about 63 percent of the requirement.

6. The Enardo 660 does not provide enough capacity for vacuum protection, depending on the tank, it supplies about 63 percent of the requirement.

The recommendations to make the system safe were:

1. For pressure
   a) increase the vapor inlet/outlet piping and nozzle nominal diameter from 2” to 10”
   b) add an Enardo 800-2” to the gun barrel and the water tanks
   c) add an Enardo 900-2” to the oil tank

2. For vacuum
   a) add an Enardo 800-2” to the gun barrel and the water tanks
   b) add an Enardo 900-2” to the oil tank

Table 7 - Project 4

	Tank 1	Tank 2	Tank 3
Service	**Gun barrel**	**Oil**	**Water**
Capacity [Barrels]	300		
Pressure/Vacuum Relief Device	Jayco JT-8	Enardo 660	
Pressure Setting [oz/in^2]	8		
Maximum Filling Rate [GPM]	42	0.44	41
Maximum Draining Rate [GPM]	42	252	252
Latitude Factor	0.32		
Insulation Factor	1		
C Factor	6.5		
Environmental Factor	1		
Normal out-breathing resulting from a maximum inflow of liquid into the tank [SCFH] API 2000	331	4	328
Thermal Outbreathing [SCFH] API 2000	387	387	387
Total Outbreathing [SCFH] API 2000	718	391	715
Initial Fire Case Temperature [Degrees F]	160	160	212
Initial Fire Case Flowrate Per API 2000 [SCFH]	407,592	352,834	121,316
Emergency Venting Per API 12F [SCFH]	378,700	378,700	378,700
API 2000 Fire Flow/API 12F Fire Flow	1.08	0.93	0.32
Additional pressure case - well blowdown [SCFH]	250,000		
Device Pressure Rate [SCFH]	111,000	31,154	31,154
Percent capacity provided for maximum case	27.23	8.83	25.68
Required Vacuum Flowrate [SCFH]	3,962	5,652	5,652
Device Vacuum Flowrate [SCFH]	2,500	3,586	3,586
Percent capacity provided for vacuum	63.10	63.45	63.45
Solution for pressure	Added Enardo 800 - 2". Increase piping from 2" to 10"	Added Enardo 900 - 2". Increase piping from 2" to 10"	Added Enardo 800 - 2". Increase piping from 2" to 10"
Solution for vacuum	Added Enardo 800 - 2"	Added Enardo 900 - 2"	Added Enardo 800 - 2"

3.5 Project No. 5 - API 2000 versus API 12F:

Project No 5 is an upstream oil production facility with two tanks: a 400 bbl Condensate Tank, and a 400 bbl Water Tank. Both tanks were equipped with a Enardo 660 Pressure / Vacuum relief device set at 12 oz/in^2. Based on the results of the study, detailed in Table 8 below, the following conclusions were drawn:

1. During normal operation, there is a significant amount of liquid flashing, resulting in vapors needing to exit the tank. Unfortunately, this is not taken into account in the API 12F method resulting in the API 12F flow rate being only about 58% of actual for the oil and 74% of actual for the water. This means that only using the API 12F method could result in a tank being underprotected during normal operation.

2. Performing final fire case provides insight into how hot the tank is going to get.

3. As the fire case proceeds, the oil tank temperature increases as the more volatile components are preferentially boiled off. Since the water is much closer to a pure component the temperature does not change significantly as the fire progresses.

4. The fire case flow rate increases as the fire proceeds on the oil tank. Comparing the generic API 12F tables with the more specific API 2000 calculations the conclusions are:

   a) The oil tank has an actual fire case relief flow rate of about 43% more than that calculated using API 12F.
   b) The water tank has an actual fire case relief flow rate of only about 37% of that calculated using API 12F.

5. Using the API 12F method and not using the API 2000 method means that the oil tanks may be underprotected and the water tanks may be overprotected. There may be an opportunity for cost savings by not placing the same devices on all tanks.

6. The Enardo 660 does not provide nearly enough capacity for overpressure protection. Depending on the tank it provides between 4 and 14 percent of what is needed.

7. The Enardo 660 does not provide enough capacity for vacuum protection. It supplies only about 64 percent of the requirement.

The recommendations to make the system safe were:

1. For the oil tank, add a nozzle and 10" Enardo 950 for pressure/vacuum

2. For the water tank, add a nozzle and 6" Enardo 960 for pressure/vacuum

Table 8 - Project 5

	Tank 1	Tank 2
Service	**Oil**	**Water**
Capacity [Barrels]	400	
Pressure/Vacuum Relief Device	Enardo 660	
Pressure Setting [oz/in^2]	12	
Maximum Filling Rate [GPM]	8	15
Maximum Draining Rate [GPM]	650	
Latitude Factor	0.25	
Insulation Factor	1	
C Factor	5	
Environmental Factor	1	
Normal Flashing Rate [SCFH] API 2000	229	30
Normal out-breathing resulting from a maximum inflow of liquid into the tank [SCFH] API 2000	59	117
Thermal Outbreathing [SCFH] API 12F	400	400
Thermal Outbreathing [SCFH] API 2000	392	392
Total Outbreathing [SCFH] API 2000	680	539
Total Outbreathing [SCFH] API 12F	400	400
% API 12F Outbreathing/API 2000 Outbreathing	58.8	74.2
Initial Fire Case Temperature [Degrees F]	452	216
Final Fire Case Temperature [Degrees F]	901	216
Initial Fire Case Flowrate Per API 2000 [SCFH]	553,445	165,755
Final Fire Case Flowrate Per API 2000 [SCFH]	640,014	
Emergency Venting Per API 12F [SCFH]	446,350	446,350
API 2000 Fire Flow/API 12F Fire Flow	1.43	0.37
Device Pressure Rate [SCFH]	24,310	24,310
Percent capacity provided for maximum case	4.39	14.67
Required Vacuum Flowrate [SCFH]	5,555	5,555
Device Vacuum Flowrate [SCFH]	3,586	3,586
Percent capacity provided for vacuum	64.55	64.55
Solution for pressure	10" Enardo 950	6" Enardo 960
Solution for vacuum	10" Enardo 950	6" Enardo 960

3.6 Project No. 6 - API 2000 versus API 12F:

Project No 6 is an upstream oil production facility with two tanks: a 400 bbl Condensate Tank, and a 400 bbl Water Tank. Both tanks were equipped with a Enardo A Pressure / Vacuum relief device set at 6 oz/in^2. Based on the results of the study, detailed in Table 9 below, the following conclusions were drawn:

1. Comparing the generic API 12F tables with the more specific API 2000 calculations the conclusions are:
   a) The oil tank has an actual fire case relief flow rate of about 47% more than that calculated using API 12F.
   b) The water tank has an actual fire case relief flow rate of only about 37% of that calculated using API 12F.

2. Using the API 12F method and not using the API 2000 method means that the oil tanks may be underprotected and the water tanks may be overprotected. There may be an opportunity for cost savings by not placing the same devices on all tanks.

3. The Enardo A does not provide nearly enough capacity for overpressure protection of the oil tank, only 62% of what is needed. It does, however, provide more than enough capacity for the water tank.

4. The Enardo A does not provide enough capacity for vacuum protection. It supplies about 43 to 66 percent of the requirement, depending on the tank.

The recommendations to make the system safe were:

1. For the oil tank, add 2 Enardo 951s

Table 9 - Project 6

	Tank 1	Tank 2
Service	**Oil**	**Water**
Capacity [Barrels]	400	
Pressure/Vacuum Relief Device	Enardo A	
Pressure Setting [oz/in^2]	6	
Maximum Filling Rate [GPM]	43.75	59
Maximum Draining Rate [GPM]	798	504
Latitude Factor	0.32	
Insulation Factor	1	
C Factor	6.5	
Environmental Factor	1	
Normal out-breathing resulting from a maximum inflow of liquid into the tank [SCFH] API 2000	351	468
Thermal Outbreathing [SCFH] API 2000	501	502
Total Outbreathing [SCFH] API 2000	852	970
Initial Fire Case Temperature [Degrees F]	200	212
Initial Fire Case Flowrate Per API 2000 [SCFH]	654,730	164,926
Emergency Venting Per API 12F [SCFH]	446,350	446,350
API 2000 Fire Flow/API 12F Fire Flow	1.47	0.37
Device Pressure Rate [SCFH]	409,554	409,554
Percent capacity provided for maximum case	62.55	248.33
Required Vacuum Flowrate [SCFH]	6,844	4,487
Device Vacuum Flowrate [SCFH]	2,993	2,993
Percent capacity provided for vacuum	43.73	66.70
Solution for pressure	2 Enardo 951s	

3.7 Project No.7 - API 2000 versus API 12F:

Project No 7 is an upstream oil production facility with four tanks: a 1000 bbl capacity Gun Barrel Tank (which is larger than the largest API 12F tank), a 300 bbl Oil Tank, a 750 bbl Water Tank and a 300 bbl Slop Tank. All tanks were equipped with an Enardo A Pressure / Vacuum relief device. Based on the results of the study, detailed in Table 10 below, the following conclusions were drawn:

1. Performing final fire case provides insight into how hot the tank is going to get. As the fire proceeds, the gunbarrel, oil and slop tank temperatures increase as the more volatile components are preferentially boiled off. Since the water is much closer to a pure component the temperature does not change significantly as the fire progresses.

2. Comparing the generic API 12F tables with the more specific API 2000 calculations

the conclusions are:

a) the oil tank has an actual fire flowrate of about 79% more than API 12F.
b) the water tank has an actual fire flowrate of only about 38% of API 12F.
c) the slop tank has an actual fire flowrate of about 79% more than API 12F.

3. Using the API 12F method and not using the API 2000 method means that the oil and slop tanks may be underprotected and the water tanks may be overprotected. There may be an opportunity for cost savings by not placing the same devices on all tanks.

	Tank 1	**Tank 2**	**Tank 3**	**Tank 4**
Service	**Gunbarrel**	**Oil**	**Water**	**Slop**
Capacity [Barrels]	1000	300	750	300
Maximum Filling Rate [GPM]	1584	984	1584	100
Maximum Draining Rate [GPM]	252	136.5	252	252
Latitude Factor	0.32			
Insulation Factor	1			
C Factor	6.5			
Normal Flashing Rate [SCFH] API 2000	4,167			
Normal out-breathing resulting from a maximum inflow of liquid into the tank [SCFH] API 2000	12698	7890	12698	802
Thermal Outbreathing [SCFH] API 12F		300	750	300
Thermal Outbreathing [SCFH] API 2000	1144	387	883	387
Total Outbreathing [SCFH] API 2000	18,009	8,277	13,581	,189
Total Outbreathing [SCFH] API 12F	-	300	750	300
% API 12F Outbreathing/ API 2000 Outbreathing	-	3.6	5.5	5.2
Initial Fire Case Temperature [Degrees F]	170		212	170
Final Fire Case Temperature [Degrees F]	327			327
Initial Fire Case Flowrate Per API 2000 [SCFH]	1,065,208	678,219	204,049	678219
Final Fire Case Flowrate Per API 2000 [SCFH]	744,370	473,942		473942
Emergency Venting Per API 12F [SCFH]		378,700	543,600	378,700
API 2000 Fire Flow/API 12F Fire Flow		1.79	0.38	1.79
Required Vacuum Flowrate [SCFH]	2,864	1,458	2,711	2384

4 Conclusions:

Based on the results of the various studies undertaken by Contek, it was determined that when using API 2000 to calculate the venting flow rate, the water tanks almost always have a lower fire case flow rate, typically about 32-38% of that calculated using API 12F. There are potential savings by putting smaller devices on the water tanks than the hydrocarbon tanks.

Using the API 2000 calculation method, the gunbarrel tanks always have a higher fire case relief flow rate, typically about 2-8% more than that calculated using API 12F. Therefore, using the API 12F method and not using the API 2000 method means that the gunbarrel tanks may be underprotected.

Using API 2000, the oil tanks almost always have a higher fire case relief flow rate, from 2-47% more than API 12F. Using the API 12F method and not using the API 2000 method means that the oil tanks may be underprotected.

By itself, one Enardo 660 thief hatch almost never provides sufficient overpressure protection and frequently is also insufficient for vacuum protection. Multiple devices are frequently needed.

The API 12F method does not take the flashing of the tank contents into account, which means that relying on the API 12F method may result in underprotected storage tanks.

Looking at different points in the fire duration provides more insight into the event. For example the maximum fire case temperature can be determined, which in turn could be used to discern potential for tank failure. Additionally, the worst case flow rate is not always at the start of the fire, but rather after the fire has been burning for some time.

The device installed on the 8" thief hatch can greatly restrict the venting capacity, and the installer should check with the suppliers to verify the actual venting capacity.

With tanks connected in network, it is recommended to determine if protection devices on other tanks can be counted.

5 References

[1] API 12F, Twelfth Edition, October 2008, Specification for Shop Welded Tanks for Storage of Production Liquids.

[2] API 2000, Seventh Edition, March 2014, Venting Atmospheric and Low-pressure Storage Tanks.

Overfilling Protection for Weak Tanks

Rahul Raman, Justin Phillips
Provenance Consulting LLC
1820 S. Mason Rd., Katy, TX 77450
Email rahul.raman@provenanceconsulting.com

Dr. Davide Moncalvo, Dr. Thomas Heidermann, Frank Heinrich
Braunschweiger Flammenfilter GmbH (PROTEGO®)
Industriestraße 11, 38110 Braunschweig, Germany
Email davide.moncalvo@protego.com

Steve Kostos
Bayer Business and Technology Services LLC
8500 West Bay Road, MS 52, Baytown, TX 77523-8727, USA

Prepared for Presentation at
American Institute of Chemical Engineers
2015 Spring Meeting
11th Global Congress on Process Relief
Austin, Texas
April 27-29, 2015

Overfilling Protection for Weak Tanks

Rahul Raman
Provenance Consulting LLC
1820 S. Mason Rd., Katy, TX 77450
Email rahul.raman@provenanceconsulting.com

Dr. Davide Moncalvo, Dr. Thomas Heidermann, Frank Heinrich
Braunschweiger Flammenfilter GmbH (PROTEGO®)

Justin Phillips
Provenance Consulting LLC

Steve Kostos
Bayer Business and Technology Services LLC

Keywords: overfilling protection, pressure vacuum relief valves, atmospheric tanks, weak tanks

Abstract

In this paper the use of pressure relief valves for protection against overfilling is discussed. Two practical approaches are herein discussed and their strengths are pin-pointed. The validation against measurements with two PROTEGO® relief valves has been successful for both methods. Nevertheless, further research should be done on this topic.

1. Introduction

The protection of near atmospheric or low pressure non-refrigerated aboveground tanks with a design pressure of 15 psig (1.034 bar-g) or below from overpressure due to overfilling by using low pressure relief valves is discussed in this paper. Low pressure relief valves are normally sized for gas flows according to API 2000 [1] and are generally installed at the top of the non-refrigerated tanks. Indeed, API 2000 suggests other solutions for liquid overfilling, like the installation of two repairable level sensing instruments and an independent actuator to close the filling valves and it refers to API RP 2350 [2] for full details.

However, electrical transducers and actuators must be maintained accurate by regular calibration and inspection as specified in API RP 2350, which may be a burden especially for remote tanks, where relief valves are usually already installed. In this context it may be conceivable to use the low pressure relief valve to discharge the liquid overfilling so that acceptable accumulation can be built up in the tank. This solution can be considered as an independent additional measure of protection for the case that the filling sensors are not recently calibrated or the electricity supply to them is interrupted. Indeed, one of the main hindrances in the application of low pressure relief valves for the prevention of liquid overfilling is the paucity of available data to estimate even roughly if an available valve would suffice to prevent overfilling.

In this paper the authors offer two simple and immediate approaches to estimate the liquid capacity of low pressure relief valves. The accuracy of the predictions is compared against measurements performed with two PROTEGO® valves at an independent testing partner. One approach is applicable when little information is available about the pressure relief valve and general literature data must be used, while the second approach may be useful when the air capacity of a selected valve is available.

2. Experimental background

PROTEGO® has several years of experience and a large database of dedicated water capacity measurements through flame arresters and low pressure relief valves.

In accordance with PROTEGO® policy the liquid flow capacity of low pressure relief valves is tested at the facility of an independent partner, Dr.-Ing. T. Bäumer GmbH. Additionally, the flow coefficient K_v is calculated from the measured capacity at each opening pressure in agreement with IEC 60534-2-3 [3] for non-choked turbulent flows using the formula in Eq. 1.

$$K_v = N_{water} \cdot Q \cdot \sqrt{\frac{\rho_1/\rho_0}{\Delta p}} \quad \text{with} \quad N_{water} = 1bar^{-0.5} \quad \text{[Eq. 1]}$$

In Eq. 1 Q is the volumetric flow rate measured at a certain difference between opening and back pressure, Δp, for a medium of density ρ_1, and ρ_0 is the density of water at 15°C (60°F). The constant N is adopted for dimensional correctness and is numerically equal to one for pressure drops given in bar. The opening pressures and the flow coefficients of end-of-line PROTEGO® valves VD/SV DN50 (2 in.) and DN80 (3 in.) at a set pressure of 20 mbar (8 in. WC) gauge and atmospheric discharge are reported in Table 1 for increased water flow capacity. From that table it can be seen that the flow coefficient K_v always reaches an asymptote.

Table 1. Measured opening pressures and calculated flow coefficients of PROTEGO® valves VD/SV DN50 and DN80 at several water capacities at a set pressure of 20 mbar-g.

	VD/SV DN50		VD/SV DN80	
Water capacity [m³/h]	Opening pressure [mbar]	Kv [m³/h]	Opening pressure [mbar]	Kv [m³/h]
10	24	64.5	25	63.2
20	38	102.6	26	124.0
30	77	108.1	29	176.2
40	135	108.9	32	223.6
50	241	101.9	38	256.5
60	298	109.9	48	273.9
70	405	110.0	64	276.7
80	539	109.0	83	277.7
90	657	111.0	105	277.7
100	812	111.0	129	278.4

The test apparatus is schematically described below.

2.1 Description of the test rig

The test facility for the flow capacity measurement of low pressure relief valves is reproduced in Fig. 1a. Water is pumped out from a 1 m³ (35.3 ft³) collection tank to the test valve from where it either flows back to the tank or is discharged in the sewers. The distances of the pressure sensors from the valve are in agreement with IEC 60534-2-3 [3]. A temperature sensor is located upstream from the valve to calculate the medium density ρ_1. The volumetric flow capacity is measured with a magnetic inductive mass flow meter in position FIR II.

The accuracy of instrumentation meets the requirements of IEC 60534-2-3, namely for the thermometers ± 1°C (± 1.8°F) and for the flow meter within 2 % of the measured value.

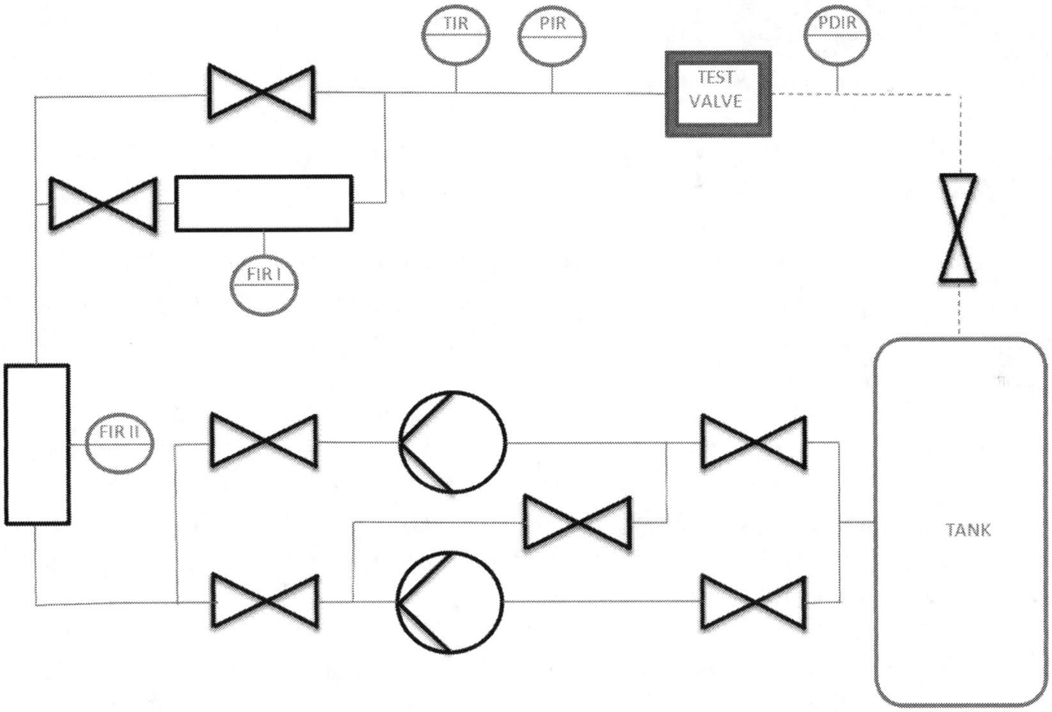

Fig. 1a. Test rig for measurement of relief valves water capacity.

Fig. 1b VD/SV tested with water

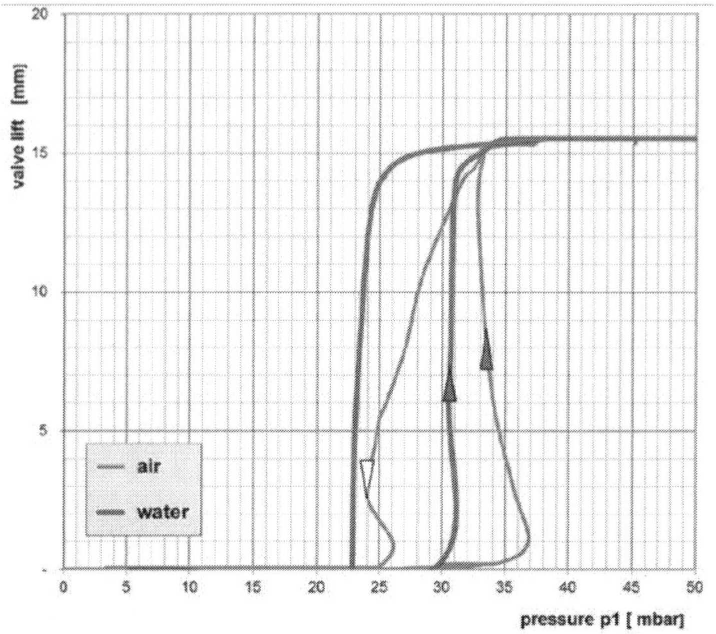

Fig. 1c. Example of Valve function test

2.2 *Valve function*

PROTEGO® relief valves start to open at the set pressure and require 10 % overpressure to achieve full lift. The full lift type technology allows the valve to be set just 10% below the maximum allowable pressure in the tank or equipment.

The opening characteristic is well defined for gas and vapors. But, does the relief valve, like the one in Fig. 1b, show the same characteristic for liquid flows? For any capacity estimation or calculation the valve characteristic curve should be known. To verify the valve characteristic curve the same test apparatus as given in Fig. 1a was used. The valve pallet was equipped with a distance sensor as a lift meter, which was used to indicate the position of the valve pallet, so that the sensor reads zero, when the valve is closed.

Figure 1c shows the lift as a function of tank pressure for air and water. The full lift type valve with a set pressure of 30 mbar-g (12 in. WC-g) shows a hysteresis, with different curves for the opening (pallet moving upwards) and closing (pallet moving downwards) of the valve. The valve opens with an overpressure of 10% for both air and water. For Newtonian liquids with a viscosity close to water, the opening characteristic of full lift relief valves is similar to that of gaseous fluids.

3. Approach using literature data

There are currently limited methods for sizing low pressure relief valves for liquid flow. The method in this section is a quick calculation to estimate the liquid flow through low pressure relief valves based on conventional techniques readily available for engineers at their disposal. A step-by-step approach is suggested and the reader is cautioned to follow all local and federal regulations, and internal company guidelines. Standard formulae are available for performing hydraulic analysis and engineers use a variety of tools at their disposal.

This method is based on first estimating the ideal flow capacity of a relief valve using available methods. The ideal flow capacity is then compared with experimental test data to correlate the calculation and test results. A correction factor is proposed as a function of the percentage overpressure and multiplied by the ideal flow capacity to obtain the corrected flow rate values. The correction factor is generated for both United States Customary System (USCS) and metric units. This is an extension of a previous work already presented at the DIERS User Group Spring 2014 Meeting [9].

3.1 Calculating flow coefficient

Relief system designers often use flow curves or software provided by vendors to size low pressure relief valves for weak tanks. PROTEGO® sizing program offers flow curves for valves in the product catalog [4]. It is suggested to use standard control valve sizing equations to back calculate C_v, the flow coefficient of the valve [3], or control valve manufacturer's method. Eq. 2 calculates the flow coefficient for a given pressure drop, upstream pressure P_{1a}, W flow rate of the pressure relief valve, and thermo-physical properties of the gas.

$$C_v = \frac{W}{39.612\sqrt{F_\gamma P_{1a}\rho_1}} \quad \text{[Eq. 2]}$$

The absolute upstream pressure, P_{1a}, should be evaluated 100% overpressure, i.e. at least two times the set pressure. At lower pressures the effect of the weighted pallets are more pronounced and the coefficients therefore need to be calculated at higher overpressures. Essentially at higher overpressure, the resistance of the weights becomes negligible and the valve can be treated as a pipefitting. However, the low pressure vents are typically used for low pressure tanks that have design pressure less than 15 psig.

In addition to the above constraints, the design pressure of the tank is defined at the tangent line and a relief system designer must consider the effects of static head especially for liquid flow through the relief valve. The resulting relief pressure is a combination of the static pressure at which sizing is done, the static head from tangent line to the seat of the atmospheric vent or top face of outlet flange on a pipe away model, and the velocity head to accelerate the fluid from tank nozzle to vent entrance. The relief pressure must not exceed the maximum allowable accumulation pressure of the tank. The capacity calculated using this method should exceed or equal the required capacity of the overpressure scenario.

The set pressure of such devices is typically well below the design pressure of the tank. Typical overpressures that are allowed for such valves range from 10% to roughly 200% of set pressure; see the paper of Moncalvo et al. on the effect of overpressure on valve performance [5].

3.2 Sizing Calculations

Hydraulic analysis of a piping system is a common practice and a day-to-day activity for engineers. CRANE Technical paper no. 410, published in both USCS and Metric units is a useful resource for engineers [6]. A basic review of the required methodology is highlighted here and is not all-inclusive. The pressure drop through valves, fittings, and pipe is calculated using,

$$\Delta P = 1.801 * 10^{-5} * \frac{K * \rho * Q^2}{d^4} \text{ USCS Units [Eq. 3]}$$

$$\Delta P = 225.2 \frac{K * \rho * Q^2}{d^4} \text{ Metric Units [Eq. 4]}$$

The total resistance to flow in the system is defined as a unit-less resistance coefficient, K. This is the sum of the valves, fittings, and pipes. For the purpose of this method, Eq. 4 is considered to be the ideal flow through the system and is next compared with experimental values to obtain a correction factor multiplier to estimate the corrected flow. The flow coefficient calculated using Eq. 1 can be converted to resistance coefficient using the following equations,

$$K_v = 0.865 * C_v \text{ [Eq. 5]}$$

$$K = 890.3 * \frac{d^4}{C_v^2} \text{ USCS Units [Eq. 6]}$$

$$K = 0.0016 * \frac{d^4}{K_v^2} \text{ Metric Units [Eq. 7]}$$

The diameter for the above equations can equal the throat diameter of the relief valve. Typically this can be considered equal to the inlet nominal pipe size of a low pressure relief valve.

3.2.1 Estimation of overpressure correction factor

The flow through the relief valve is affected by the weights of the pallet. At lower pressures the flow is lower than calculated ideal flow through a short piece of pipe. The flow coefficient is added as a resistance coefficient to find the actual resistance caused by the valve without additional resistance created by the weights. In addition to the above, a sharp edged conservative entrance plus the length from the vessel nozzle to the throat of the vent should be considered. In the example calculations shown below, the characteristic nozzle length of 12 inches was considered. The relieving fluid was water and its properties at standard conditions were incorporated for the calculations. The example calculations are shown for a DN80 (3-inch) VD/SV PROTEGO® relief valve in USCS units and the same approach was repeated in Metric units. Eq. 8 calculates K_p, the overpressure correction parameter. See Table. 2 for value of K_p for the given data obtained from flow test. Q_{est} for Table. 2 are calculated using Eq. 3 and Eq. 4 using an iterative calculation.

$$K_p = \frac{Q_{est}}{Q_{data}} \text{ [Eq. 8]}$$

Table 2. Estimated flow rate from Eq. 3 & Eq. 4 for a DN80 (3 inch) VD/DV relief valve

ΔP	% OP	Qest	Qdata	Kp	Qest	Qdata	Kp
[psi]	0%	[gpm]	[gpm]	[--]	[m^3 / hr]	[m^3 / hr]	[--]
0.363	25%	129	44	0.343	10	28.98	0.345
0.377	30%	131	88	0.672	20	29.58	0.676
0.421	45%	138	132	0.954	30	31.32	0.959
0.464	60%	145	176	1.211	40	32.88	1.216
0.551	90%	159	220	1.389	50	35.88	1.395
0.696	140%	178	264	1.482	60	40.32	1.489
0.928	220%	206	308	1.497	70	46.56	1.504
1.204	315%	234	352	1.502	80	53.04	1.509
1.523	425%	264	396	1.502	90	59.64	1.509
1.871	545%	292	440	1.506	100	66.12	1.513

Overpressure correction factor, K_p as a function of percent overpressure is shown in Figure 2 for both valves. A sum of least square method with a third order polynomial was regressed with a maximum constraint of 1.3 for the correction factor. This maximum constraint of 1.3 is chosen in order to obtain conservative flow rates, i.e. lower flow rates and is the lower bound. For higher overpressures the factor does not exceed 1.3 and is considered to be a conservative estimate. Figure 2 shows the Eq. 9 obtained by a sum of least squares method.

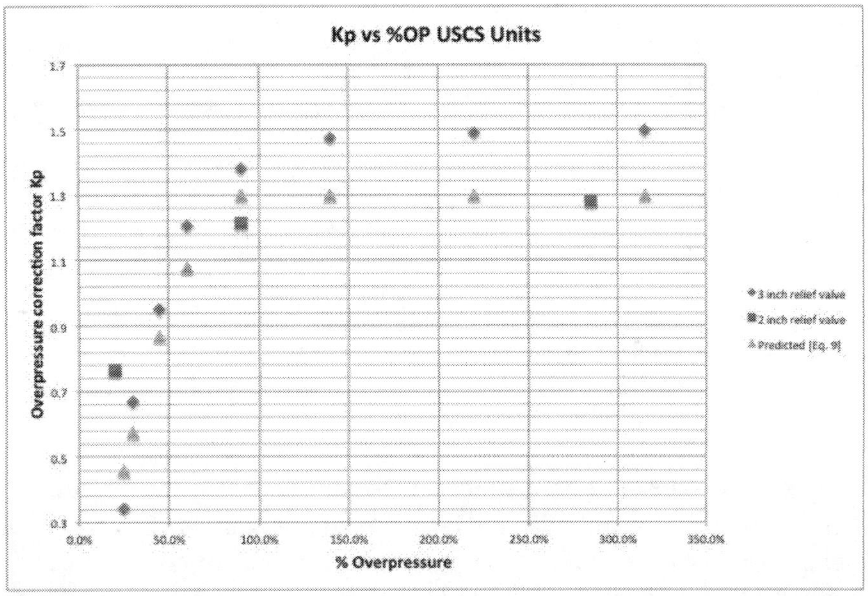

Figure 2. Overpressure Correction Factor Vs % Overpressure USCS Units

$$K_P = 0.577 * \%OP^3 - 2.595 * \%OP^2 + 3.645 * \%OP - 0.299 \text{ USCS Units [Eq. 9]}$$

$$K_P = 0.559 * \%OP^3 - 2.517 * \%OP^2 + 3.535 * \%OP - 0.25 \text{ Metric Units [Eq. 10]}$$

3.2.3 Equations for sizing low pressure relief valves

The correction factors are multiplied to Eq.3 and Eq. 4 to obtain the following equations that can calculate the flow through the relief valve as a function of pressure drop across the valve, throat diameter of the valve, density of fluid, and resistance coefficient.

$$Q = \sqrt{\frac{\Delta P * d^4}{1.801 * 10^{-5} * K * \rho}} * K_p \text{USCS Units [Eq. 11]}$$

$$Q = \sqrt{\frac{\Delta P * d^4}{225.2 * K * \rho}} * K_p \text{ Metric Units [Eq. 12]}$$

The flow capacities of the valves are calculated and compared with test data. Table 3 shows the values for both DN50 (2") and DN80 (3") valves along with error percentages. The Kp curve is modeled using a constraint of 1.3 to ensure that we do not over predict the capacity of the valve. The error percentages can be within a ± 7.5 % error if an unconstrained sum of least square method is performed. .

Table 3. Percentage flow error against measured data for PROTEGO® VD/SV valves

		VD/SV DN50 (2 inch)			VD/SV DN80 (3 inch)		
Qdata (gpm)	Qdata (m3 / hr)	% OP	Qerror (USCS)	Qerror (Metric)	%OP	Qerror (USCS)	Qerror (Metric)
44	10	20%	-57%	53%	25%	35%	40%
88	20	90%	7%	-6%	30%	-14%	-11%
132	30	285%	2%	-1%	45%	-8%	-8%
176	40	575%	1%	0%	60%	-11%	-11%
220	50	1105%	8%	-7%	90%	-6%	-7%
264	60	1390%	0%	1%	140%	-12%	-13%
308	70	1925%	-0.1%	1%	220%	-13%	-14%
352	80	2595%	1%	0%	315%	-13%	-14%
396	90	3185%	-1%	2%	425%	-13 %	-14%
440	100	3960%	-1%	2%	545%	-13%	-14%

The method proposed here seems to be a quick, accurate, and easy technique that can be easily adopted by process engineers for calculating liquid flow capacities through low pressure relief valves. The reader is cautioned not to use the above method in the range of overpressures below 30% as less conservative values are obtained. The estimated capacities for overpressures above

30% are within reasonable tolerances. The user is cautioned that this provides the relief capacity of the valve only and effects of static and velocity head must be accounted for in determining the adequacy of the entire relief system.

4. Approach using manufacturer's relief valve air capacity data

This experimental rule-of-thumb consists of finding a relationship between the capacity of air and water flowing through the valve.

In order to develop such a relationship a connection between the flow rates and the pressure drop through the device for both media must be derived. By a cross comparison among several manufacturers a polynomial relationship like the one in Eq. 13 between air capacity and pressure drop seems to fit the curves at best. In that equation the parameter K_{air} can be treated as the flow coefficient for air flows and the safety factor φ is a fitting parameter to consider the ratio between the certified and the measured flow capacity. If no exact value for φ is given by the manufacturer for the specific valve, it can be assumed equal to 1.1.

$$Q_{air} = \varphi \cdot N_{air} \cdot K_{air} \cdot \Delta p^{a} \quad \text{with } N_{air} = 1\, bar^{-a} \quad \text{[Eq. 13]}$$

For water flows a similar polynomial relationship can be derived by adjusting Eq. 1 as shown in Eq. 14. For sake of simplicity, the flow coefficient can be assumed roughly constant and equal to the asymptotic value extracted from Table 1.

$$Q_{water} = N_{water} \cdot K_{water} \cdot \Delta p^{0.5} \quad \text{with } N_{water} = 1 bar^{-0.5} \quad \text{[Eq. 14]}$$

In this case the flow coefficient for water is measured but that is not the case in the normal practice. The simplest method to estimate it is to consider that at near atmospheric pressures air tends to behave as a weakly compressible or almost incompressible medium. Therefore, on behalf of the ideal nozzle theory applied to safety valves as in ISO 4126 [4] or API 520 [8] the ratio of the flow coefficients is equal to the square root of the inverse of the density ratio, Eq. 15.

$$\begin{aligned} K_{water}/K_{air} &\approx N_{air}/N_{water} \cdot \varphi \cdot Q_{water}/Q_{air} \cdot \Delta p^{a-0.5} \\ &\approx N_{air}/N_{water} \cdot \varphi \cdot \sqrt{\rho_{air}/\rho_{water}} \cdot \Delta p^{a-0.5} \\ &\approx (\Delta p/bar)^{a-0.5} \cdot \varphi \cdot 0.03165 \quad \text{[Eq. 15]} \end{aligned}$$

For the two valves VD/SV DN50 and DN80 based on Eq. 15 the approximated water flow coefficients obtained from the TÜV certified air flow curves are respectively 105.7 and 231.2 Nm^3/h, assuming the parameter *a* close to 0.5. The roughly estimated water flow coefficients are rather close to the measured ones, respectively 111 and 278 Nm^3/h. based on the obtained water flow coefficients the water flow capacities are calculated using Eq. 3 and compared against the measured values of Table 1 in Table 4.

From Table 4 it can be easily seen that the proposed rule of thumb results in reasonable agreement with the measured water flow rates except for the range of very low opening pressures. The authors recommend for very low capacities to modify the control valve equation in Eq. 3 by considering a low capacity correction factor K_v, postulated in API 520 [8] or by Darby and Molavi [8] as shown in Eq. 16. A validation of this formula is still ongoing.

$$Q_{water} = K_{water} \cdot K_v \cdot \Delta p^{0.5} \quad \text{with} \quad N_{water} = 1bar^{-0.5} \quad \text{[Eq. 16]}$$

Table 4. Measured vs. calculated water flow rates at several opening pressures for PROTEGO® valves VD/SV DN50 and DN80 at the set pressure of 20 mbar-g (8 in. WC-g).

	VD/SV DN50			VD/SV DN80		
Measured Water capacity [m³/h]	Opening pressure [mbar]	Calculated Water capacity [m³/h]	Error [%]	Opening pressure [mbar]	Calculated Water capacity [m³/h]	Error [%]
10	24	16.37	63.7%	25	36.56	265.6%
20	38	20.60	3.0%	26	37.28	86.4%
30	77	29.33	2.2%	29	39.37	31.2%
40	135	38.84	2.9%	32	41.36	3.4%
50	241	51.89	3.8%	38	45.07	9.9%
60	298	57.70	3.8%	48	50.65	15.6%
70	405	67.27	3.9%	64	58.49	16.4%
80	539	77.60	3.0%	83	66.61	16.7%
90	657	85.68	4.8%	105	74.92	16.8%
100	812	95.25	4.8%	129	83.04	17.0%

	VD/SV DN50			VD/SV DN80		
Measured Water capacity [gpm]	Opening pressure [psi]	Calculated Water capacity [gpm]	Error [%]	Opening pressure [psi]	Calculated Water capacity [gpm]	Error [%]
44.03	0.348	72.07	63.7%	0.363	160.97	265.6%
88.06	0.551	90.70	3.0%	0.377	164.14	86.4%
132.09	1.117	129.14	2.2%	0.421	173.34	31.2%
176.11	1.958	171.01	2.9%	0.464	182.10	3.4%
220.14	3.495	228.46	3.8%	0.551	198.44	9.9%
264.17	4.322	254.05	3.8%	0.696	223.01	15.6%
308.20	5.874	296.18	3.9%	0.928	257.52	16.4%
352.23	7.818	341.66	3.0%	1.204	293.28	16.7%
396.26	9.529	377.24	4.8%	1.523	329.86	16.8%
440.29	11.777	419.37	4.8%	1.871	365.61	17.0%

5. Conclusion

In this paper two approaches for the prediction of water capacity in case of overfilling are presented, either assuming the availability of certified or published values for the air capacity of pressure vacuum valves or using literature data. Both methods have been compared with measurements of water capacity done with two PROTEGO® pressure relief valves and both approaches delivered reasonable agreement. The user is cautioned that calculation of the venting capacity at the maximum tank venting pressure must include the static and velocity head due to the flow of the fluid. The limits of application of the two methods proposed here are at the moment under investigation.

Future work would be to further test both approaches with additional types of pressure relief valves to broaden the range of applications. Nevertheless, the authors still recommend to contact the valve manufacturer for dedicated certified water capacity measurements.

6 Notation

Notation	USCS Units	Metric Units
Mass flow rate, W	Lbs_m/hr	Kg/h
Fγ, Specific heat ratio factor	Dimensionless	Dimensionless
X_T, Pressure differential ratio factor	Dimensionless	Dimensionless
Upstream pressure, P_{1a}	Psia	Pa
Static pressure at valve inlet, P_1	psig	Pa-g
Flow coefficient	Cv = GPM of water per psi pressure drop	Kv = cubic meters per hour at a pressure drop of one kilogram per square centimeter.
Pressure drop, ΔP	Psi	Pa
Resistance Coefficient, K	Dimensionless	Dimensionless
Volumetric flow rate, Q	Gpm (US)	Liters per minute (lpm)
Density of fluid, ρ	Lb_m/ft^3	Kg/m^3
Diameter of piping system, d	Inches	Millimeter
Overpressure correction factor	Dimensionless	Dimensionless

7 References

[1] API 2000 Seventh Edition, Venting Atmospheric and Low pressure Storage Tanks, 2014.
[2] API Recommended Practice 2350 Third Edition, Overfill Protection for Storage Tanks in Petroleum Facilities 2005.
[3] IEC 60534-2-3: Industrial Process Control Valves. Part 2-3: Flow Capacity. Test Procedures, 1997.
[4] PROTEGO® sizing program, Available at http://www.protego.com/engineering/design.html, Accessed on February 8, 2015.
[5] D. Moncalvo, M. Davies, M. Barfield, C. Mason, Influence of Overpressure in Pressure Vacuum Safety Valves on Emission Reduction and Explosion Risk Minimization of Atmospheric Storage Tanks, Proc. 11th GCPS, Auston, TX, 2015.
[6] CRANE Technical Paper No. 410, Flow of fluids through valves, fittings and pipe, 2014.
[7] ISO 4126-1, Safety devices for protection against excessive pressure -- Part 1: Safety valves, 2013.
[8] R. Darby and K. Molavi, Viscosity correction factor for emergency relief valves, Proc. Saf. Prog. 16 (1997), 80 -82.
[9] Raman - Overflow Protection for Weak Tanks, DIERS User Group Spring 2014 Meeting.

Changes Between API STD 521 6th Ed and 5th Ed Cataloged

Dustin Smith
Smith & Burgess LLC
7600 W Tidwell, Houston, TX
Dustin.Smith@smithburgess.com

John Burgess
Jarrett Doe
Smith & Burgess LLC
7600 W Tidwell, Houston, TX

Prepared for Presentation at
American Institute of Chemical Engineers
2015 Spring Meeting
11th Global Congress on Process Safety
Austin, Texas
April 27-30, 2015

UNPUBLISHED

Changes Between API STD 521 6th Ed and 5th Ed Cataloged

Dustin Smith
John Burgess
Jarrett Doe
Smith & Burgess LLC
7600 W Tidwell, Houston, TX
Dustin.Smith@smithburgess.com

Keywords: API STD 521, PSM Compliance, Relief Systems Design, ASME Pressure Relief Device Design, Analysis Overpressure

Abstract

API STD 521, Pressure-relieving and Depressuring Systems, is perhaps the most thorough published guideline on how to size relief devices and flare systems. This paper reviews the differences between the 6th Edition of API STD 521 (Published in 2014) and the 5th edition (Published in 2007). The authors will review changes and highlight several new requirements that are deemed critical to relief systems design. The paper is based on a comprehensive review of the two editions and is a valuable resource for any engineer or process designer who needs to be knowledgeable of the latest relief systems recommendations.

1. Background

This paper provides the reader with a summarized listing of the changes between the current (6th edition) and past (5th edition) of *API Standard 521, Petroleum and natural gas industries—Pressure-relieving and depressuring systems* (hereafter referred to as "521") [1, 2]. This API document provides guidance for determining overpressure scenarios, determining the required relief rate for sizing a relief device, and criteria and considerations for sizing a disposal system.

As Table 1 shows, a considerable amount of guidance has been added to the 6th edition of 521.

Table 1. Summary of Guidance in the 6th/5th Editions of API STD 521

Items	5th Edition (2007)	6th Edition (2014)
"Should" Appears	515	611
"Shall" Appears	137	154
Pages	196	248
ISO Co-branded	Yes	No

The 5th edition was co-branded as an International Organization for Standards (ISO 23251) and American Nation Standards Institute (ANSI) standard. The current (as of the time of writing) 6th Edition is neither an ISO nor ANSI standard.

Note: Do not add page numbers. Do not refer to page numbers when referencing different portions of the paper

In addition, a special note was added to ensure that the users knew relief systems design and engineering problems are installation specific, and need to be designed/engineered by a competent person knowledgeable in the art of pressure relief system design.

1.1 Terms and usage

The following familiarizes the reader of this paper with acronyms and terms that may be encountered.

1.1.1 SCPRS

The API Committee on Refinery Equipment (CRE) subcommittee that is responsible for 521 is the Subcommittee on Pressure-Relieving Systems (SCPRS).

1.1.2 Section References

Section numbers that are prefaced by "§" reference the API standard and not this paper. Unless otherwise mentioned, the section numbers referred to are for the 6th ed. of 521.

2. Summary of changes

The changes are broken down into the following categories. The remainder of this paper will follow the order shown in the following table.

Table 2. Organization of the Summary of the Changes.

Paper Section	Topic	Items
2.1	Moved Material	104
2.1	Clarified	12
2.2	New Definitions	14
2.3	Deleted	9
2.4	Equation Modifications	2
2.5	New Sections	24
2.6	Additional Requirements or Guidance	29
	Total	*194*

2.1 Moved Material / Clarified

The SCPRS spent a lot of time trying to organize and clarify this standard. This work was performed to ensure a more useful resource for the end users. As a result a majority of the changes are text moving from one section to another. These changes are numerous, pervasive and do not change the guidance. Therefore, they are not further elaborated in this paper.

2.2 New Definitions

The 6th edition of API STD 521 has had additional definitions added into §3.1 Terms and Definitions. The following terms have been defined.

§3.1.5 availability	§3.1.49 molecular mass
§3.1.8 blowdown drum	§3.1.51 open fire
§3.1.16 choke valve	§3.1.58 pressure-relief device
§3.1.19 combustion efficiency	§3.1.60 pressure system
§3.1.20 confined fire	§3.1.69 required relief rate
§3.1.36 fuel-controlled fire	§3.1.70 risk
§3.1.44 lift-off (Substituted for blow-off)	§3.1.85 ventilation-controlled fire

A few of the definitions have been further discussed in this paper. The reader of this article is encouraged to further read the standard to get the official definition which is not included herein.

2.2.1 Confined Fire / Open Fire

The SCPRS introduced a new definition, concept, and requirements. These specify that if a fire is confined, the heat flux requirements are higher than an open pool fire (see §4.4.13.3). The open pool fire is the historical fire method in 521.

2.2.2 Risk

In the 5th edition, risk was addressed throughout the document and specifically in Annex E. The 6th edition provides the first definition for use in 521 and related relief device decisions.

2.3 Deleted

The following items were addressed in the 5th edition of 521, but have not been included in the 6th edition. The section numbers refer to the 5th edition.

§3.7	blowdown (Definition)	§3.46	marked burst pressure (Definition)
§3.9	breaking-pin device (Definition)	§3.53	pin device (Definition)
§3.17	closed disposal system (Definition)	§3.71	shear pin device (Definition)
§3.37	huddling chamber (Definition)	§7.3.1.2	design of relief device inlet piping

The SCPRS deleted the design of relief systems inlet piping from API 521 as it is covered in API STD 520 Part II.

2.4 Equation Modifications

The following equations were modified between the 5th and 6th editions.

2.4.1 Estimation of Purge Gas Requirements (Eq. 46/47 6th Ed.)

The Ci term is for the volume fraction of a component in the purge gas and has been changed from a percentage to a fraction. This is for Equations 46/47 in §5.7.6.2 in the 6th edition (Equations 59/60 in §7.3.3.3.3 in the 5th edition).

2.4.2 Equation determining the droplet Velocity in a Flare Knockout Drum (Eq. 55-57, 6th Ed.)

In the 6th edition, the density of the vapor term has been added to the equation determining the droplet velocity in a flare knockout drum. This applies to Equations 55 - 57 in §5.7.9.5 in the 6th edition (Equations 37 - 39 in §7.3.2.1.2 in the 5th edition). Note that these this change was included in the 5th edition with the May 2008 Addendum.

2.5 New Sections

The following sections were added or significantly modified between the 5th and 6th editions.

2.5.1 Additional Considerations Involving Pumps (§4.4.2.2)

The 6th edition provides new guidance on overpressure protection for pumps. In general, centrifugal pumps don't require overpressure protection.

2.5.2 Additional Considerations Involving Reciprocating Compressors (§4.4.2.3)

The 6th edition provides new guidance on overpressure protection for reciprocating compressors.

2.5.3 Choke Valve Failure (§4.4.8.6)

The 6th edition provides new guidance on overpressure protection specifically related to choke valves failing.

2.5.4 Estimating the Required Relief Rate from Check Valve Failure (§4.4.9.3.3)

The 6th edition provides new guidance on estimating the required relief rate from a failed check valve. One method is based on the size of the valve and the other is based on the normal forward flow.

2.5.5 Reciprocating Compressor Rod Packing Failure (§4.4.9.4)

The 6th edition provides new guidance on overpressure protection for reciprocating compressors specifically related to the failure of rod packing and how to estimate the required relief rate.

2.5.6 Alternative Method to Estimate Overpressure from External Fire (§4.4.13.2.4.4)

The 6th edition provides new guidance on overpressure protection from fire in the event that a fluid phase change does not occur.

2.5.7 Confined Pool Fires (§4.4.13.3)

The 6th edition provides new guidance on estimating the required relief rate from a fire inside a building or in a confined area. *The methods in the previous (5th and older) editions of 521 predict a lower relief rate for these cases.*

2.5.8 Additional Considerations for Plate and Frame Exchanger (§4.4.14.4)

The 6th edition provides new guidance on overpressure protection for plate and frame exchangers. In general, full plate failure is not considered likely but a failure from a hole caused by corrosion should be considered.

2.5.9 Overpressure Prevention during Maintenance (§4.4.16)

The 6th edition includes a new section providing guidance for considering overpressure protection during maintenance activities.

2.5.10 Initiation of Depressuring (§4.6.2)

The 6th edition provides a new section with descriptive information and considerations for initiating an emergency depressuring system including a fire and gas detection system.

2.5.11 Depressuring Low Temperature Considerations (§4.6.3)

The 6th edition provides a new section with descriptive information on the potential for auto-refrigeration from depressuring some process systems.

2.5.12 Flare and Outlet Piping Acoustic Fatigue (§5.5.12)

The 6th edition provides a new requirement with the addition of substantive new guidance and requirements on evaluating acoustic fatigue in flare headers primarily based on the work of Carucci and Mueller [3, 4]. *This section adds an additional requirement for flare systems design based on past incidents.*

2.5.13 Flame Stability (§5.7.2.4)

The 6th edition provides a new section with descriptive information on flare flame stability and common language to describe flame instability.

2.5.14 Noise from Smokeless Flaring (§5.7.3.2.11)

The 6th edition provides a new section with descriptive information on the issues associated with noise caused when adding assist fluid to attempt smokeless flaring.

2.5.15 Pilot Ignition Systems and monitoring (§5.7.7.3 and §5.7.7.4)

The 6th edition provides new sections with descriptive information on pilot ignition systems and methods for monitoring pilots. These sections refer the users to API STD 537 for additional guidance. [5]

2.5.16 Flare Knockout Drum (§5.7.9.1)

The 6th edition provides a new section with descriptive information in regards to the location of a flare knockout drum.

2.5.17 Risk of Overfilling Flare Knockout Drum (§5.7.9.6)

The 6th edition provides new requirements that *shall* be considered (i.e. are required) in reference to flare knockout drums being overfilled (or those that have the potential to be overfilled).

2.5.18 Production Facility Flare Knockout Drums (§5.7.9.8)

The 6th edition provides a new section giving guidance for knock out drum sizing due to the special nature of production facilities.

2.5.19 Siting Considerations for Flares (§5.7.10)

The 6th edition provides a new section with guidance associated with facility siting concerns for flares (topography and dispersion).

2.5.20 Alternative Analytical Method (Fire, Annex A.3)

The 6th edition provides a new section that provides an alternative means to estimate the heat loads from an external fire. The analytical method can apply to situations where the guidance in the main body may not be applicable (e.g. vapor/jet fires).

2.6 Additional Requirements or Guidance

The following existing sections were modified in ways that, in the opinion of the Authors, provide additional requirements or guidance. If the reader of this paper designs or audits relief systems, then he or she is encouraged to purchase a copy of the 6th edition of API STD 521 to ensure designs are per the most recent guidance. The information contained herein is insufficient for design.

2.6.1 Operator Error/ Response Criteria (§4.2.5)

The Operator Error and Effect of Operator Response sections (§4.2.3 and §5.4) in the 5th edition combined to create §4.2.5 with additional recommendations for operator response in overpressure protection. Identifies criteria for deciding to consider operator intervention for relief systems design.

2.6.2 Inadvertent Valve Opening (§4.4.9.2)

The 5th edition sections on Inadvertent Valve Opening and Opening Manual Valves (§4.3.3 and §5.17) were combined and new guidance on relief rate determination provided. Clarification that opening of multiple valves may be credible under certain circumstances was added.

2.6.3 Check Valve Leakage or Failure (§4.4.9.3)

The 6th edition moved this section and provides new guidance on determining the required relief rate. The wording is less strong (as compared to the 5th Edition) and additional guidance on inspection and testing of check valves is provided.

2.6.6 Relief Systems Documentation Requirements (§4.7)

In the 6th edition, the Relief System Design Documentation section was overhauled. The language from the 5th edition was softened to ensure the guidance was not confused as a requirement. Additional suggestions on nature, type, and goal of relief system design documentation were included.

2.6.7 Level Instrumentation (§4.4.7.3)

The section providing guidance on level instrumentation was moved and four additional criteria for consideration of level valve independence were added.

2.6.8 Depressuring Systems Design Criteria (§4.6.5)

The 6th edition provides a new requirement that the discharge location of a depressuring system *shall* be a safe location.

2.6.9 Liquid Overfilling (§4.4.7)

The 6th edition provides more concrete language around evaluating the design of systems for overpressure protection from liquid overfilling.

2.6.11 Header Segregation (§5.4.4)

The 6th edition consolidated the guidance related to header segregation and added a list of criteria to consider when designing a flare system with for multiple relief devices, all of which are recommendations (should).

2.6.12 Pipe Anchors, Guides, and Supports (§5.5.16)

The 6th edition updates and consolidates information on *Pipe Anchors, Guides, and Supports* into a section of the same name providing additional guidelines to design effective pipe anchors and supports.

2.6.13 Flaring Guidance on Smokeless Requirements (§5.7.3.1)

The 6th edition provides additional guidance as to when the user is to and how the user should specify the smokeless requirements for a flare.

2.6.14 Flares with Smoke Suppression (§5.7.3.2)

The 6th edition provides additional guidance on how flares achieve smokeless operation and/or suppress the generation of smoke.

2.6.15 Ground Flares (§5.7.4.1)

Additional guidance provided on the design, specification, and operation of ground flares including a reference to API STD 537 for more detailed description.

2.6.16 Steam Injection for Smokeless Operation (§5.7.3.2.2)

The notes for Table 14 (Table 11 in 5th edition) are updated to ensure that the user is properly specifying and designing the steam injection for smokeless flare operation.

2.6.17 General Purge Requirements (§5.7.6)

The section is updated and organized to include all design aspects of purging a disposal system to reduce the possibility of flashback.

2.6.18 Seal Drum Design (§5.7.8.7)

The 6th edition provides new equation / criteria to determine the minimum required diameter for a vertical seal drum and minimum free liquid surface area of a horizontal seal drum (Equations 50 and 51).

2.6.19 Flare Knockout Drum (§5.7.9.2)

The 6th edition provides new guidance "In general, vapor outlet nozzles should *not* be fitted with any devices (e.g. deflection plates)." which is *opposite* of previous guidance. This is due to an incident where a flare drum internal blocked the flare system of a platform on the Visund field. For more information on this incident see the PSA investigation online [6].

2.6.20 Flare Liquid Separation Droplet Size (§5.7.9.4)

The 6th edition provides additional guidance for different types of flares that can handle different liquid loading and droplet sizes.

2.6.20 Vent Stacks (§5.8.8)

The 6th edition provides additional guidance given for vent tips in applications where existing space and other restrictions have to be considered.

2.6.21 Sizing of Vent Stacks (§5.8.10.3.1)

The 6th edition provides updates to the calculation of noise level for vent stacks. If the reader is designing a vent stack they should be warned that the Vent Stacks section (§5.8.10) contains *shall* 8 different times.

2.6.23 Sample Calculations (Annex C)

The 6th edition provides all the sample calculations in a single location, Annex C.

2.6.24 High-integrity Protection Systems (Annex E)

The 6th edition changes the ASME Boiler and Pressure Vessel Code exception that allows instrumentation in lieu of relief devices from Code Case 2211-1 to UG-140. This was adopted in the 2008 addendum to the 2007 publication of ASME B&PVC Section VIII Division I. [7] The author has previously written more information on the specific requirements to use instrumentation / system design instead of relief devices in accordance with UG-140. [8]

3. Conclusion

The 6th Edition of API STD 521 has been reformatted and updated. This document has changed extensively from the previous addition. The following lists the largest changes (in the view of the authors) for this document:

- Guidance on Flare Knockout Drum internals is ***opposite*** of previous editions of this standard.
- New requirements for acoustically induced vibrations are included.
- Confined external fires are defined and guidance is provided.

4. References

[1] API Standard 521, FIFTH ED., JANUARY 2007 (w/ ERRATA JUNE 2007) ADDENDUM, MAY 2008 ISO 23251 (Identical), Petroleum and natural gas industries—Pressure-relieving and depressuring systems

[2] API Standard 521, SIXTH ED., JANUARY 2014, Petroleum and natural gas industries—Pressure-relieving and depressuring systems

[3] V. A. Carucci and R. T. Mueller, "Acoustically Induced Piping Vibration in High Capacity Pressure Reducing Systems," ASME 82-WA/PVP-8; 1982.

[4] J. Cowling, "Design Strategies for Acoustically Induced Vibration in Process Piping," INTER-NOISE 2012 Proceedings, August 9–22, 2012, New York, available from Institute of Noise Control Engineering of the USA (INCE/USA)

[5] API Standard 537, Flare Details for General Refinery and Petrochemical Service

[6] "PSA completes investigation of Visund incident," June 2006, http://www.psa.no/news/psa-completes-investigation-of-visund-incident-article2733-878.html (accessed March. 15, 2015)

[7] American Society of Mechanical Engineers. 2007 ASME Boiler & Pressure Vessel Code: Section VIII,Division I. with 2008 Addendum

[8] Smith, D. J. (2014), Safety instrumented systems in lieu of pressure relief valves. Proc. Safety Prog., 33: 345–349. doi: 10.1002/prs.11681

Changes Between API STD 520 – Part II 6th Ed and 5th Ed Cataloged

Dustin Smith
Smith & Burgess LLC
7600 W Tidwell, Houston, TX
Dustin.Smith@smithburgess.com

John Burgess
Jessye Palladino
Smith & Burgess LLC
7600 W Tidwell, Houston, TX

Prepared for Presentation at
American Institute of Chemical Engineers
2015 Spring Meeting
11th Global Congress on Process Safety
Austin, Texas
April 27-30, 2015

UNPUBLISHED

Changes Between API STD 520 – Part II 6th Ed and 5th Ed Cataloged

Dustin Smith
John Burgess
Jessye Palladino
Smith & Burgess LLC
7600 W Tidwell, Houston, TX
Dustin.Smith@smithburgess.com

Keywords: API STD 520 – Part II, PSM Compliance, Relief Device Installation, Relief Device Inlet Pressure Losses, Relief Device Backpressure Limits

Abstract

American Petroleum Institute (API) STD 520 – Part II, Sizing, Selection, and Installation of Pressure-Relieving Devices in Part II—Installation, is perhaps the most thorough published guideline on how to install relief devices. This paper reviews the differences between the 6th Edition of API STD 520 Part II (Published in 2015) and the 5th edition (Published in 2003, Reaffirmed in 2011). The authors will review changes and highlight several new requirements that are deemed critical to relief system installations. The paper is based on a comprehensive review of the two documents and is a valuable resource for any engineer or process designer who needs to be knowledgeable of the latest relief systems recommendations.

1. Background

This paper provides the reader with a summarized listing of the changes between the current (6th edition) and past (5th edition) of *API Standard 520 Part II- Sizing, Selection, and Installation of Pressure-Relieving Devices in Refineries Part II—Installation* (hereafter referred to as "520-Part II") [1, 2]. This API document provides guidance for the installation of pressure relief valves, rupture disks, and inlet and outlet piping considerations for pressure relief devices.

As Table 1 shows, a considerable amount of guidance has been added to the 6th edition of 520-Part II.

Table 1. Summary of Guidance in the 6th/5th Editions of API STD 520-Part II

Items	5th Edition (2011)	6th Edition (2015)
"Should" Appears	102	172
"Shall" Appears	38	58
Pages	29	55

In addition, note that the 5th edition was issued as an API recommended practice. The current (as of the time of writing) 6th Edition is issued as a API standard.

Note: Do not add page numbers. Do not refer to page numbers when referencing different portions of the paper

In addition, the scope of the document was further clarified to ensure that the users knew that the standard covers the installation methods of pressure relief devices (PRDs) for systems with maximum allowable working pressure (MAWP) equal to or greater than 15 psig. This edition of API 520-II discusses the installation of PRDs for the use with compressible, incompressible, and 2-phase systems. However, it is noted that the standard does not cover systems with special applications that need unique installations. Additionally, this standard is stated to be used in tandem with sound business, scientific, engineering, and safety judgments.

1.1 Terms and usage

The following helps the reader of this paper with some of the acronyms and terms.

1.1.1 SCPRS

The API Committee on Refinery Equipment (CRE) subcommittee that is responsible for this document is the Subcommittee on Pressure-Relieving Systems (SCPRS).

1.1.2 PRDs

Pressure relieving devices, that refers to pressure relief valves (PRVs) and/or rupture disks (RDs).

1.1.3 Section References

Unless otherwise mentioned, the section numbers referred to are for the 6th ed. of 520-Part II. Section numbers that are prefaced by "§" reference the API standard and not this paper.

2. Summary of changes

The changes are broken down by categories in Table 2. The remainder of this paper will follow the order shown in the following table.

Table 2. Organization of the Summary of the Changes

Paper Section	Topic	Items
2.1	Moved Material	64
2.1	Clarified	53
2.2	New Figures	10
2.3	Deleted	8
2.4	Equation Modifications	2
2.5	New Sections	23
2.6	New Requirements	4
	Total	164

2.1 Moved Material / Clarified

The SCPRS spent a lot of time on this document to try and organize and clarify this standard. This work was performed to ensure that it was more useful to the end users. As a result a majority of the changes are text moving from one section to another. These changes are numerous, pervasive and do not change the guidance. Therefore, they are not further elaborated in this paper.

2.2 New Figures

The 6th edition of API STD 520-Part II has added additional figures throughout the standard. The following figures have been added:

Figure 3	Typical Rupture Disk Device Installation: Atmospheric (Open) Discharge Disk
Figure 3B	Typical Rupture Disk Device Installation: Closed System Discharge
Figure 19	Bonnet Vent for Bellows Valves with Vent Located at the PRV
Figure 20	Bonnet Vent for Bellows Valves Handling Vapor with Remote Vent Location
Figure 21	Bonnet Vent for Bellows Valves Handling Liquids Where a Leak Needs to Be Routed Away from the PRV
Figure 22	Bonnet Vent for Bellows Valves Handling Liquids or Flashing Liquid and Vapor with Remote Vent Location
Figure A.1	Typical Configuration of Companion Flanges, Gaskets and Rupture Disk Assembly
Figure A.2	Proper Handling of a Rupture Disk
Figure A.3	Improper Handling of a Rupture Disk
Figure A.4	Proper Alignment of Rupture Disk indicated by Tag Arrows
Figure 8	Typical Rupture Disk Device in Combination with Relief Valve: Inlet Side Installation

2.3 Deleted

The following items were in the 5th edition of 520-Part II and have not made it to the 6th edition. The section numbers refer to the 5th edition.

§4.8	Turbulence in Pressure Relief Device Inlets (See section 9.3 in 6th edition)
§6.3.1a*	Inlet isolation valves full bore requirement
§6.3.2a*	Outlet isolation valves full bore requirement
§Appendix C	Technical Inquiries

* There is still the requirement that the inner diameter all fittings be at least as large as the inlet/outlet diameter of the relief device. Thus while the verbiage changed, the logic was to allow for the use of non-full bore valves are of a larger diameter such that the minimum flow diameter of piping and fittings are all greater than the associated relief device connection. A

requirement of a full bore relief device would not allow for this installation type in the 5th edition of 520 Part 2.

2.4 Equation Modifications

The following equations were modified between the 5th and 6th editions. The updates correspond to unit updates only.

2.4.1 Reaction Force Due to Vapor Discharge §5.8.2.1 Equation 1, 6th edition (§4.4.1.1 Equation 1, 5th edition)

For the SI units equation the units for the outlet pressure and the associated constants were updated from bar(g) in the 5th edition to kPa(g) in the 6th edition.

2.4.2 Reaction Force Due to 2-Phase Discharge §5.8.2.2 Equation 1, 6th edition (§4.4.1.2 Equation 1, 5th edition)

The mass flow units were updated from kg/hr in the 5th edition to kg/s in the 6th edition. Additionally, in the 5th edition the static pressure at the pipe is depicted as absolute pressure minus atmospheric pressure. In contrast, the static pressure at the pipe exit is given solely in gauge pressure.

2.5 New Sections

The following equations were added or significantly modified between the 5th and 6th editions

2.5.1 Operating Environment (§4.5)

The 6th edition provides a new recommendation for installing rupture disks. The location of rupture disk should be taken into consideration, as different process conditions can affect rupture disk reliability. Cleaner or cooler areas of a process may be preferred.

2.5.2 Free Draining (§4.6)

The 6th edition provides a new recommendation for the inlet and outlet piping of a PRD. Both the inlet and outlet of the PRD should not be pocketed and be free draining away from the PRD.

2.5.3 Maintainability (§4.7)

The 6th edition provides a new recommendation that a PRD should be located in an area that enables access for proper maintenance.

2.5.4 Backpressure Limitation and Sizing of Pipe - Thermal Relief Valves (§6.3.2)

The 6th edition provides new guidance on the applicability of backpressure limits for thermal relief valves. This is mirrored in §7.3.7.3 for inlet pressure requirements.

2.5.5 PRV Stability (§7)

The 6th edition provides a comprehensive clarification to the 3% rule along with suggests to the components of a proper engineering analysis. The sub-sections of §7 have also been thoroughly re-written these subsections are:

7.1 General [PRV Stability]
7.2 Potential Causes of PRV Instability
7.3 PRV Inlet Pressure Drop Limitations
7.4 PRV Trim Selection

In 7.3.6 the standard provides more guidance to the requirements for an engineering analysis. This combined with Annex C add specificity to the engineering analysis which in the 5th edition was lacking.

2.5.6 Rupture Disk Installations (§9)

The 6th edition provides new guidance on various rupture disk installation configurations. This new section incorporates the 5th edition §4.6.

2.5.7 Rupture Disk in Series (§9.2)

The 6th edition provides new guidance on using rupture disks in series.

2.5.8 Bonnet Vent for Bellows Valve Handling Non-Hazardous Vapors (§10.3.2)

The 6th edition provides new guidance on venting a bonnet vent when non-hazardous vapors are relieved.

2.5.9 Bonnet Vent for Bellows Valve Handling Hazardous Vapors (§10.3.3)

The 6th edition provides new guidance on venting a bonnet vent when hazardous vapors are relieved.

2.5.10 Bonnet Vent for Bellows Valve Handling Non-Hazardous Liquids (§10.3.4)

The 6th edition provides new guidance on venting a bonnet vent when non-hazardous liquids are relieved.

2.5.11 Bonnet Vent for Bellows Valve Handling Hazardous Liquids (§10.3.5)

The 6th edition provides suggestion that bonnet vents for hazardous liquids be consider on a case by case basis to ensure safe routing.

2.5.12 Annex A.2 – Companion Flanges

The 6th edition has a new annex that provides new guidance for the pipe flanges that a rupture disk is to be installed into.

2.5.13 Annex A.3 – Gasket Selection

The 6th edition has a new annex that provides guidance on the gasket selection for rupture disks.

2.5.14 Annex A.4 – Rupture Disk Holder Serviceability

The 6th edition has a new annex that provides guidance about the serviceability of the rupture disk holder to maintain proper function.

2.5.15 Annex A.5 – Rupture Disk Suitability for Application

The 6th edition has a new annex that provides guidance on the suitability of a rupture disk for the given application and the importance of the compatibility between a rupture disk and rupture disk holder.

2.5.16 Annex A.6 – Preparing the Rupture Disk for Installation

The 6th edition has a new annex that provides guidance on how to prepare the rupture disk for the given installation.

2.5.17 Annex A.7 – Installation of the Rupture Disk into the Rupture Disk Holder

The 6th edition has a new annex that provides guidance on installing the rupture disk into the holder.

2.5.18 Annex A.8 – Installation of the Rupture Disk Device into a Piping/Pressure-Relief System

The 6th edition has a new annex that provides guidance on installing the rupture disk into a piping/pressure-relief system.

2.5.19 Annex A.9 – Rupture Disk Life-Cycle and Maintenance

The 6th edition has a new annex that provides guidance on the life cycle and maintenance of a rupture disk.

2.5.20 Annex C – PRV Acoustic Interactions

The 6th edition has a new annex that is an informative section about a type of acoustic interaction with direct spring loaded PRVs. Research still being done in this area.

2.6 New Requirements

The following sections were added that, in the opinion of the Authors, provide additional requirements or guidance. If the reader of this paper designs or audits relief systems, then he or she is encouraged to purchase a copy of the 6th edition of API STD 520-Part II to ensure designs are per the most recent guidance. The information contained herein is insufficient to design by.

2.6.2 Pilot-Operated Valves (§10.5)

Provides an additional recommendation that if a pilot valve is a balanced bellows type the vent of the pilot valve bonnet should follow 10.3 bonnet venting guidelines.

2.6.3 Gate Valves (§8.3.2(c))

The 6th edition provides guidance on the installation of gate valve related to the position of the stems when the gate valves are used as an isolation valve from recommendation to requirement.

2.6.4 Gate Valves (§8.3.2(d))

The 6th edition provides guidance on the installation of bleed valve between isolation valve and PRD.

3. Conclusion

The 6th Edition of API STD 520-Part II has been reformatted and updated. This document has changed extensively from the previous addition.

The expansion of the inlet pressure loss and stability guidance along with provide information on valve stability really increases the usability of this document.

4. References

[1] API Recommended Practice 520-Part II, FIFTH ED., AUGUST 2003, REAFFIRMED FEBRUARY 2011, Sizing, Selection, and Installation of Pressure-Relieving Devices in Refineries – Part II Installation

[2] API Standard 520-Part II, SIXTH ED., MARCH 2015, Sizing, Selection, and Installation of Pressure-relieving Devices-Part II—Installation

11TH GLOBAL CONGRESS ON PROCESS SAFETY

Evolution of Relief Sizing at an Operating Company

Mike J. Maness
Eastman Chemical Company
P.O. Box 511
Kingsport, TN 37662
mjmaness@eastman.com

J.F. Mize, J.W. Chastain, J.J. Falin
Eastman Chemical Company
P.O. Box 511
Kingsport, TN 37662
jfmize@eastman.com, chastain@eastman.com, jfalin@eastman.com

Prepared for Presentation at
American Institute of Chemical Engineers
2015 Spring Meeting
11th Global Congress on Process Safety
Austin, Texas
April 27-29, 2015

UNPUBLISHED

Evolution of Relief Sizing at an Operating Company

Mike J. Maness
Eastman Chemical Company
P.O. Box 511
Kingsport, TN 37662
mjmaness@eastman.com

J.F. Mize, J.W. Chastain, J.J. Falin
Eastman Chemical Company

Keywords: emergency relief systems, recognized and generally accepted good engineering practice, RAGAGEP

Abstract

Codes, Standards, and Recognized and Generally Accepted Good Engineering Practices (RAGAGEP) regarding the engineering design of pressure relief systems have undergone significant changes over time. This paper traces the development of the approach to relief system design by one operating company as the codes, standards, and RAGAGEP have changed and as the awareness of the recommendations of these various guidelines have penetrated the consciousness of the organization. Initially, relief design consisted of simple sizing of orifice areas based solely on the external fire case. Over time, relief design evolved to include additional relief cases being considered in the design, the evaluation of inlet and outlet pressure drops for reclosing devices (both pressure relief valves and low pressure devices), the consideration of reactive relief systems and different approaches to reactive cases, the evaluation of atmospheric venting of relief systems from a consequence assessment standpoint, the elimination or mitigation of relief sizing contingencies based on system limitations or active protective systems, and appropriate protection of gas-filled vessels with depressuring systems. In addition, new topics under consideration including relief valve stability will also be addressed.

1. Introduction

Tennessee Eastman Corporation was founded in Kingsport, TN in 1920 by George Eastman to produce chemicals for Eastman Kodak Company. The facility has been in continuous operation since that period, becoming an independent operating company in 1994. Eastman's Tennessee Operations employs approximately 7000 people and is one of the largest chemical manufacturing sites in North America. The facility occupies more than 550 buildings and approximately 4000 acres of land with the main plant site occupying approximately 900 acres.

Eastman's Tennessee Operations has more than 9,800 relief valves, 1,700 rupture disks, and 2,600 low pressure conservation vents in service. Some of these relief systems are protecting facilities as old as the site itself (95 years). With such a large site and extensive history, the pressure relief design practices have gone through significant revision over the life of the facility. This paper explores the changes in the relief design practices at the Tennessee Operations site over time including the perception and approach to Recognized and Generally Accepted Good Engineering Practice (RAGAGEP). It is hoped this review can assist other organizations that may be at different stage of development along the path of relief design practice.

2. Nascent Relief Design at the Tennessee Operations Site

Information related to relief systems designed before the 1980s is limited at the site. No uniform procedure existed for the evaluation of relief systems or the appropriate documentation to be produced by the relief evaluation. Relief calculations were either performed by hand or performed by vendors who then provided relief devices with limited documentation. A review of the limited documentation prior to this period revealed that most relief designs assumed that the worst case was pool fire and designed a relief system on this basis. Subsequent inspection of relief designs completed prior to the 1980s have shown that many did not explicitly evaluate the pressure drop associated with the inlet or outlet piping in the relief system. Documentation of relief design bases prior to this period is also limited. Often, a specification sheet for the relief device indicating the sizing basis may be the only documentation maintained from the original design. There are also notations on vessel drawings that indicate equipment vendors were often requested to supply relief devices for the vessels they were designing and constructing for the site.

3. 1980's and 1990's

In the 1980's and early 1990's, process engineers began to recognize that the approaches taken for relief design in previous years had not been as comprehensive as necessary. Prior to this period, evaluation of the full range of potential relief scenarios other than external fire had not been typically done. Recognition of the broader possibilities for design cases for relief systems eventually became more common. The advent of computers in process engineering allowed more complex calculations to be conducted, even though this often required the development of a program to execute the calculations. For Tennessee Operations, the calculations conducted during this period included evaluations of relief header systems for simultaneous relief cases, early reactive relief sizing based on scaling from calorimetry, and dynamic modeling of reactive systems using specifically generated Fortran code.

In 1993, Tennessee Operations' Engineering Division issued its Relief and Venting Systems Design Handbook, a comprehensive guidance document for relief evaluation methodology and documentation. This document established standardized practices and the appropriate documentation to be developed and maintained for relief device designs. The establishment of a standardized methodology, which included evaluation of all appropriate cases for consideration, led to recognition that in some cases pool fire was not the governing case. In such situations, this sometimes resulted in an increase in the size of the relief system required to adequately address the actual governing relief case. When this increase in size was deemed to result in an undesirable consequence due to activation of the relief device, safety interlocks were designed and applied to the process to remove relief design cases from consideration. The application of safety interlocks in this manner not only reduced the cost of relief systems but also provided the additional benefit of preventing a loss of containment.

4. 29 CFR 1910.119--OSHA Process Safety Management of Highly Hazardous Chemicals

By the early 1990s, promulgation of the OSHA PSM Standard prompted the Eastman Kingsport Site (EKS) Engineering Division to develop a more uniform method of performing and documenting relief calculations as part of the required process safety information (PSI) for PSM-covered processes at the Kingsport site, as well as all capital projects managed and executed at that location [OSHA, 1992]. This revised approach was spreadsheet-based and consisted of a series of macro-driven files that utilized physical property data from AIChE Design Institute for Physical Properties (DIPPR), user input data, and algorithms for performing calculations related to relief valves, rupture disks, low-pressure vents, and open pipe systems [AIChE, 2002].

The original intent of this methodology was to provide a way to quickly evaluate relief systems for the wide variety of chemicals used in Eastman's batch manufacturing units. However, use of the spreadsheets was soon expanded to the other continuous operations located in Kingsport, as well as new capital projects being executed at the time. Eastman design engineers as well as engineering contractors were given training on the use of the tool which allowed both groups to produce consistent calculation results that could be stored for future reference, along with input files that could be used as a starting point for the next evaluation of the system. The user input forms allowed clear indication and discussion of all cases considered, input of device-specific coefficients of discharge, selection of fire heating correlations, a choice of allowable accumulation options, and input of piping system details (linear piping and fittings). The resulting output included a summary of the design basis (worst case) as well as individual sections with detailed results for each credible case evaluated.

While the uniform calculational approach and documentation offered by the spreadsheet method was an improvement over previous hand calculations, there were still significant limitations present in the system. Although large lists of chemicals could be evaluated quickly, the calculations performed were done on a single (pure) component basis. Inclusion of mixtures and multicomponent vapor liquid equilibrium (VLE) data were not possible using the spreadsheet approach alone. Vapor phase calculations used only ideal gas equations. Any departure from ideality was not adequately captured with this method. There was also no way to effectively evaluate two-phase flow. At best, designers would use the previously recommended API method of determining the vapor/liquid composition split, size the required area for each phase separately, and then add the required areas together to obtain the total required area for the two phase stream [API, 1990]. In addition, pressure drop through the associated piping system was based on required flow rather than rated flow for PSVs, which ran the risk of under-predicting inlet and discharge pressure drops for the purposes of evaluating the ASME 3% and 10% guidelines. Effective characterization of low pressure (sub-critical flow) vents in a relief system was also problematic. There was an additional spreadsheet component for evaluating header systems, but it was rather difficult to use and, as a result, its accuracy was somewhat suspect.

5. Standardization of Methodology and Documentation

5.1 Development of a core group of Subject Matter Experts

The implementation of the spreadsheet tool for relief system evaluation allowed engineers with limited experience in the subject matter to evaluate existing relief systems and specify new installations. The ease of use of the spreadsheet and widespread access to Excel ensured that any contracting firm could perform relief calculations on behalf of EKS. The majority of relief calculations were assigned to contractors, who had the manpower available for development of PSI for existing installations and specification of new devices for capital projects. Eastman process engineers reviewed the work of the contractors but were not tasked with further developments in the technology associated with relief system evaluation. However, as the limitations of the spreadsheet tool became more apparent and the need for more sophisticated analysis was identified, it became increasingly evident that a core group of subject matter experts (SMEs) in relief system design should be developed. This core group was formed in 2005 and consisted of three chemical engineers with several years of relief design experience in the EKS Engineering Division. Additional members have been added over time such that, currently, there are eight chemical engineers on the Relief and Venting Team who are charged with managing the relief system efforts at EKS. These efforts include review of all relief calculations performed on behalf of EKS in addition to technology developments in relief system evaluations. Additionally, this team is responsible for training all relief designers prior to them performing work for EKS.

5.2 Reactive Relief Design

EKS had started practicing reactive relief using the established Design Institute for Emergency Relief Systems (DIERS) methodologies in the 1990's [DIERS, 1992]. EKS engineers were trained in the DIERS technology through the AIChE DIERS course and began applying the principles primarily in batch process applications. Subsequent training in the approach to reactive relief developed by Hans Fauske led to EKS procuring an Advanced Reactive System Screening Tool (ARSST) and conducting reactive relief via direct scaling of data obtained from this calorimeter [Fauske, 2000].

The group of EKS relief design SMEs continued to develop depth of knowledge in the area of reactive relief design through participation in the DIERS Users Group. As this group developed additional expertise in the subject matter, they determined that other

DIERS methodologies, such as dynamic simulation of runaway reaction systems using kinetic models, could be successfully applied to evaluate relief systems. In certain applications this approach would allow less conservative answers than those developed from scaling methodologies. Early efforts at EKS in this arena focused on using the AIChE SAFIRE code or developing a dynamic model in a spreadsheet to conduct the evaluation [DIERS, 1992].

It was determined that the number of applications required for rigorous reactive relief sizing required a better solution. Software was needed which could allow quick development of a dynamic reactive model.

5.3 Selection of Third Party Software

Based on the limitations of the existing approach and methodology given above and the need to have a tool to conduct dynamic relief simulation, a comprehensive evaluation of several commercial software packages available from various vendors was conducted by the EKS relief design SMEs. The software packages were evaluated based on their capabilities to handle multicomponent mixtures and vapor liquid equilibrium, to address the addition of new chemicals to the database, to perform detailed piping analysis for pressure drop including two phase flow evaluation, to evaluate compliance with key aspects of RAGAGEP, to conduct dynamic modeling, to rigorously address reactive relief evaluation through the applications of kinetic models, and finally on the cost of the software and the availability of ongoing product support.

Upon completion of this evaluation effort, SuperChems™ by ioMosaic Corporation was selected as the standard software [ioMosaic, 2015]. SuperChems™ was found to meet all of the key criteria evaluated. An additional consideration was the fact that SuperChems for DIERS is the official DIERS software of AIChE. As a result, the software has a large user base, an active users group, and significant ongoing development.

After the selection of a standard software package, EKS personnel highlighted several areas which could be improved for ease of use. Eastman Chemical worked with the vendor to address some of these areas including the development of relief templates to allow efficient use by non-expert users, the incorporation of a graphical interface to allow isometrics to be constructed visually for the evaluation of relief capacity and piping pressure drop, and an interface to allow the transcription of the latest DIPPR database into the software on a regular basis.

Use of a standard software package has greatly improved the consistency of documentation of relief calculations at EKS. One goal of the selection of the third party software package had been the capability of producing all relief documentation using one program. This goal has not yet been fully realized, but the vast majority of documentation is produced by SuperChems in a standard report format. The standardization of methodology and documentation by all relief designers, whether internal SMEs, contractors assisting with the evaluation of existing systems, or detailed design firms completing capital projects, helps to speed relief reviews and ensure that archived relief calculations will be consistent and complete.

5.4 Technology Development

In addition to implementing a standard software package at EKS, significant technology developments were occurring in relief system design. As the Relief and Venting Team increased in knowledge and expertise, solutions were devised to address these developments.

5.4.1 Fire Case Evaluation

The Relief and Venting Team pursued several issues related to vessels impacted by external fire. Historically, gas filled vessels had been protected with process safety valves. Reclosing devices do not provide adequate protection for gas filled vessels that can be exposed to a pool fire due to overheating of the vessel walls [API, 2014]. Recognition of this issue resulted in establishing alternate means of protection for such systems. The next approach taken was installation of a rupture disk to provide a non-reclosing depressurization path for the vessel. However, in many cases, excessive heating of the vessel wall may occur well before pressure increases sufficiently to activate the rupture disk. The recommended protective measures were revised to include installation of fire resistant insulation in combination with approved deluge per NFPA 15, a properly sized non-reclosing device initiated on (pressure provided that pressure is an effective initiator), or a properly designed depressurization system initiated by temperature in the fire zone [NFPA, 2012].

Another concern related to temperatures experienced during a pool fire is associated with materials with high boiling points (>260°C, 500°F). Equipment containing such materials can also fail due to overheating of the vessel walls when exposed to a pool fire. Emergency pressure relief may not be possible for these vessels since they may fail before any appreciable internal pressure is generated; therefore, EKS has taken an approach that fire protection and emergency response are the appropriate mitigations for this equipment.

Fire impingement on vessels that contain thermally sensitive or reactive materials can result in the need for very large relief devices, particularly if significant off-gassing is expected during the relief event. EKS has established that the application of foam fire systems to extinguish a pool fire and prevent heating of these vessels is an effective mitigation when relief systems are not practical or when activation of the relief system and release of effluent presents a significant hazard.

5.4.2 Fire Duration/Heat-up Time

For some large storage tanks or vessels with significant liquid inventory, the heat input from fire exposure may not raise the bulk material to the boiling point for a very long time. Under such circumstances, it may be appropriate to consider a time limit for duration of the pool fire scenario [CCPS, 1998]. A duration limit for a pool fire should consider aspects such as the amount of fuel which could be released to feed the fire, the time for emergency response, and the capability of the emergency response to address the scenario.

5.4.3 Supercritical System Evaluation

Fire exposure for systems with very high Maximum Allowable Working Pressures (MAWPs) of several hundred psig or higher can result in a situation very similar to relief of gas-filled vessels. This is due to the combination of pressure and temperature that occurs during such an event, which can exceed the critical point of liquids present in the system. Beyond the critical point, there is no heat of vaporization as the contents transition to a supercritical fluid. Not only do the very high temperatures challenge the vessel integrity, but also the supercritical fluid does not remove heat from the system through vaporization as liquids will. Fire protection methods such as insulation, water spray (or foam), and depressurization are extremely important measures to consider for such systems [CCPS, 1998].

5.4.4 Effluent Handing Analysis

Relief events are typically very low frequency events. The historical precedence at EKS was to relieve vessels directly to the atmosphere through outlet piping adequate to ensure a relief event would not expose personnel in the area to the release. As the sophistication of the relief program at EKS increased and the tools for conducting dispersion analysis became accessible, the concept of modeling the effluent from a relief system to determine the post release impact became more common. CCPS recommends this practice to determine if the addition of effluent handling equipment and systems is appropriate [CCPS, 1998].

Dispersion and effluent handling analysis was first applied to toxic streams to ensure that released materials, in the rare event of a relief device operating, would not put personnel or the community at risk. Later, the approach was expanded to encompass releases of flammable materials. Broader use of this approach increased the application of relief effluent mitigation strategies such as vapor - liquid separators, vent stacks, scrubbers, and flares. The broader knowledge of the possible impact of relief effluent systems currently results in Process Hazard Analysis teams considering the potential consequences of relief scenarios and often making recommendations for evaluation and mitigation of these scenarios where necessary.

5.4.5 Vacuum Evaluation of Loss of Steam Case for Columns (Improvements)

Although not specifically addressed by ASME Code, protection of equipment against vacuum conditions is another area of relief design that has evolved over time at EKS [ASME, 2010]. API STD 2000 provides guidance for low pressure storage tanks based on pump-out and thermal inbreathing, but process equipment with other sources of vacuum can present significant challenges [API, 2014]. Process equipment can also be subjected to vacuum due to process upsets, failure of utilities, or other means. If process equipment is not rated for full vacuum, these conditions can result in damage to the equipment and the potential for a release. One unit operation of particular concern is distillation. The condenser in a distillation system can be a very effective form of vacuum generation. The energy balance in a distillation system is in equilibrium during normal operation with heat input by the reboiler and heat removed by the condenser. If this balance is disrupted by a sudden loss of reboiler heating (e.g., utility failure, ESD, etc.), the continued operation of the condenser (since cooling water is typically supplied continuously) can cause a collapse of the vapor inventory in the system and result in sub-atmospheric pressure which may challenge the mechanical integrity of the system if the various components (column, condenser, reflux drum) are not rated for the vacuum condition. A method for addressing this scenario was developed by Eastman engineers [Lodal, 1995]. This method was later modified to reduce the required size of the vent based on the effectiveness of non-condensables reducing the condensing capacity in the system [Stark, 2000]. Both the original method and the modified version have been used by relief design practitioners at EKS as part of the overall relief design approach.

5.4.6 Low Pressure Vent Piping Analysis

Flow charts and capacity tables provided by manufacturers of low pressure vents (also called conservation vents or breather vents) do not account for pressure drop that may occur through inlet or discharge piping connected to such devices. This pressure drop can significantly impact the actual capacity of such devices (both pressure and vacuum vents) since it reduces the pressure differential available for pallet lift. The methods for evaluating venting systems that utilize these devices at EKS have been revised to account for this effect.

5.4.7 Incorporate Relief as Part of the Process Design

Relief system design was typically a task that occurred during detailed design when the necessary process safety information (PSI) had been developed to allow relief designers to fully document and specify relief devices and associated piping. This activity was often viewed as an "add-on" to the main process and equipment design efforts. However, vessel ratings were often already selected prior to the start of the relief design efforts, effectively preventing the use of inherently safer design practices, such as selecting vessel pressure and vacuum ratings to eliminate potential relief scenarios from consideration. Vessel ratings were often selected only on the basis of providing a sufficient margin between operating pressure and the set pressure of the relief device. As a result, the requested vessel ratings were often different for each individual vessel. Groups of vessels protected together as a system were then limited by the MAWP of the lowest rated vessel in the system. In many cases, this resulted in larger relief systems or the need for mitigations to ensure that the system could be appropriately protected during overpressure events. Failing to request a full vacuum rating for vessels resulted in the need for vacuum relief, which adds additional complexity in the relief system design along with the risk associated with introduction of air into process equipment via vacuum vents.

Incorporating relief as part of the overall process design allows for the relief designer to identify credible relief cases and perform preliminary relief sizing before the equipment specifications are finalized. They can evaluate the use of inherently safer design options and recommend appropriate vessel ratings to eliminate certain relief cases and minimize the required relief size for those that cannot be eliminated by system design. The relief design is then integrated with the equipment design process and the process hazard and risk analysis that occurs during the capital project.

5.4.8 Integrate Relief System Analysis with Process Hazard and Risk Analysis

As relief system expertise was concentrated into the group responsible for process safety at EKS, the process safety engineers also started serving as the "safety lead" on all capital projects. The safety lead manages the efforts of the relief design lead, PHA study leader, and Layer of Protection Analysis (LOPA) lead on all capital projects (one individual may serve all three roles on a project if they have the expertise). Having all process safety expertise in one group allows for relief system analysis to be integrated with process hazard analysis and risk assessment. This approach has been applied to the revalidation of PHAs for existing covered processes in addition to all capital projects managed by EKS.

Preparation for a process hazard analysis requires compilation of process safety information, including relief system design and basis. In order to evaluate relief system design and basis, many other pieces of PSI are needed, including process chemistry, piping and instrumentation diagrams (P&IDs), design codes and standards employed, and information on associated safety systems [OSHA, 1992]. The relief design expert is therefore able to assess the health of numerous elements of PSI as part of the evaluation and documentation of the relief system design and basis. Performing a comprehensive assessment of the relief system design and basis for all equipment and vessels within the process boundary ensures that the PHA team has the information they need to evaluate the adequacy of controls for scenarios that result in the potential for overpressure and failure of vessels. The relief system is often the last line of defense against such overpressure scenarios.

5.4.9 Mitigation of Relief Scenarios through the Application of Safety Instrumented Systems

Elimination of relief cases from consideration through the application of interlocks and emergency shutdown systems is a practice that has been well established at EKS. This practice was established as an industry norm in API 521 and in the ASME Code, first through the development of Code Case 2211, and then through adoption of this practice into the code as UG-140 [API, 2014], [ASME, 1995], [ASME, 2010]. Consensus standards for safety instrumented systems (SIS) were developed by the ISA and then further developed by the IEC [ISA, 1996], [IEC, 2004]. These consensus standards established a life cycle for the design, installation, operation, maintenance, and decommissioning of Safety Instrumented Systems (SIS). One application of SIS can be the mitigation of relief cases to remove them from consideration for design.

Before relief system evaluation was integrated with the PHA process, relief designers often noted the need for mitigation but were not responsible for ensuring effective mitigation was provided by existing instrumentation or other protective systems. Review of relief design and basis documentation in preparation for a PHA often identified the need for additional risk assessment to evaluate the adequacy of existing controls or to determine the required reliability for proposed SIS.

In order to provide the documentation required by UG-140 when SIS is used to provide overpressure protection, a formalized process was developed for the mitigation of relief scenarios. This process is based on Layers of Protection Analysis (LOPA) for all relief scenarios that need to be mitigated. A LOPA practitioner will perform the required assessments to evaluate the risk associated with a given overpressure event and will provide recommendations to the relief designer that are included in the relief documentation.

5.4.10 Heat Exchanger Relief Design

Relief valves have typically been the protective device of choice applied to shell-and-tube exchangers. However, as knowledge and understanding grew among EKS relief designers with respect to the hazards and difficulties associated overpressure protection for exchangers with large pressure differentials between the two chambers, criteria were developed to provide guidance on when reclosing devices were an appropriate choice, and when non-reclosing devices should be used instead due to the faster opening response required when large pressure differentials are present [CCPS, 1998]. The design criteria take into consideration the pressure ratio between the two sides, as well as the process fluid state (gas or liquid) present on each side.

In addition to establishing criteria for proper relief device selection for the exchanger tube rupture scenario, relief designers at EKS have incorporated the use of LOPA in the relief design process for this case, which has led to a change in how tube rupture scenarios are evaluated. Vessel repair histories and input from equipment inspectors have allowed the use of LOPA to account for the relatively low occurrence of full tube rupture events as compared to pinhole leaks for shell and tube exchangers at EKS. As a result, the potential for a full tube rupture, which can present a difficult relief problem, can often be shown to be well below the risk threshold of concern.

5.5 Benefits of Improvements in Relief Design

By integrating relief design with PHA and LOPA, the overall health of relief systems, safety systems, and vessels is documented and understood so that risk can be effectively managed. Understanding the options for providing overpressure protection through a combination of relief devices, SIS, and inherently safer design practices allows for selection of the most cost-effective solution for a given installation. This approach is applicable to capital projects as well as existing installations.

Relief system evaluation is an important part of facility siting and emergency response planning. Steady-state and dynamic modeling of relief cases provide source terms for dispersion modeling to assess the adequacy of effluent handling systems and the potential impact on personnel and the surroundings. Dynamic modeling of vessels subjected to pool fire impingement evaluates the changes that occur in the vessel as the relief event progresses, including determination of time to vaporize all of the contents (boil dry). After the vessel boils dry, the vessel could fail due to weakening of the walls due to exposure to high temperature. Emergency responders could be injured due to failing vessels. Understanding the time available to effectively fight a fire before vessel failure occurs is essential to emergency response planning.

6. Future Technology Development

While EKS has gone through a significant amount of development in its approach to relief system design and the sophistication of its understanding and methods, as an organization it continues to learn more about the topic through continued engagement with peers in other operating companies and subject matter experts in consulting firms in the DIERS organization. Relief systems are often thought of as the last line of defense against loss of containment in overpressure events. The challenge frequency for these systems has thankfully been very low at EKS and across the industry. While this speaks highly of our industry's safety record, it also means that we must be very thoughtful about all contingencies that the relief device may need to protect against. In this arena, there are several issues which we are still working on incorporating into our relief design practices.

6.1 Shock Analysis

Above, the selection of the appropriate type of relief device for tube rupture scenarios was discussed with the governing criteria being the circuit pressure ratio. A high pressure ratio results in the selection of a non-reclosing device because of the required speed of response. For very high circuit differentials where the high pressure circuit is a gas or vapor and the low pressure circuit is a liquid, the failure of a tube can result in a localized shock to the wall of the exchanger. Indications are that this shock wave could result in failure of the exchanger wall before the opening of a non-reclosing device can protect the exchanger. Analysis of this phenomenon is beyond most SMEs and requires specialized knowledge of the vessel walls response to the event.

6.2 Vessels Transitioning to Gas/Vapor Filled Vessels during Fire Case

The issues with vapor or gas filled vessels in the event of a pool fire were discussed above and the approach to protecting these vessels has been established at EKS. Likewise, the approach taken for vessels with liquid present during a pool fire is established. One issue still under discussion and development at EKS is an approach that accounts for the transition of a vessel during a pool fire from a liquid containing vessel to a gas or vapor filled vessel due to boiling the vessel dry. If this occurs and a reclosing relief device has been used to protect the vessel from overpressure, the weakening of the wall due to fire on dry metal can result in failure of the vessel at high pressure. Development of a comprehensive plan for these situations should account for emergency response, fire duration, presence of deluge and fire resistant insulation, and other factors. This is also still under discussion and development.

6.3 Advanced Issues with Relief Valve Stability

One area under discussion across the industry is relief valve stability. ASME Code indicates that inlet pressure drops are to be maintained below 3% of the set pressure of the device to prevent valve chattering which can result in damage to the relief device [ASME, 2010]. The area is currently seeing a significant amount of research and presentations at DIERS meetings have indicated that inlet pressure drop may not be the primary driver causing relief valve instability issues. EKS SMEs continue to monitor this developing area to determine an appropriate response in our designs for stable operation of relief valves.

7. Conclusion

The evolution of a comprehensive approach to relief design and evaluation at EKS has included the development of a team of subject matter experts to shepherd the evaluation and application of relief technology. This development has included recognition of the broad scope of RAGAGEP that applies to relief design and the implementation of these practices for new designs and in the evaluation of existing systems. The integration of concepts into relief design has been a slow process occurring over many years as the depth of knowledge increased through participation in industry groups such as the DIERS Users Group and the Global Congress on Process Safety. The development of consistent relief design practices has also allowed standardization of relief documentation and review of relief calculations for both internally developed calculations as well as those completed by contractors.

8. References

[1] American Institute of Chemical Engineers, DIPPR 801 Database, 2002.

[2] American Petroleum Institute, *Recommended Practice 520: Sizing, Selection, and Installation of Pressure-Relieving Devices in Refineries – Part 1: Sizing and Selection*, 1990.

[3] American Petroleum Institute, *Recommended Practice 521: Pressure Relieving and Depressuring Systems*, 2014.

[4] American Petroleum Institute, *Standard 2000: Venting Atmospheric and Low-Pressure Storage Tanks*, 2014.

[5] American Society of Mechanical Engineers, *Pressure Vessels with Overpressure Protection by System Design, Section VIII, Divisions 1 and 2, ASME Code Case 2211, The 1995 Boiler Pressure Vessel Code*, 1995.

[6] American Society of Mechanical Engineers, *Boiler and Pressure Vessel Code - 2010 Edition, Section VIII, Division 1*, 2010.

[7] Center for Chemical Process Safety, *Guidelines for Pressure Relief and Effluent Handling Systems*, 1998.

[8] Design Institute for Emergency Relief Systems of the American Institute of Chemical Engineers, *Emergency Relief System Design Using DIERS Technology*, 1992.

[9] Fauske, H.K., Properly Size Vents for Nonreactive and Reactive Chemicals, Chem Eng Progress, (2000), 17-29.

[10] International Electrotechnical Commission (IEC), *Functional Safety: Safety Instrumented Systems for the Process Sector*, Geneva, Switzerland, 2003.

[11] ioMosaic Corporation, ***SuperChems***^{TM}, 2015.

[12] The International Society of Automation (ISA), *Application of Safety Instrumented Systems for the Process Industries,* Research Triangle Park, NC, 1996.

[13] Lodal, P.N., Mahanes, J. L., Calvert, J. I., and Keel, J.M., Revised Emergency Vacuum Relief Device Sizing for Atmospheric Distillation Systems, J. Loss Prevention Process Ind., Vol. 8, No. 6, (1995), 331-341.

[14] National Fire Protection Association, *NFPA 15: Standard for Water Spray Fixed Systems for Fire Protection*, 2012.

[15] Occupational Safety and Health Administration, *Process safety management of highly hazardous chemicals, 29 CFR 1910.119*, 1992.

[16] Stark, D.E., Shonnard, D. R., and Crowl, D. A., Safety Aspects During Transient and Steady-State Operation of a Pilot Scale Distillation Column, Process Safety Progress, Vol. 19, No. 3, (2000), 179-188.

Product Failure Rates vs Total Failure Rates at Specific Sites: Implications for Safety

Julia V. Bukowski, Ph.D.
Dept. of Electrical & Computer Engineering
Villanova University, Villanova, PA 19085
Julia.Bukowski@villanova.edu

William M. Goble, Ph.D., CFSE
exida
64 N. Main St., Sellersville, Pa 18960
wgoble@exida.com

Iwan van Beurden, MSc
exida
64 N. Main St., Sellersville, Pa 18960
vanbeurden@exida.com

Prepared for Presentation at
American Institute of Chemical Engineers
2015 Spring Meeting
11th Global Congress on Process Safety
Austin, Texas
April 27-29, 2015

Product Failure Rates vs Total Failure Rates at Specific Sites: Implications for Safety

Julia V. Bukowski, Ph.D.
Dept. of Electrical & Computer Engineering
Villanova University, Villanova, PA 19085
Julia.Bukowski@villanova.edu

William M. Goble, Ph.D., CFSE
exida

Iwan van Beurden, M.Sc.
exida

Keywords: total failure rate, product failure rate, site-generated failure rate, FMEDA, PFDavg

Abstract

To quantify the safety achieved through the use of automatic safety equipment, one must be able to determine the failure rates of the equipment which are then used to compute average probability of failure on demand (PFDavg). This is usually done by predicting failure rates via a failure modes, effects and diagnostics analysis (FMEDA) or by estimating the failure rates from failure data obtained from one or more specific end-user sites. When the predicted failure rates are compared to the estimated failure rates, there are occasions where they differ considerably. Furthermore, studies of field failure data (FFD) from different sites have shown the estimated total failure rate for a particular product model number has varied by up to four times between best and worst sites. An understanding of why this number varies from site to site has led to the conclusion that some random failures are caused by end-user practices at a particular site. This means that total failure rates computed from site-specific FFD consist of two parts: the inherent product failure rate and the specific site-generated failure rate. Inherent product failure rates which are predicted from failure rate and mode databases assume that the site end-user follows a maintenance process per IEC 61511 [1] and that actions to properly store, install, maintain, test and repair safety related equipment are performed accurately. Site(s)-generated total failure rates account not only for the inherent product failure rate but also, account for the effects of site-specific end-user practices that cause failures.

This paper

- Provides background information sufficient so that the paper is relatively self-contained
- Presents convincing evidence that the total failure rate estimated from FFD for a specific piece of mechanical equipment from a specific site in fact consists of an inherent product failure rate and a specific site-generated failure rate and that the site-generated failures may account for half or more of the total failures observed

Note: Do not add page numbers. Do not refer to page numbers when referencing different portions of the paper

- Shows that estimated total failure rates based on site-specific FFD may include failures from any life phase, i.e., initial failures, infant mortality failures, useful life failures, and end of life or aging failures and that all of these must be included in assessing the safety performance of a SIF
- Discusses the safety implications of omitting site-generated failure rates when assessing safety performance at a specific site and of using total failure rates based on multi-site FFD
- Proposes a method for extending prediction techniques to predict specific site-generated failure rates as well as inherent product failure rates

1. Introduction

Every electrical, electronic, mechanical or electromechanical product, including any equipment used to implement safety instrumented functions (SIF), has an inherent product failure rate due to the relative strengths and weaknesses of the product (as determined by the product's design and manufacture/assembly) and due to the environmental stresses in which the product is deployed. The authors contend that the inherent product failure rate is only one part of the total failure rate as experienced by the product's end user. In addition to the inherent product failure rate, there is an additional specific site-generated failure rate due to the actions (or inactions) of the end user. Since this additional failure rate is determined by human activities, both accidental and the result of deliberate decisions reflecting institutional safety culture, for the same product this additional failure rate may vary from one end-user site to another even within the same company.

For an end user to determine accurately the true safety performance of a SIF at a specific site, it is not sufficient to consider only the inherent product failure rates of each piece of equipment used to implement the SIF. The end user must also consider the site-generated failure rate for each piece of equipment which is determined by the actions or inactions taken at the site where the equipment is deployed.

As there is significant evidence that a product's total failure rate as experienced by the end user at a specific site is often two times or more the inherent product failure rate, it is clear that the site-generated failure rate is a significant contributor to the total failure rate and, consequently, to the safety achieved. The end user who does not include the site-generated failure rate in safety calculations risks seriously overstating the safety being achieved at a specific site.

2. Notation

exp[y]	notation equivalent to e^y
FFD	field failure data
FIT	one failure per 10^9 hours of operation
FMEDA	failure modes, effects and diagnostic analysis
FTO	fail to open
IEC	International Electrotechnical Commission
NBIC	National Board Inspection Code

PFD(*t*)	time varying probability that SIF is in a state of dangerous failure
PFDavg	average probability of dangerous failure on demand
PIF	probability of initial failure (may include infant mortality failure as explained in the paper)
PIF_{PROD}	inherent product PIF
RCA	root cause analysis
SIF	safety instrumented function
SIL	safety integrity level
SOPRV	spring operated pressure relief valve
T_P	time to proof test; time to periodic maintenance and inspection
λ_D	constant dangerous failure rate
λ_{DPROD}	inherent product constant dangerous failure rate
λ_{DSITE}	site-generated constant dangerous failure rate
λ_{DTOTAL}	total constant dangerous failure rate based on field failure data
<<	very much less than
≈	approximately equal to

3. Background

3.1 What Is a SIF?

A SIF is a collection of equipment generally consisting of one or more input sensors, a logic solver, and one or more final elements such as switches or valves. SIFs are used to monitor processes such as power production, pharmaceutical manufacturing, or gasoline production and to shut down these processes safely in the event that such processes cease to operate within normal operating parameter ranges and enter operational regions known to be dangerous. When a process moves from a safe operational range into an unsafe operational range it is said that the process places a demand on the SIF or, more simply, that a demand has occurred.

A correctly operating SIF functions as follows. If the information from input sensors such as pressure transmitters or thermocouples indicates that the process is functioning normally, the SIF allows the process to continue. If, however, based on information from the input sensors, the logic solver within the SIF determines that a demand has occurred, then the logic solver instructs the final elements to safely shut down the process. For example, this shutdown may involve a switch opening to deprive the process of energy which would be of the "de-energize to trip" design.

The SIF itself can fail in two ways. In the event of a safe failure, the SIF unnecessarily causes a safe shutdown of a correctly operating process. If the SIF is in a state of safe failure, the consequences are primarily economic due to loss of production time; with a safe failure there is generally no damage to equipment or the environment nor injury to personnel. In the event of a dangerous failure the SIF is no longer able to intervene and shut down the process should a process demand arise. If the SIF is in a state of dangerous failure and no demand occurs, no harm is done. However, if the SIF is in a state of dangerous failure and a demand occurs, the possible consequences include equipment damage, environmental damage, injury, and/or loss of life.

Clearly, from a safety perspective, dangerous failures of the SIF are of prime concern; to maximize safety one needs to minimize the time that the SIF is unable to respond to demands. These dangerous failures can be further subdivided into two categories: dangerous failures detected by internal diagnostics and dangerous failures undetected by internal diagnostics. If a dangerous failure is detected by internal diagnostics and is promptly repaired, the time during which the SIF was in a state of dangerous failure is relatively short and does not significantly impact the safety performance of the SIF. If, however, the SIF is in a state of undetected dangerous failure, it is likely to remain in that state unless the dangerous failure is discovered and repaired at a time of periodic maintenance.

3.2 Quantifying the Safety Performance of a SIF via PFDavg

Safety standards, such as IEC 61511 [1], IEC 61508 [2] and ANSI/ISA SP84.00.01 [3], define a number of metrics which can be used to quantify the safety performance of a SIF. The principle metric used for equipment deployed in processes with low demand rates is PFDavg, the average probability that the SIF is in a state of dangerous failure when a demand occurs. PFDavg is defined as

$$\text{PFDavg} = (1/T_P)\ \int_0^{T_P} \text{PFD}(t)\,dt \qquad \text{[Eq. 1]}$$

where T_P is the time between periodic proof tests and PFD(t) is the probability that the SIF is in a state of dangerous failure as a function of time. Clearly PFDavg can vary depending on the reliability model that generates PFD(t). The standards allow PFDavg to be computed by any method generally accepted by the reliability and safety community as long as the computations can be justified. However, IEC standards require only that the failure models used account for failures which can realistically be modelled by a constant failure rate.

Approximately 70% of all process safety systems consist of a single channel, i.e., the components comprising the SIF are all connected in series. In this case, PFDavg is often approximated in practice as

$$\text{PFDavg} \approx \text{PIF} + (1 - \text{PIF}) \times \lambda_D \times T_P/2 \qquad \text{[Eq. 2]}$$

where λ_D is the total constant failure rate for dangerous failures computed by summing the constant failure rates for dangerous failures for each of the SIF components (sensors, logic solver, final elements), T_P is the proof test interval, and PIF is the probability of initial failure. When PIF equals zero, Eq. (2) reduces to the more familiar form

$$\text{PFDavg} \approx \lambda_D \times T_P/2. \qquad \text{[Eq. 2a]}$$

The accuracy of this approximation depends on several important assumptions, specifically, that $\lambda_D T_P << 1$ (which is usually the case) AND that the SIF reliability with respect to dangerous failures is exponential in form. This is true only if the SIF consists of a series combination of components AND if each component's dangerous failures are adequately modeled solely by a constant failure rate, i.e., that the failures are due to random stresses and are equally likely to occur during any operational time interval. Therefore, Eqs. (2) and (2a) do **not** apply to redundant architectures. Further, even for SIF's configured in series, Eqs. (2) and (2a) will **not** apply if any of the component failures are due to infant mortality failures, or failures due to aging or end of life. Also, Eq. (2) will not apply if any component failures are due to initial failures.

While it may be the case that most electrical/electronic components generally experience failures that are reasonably modeled by constant failure rates, this is not usually the case for mechanical components which often comprise the final element(s) of a SIF. Mechanical failures can generally include initial failures, infant mortality failures, and aging failures in addition to failures due to constant failure rates and there is significant evidence that they do [4, 5, 6, 7]. Thus, in this paper, mechanical equipment, specifically, spring operated pressure relief valves (SOPRV) form the basis of the data example and illustrate the various failure types.

3.3 Basics of Spring Operated Pressure Relief Valves

Many industrial processes use a SOPRV as a SIF to mitigate the hazards of a process overpressure event. The basic mechanics of a typical SOPRV are illustrated in the conceptual representation shown in Figure 1. While there are many possible design variations, Figure 1 along with its description provides sufficient background to understand the research presented in this paper.

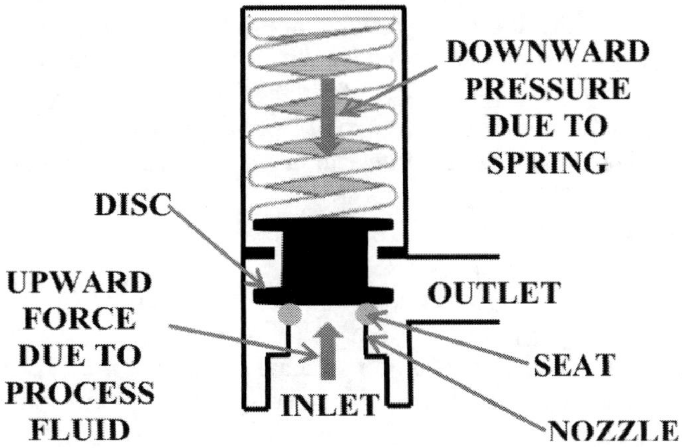

Fig. 1 Conceptual representation of a typical SOPRV

In a properly operating SOPRV a spring exerts a downward force/pressure on the disc pressing the disc against the seat. The seat base is generally the top surface of the wall of the nozzle and the seat itself is often a small raised structure of specific width on the seat base. The green circles in Figure 1 are not intended to describe the shape of the seat but merely to indicate its position. The orifice diameter (area) is the diameter (area) of the circular opening formed by the nozzle. The spring pressure on the disc results in the formation of a fluid tight seal preventing process fluid, which reaches the nozzle through the inlet, from leaving the nozzle through the orifice and outlet. The process fluid exerts an upward force/pressure on the disc. However, since the process *pressure* is nominally about 80-90% of the spring "*set pressure*" the disc remains closed.

During normal plant operation the SOPRV is in the closed position. If the process pressure increases beyond that of the spring set pressure, the disc will lift allowing process fluid to flow through the outlet thereby relieving excess process pressure. When the process pressure returns to the closing pressure of the SOPRV the disc once again seals the SOPRV and the process proceeds normally.

The SOPRV can fail in a number of ways. If the SOPRV either opens slightly or fails to form a fluid tight seal when the process pressure is within normal ranges, the valve is said to leak. This is usually a safe failure, provided that the unintended pressure relief and fluid release does not itself induce a safety hazard. On the other hand, if the SOPRV does not open under conditions of excessive process pressure, the valve is said to fail to open (FTO), or to be "stuck shut," and this is a dangerous failure. Because the SOPRV is normally closed while the SOPRV is in service, the FTO state is a state of undetected dangerous failure. This dangerous failure can only be observed in two ways: as the result of an accident when the SOPRV fails to respond to an overpressure event or in proof testing.

4. Evidence for the Existence of Site-Generated Failure Rates

Although it is easy to imagine site-specific actions or inactions that would increase the SIF total failure rate beyond that of the inherent product failure rate alone, it is nevertheless incumbent upon the authors to offer some actual evidence of instances where such practices clearly impact total failure rate and create a site-generated failure rate. In the following sections anecdotal evidence, some hypothetical scenarios, and actual field data evidence from proof test results are offered that clearly indicate that site-generated failure rates are a real phenomenon.

4.1 Anecdotal Evidence and Hypothetical Scenarios

In examining warranty data for a particular product, it was noted that when the data for that product were sorted by site there was a difference of greater than two times between the failure rate computed from failure data from the best site and the worst site [8]. Site visits provided an explanation. At the worst site some bad maintenance practices were routinely used. However, even when these practices were corrected, there still remained failure rate differences, albeit smaller ones, between sites. Even sites with excellent maintenance practices make mistakes that can result in dangerous failures and a site-generated failure rate.

This is not the only story of its kind and maintenance is not the only issue. For example, site-specific practices of delaying proof testing beyond the prescribed interval can lead to safety systems entering timeframes where aging failures rapidly accumulate, thereby increasing the number of failures of the equipment disproportionately to the constant failure rate assumption of equal likelihood of failure on all time intervals. This practice may not be under the control of maintenance but may depend on decision making in other areas of the company such as operational scheduling. Site-specific practices of failing to proof test equipment immediately after installation can result in initial failures if the installed equipment was in a state of dangerous failure or if installation was not performed correctly resulting in a dangerous failure. Sites that perform less than complete proof tests can generate initial failures that affect the safety performance of the equipment during the next in-service interval. End-users who do not calibrate equipment on schedule may induce failures. End-users may have official policies that support good practices but these may not always be followed. In short, end-user practices can affect the total failure rates of a SIF and create site-generated failure rates.

But it should also be noted that site-generated failure rates may relate not to particular practices or policies but simply to the fact that human beings make random mistakes. The technician that correctly calibrates equipment 999 times may perform the task incorrectly the next time. Sites that faithfully proof test equipment on schedule may occasionally miss a piece of equipment during a proof test cycle. Maintenance personnel may correctly place equipment in maintenance mode and correctly perform a maintenance task but fail to return the device to its operational mode. People make mistakes; it is impossible to prevent them all; some of these mistakes can result in dangerous failures.

4.2 Data Evidence That Site-Generated Failure Rates are Real Phenomena

Normally, a single site does not have sufficient failures on a particular equipment to permit useful estimation of total failure rates. However, a recent study [9] reported that a single site had a total of 13 dangerous failures on a total of 195 proof tests on very similar SOPRV. All the SOPRV in the study had stainless steel seats and discs, relatively low set pressures (less than 150 pounds per square inch gauge), and orifice diameters less than or equal to 1.0 inch. While not identical, they represent a remarkably similar group of equipment. The average proof test interval for this type of SOPRV at the data site was 3.7 years. Furthermore, all dangerous failures were subjected to root cause analysis (RCA) so that the underlying cause of each failure was known. This provided excellent data to compare the predicted inherent product failure rate to the site's estimated total failure rate.

Using existing failure rate and failure mode databases, FMEDA's were performed on the two most populous valve series in the study population. Together these two valve series accounted for 60% of all SOPRV in the study. For both valves the FMEDA's predicted an inherent product PIF (PIF_{PROD}) of 1.1% and an inherent product constant dangerous failure rate, λ_{DPROD}, of 437 FITS. (A FIT is one failure per 10^9 hours of operation.) Thus the FMEDA predicted

$$\lambda_{DPROD} = 437 \text{ FITS}$$

$$PIF_{PROD} = 0.011.$$

It should be noted that the safe failure rates of the two valve series differed by 5% and the residual rates (computed for failures that do not affect safety) differed by 9%. The similarity of the two inherent product failure rates across all types of failure rates (dangerous, safe, residual) confirms that the study population is sufficiently similar to be treated as a nearly identical group. Based on the average proof test interval of 3.7 years for this type of SOPRV at the data site, it is possible to project the expected number of dangerous failures at the proof test interval. The site does pre-installation testing on all SOPRV as post-installation testing is not possible for this equipment. Thus, it would be expected that there would be no initial failures and PIF would equal zero. However, as is explained in Appendix A, infant mortality failures which happen in-service can be conservatively modeled as PIF. Therefore, the FMEDA prediction accounts for infant mortality failures via PIF. The expected number of dangerous failures is

E[# of dangerous failures] =

$=195\text{x}\{0.011+(1-0.011)\,[1 - \exp(-437\text{x}10^{-9}\text{failures/hr x } 3.7 \text{ yrs x } 8760 \text{ hr/yr})]\}$

$= 4.86 \approx 5$ SOPRV dangerous failures due to product failure rates and PIF. [Eq. 3]

Of these five failures, approximately two (1.1% x 195 SOPRV) would be expected to be infant mortality failures. But the actual number of dangerous failures was 13!

The most frequently used technique for estimating a total constant failure rate from FFD is to compute

λ_{DTOTAL} = # of failures / total # of operating hours. [Eq. 4]

Based on the data from the research study, this means that one would estimate λ_{DTOTAL} as

λ_{DTOTAL} = 13 failures/(195 SOPRV x 3.7 yrs/SOPRV x 8760 hrs/yr) = $2.057 \text{ x } 10^{-6}$ failures/hr

= $2057 \text{ x } 10^{-9}$ failures/hr = 2057 FITS. [Eq. 4a]

Note that the estimated total failure rate is more than four times that of the predicted inherent product failure rate but in assessing that difference one must remember that the estimated total failure rate does not include an estimate of initial failure whereas the FMEDA method also predicts a PIF for the product. Now, why is there such a difference?

Table 1 lists the dangerous failures and their causes as well as classifies the failures as inherent product failures or site-generated failures and identifies the possible life phases when the failures occurred. Note that there are six inherent product failures. However, the two manufacturing defect product failures were on the same physical valve. Although this specific SOPRV failed its first proof test after being in service because the first lift exceeded 50% of set pressure, subsequent lifts during the same proof test were normal and it was returned to service which was permitted by SRS maintenance program protocol. It failed its next proof test after additional in-service time by lifting for the first time at pressures exceeding 50% of set pressure and again returned to normal pressure lifts on subsequent lifts during the same proof test. Despite the fact that the SOPRV again met the SRS maintenance program protocol for being retagged and returned to service, the site decided to rebuild the valve and found that in the original assembly the wrong spring had been used. Thus one of the six SOPRV inherent product failures was a "duplicate" failure and five were individual inherent product failures. Note further that two of these product failures were due to sticking which is typically an infant mortality failure mechanism [10]. So the FMEDA's predicted λ_{DPROD} and PIF_{PROD} accounted appropriately for the number of inherent product failures and correctly divided those product failures between initial/infant mortality and useful life failures. This provides further validation of the FMEDA method and the associated databases.

Table 1. Categorization of Spring Operated Relief Valve Dangerous Failures

Failure #	Cause of Failure	Product or Site Failures	Possible Failure Life Phase
1	Sticking (Stainless Steel Adhesions)	Product	Infant mortality
2	Sticking (Stainless Steel Adhesions)	Product	Infant mortality
3	Galvanic corrosion cell	Product	Useful Life
4	Galvanic corrosion cell	Product	Useful Life
5	Manufacturer Defect	Product	Useful Life or Aging
6	Manufacturer Defect	Product	Useful Life or Aging
7	Corrosion; inappropriate design choice	Site	Useful Life or Aging
8	Corrosion; inappropriate design choice	Site	Useful Life or Aging
9	Corrosion; unprotected exposure to weather	Site	Useful Life or Aging
10	Corrosion; in service too long	Site	Useful Life or Aging
11	Corrosion; in service too long	Site	Useful Life or Aging
12	Corrosion; in service too long	Site	Useful Life or Aging
13	Sticking; in service too long	Site	Useful Life or Aging

The total failure rate estimate from Eq. (3) differs from the predicted product failure rate due to the site-generated failure rate which, based on seven site-generated failures, can be now be estimated as

$$\lambda_{DSITE} = 7 \text{ failures}/(195 \text{ SOPRV} \times 3.7 \text{ yrs/SOPRV} \times 8760 \text{ hrs/yr}) = 1.108 \times 10^{-6} \text{ failures/hr}$$

$$= 1108 \times 10^{-9} \text{ failures/hr} = 1108 \text{ FITS.} \qquad \text{[Eq. 5]}$$

Note that λ_{DSITE} is a little more than half the size of λ_{DTOTAL}, which is expected because the number of site-generated dangerous failures account for a little more than half of all the dangerous failures. These site-generated failures are important because they represent failures over which the site *may* have some control and *may* be able to eliminate by changes in site practices.

In this example, none of the site generated failures were due to initial failures or to infant mortality failures. However, it is entirely possible that these types of failures could be generated by certain kinds of end-user activities and therefore may need to be accounted for in particular cases.

4.3 Why All Failures Must Be Accounted for in Estimating Failure Rates and Computing PFDavg

Safety standards such as [1, 2, 3] require that the assessment of safety performance via failure rates include failures due to failure mechanisms which can be described by a constant failure rate. The standards do not require inclusion of failures due to mechanisms that are not described by constant failure rates. However, while the standards do not require the inclusion of all failures in the computation of PFDavg, it is also true that they do not prohibit inclusion of all failures either. Dangerous failures that exist when a demand occurs lead to significant consequences. If the consequences of a dangerous failure are the same regardless of how the corresponding dangerous failure rate is *modeled* then there is little justification for eliminating certain failure types from the analysis of failure rates and the computation of PFDavg. After all, if one is a resident whose home is lost to an uncontrolled oil refinery fire, that individual does not care what type of dangerous failure caused the fire. If a family member is disabled or dies as the result of an industrial accident unmitigated by safety equipment that was in a state of dangerous failure, again, the family will not care what type of dangerous failure failed to prevent the accident. Further, most FFD does not contain sufficient information to correctly distinguish between different underlying causes of failure. Thus, when failures are eliminated from analysis and computation, safety *appears* to be better but in reality is not. Consequently, it is the opinion of the authors that it is important to include all failures discovered in FFD or reasonably anticipated/expected/predicted to occur (even if not observed in FFD) in the computation of PFDavg.

4.4 Safety Implications of Omitting Site-Generated Failure Rates When Assessing Safety Performance

If no FFD were available for the SOPRV subpopulation above, then the computation of safety performance would have to be based on predictions about λ_D. If an analyst chose to limit the analysis only to the predicted constant failure rate, λ_{DPROD}, (and therefore omitted the predicted PIF_{PROD} as permitted by standards [1, 2, 3]), the PFDavg at the time of proof test would be computed from Eq. (2a) as

$$PFDavg = 0.5 \times 437 \times 10^{-9} \text{ failures/hr} \times 3.7 \text{ yrs} \times 8760 \text{ hr/yr} = 0.007 \quad \text{[Eq. 6]}$$

whereas PFDavg based on λ_{DTOTAL} would be

$$PFDavg = 0.5 \times 2057 \times 10^{-9} \text{ failures/hr} \times 3.7 \text{ yrs} \times 8760 \text{ hr/yr} = 0.033 \quad \text{[Eq. 7]}$$

and finally PFDavg based on PIF_{PROD}, λ_{DPROD}, and λ_{DSITE} would be

$$PFDavg = 0.011 + (1-0.011) \times 0.5 \times (437+1108) \times 10^{-9} \text{ failures/hr} \times 3.7 \text{ yrs} \times 8760 \text{ hr/yr} = 0.036. \quad \text{[Eq. 8]}$$

Table 2 shows the correspondence between PFDavg and safety integrity levels (SIL) for low demand operation which is characteristic of processes using SOPRV. Using the value of PFDavg from Eq. (6) which is based on inherent product failure rate alone provides a rating of SIL 2 for the SOPRV. Using the value of PFDavg computed from either Eqs. (7) or (8), which both include the site-generated failure rate (though in different ways), provides a rating of SIL 1 for the SOPRV. Excluding site-generated failure rates can cause an order of magnitude difference in assessed SIL level.

Table 2 Correspondence Between PFDavg and SIL

SIL per IEC61508[1]	PFDavg
1	$\geq 10^{-2}$ to $< 10^{-1}$
2	$\geq 10^{-3}$ to $< 10^{-2}$
3	$\geq 10^{-4}$ to $< 10^{-3}$
4	$\geq 10^{-5}$ to $< 10^{-4}$

As was previously mentioned, it is rare for a single site to have sufficient FFD on a particular piece of equipment to produce accurate estimates of dangerous failure rates. Furthermore, if multiple-site FFD is pooled in order to acquire sufficient data for an estimate, the resulting λ_{DTOTAL} will containing the product failure rate along with an *average* site-generated failure rate. This will benefit the assessed safety performance of sites that engage in practices that are likely to increase site-generated failure rates while penalizing sites that engage in practices that are likely to decrease site-generated failure rates.

5. Incorporating All Failure Mechanisms into the Computation of PFDavg

Clearly the above example demonstrates that an accurate evaluation of safety performance requires inclusion of all failures and that the different failure mechanisms be accounted for appropriately at both the product and site levels. Based on the RCA for the 13 failures in Table 1, it is clear that both inherent product and site-generated failures were not limited to failure mechanisms consistent with constant failure rates. Failures were identified as being consistent with infant mortality failure, useful life failure, and failure due to aging. At a site that does not perform pre-installation proof testing of SOPRV (or immediate post-installation proof testing of equipment that can be tested after installation) the possibility of initial failure also exists.

If an analyst is to appropriately include in the computation of PFDavg initial failures (which require one parameter), infant mortality failures (which require a minimum of two parameters), useful life failures (which require one parameter) and failures due to aging (which require a minimum of two parameters), the task seems overwhelming. Further, if there is insufficient information in failure data to classify the failures, how can these different failure types be accounted for and appropriately incorporated into the computation of PFDavg? Fortunately, PFDavg can be *conservatively* approximated by Eq. (2) using only two parameters while still including all failure mechanisms. This is accomplished by including in PIF the fraction of equipment population expected to exhibit infant mortality failure and by including in the predicted constant failure rate the failures expected due to aging. These predictions can be made separately for inherent product failures (as is already done with the FMEDA technique) and for site-generated failures (as is proposed below as an extension to the FMEDA technique). The reasons why all failure mechanisms can be conservatively modelled with only two parameters is outlined in the Appendix.

6. Extending the FMEDA Technique to Predict Site-Generated Failure Rates & PIF

Most sites do not have sufficient FFD of their own for a given model of equipment to estimate a λ_{DSITE} for that specific site. However, the FMEDA technique [11] which is a proven method for product failure rate prediction, and for some equipment product PIF prediction also, offers a practical solution to the problem of insufficient data. The FEMDA technique bases prediction of a constant failure rate on the assumptions that the SIF components are configured in series with each component's failure rate modeled as a constant. A validated database [12, 13] which includes component failure rates for each component failure mode serves as a basis for predicting the failure rates by mode of equipment comprised of various components in the database. The component failure rates are adjusted for specifics of equipment usage and environment.

There is no reason why the FMEDA component databases cannot be extended to include probabilities of initial and infant mortality failures for additional equipment and to include adjustments to the component constant failure rates that would account for different site practices as these impact useful life and aging failures. In computing PFDavg based on FMEDA predictions, some site practices, such as performing incomplete proof testing, are already included.

The current FMEDA component databases were constructed by initializing the failure rate values of a large number of frequently used components for each of the various component failure modes. Equipment failure rates were then predicted using the component database. As FFD became available for the equipment with failure rate predictions, the predicted equipment failure rates were compared to the FFD equipment rates, component failure rates were adjusted if necessary, and new predictions were made. The databases were also supplemented with additional component failure modes as these were discovered. This feedback process was repeated over many different equipment products until the component failure rate and failure mode database stabilized.

It is proposed to proceed in a similar fashion to extend the FMEDA component database to include the effects of site practices. Five levels of site practices are being defined from the lowest level where little or no effort is made to avert failures chargeable to site actions to the highest level where site practices include documentation and analysis of all field failures with understanding of root cause and a written, auditable plan for improvement. Each level of site practice will result in a site-generated constant failure rate for each component's failure mode. Additionally for each level of site practice, each component's failure mode may have a non-zero probability of initial and infant mortality failure. Using this database, new predictions will be made for various equipment and these predictions will be compared to existing FFD on a site by site basis because site practices can vary from location to location. Adjustments will be made to the extended database until it stabilizes.

This procedure will allow sites not only to predict the failure rates of equipment at their particular locations but also to quantify how much improvement in safety can be realized by changes in specific site practices.

7. Conclusions

This paper clearly shows that the safety performance of equipment is a function not only of the inherent product failure rates and PIF but also of site-generated failure rates determined by end-user practices as well. This means that for an end user to assess realistically the safety being achieved by a particular piece of safety equipment at a specific site, λ_{DPROD}, as provided by the manufacturer or assessed by other means meeting IEC minimum requirements, is insufficient. Information about the site-generated failure rate must also be included. The methods presented in this paper provide a framework for accomplishing this task.

8. References

[1] IEC 61511, *Application of Safety Instrumented Systems for the Process Industries*, Geneva, Switzerland, 2nd edition, 2004.

[2] IEC 61508, *Functional Safety of electrical / electronic / programmable electronic safety-related systems*, Geneva, Switzerland, 2010.

[3] ANSI/ISA SP84.00.01 – 2004 (IEC 61511-1 Mod.), *Application of Safety Instrumented Systems for the Process Industries*, Raleigh, NC, ISA, 2004.

[4] J. V. Bukowski, R. E. Gross and W. M. Goble, "Probability of Initial Failure for Spring Operated Relief Valves," *Proceedings of the ASME 2011 Pressure Vessels and Piping Division Conference*, Baltimore, MD, July 2011.

[5] J. V. Bukowski, R. E. Gross and W. M. Goble, "The Adhesion Failure Mode in Stainless Steel Trim Spring Operated Pressure Relief Valves," *ASME Journal of Pressure Vessel Technology*, Vol 135 No 6, December 2013.

[6] L. Stewart, J. V. Bukowski and W. M. Goble, "Improving Reliability & Safety Performance of Solenoid Valves by Stroke Testing," *Proceedings AIChE 9th Annual Global Conf on Process Safety – Process Plant Safety Symposium*, San Antonio, TX, April-May 2013.

[7] J. V. Bukowski, R. E. Gross and W. M. Goble, "Investigation of Adhesion Formation in New Stainless Steel Trim Spring Operated Pressure Relief Valves," *ASME Journal of Pressure Vessel Technology*, Paper 061602 1-9, Vol 136 No 6, December 2014.

[8] W. M. Goble, "Maintenance capability: Does it matter in the long term?" Hydrocarbon Processing, June 1, 2013.

[9] J. V. Bukowski, W. M. Goble, R. E. Gross and S. P. Harris, "The Effects of Maintenance Actions on the PFDavg of Spring Operated Pressure Relief Valves," *Proceedings of the ASME 2014 Pressure Vessels And Piping Division Conference*, Anaheim, CA, July 2014.

[10] Z. Zhao and B. Bhushan, "Effect of Lubrication Thickness and Viscosity and Rest Time On Long-Term Stiction in Magnetic Thin-Film Rigid Disks," *IEEE Transactions On Magnetics*, Vol 34. No 4, July 1998.

[11] W. M. Goble and A. C. Brombacher, "Using a Failure Modes, Effects and Diagnostic Analysis (FMEDA) to Measure Diagnostic Coverage in Programmable Electronic Systems," *Reliability Engineering and System Safety*, Vol. 66, No. 2, Nov 1999, pp. 145-148.

[12] W. M. Goble and J. V. Bukowski, "Development of a Mechanical Component Failure Database," *Proceedings of the 2007 Annual Reliability and Maintainability Symposium*, Orlando, FL, January 2007, pp. 451-455.
[13] J. V. Bukowski and, W. M. Goble, "Validation of a Mechanical Component Constant Failure Rate Database," 2009 *Proceedings Annual Reliability and Maintainability Symposium*, Fort Worth, TX, January 2009, 338-343.

9. Acknowledgement

The authors wish to acknowledge with gratitude the following individuals who have been involved in and/or have supported the Savannah River Site SOPRV data collection and RCA analyses: Robert E. Gross, PE, Site Infrastructure Engineering; James Fulmer and Chester Enlow, SRS Valve Shop Mechanics; Duane Edington and Albert (Chip) Jenison, SRS Valve Shop Management; Bob Davis and Jim Broome, Site Services SOPRV Coordinators; SRS Pressure Protection Committee Members.

10. Appendix

In order to show that all four failure mechanisms can be *conservatively* modeled using only two parameters, first consider computing PFD(t) assuming that the times to failure of the failed units **are known**. Figure A1a shows PFD(t) data used for computing PFDavg of a hypothetical population of 1000 identical SIF. In Figure A1a, the state (successful or failed) of ten units of the total population is displayed as a function of time over the time interval [0, T_P] where T_P equals 5 years. Note, for each unit's timeline, the dashed line indicates that the SIF is successful. If the line ends in S at t equals 5 years this indicates the system was successful on the entire time interval. A dashed line that changes to a series of F's indicates that a dangerous failure occurred at the time of the change and that the SIF remained in the dangerous failure state until T_P. Figure A1a shows that SIF units 9 and 10 have always been successful on the time interval [0, T_P] and for the purposes of this illustration it is also assumed that 990 other SIF units not shown in Figure A1a were also successful over the entire time interval. Figure A1a also shows eight other SIF units that were discovered failed in proof test.

Based on the information in Figure A1a it is possible to construct an empirical PFD(t) as follows: for any point in time between 0 and 5 years, PFD(t) is the percent of units failed dangerously recalling, of course, that there are a total of 1000 units. At t equal 0, PFD(0) equals 0 because there are no failed units. At t equals 1 year, PFD(1 year) = 0.001 because there is one failed unit (unit 7) out of the total population of 1000 SIF units. Continuing in this manner one can construct the empirical PFD(t) associated with Figure A1a which is displayed in Figure A1b. Now to find PFDavg from Figure A1b, PFD(t) is integrated over the interval [0, 5] and divided by 5 years. The area under the step function is easily computed noting that the discontinuities occur at times 0.77, 1.23, 2.21, 2.65, 2.98, 3.20, 4.10, and 4.44 years, respectively. This results in an empirical value of PFDavg of 0.0037.

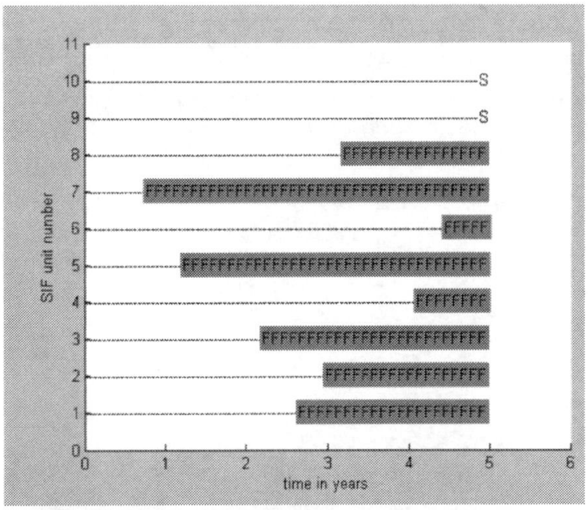

Figure A1a. Timelines of hypothetical SIF states for 10 SIF units consistent with a constant failure rate model.

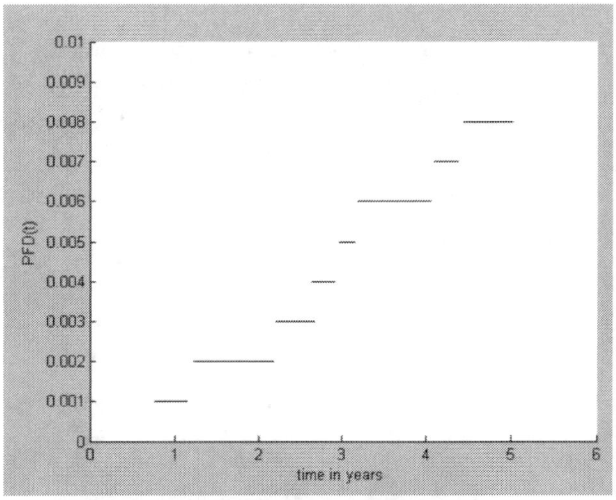

Figure A1b. Empirical PFD(*t*) corresponding to the hypothetical timelines in Figure A1a.

In general, the times to failure will **not** be known. Therefore, some assumptions must be made about the failure rate describing the data so that mathematical models can be used to compute PFDavg. Assuming that the failures are governed by a constant failure rate, and knowing that there were a total of 8 failures in 1000 units with about 5 years of operating time on each unit, λ_D would commonly be estimated as

λ_D = total # failures / total operating time [Eq. A1a]

= 8 / (1000 units * 5 yrs * 8760 hrs/yr) = $1.83 * 10^{-7}$ failures/hr. [Eq.A1b]

Then, from Eq. (2a), PFDavg $\approx 1.83 * 10^{-7}$ * (5 yrs*8760 hrs/yr)/2 = 0.004. The empirical PFDavg from Eq. (A1b) and the PFDavg calculated from Eq. (2a) match well because the failure data exhibited in Figure A1a were generated under the assumption of a constant failure rate. Note that PFDavg is directly related to the amount of time the SIF population is in a state of dangerous failure so that PFDavg is related to the amount of red shown in Figure A1a. Clearly it is true that the more red shown in a figure of timelines, the higher the value of PFDavg and, correspondingly, the less red, the smaller the value of PFDavg.

Because, in general, the times to failure are **not** known and based on proof test data of 8 dangerous failures in 1000 proof tests, the failure times *could* just as well have occurred at the times shown in Figure A2. In this figure it is clear that the times-to-failure pattern is more consistent with an increasing failure rate mechanism, i.e., failures due to aging. In this case, the empirical PFDavg would be substantially smaller than the value of 0.0037 computed above, consistent with the substantially smaller amount of time the SIF population is in a state of dangerous failure and the, consequently, smaller amount of red in Figure A2. However, without knowing the times to failure, if a constant failure rate were assumed to describe the data, then the mathematical model for λ_D would be identical to that of Eq. (A1b) above and again the calculated PFDavg would have been 0.004. This would lead to a conservative value of PFDavg

Note: Do not add page numbers. Do not refer to page numbers when referencing different portions of the paper

for the failures due to aging. Thus it can be concluded that, in general, combining failures that are due to mechanisms described by a constant failure rate with failures due to aging and modeling them together by an equivalent constant failure rate, λ_D, leads to an acceptably conservative approximation for PFDavg when Equation (2a) is then used.

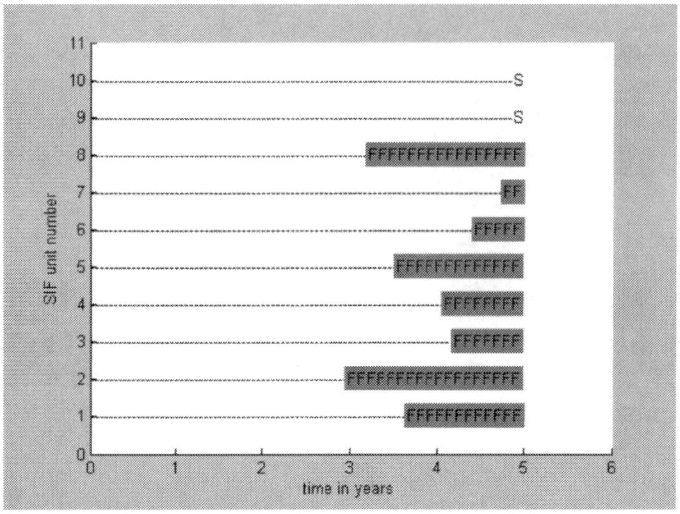

Figure A2. Timelines of hypothetical SIF states for 10 SIF units consistent with an increasing failure rate model due to aging.

Now consider Figure A3 which shows 8 failed units where the failures were all due to initial failures. Clearly in this case, PFDavg is equal to 0.008. Lastly, consider Figure A4 which shows the 8 failed units to have failed at times consistent with a combination of initial failures and infant mortality failures. In this case it is clear that the empirical PFDavg would be less than the 0.008 from Figure A3 but greater than the 0.004 that would be computed under the assumption of a constant failure rate model. Further, it is clear that the empirical value would be closer to 0.008 because the SIF population spends only slightly less time in a state of dangerous failure in Figure A4 when compared to Figure A3, i.e., there is only slightly less red in Figure A4 than in Figure A3. Therefore, if all initial failures and infant mortality failures are combined and modeled as initial failures this leads to a conservative approximation for PFDavg for those two types of failures.

Since all failure mechanisms can be conservatively modelled using only two parameters, PFDavg can be conservatively approximated using only two parameters. This makes the problem of including all failures in the computation of PFDavg both tractable and practical. It requires only that all failures be divided into one of two categories: initial or infant mortality failures vs constant failure rate or end-of-life/aging failures.

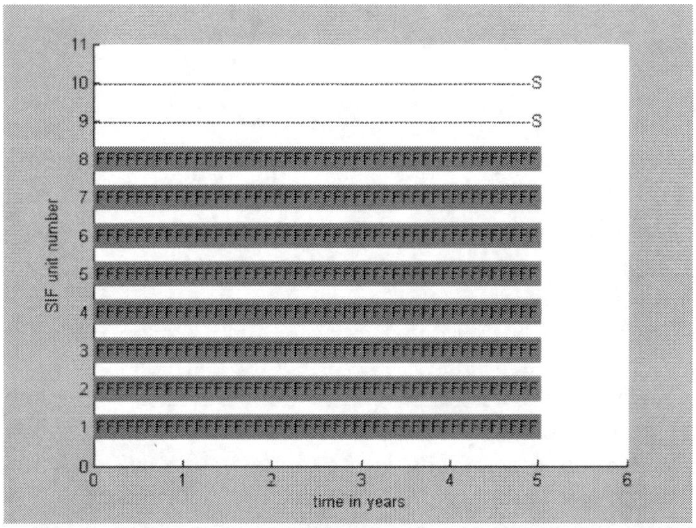

Figure A3. Timelines of hypothetical SIF states for 10 SIF units consistent with initial failures.

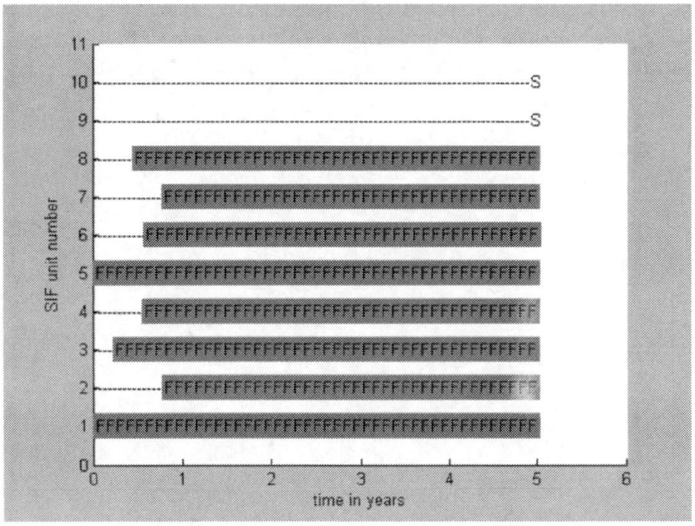

Figure A4. Timelines of hypothetical SIF states for 10 SIF units consistent with initial failures and infant mortality failures.

397151 Exploring Dependencies of Ignition Locations on Probabilistic Explosion Analysis

Abstract:

CFD based Probabilistic Explosion Analysis has become a common method for estimating structural design accidental load (DAL) specifications for offshore capital projects. This approach has been shown to yield more reasonable results than traditional analysis i.e. solely consequence based which generally result in highly conservative values. As expected, results from probabilistic based approach may have dependencies based on chosen methodologies. Previous research and studies have generally focused on exploring methodology changes which are considered to result in significant differences i.e. transient vs. non-transient dispersion modeling, different levels of 3-D model detail, etc. The purpose of this paper is to explore the dependency that can come from changes to ignition locations (often viewed as small adjustment) within a probabilistic explosion analysis. Commonly a limited number of ignition points are chosen for each simulated cloud location/size, per NORSOK Standard Z-013 [2] *"Ignition can in principle take place anywhere in the gas cloud. Explosion simulations shall include at least three different ignitions: central, edge and other."* This paper will be broadly divided in two parts where the first part will explore the use of synthesized ignition locations based on data from CFD simulations conducted using a fewer number of ignition points. This will be compared to actual simulation data from a larger set of ignition locations for establishing convergence. Having established criteria from the first part, the second part of the paper will explore the use of different ignition probability distribution functions (equal, linear, Gaussian, etc.) and compare the exceedance results on specified targets. It is the intent that by understanding how the locations of ignition sources affect the sensitivities of the forthcoming QRA work performed to internally or externally demonstrate ALARP for most facility design.

397853 Process Safety Near-Miss Management System in Takreer Ruwais Refinery: An Effective Tool for Applying Prevention Strategies

Abstract:

The primary focus of any process safety management system is always to reduce the risk of a major incident by controlling major accident hazards. Process safety events (which include near misses and minor & major incidents) are key indicators of a company's process safety performance that demonstrate how effectively a system is working to deliver its objective of abating major accident hazards. While analysing various process safety events, we observe that there is a close relationship between near-miss events and major incidents as both have often common causes. A near-miss could be defined as a hazardous condition where the event sequence could lead to an accident if it had not been interrupted. An effective Process Safety Near-miss Management System aims to quickly recognize signals from the operational fields in order to apply prevention strategies.

TAKREER Ruwais Refinery has been reviewing process safety near miss events for many years but has developed within the last two years a detailed and systematic Process Safety Near Miss Program which is modelled after API and CCPS guidelines. This paper introduces some practical strategies adopted in Ruwais Refinery for implementing a model near miss reporting system which has contributed to excellence in health and safety. It is hoped this paper will initiate further consideration and conversation about ways to enhance involvement in Near-miss reporting and thereby reduce the occurrence of unintentional injuries, fatalities, and property-damage incidents.

How Maintaining Current Heat & Material Balances Commits to Process Safety

Dustin Smith
Smith & Burgess, LLC
7600 W. Tidwell Rd., Ste. 600, 77040
Dustin.Smith@smithburgess.com

Dick Baum
Smith & Burgess, LLC
7600 W. Tidwell Rd., Ste. 600, 77040
Dick.Baum@smithburgess.com

Nicholas Cristea
Smith & Burgess, LLC
7600 W. Tidwell Rd., Ste. 600, 77040
Nicholas.Cristea@smithburgess.com

Juan Lopez
Smith & Burgess, LLC
7600 W. Tidwell Rd., Ste. 600, 77040
Juan.Lopez@smithburgess.com

Prepared for Presentation at
American Institute of Chemical Engineers
2015 Spring Meeting
11th Global Congress on Process Safety
Austin, Texas
April 27-30, 2015

UNPUBLISHED

How Maintaining Current Heat & Material Balances Commits to Process Safety

Juan Lopez
Smith & Burgess, LLC
7600 W. Tidwell Rd., Ste. 600, 77040
Juan.Lopez@smithburgess.com

Nicholas Cristea
Smith & Burgess, LLC

Keywords: LOPA, PHA, Relief Systems Design, Heat and Material Balance.

Abstract

A Heat & Material Balance (H&MB) is a valuable component of Process Safety Management (PSM) because it helps to follow the four pillars of risk-based process safety. It is important to ensure H&MBs are updated on a regular basis to accurately represent how the process is running. An H&MB is a valuable source of information for many aspects of PSM such as; Relief Systems Analysis, Process Hazard Analysis (PHA), and Layers of Protection Analysis (LOPA). If the information is not accurate and current, the analysis may lead to inaccuracies and inconsistencies that could result in: increased expense (rework), increased exposure (citations), and greater potential for incidents. This paper will look at some of the possible consequences of not having up-to-date H&MB in regards to a Relief System Analysis, PHA, or a LOPA.

1. Introduction

OSHA's Process Safety Management standard cites Heat and Material Balances (H&MBs) for any process built after May 26, 1992 as a requirement for PSM Compliance for the oil and gas industries.[1] Any modification to an existing facility that requires installing a new line, unit, or train to supplement the process could be considered as a new process and therefore requires H&MBs.[2]

Even though a plant may have been built before 1992, the plant owner should still actively attempt to keep updated documentation at least as a means to decrease the probability of incidents, which in turn may result in additional expenses or citations. Chances are even if the process was built before 1992, a small change or the installation of additional equipment may both be interpreted by the authorities as a change in the process requiring updated H&MBs. In many cases, the owner of the plant may not even be aware of these small changes that may occur over time, hence, highlighting the importance of updated and organized documentation.

It is a common occurrence in today's oil and gas industry that an operating facility does not have updated or readily available documentation regarding its process or equipment. Over time, changes in the process may occur, additional pieces of equipment may be added, or modifications may be made to existing pieces of equipment. All of these changes have the potential to adversely affect previous analyses. Not only is it a requirement by OSHA to regularly update the applicable PSM Compliance documentation, but inaccuracies in the documentation may also lead to inaccurate results in other aspects of PSM such as relief systems analysis, LOPA, and PHA. Because they normally provide compositions, temperature, pressure, and flow rates related to the process, H&MBs are a major part of PSM Compliant documentation that can help in obtaining accurate results in the aforementioned areas of PSM. Keeping updated H&MBs speaks volumes about the commitment of a plant in keeping process safety as a priority.

2. Discussion

2.1 PHA and LOPA

A process hazard analysis (PHA), also referred to as process hazard evaluation, is a major part of the PSM program. A PHA is performed by a group of qualified engineers that leads to identifying the potential of hazards associated to the process or handling of a material. It is used as a tool to make risk based decisions on their process. A Layer of Protection Analysis (LOPA) is often used as an addition to a PHA in order to facilitate the decisions to be made for a plant. Both of these types of analyses are risk based and attempt to make important modifications or installation of additional layers of protection based on the probability and consequence of any given scenario.

OSHA's standard 1910.119 App. C highlights the importance of having knowledge about the process and how innovation or change may affect the methodology:

> *"The selection of a PHA methodology or technique will be influenced by many factors including the amount of existing knowledge about the process. Is it a process that has been operated for a long period of time with little or no innovation and extensive experience has been generated with its use? Or, is it a new process or one which has been changed frequently by the inclusion of innovative features [?]..."* [3]

Often times an H&MB is not available at the time of performing a Process Hazard Analysis or Layer of Protection Analysis. However, due to the nature of these analyses, which attempt to measure probability and consequence as a means to assess risk, the lack of an H&MB during the analysis could contribute to inaccuracies when measuring the consequence of an event. An H&MB includes information such as temperature, pressure, and composition of the fluid within a system; these variables can make a significant difference when measuring the consequence of a release.

For example, a large enough leak in a pipeline containing a liquid hydrocarbon at high pressure and temperature may result in an immediate loss of pressure and lead to a

Boiling Liquid Expanding Vapor Explosion (BLEVE). If one does not have the pressure, temperature, and composition of the fluid, the consequence of such an event could be miscalculated. The PHA analyst may reach the conclusion that a leak in the pipeline may only result in a vapor cloud that dissipates quickly. Having an H&MB during the PHA or LOPA could be the difference between assessing the consequence of the leak as a BLEVE, vapor cloud release, or the release of toxic fluid.

In a LOPA, the consequences of events are categorized depending on their severity. Examples of consequence categorization that can be used in a LOPA are shown in Table 1 (shown below). In the table, the event is categorized depending on the characteristics and the quantity of the fluid being released.[4] As can be seen from the example, the quantity and characteristics of a certain release can affect the categorization of an event, and can directly influence whether a decision is made to add additional safety measures to reduce the probability of an incident.

Table 1. Example Consequence Categorization[4]

Release Characteristic	Size of a release (beyond a dike)					
	1 to 10 pound release	**10 to 100 pound release**	**100 to 1,000 pound release**	**1,000 to 10,000 pound release**	**10,000 to 100,000 pound release**	**>100,000 pound release**
Extremely toxic above BP*	Category 3	Category 4	Category 5	Category 5	Category 5	Category 5
Extremely toxic below BP or highly toxic above BP	Category 2	Category 3	Category 4	Category 5	Category 5	Category 5
Highly toxic below BP or flammable above BP	Category 2	Category 2	Category 3	Category 4	Category 5	Category 5
Flammable below BP	Category 1	Category 2	Category 2	Category 3	Category 4	Category 5
Combustible liquid	Category 1	Category 1	Category 1	Category 2	Category 2	Category 3

*BP = atmospheric boiling point

Consequence Characteristic	Magnitude of Loss					
	Spared or non-essential equipment	**Plant Outage <1 month**	**Plant Outage 1-3 months**	**Plant Outage >3 months**	**Vessel rupture 3,000 to 10,000 gal 100-300 psi**	**Vessel rupture >10,000 gal >300 psi**
Mechanical damage to large main product plant	Category 2	Category 3	Category 4	Category 4	Category 4	Category 5
Mechanical damage to small by-product plant	Category 2	Category 2	Category 3	Category 4	Category 4	Category 5

2.2 Relief Systems Design

Relief systems design is also recognized as part of PSM compliant documentation. During relief systems design, the entire plant, one piece of equipment at a time, is analyzed for the potential scenarios that could lead to overpressuring a piece of equipment over the safe limit. All potential applicable scenarios are considered, and calculations are performed in order to determine the rate, temperature, and thermodynamic properties of the fluid that needs to be relieved, so as to keep the vessel from being pressurized over the acceptable limit. A relief device is then sized and recommendations can be made if any changes should be performed. Any change in the conditions of the process, including temperature, pressure, composition, or process rate, may lead to a relief device of different size.

As part of this research paper, a blocked vapor outlet scenario was simulated for a depropanizer column. The purpose of the simulation was to find the effect of changing feed compositions in relief valve sizing. The column was simulated with different propane compositions; the other hydrocarbons in the feed stream were ethane, isobutane, n-butane, isopentane, n-pentane, n-hexane, and n-heptane in the same ratio. The composition of the aforementioned components was lowered equally with increasing compositions of propane. The diagram of the Aspen HYSYS Simulation © is shown in Figure 1.

The following specifications were used in order to simulate the column:

- Feed charge of 200,000 lb/hr
- Relieving pressure of 300 psig
- Constant reflux rate of 80,000 lb/hr
- Reboiler duty of 20 MMBtu/hr

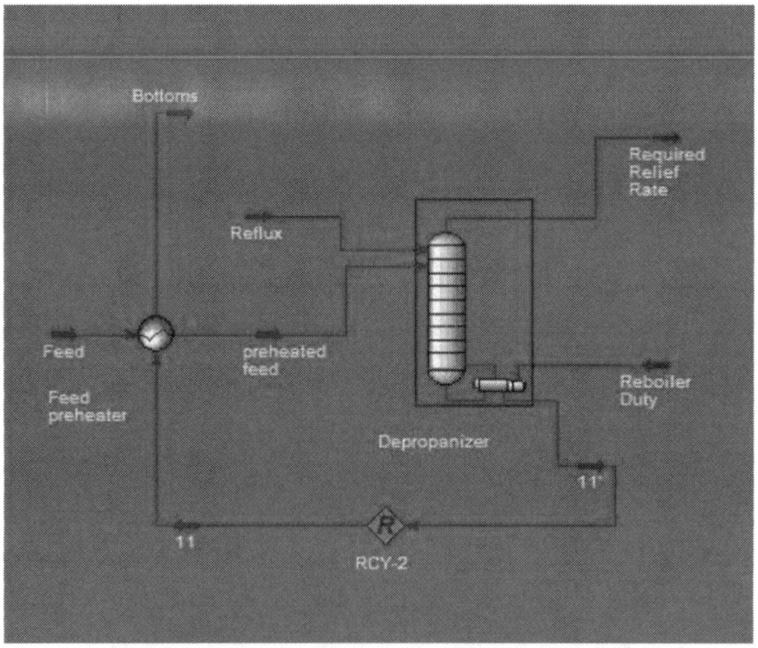

Figure 1. Depropanizer column simulation

The resulting effective discharge area required to achieve the required flowrate was found for each composition of propane using the derived critical flow sizing equations from API Standard 520 Part I, shown as Equation 1 below.[5]

$$w = A \times C \times K_d \times K_B \times P \times \sqrt{\frac{M}{TZ}} \qquad \text{Eq. (1)}$$

Where

w = mass flow through the orifice
A = Effective orifice area
C = function of the ratio of specific heats
K_d = Effective coefficient of discharge
K_B = Backpressure correction factor
P = Relieving pressure
M = Molecular Weight
T = Relieving Temperature
Z = Compressibility Factor

The function of the ratio of specific heats, *C*, is defined by Equation 2.

$$C = \sqrt{k\left(\frac{2}{k+1}\right)^{\frac{(k+1)}{(k-1)}}} \qquad \text{Eq. (2)}$$

Where
k = ideal ratio of the specific heats

Figure 2 shows the effective discharge area as a function of propane compositions. As shown in the figure, the effective discharge area has a direct correlation with increasing compositions of propane. While the composition of a feed stream is only a portion of all the information contained in an H&MB, it can make a difference in the results of the design.

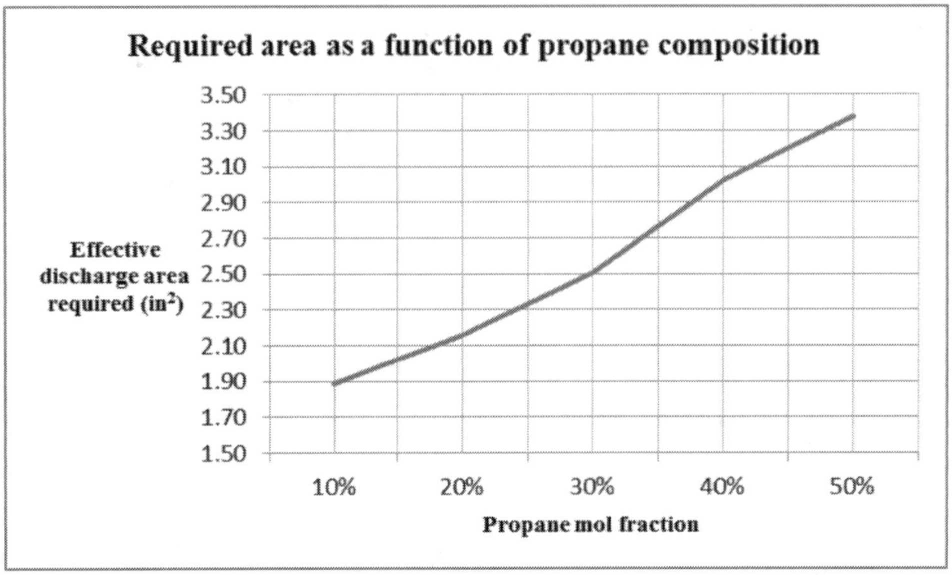

Figure 2. Depropanizer relief valve effective discharge area vs propane composition.

It is common in the industry that the process safety analyst who performs the relief systems design is provided with an outdated H&MB or no H&MB at all as supporting documentation. If this situation occurs, the analyst is forced to make conservative assumptions for the process that can often result in oversizing a relief device; this can lead to relief device chattering and additional cost.

If the relief systems analysis is performed with an inaccurate H&MB, the results of the work performed may be unreasonable to the plant owner. When the analysis is performed based on outdated information, the result of the analysis may not be what the client initially expects due to the conservative measures that need to be assumed by the engineer. A significant number of action items may be generated due to the conservative measures, and the plant may have to choose to perform the entire analysis again after they have updated the corresponding documentation; this decision may depend on the cost associated with mitigating the extensive list of action items versus the combined cost of updating H&MBs, performing the analysis for a second time, and dealing with any leftover action items. Therefore, in order to avoid rework and additional expenses, keeping updated H&MBs should be a priority.

3. Conclusion

The main reason to maintain current H&MBs is to make the plant safer. It is common in the process safety industry that the H&MBs, provided as supporting documentation for performing PSM related analyses, are outdated or not provided at all. Flow rates, compositions, pressures, and temperatures shown in H&MBs can affect the sizing during relief systems design, decision making during a PHA, and consequence categorization during a LOPA, all of which can contribute to process safety.

Current H&MBs are an important aspect of PSM documentation that relates directly to the commitment to process safety. Any process built after 1992 must have updated H&MBs in order to comply with the PSM standard. In addition, modifications made to a plant could be interpreted by OSHA as a change to the process requiring updated H&MBs.

In addition to being an integral part of process safety, maintaining current Heat and Material Balances can help in:

- Providing accurate sizing for relief devices
- Avoiding rework related to PHA, LOPA, and relief systems design
- Reducing action items from relief systems design
- Preventing citations related to outdated documentation
- Reducing the potential for incidents
- Accurately measuring the consequence of incidents

4. References

[1] "Process Safety Management of Highly Hazardous Chemicals -- Compliance Guidelines and Enforcement Procedures", 29 CFR 1910.119, 09/28/1992

[2] "Clarification of the Process Safety Management standard with regard to material and energy balances", 29 CFR 1910.119, 09/25/1995

[3] "Compliance Guidelines and Recommendations for Process Safety Management (Nonmandatory)", 29 CFR 1910.119 App C

[4] "Layer of Protection Analysis – Simplified Process Risk Assessment", Center for Chemical Process Safety of the American Institute of Chemical Engineers

[5] Sizing, Selection, and Installation of Pressure-relieving Devices Part I – Sizing and Selection, API Standard 520, 9th Edition, July 2014

397930 Safety Culture and Leadership- Independent of Language

Abstract:

The existence of safety in the workplace is mandatory for a sustainable business. Strong safety cultures are characterized by workforce engagement and fully committed safety leadership. With that being said, increased diversity in work environments and international recruitment can oftentimes complicate the process of implementing a common safety culture.

More specifically, those who come from different types of work environments than their current environment may have trouble interpreting the procedures and guidelines that are enforced at their new place of employment. Differences in languages and dialects may also lead to misinterpretations of intended safety guidelines. Lastly, there may be some instances where the workplace follows a non-flexible and hierarchical structure that is not very open to change.

In order to best develop safety practices in the workforce, the senior leaders need to first develop an environment based on the understanding of the different rationales and expectations of a multi-cultural workforce. Once that has been established, communicating with the workforce on safety expectations of the organization and optimizing interpretation of this message is critical.

This paper will continue the discussion from a previous publication; how safety leadership can consider linguistic differences when implementing a safety plan for a diverse workforce. The discussions will emphasize the learnings of the Bradley Curve, showing how important it is to follow safe practices to reach the proximity of an almost 100% incident free area, and will provide real examples from international work assignments and diverse work teams, highlighting this need and offering potential approaches.

Best Practice in Application of CAA CAP 437 Standards for Offshore Helicopter Landing Areas

Chris Robinson PhD, CEng, MIMechE.
Principal Consultant. MMI Engineering, Apollo House, Eboracum Way, York, YO31 7RE

Tim Jones, MEng, CEng, MIMechE.
Principal Consultant. MMI Engineering, The Brew House, Wilderspool Park, Greenall's Ave, Warrington, WA4 6HL

Andrew Grand, PhD.
Senior Engineer. MMI Engineering, Corum Office Park, Crown Way, Warmley, Bristol, BS30 8FJ.

Morgan Reed, D. Eng.
Process Safety Engineer. MMI Engineering, Houston, Texas, USA.

A review has been carried out on methods for calculating turbulence with regard to the UK Civil Aviation Authority CAP437 standards for offshore helicopter landing areas. Information is provided on the implementation of CAP437 limits for turbulence conditions and appropriate assumptions which may be made. A discussion of turbulence generation and transport is included to demonstrate that these are not linear processes; and a best practice guideline for generating results to demonstrate CAP437 compliance is provided.

Keywords: CAP437, CFD, turbulence, helideck, helideck availability

Introduction

The UK Civil Aviation Authority (CAA) publishes a set of guidelines for offshore helidecks – *"CAP 437 Standards for Offshore Helicopter Landing Area"*. The first publication was released in 1981 and is now in its 7th edition; as well as being used as the standard for offshore installations in UK waters. The standard is also used by US based companies to provide guidance on their worldwide projects. It provides guidance on helideck layout and equipment, placing limits on environmental variables over the helideck for turbulence, temperature and hydrocarbon gas concentrations. CAP 437 stipulates that modeling should be used to demonstrate these limiting criteria are met, generally carried out using computational fluid dynamics (CFD).

The specific requirement for turbulence over the helideck is that the standard deviation of the vertical airflow velocity should be less than 1.75 m/s. Some practitioners struggle to interpret this requirement properly and give poor quality or incorrect advice to duty holders regarding the ability to meet the criterion in different wind speeds and hence the availability of the helideck for operations. There are several points which are commonly misunderstood. Firstly, this applies to turbulent fluctuations in the wind, not the mean velocity. Secondly, the ambient wind itself will be turbulent with the turbulence intensity dependent on the atmospheric stability conditions and wind speed; this must be taken into account along with any additional turbulence generated over the platform topsides and helideck. Finally, the amount of turbulence over the helideck is not linearly dependent on the wind speed. In some instances, practitioners have used CFD to determine the degree of turbulence at one or two wind speeds and simply scaled these over a wide range of wind velocities to determine the helideck availability. This would be appropriate if the turbulence over the helideck were linearly dependent on the mean wind velocity; however, some of the turbulence generating effects over a helideck are non-linear with mean wind and this assumption cannot be made.

The International Standard "*ISO 15138 Petroleum and natural gas industries - Offshore production installations - Heating, ventilation and air-conditioning*" also discusses requirements for turbulence modeling over helidecks. It requires that a Differential Stress Model, DSM (sometimes also referred to as a Reynolds Stress Model, RSM) is used for the CFD model calculation of turbulence over the helideck. This is an additional requirement to CAP437 but in keeping with the spirit of the CAP437 standard.

The following sections provide a review of turbulence, how turbulence is calculated in computational fluid dynamics (CFD) models and the best practices for applying these in CAP437 offshore helideck studies. Since this is a very specific and important aspect of the CAP437 standards, information is provided on the potential consequences when incorrect methods are employed.

Turbulence

The term "turbulence" is generally understood to mean flow that doesn't follow straight lines, or recirculates or swirls in some manner. Lewis Fry Richardson (1881 – 1953) a Fellow of the Royal Society and early pioneer of flow modeling particularly for weather forecasting described this in an aphorism: "Big whorls have little whorls that feed on their vorticity and little whorls have lesser whorls and so on to viscosity." This simple statement describes some important features of turbulence – that there exist turbulent "whorls" or "eddies" of different scales; and that flow energy is transferred for the largest eddies to the smallest eddies and is ultimately dissipated as heat by the action of viscosity. Large turbulent eddies typically have a length scale the same order of magnitude as the objects that create them. So in atmospheric physics turbulent eddies in weather patterns may be kilometers across – though for engineering applications the largest eddies are typically associated with the wake from buildings or the order tens of meters. The smallest eddies are defined by the Kolmogorov length scale at which the energy from turbulent motion is in balance with the energy dissipated by viscosity (Davidson, 2004). In the atmosphere this is typically in the range 0.1 to 10 mm.

In engineering applications, Reynolds number provides the definition for whether a flow is turbulent or laminar. This is a non-dimensional term which expresses the ratio of momentum to viscous forces:

$$Re = \frac{\rho U L}{\mu} \qquad (1)$$

ρ – fluid density; U – mean velocity; L – characteristic length scale; μ - fluid viscosity

For boundary layer flows, transition from laminar to turbulent flow is generally understood to start at a critical Reynolds number of the order 1000, although this depends on local conditions such as roughness and vibration. For free shear flow the transition to turbulence begins to occur much sooner – for jets the order of magnitude for the critical Reynolds number is typically 10 (Tritton, 1988).

For analysis it is often useful to decompose a velocity to its mean and turbulent components:

$$\widehat{U} = U + u' \qquad (2)$$

$\widehat{U}$ – instantaneous velocity; U – mean velocity; u' – fluctuating (turbulent) component of velocity.

In this way, the contributions to the flow of the mean and turbulent fluctuations can be separated which is often useful for calculation, computation and measurement techniques. An example is shown in Figure 1 for a velocity measured over time at a single point. The turbulent fluctuation appears to provide a random variation to the mean velocity. However it is not truly random as the turbulent velocity fluctuations are associated with the energy of individual eddies which are carried by the flow. Note that in the figure, there is also a periodic, non-turbulent variation in the mean flow which is a separate phenomenon to the turbulent fluctuation.

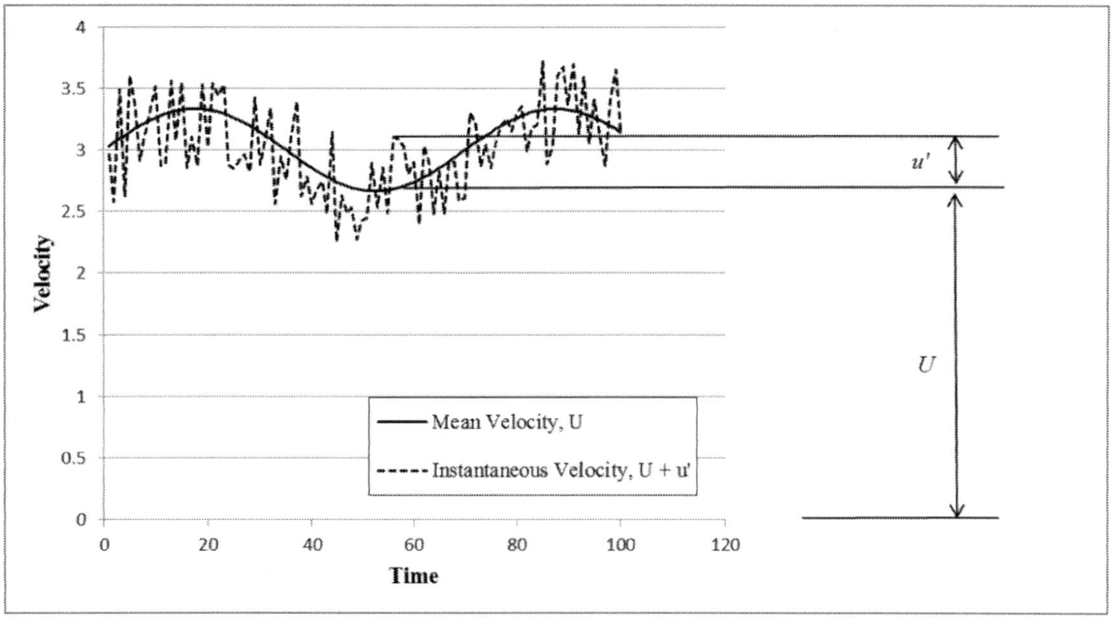

Figure 1. Example of a flow velocity measured at a single point showing the separate mean and turbulent fluctuation contributions to velocity

This breakdown of the flow into component part is important when carrying out CFD model calculations to determine whether the limits imposed by CAP437 have been achieved. In CAP437 this is stated: "*As a general rule, a limit on the standard deviation of the vertical airflow velocity of 1.75 m/s should not be exceeded*". The standard deviation of the vertical airflow is the root-mean-square value of v' (where the vertical components of velocity are normally represented by $\widehat{V} = V + v'$).

Modeling Turbulence

Computational Fluid Dynamics

Modeling turbulence is normally carried out within the framework of a computational fluid dynamics (CFD) analysis. Both CAP437 and ISO 15138 describe the use of CFD modeling for analyses of helideck turbulence. The basic principal of CFD software is to calculate flows using the conservation of momentum equations – these are effectively an expression of Newton's second law of motion which when expressed for fluids includes viscous action. The equations are more commonly referred to as the Navier-Stokes equations, after Claude-Louis Navier (1785-1836) and Sir George Gabriel Stokes (1819–1903) who, working independently, were the first to define the equations.

The Navier-Stokes equations are written:

$$\rho\frac{\partial \hat{U}_i}{\partial t} + \rho\hat{U}_j\frac{\partial \hat{U}_i}{\partial x_j} = -\frac{\partial P}{\partial x_i} + \mu\frac{\partial^2 \hat{U}_i}{\partial {x_j}^2} + \rho g_i \qquad (3)$$

t – time; x – distance; P – pressure; ρ – density; μ - viscosity; g – gravity

$i, j = 1, 2, 3$ represent the three coordinate directions along the primary orthogonal x-,y- and z-axes where the Einsteinian convention is followed that repeated indices indicate summation over the three coordinate directions. Hence equation 3 is in fact three equations for the x,y and z velocities, $\hat{U}, \hat{V}, \hat{W}$..

These provide a complete description of the flow through the entire range of turbulent energy length scales. Using the Navier-Stokes equations in this form in a CFD model is called Direct Numerical Simulation (DNS). For engineering purposes this is not particularly practicable as the model would need to resolve all turbulent eddies. In the atmosphere these are around 0.1 - 10 mm in size and have characteristic times of around 0.1 seconds. For a typical helideck on an offshore platform this would need a computational mesh size of more than 1,000 billion cells and 10,000 time steps, clearly beyond the capabilities of today's computers.

A more useful approach is to express the Navier Stokes equations in terms of the mean and fluctuating velocity components. Essentially the process is to substitute equation (2) for the three coordinate directions into equation set (3). The result is known as the Reynolds Average Navier Stokes equations (RANS).

$$\rho\frac{\partial U_i}{\partial t} + \rho\frac{\partial U_j U_i}{\partial x_j} = -\frac{\partial P}{\partial x_i} + \frac{\partial}{\partial x_j}\left[\mu\left(\frac{\partial U_i}{\partial x_j} + \frac{\partial U_j}{\partial x_i}\right) - \rho\overline{u_i' u_j'}\right] + \rho g_i \qquad (4)$$

In any averaging process information is lost and to complete the model this information must be accounted for by some means. In the RANS equations, this is manifested by the appearance of the term: $-\rho\overline{u_i' u_j'}$. This has the same units as viscous stress and acts on the mean flow velocity in the same way as viscosity μ - it is usually called the turbulent stress or Reynolds stress.

The aim of a turbulence model is to provide values for the turbulent stresses – note that when taking the indices into account for the three coordinate directions, $-\rho\overline{u_i' u_j'}$ describes six separate "stresses": $-\rho\overline{u'u'}$, $-\rho\overline{v'v'}$, $-\rho\overline{w'w'}$, $-\rho\overline{u'v'}$, $-\rho\overline{v'w'}$, $-\rho\overline{w'u'}$. For CAP437 analysis the term $-\rho\overline{v'v'}$ is the most useful as noted earlier v' is the standard deviation of vertical velocity which should be limited to 1.75 m/s.

Closure Models

There are a large number of turbulence models which have been developed for specific applications and general purposes since the 1970's. These have different levels of "closure" – i.e. use different levels of detail to calculate the turbulence stresses. This paper does not set out to review a wide range of models only to comment on the most commonly used ones for helideck airflow modeling.

Eddy Viscosity Modeling

Eddy Viscosity Models (EVM) use the Boussinesq assumption that as the turbulent stresses act in the same way as viscous stresses, they can also be calculated from the mean velocity gradients in much the same way as viscous stresses act (Versteeg & Malalasekera, 1995):

$$-\rho\overline{u_i' u_j'} = \mu_t\left(\frac{\partial U_i}{\partial x_j} + \frac{\partial U_j}{\partial x_i}\right) - \frac{2}{3}\rho k\delta_{ij} \qquad (5)$$

μ_t – turbulent viscosity; g – gravity; k – turbulent kinetic energy

The turbulent viscosity is defined from turbulent kinetic energy and the dissipation rate of turbulent kinetic energy, ε:

$$\mu_t = \rho C_\mu \frac{k^2}{\varepsilon} \qquad (6)$$

C_μ - a constant

Partial differential equations (PDE) can be formulated for the transport of k and ε through the flow and these are solved alongside the RANS momentum equations (4). This is the basis for all EVMs which include the "standard" k-ε model, RNG k-ε model, k-ω model, SST model and other similar derivations.

The turbulent kinetic energy is itself defined from the normal turbulent stresses as:

$$2k = \overline{u'u'} + \overline{v'v'} + \overline{w'w'} \qquad (7)$$

Once the PDE for k has been calculated it is possible to make an estimation of the magnitude of v' over the helideck. Typically, around half of the turbulent kinetic energy will be associated with the principal flow direction; hence $k = \overline{u'u'}$. If the remaining turbulent kinetic energy can be assumed to be evenly split between the two cross-stream directions, then $0.5k = \overline{v'v'} = \overline{w'w'}$ and the required value of v' can be extracted from the CFD model calculations.

However, these are not robust assumptions and there are a number of arguments that can be made against them – e.g. the stream-wise direction is not always in the horizontal plane and may have a vertical component – in which case the amount of turbulent kinetic energy that should be assigned to $\overline{v'v'}$ should be greater. A more conservative approach is to set $(2/3)\,k = \overline{u'u'} = \overline{v'v'} = \overline{w'w'}$, but again, this is an assumption that cannot readily be tested.

In the absence of any other method, this does provide a fair means to estimate v' and in the past it has been used reasonably widely in CAP437 helideck assessments. However, sufficient computing power is common place for more detailed approaches with higher levels of turbulence closure – this is the basis of the requirement in ISO 15138 that a Differential Stress Model of turbulence should be used.

Differential Stress Modeling

Differential Stress Models (DSM), alternatively referred to as Reynolds Stress Models (RSM) use the same RANS framework as EVMs to calculate the mean velocity in a CFD model. The conservation of momentum equation shown in equation (4) is used, but instead of calculating the turbulent stress $-\rho\overline{u_i'u_j'}$ by proxy with models for k and ε, a separate PDE for the transport (convection, diffusion, generation and destruction) of each of the six turbulent stresses is solved (Versteeg & Malalasekera, 1995).

This requires more computing power than the EVM approach, as there are more PDEs to solve within the CFD model and the CFD model will be more numerically "stiff" with dependences in each of the PDEs on the values calculated by the other PDEs. A consequence of this is that there may be poor convergence of the CFD model, particularly the convergence of the numeric for the turbulence equations. If this is the case then the DSM result for v' will still contain inaccuracies.

However, the principal benefit is that no additional assumption for the distribution of turbulent kinetic energy amongst the turbulent stresses is required; values for v' can be extracted directly from the CFD model solution.

Large Eddy Simulation

Large Eddy Simulation (DES) has the potential to become the new preferred method where turbulent fluctuations need to be resolved in cases such as CAP437 helideck studies. LES is based not on the RANS equations, but the Navier-Stokes equations (3) which calculate instantaneous velocities, rather than mean velocities. Recall from above that this is referred to as Direct Numerical Simulation if all the turbulent eddies are resolved down to the smallest length and time scales. If only the largest eddies are resolved, the technique is called Large Eddy Simulation (LES). This is viable as the largest turbulent eddies carry the greatest proportion of the turbulent energy. A filter or "grid" is used in LES to define which eddies are included explicitly in the CFD flow calculation and a "sub-grid-scale" model is used to account for the lesser action of the smaller eddies on the velocity field.

LES can be carried out reasonably efficiently in free-stream regions away from walls. However at walls, turbulent eddy size decreases linearly with distance and a finer computational grid is required to calculate their effects and this increases the computational expense of the calculation. A number of approaches are possible to improve efficiency which are based on the concept of using LES to calculate the free-stream regions away from walls and RANS models to calculate the wall-affected flows. Two such approaches are Detached Eddy Simulation (DES) and Wall-Modeled LES (WMLES) and these are now incorporated in some commercial software (Menter, 2011).

Turbulence Relationship with Mean Velocity

The relationship between mean velocity and turbulence is important in a CAP437 study as it determines both the minimum standard of turbulence model that should be used and the number of different wind speeds that need to be calculated for each wind direction. The number of wind speeds is important as it is likely that the availability for the helideck will be a key output from the study: to calculate this, the turbulence over the helideck must be determined for a large number of wind speeds. It is impracticable to calculate all the required wind speeds explicitly, yet reasonable to infer these from a limited number of calculations. If the relationship between mean velocity and turbulence is linear, then only two wind speeds are required to determine the relationship between mean velocity and turbulence. However, if the relationship is non-linear then at least three wind speeds must be calculated for each wind direction.

Homogeneous shear flow occurs in simple boundary layers without cross flow or separation – this can be found in straight pipe flow of flow over a plane surface. The only velocity gradient in such a flow is $\partial U/\partial y$ and there are no non-linear dependencies between mean velocity and turbulence. A curved flow– which will occur at an offshore installation as the wind is "bent" around the structures – does not contain only homogeneous shear. There is likely to be an asymmetric velocity profile across the flow which will lead to non-linear dependence of turbulence on the mean velocity (Ellis & Joubert, 1974). Similarly experiments (Cheah, et al, 1993) show that in swirling shear flows – e.g. where any vortices are developed around or over the helideck – the mean swirl velocity varies non-linearly with distance from the center of the vortex. Others (Craft et al 1996) have shown that linear EVMs as described above cannot calculate this non-linear variation.

To summarize these points, the non-linear dependencies between mean velocity and turbulence values mean that multiple wind speeds (at least three) should be calculated for each wind direction, if results are to be interpolated for other wind speeds in the availability study. Also that an EVM (such as k-ε, k-ω, SST models) will not be able to calculate the non-linear dependencies in the mean velocity over the helideck, and a model that can calculate non-linear turbulence effects such as a DSM is required. It is useful to know that the k-ε model (Launder & Spalding, 1974) was originally only intended for simple shear flows (e.g. flow without separation or cross-wind over a flat plate) and to expect it to calculate accurately the swirling and curved flows over a helideck is unreasonable. Any model using the same stress-strain relationship as k-ε (e.g. k-ω, SST models) will have the same limit on performance (Robinson, 2001)

Example of Helideck Model for Turbulence

A typical CFD model for a helideck analysis is shown in Figure 2 - in this particular example the helideck is located on a Living Quarters platform. Incorporated in the model are all the weather deck features which will cause flow disturbances and hence generate turbulence. This includes the crane, crane rest, telecoms mast, stairs and AHU units. The items shown in pale blue are represented in the model as porous distributed resistances as the geometry features are too small to be included in the model explicitly, but the sum effect of the small components is still significant to the flow field. For reference the mesh lines on the significant surfaces are shown; note that although these appear as tetrahedral elements; prism cells are used over the sea surface and helideck surface.

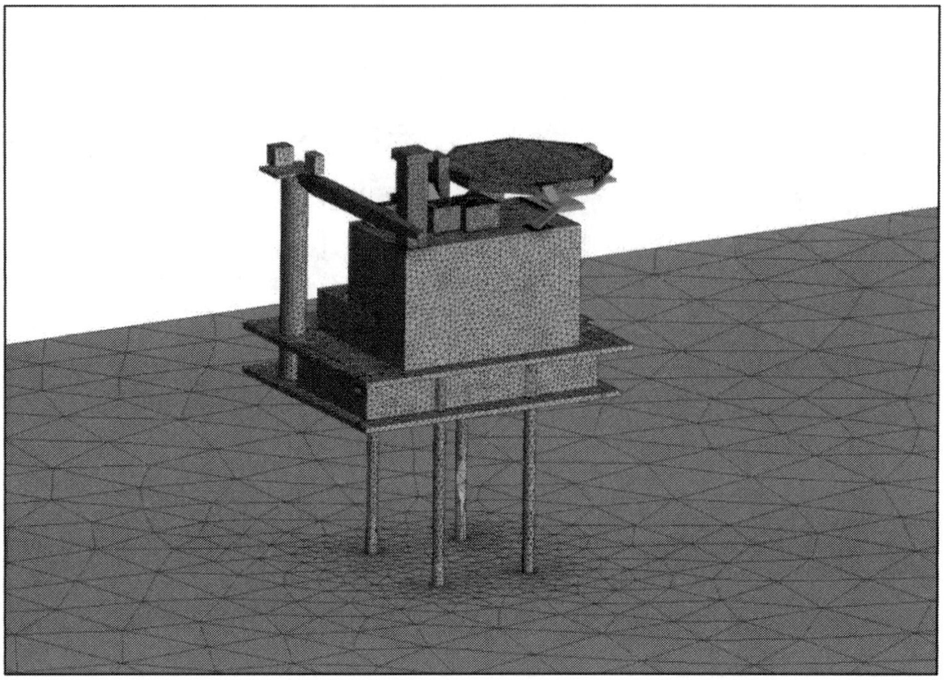

Figure 2. Example of the geometry and mesh used in a CFD model for a CAP437 helideck study on an offshore Living Quarters platform.

Results for 15 m/s wind flow in neutrally stable conditions are provided for the most onerous direction; this occurs when the wind passes over the platform before reaching the helideck and there is the largest opportunity for flow disturbance. Two calculations have been carried out using different turbulence model approaches: (i) firstly, with the SST model (Menter, 1994) which blends a k-ε model in the free stream with a k-ω model near walls and calculates turbulence with the same limitations as the EVM technique as described in this paper; (ii) secondly, with a DSM, in this case the "SSG" model (Speziale et al, 1991).

Figure 3 shows results from the CFD model calculations – the wind is passing from left to right as shown on the Figure and streamlines are used to demonstrate the degree of curvature in the wind passing over the center line of the helideck. To demonstrate the degree of turbulence calculated, an iso-surface is shown in both results for $v' = 1.75$ m/s – i.e. the limiting value in a CAP437 study. It is apparent that the EVM approach (to the left, here using the SST model) produces a much larger region of turbulence in excess of the CAP437 limit than the DSM approach (to the right, here using the SSG model).

Figure 4 shows values of v' for the same cases on a plane at 5 m above the helideck. It is clear that when using the EVM approach large areas over the helideck would be deemed non-compliant with CAP437 standard with the v' value in excess of the 1.75 m/s limit. However, the DSM approach shows that the helideck would compliant to CAP437 standard. Note that there are no differences in the cases other than the turbulence models. Figure 5 shows the plots of v' on a vertical line through which is located at the position of the maximum value of v'.

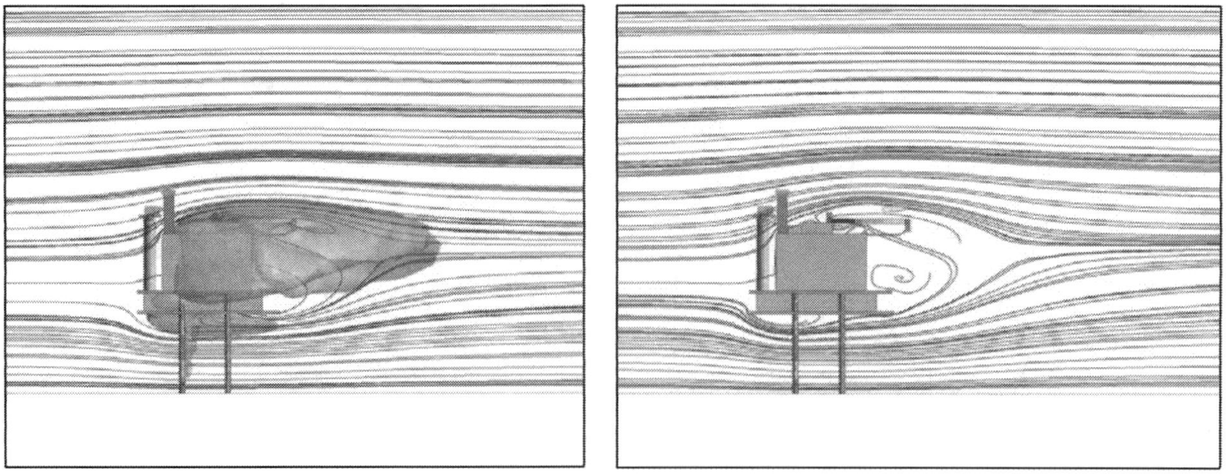

Figure 3. CFD calculations showing streamlines and an iso-surface for the value v' = 1.75 m/s.
Left – Image from SST (EVM) model; Right – Image from SSG (DSM) model.

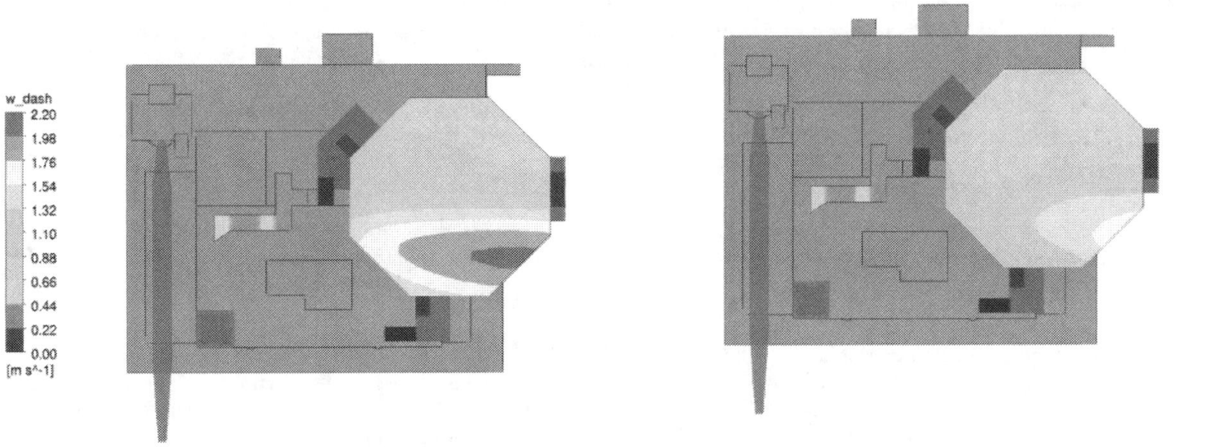

Figure 4. CFD calculations showing contours of *v'* on a plane 5 m above the helideck
Left – SST (EVM) model; Right – SSG (DSM) model.

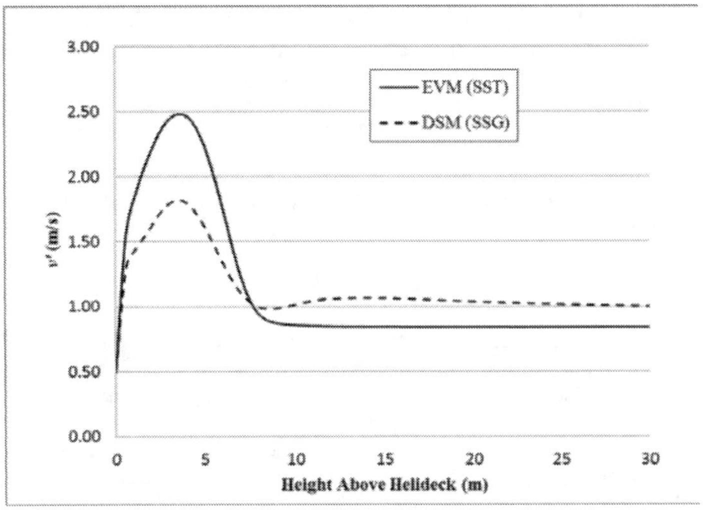

Figure 5. Plots of standard deviation of vertical velocity (*v'*) in vertical line above its maximum position on the helideck

An important point to note with these results is that independent of any measured data to validate the calculations, it is not possible to state definitively that one is more accurate than the other. However, the detail provided in this paper for calculating curved and swirling flow effects describes how DSM methods have better accuracy than EVM; there is also a large amount of published literature for curved and swirling flows which supports this.

Best Practice Guidelines

The following best practice guidelines are intended not as instructions for how to carry out CFD modeling for offshore structures in general – this is available elsewhere. In particular ISO 15138 provides good advice on CFD calculations for offshore structures and NAFEMS provides information on checking quality in CFD calculations. The purpose of the guidelines is to provide information on how a CFD model can be made suitable for CAP437 helideck assessment and how the results are used.

1. Aim of calculation

The aim of the calculation should be considered during commencement of the work. CAP437 describes limits on hydrocarbon and thermal plumes over the helideck in addition to the turbulence limits. If all these are to be included in the analysis, it may have implications for the model definition, number of cases and winds (speed and/direction) considered.

For the assessment of turbulence common requests from the duty holder will be the "availability" of the helideck and the maximum wind speed which is permissible before turbulence limits are exceeded. It is normally necessary and sufficient to calculate winds from eight directions - usually selecting N, NE, E, SE, S, SW, W, NW. If however, the wind rose for the installation is skewed such that very little of very significant amounts of wind come from a particular quarter, then fewer or more wind directions may be considered, or greater detail concentrated on a particular range of directions.

2. Turbulence Model and Software Selection

   ISO 15138 states that a DSM turbulence model should be used for proper resolution of the turbulence stress and determination of the standard deviation on vertical air velocity relative to the 1.75 m/s limit in CAP437. This is supported by the description of turbulence models in this paper, and important to reduce the reliance on assumption on the distribution of turbulence energy between the turbulent stresses.

   It is important to note however, that not all commercial CFD software have a DSM capability. For example, FLACS (GexCon) and KFX (ComputIT) are both commonly used in offshore analysis; these are specialized CFD codes which are optimized for efficient explosion and fire analysis respectively. Neither of these codes currently employ a DSM capability and if the CAP437 helideck analysis is to be carried out with either of them, the EVM approach must be used as described above.

3. Background Turbulence

   The wind approaching the helideck over the sea will contain "naturally occurring" turbulence due to physical processes in the atmosphere. The sea surface acts as a wall with low wind velocities in the boundary layer immediately above the sea level and higher velocities at higher elevations. This velocity gradient induces turbulence. Additionally, sea surface temperatures greater than the ambient atmosphere promotes convection and rising air above the sea to various degrees depending on strength of the temperature difference, relatice humidity, cloud cover, etc. These rising air currents modify the wind profile and turbulence it contains as it interacts with the helideck.

   As an example over open sea with some waves, in a neutrally stable atmosphere with a 10 m/s (approximately 20 knots) wind measured at a reference height of 40 m, the standard deviation of the vertical component of wind, $v' = 1.2$ m/s.

   This demonstrates how important it is to include the background turbulence in the model as under this particular condition it is only permissible for an additional $v' = 0.55$ m/s to be induced by the installation and helideck construction.

4. Wind Speeds

   At least three non-zero wind speeds should be determined for the CFD calculation in each wind direction. As demonstrated previously, curvature and swirl effects in the wind have a non-linear relationship with mean velocity. If only two wind speeds are calculated and the turbulence at other wind speeds is extrapolated from these, it will not be possible to quantify the non-linear effects. With three wind speeds calculated, curves can be fitted for the relationship between mean wind velocity and turbulence values to allow the vertical velocity fluctuation to be determined at intermediate wind speeds which have not been calculated in the CFD model.

   A well-designed helideck, which has a good air gap and is sufficiently remote from other structures, will sit in the undisturbed "free shear" part of the wind profile. In this instance the turbulence will vary linearly with mean velocity. However, if this is assumed from the onset of the analysis, then it reduces the likelihood of finding deficiencies in the helideck design through modeling.

5. Result Heights above Helideck

   The highest turbulence values, including v', are typically found immediately above the helideck surface due to wall shear effects. However these are not significant to the CAP437 analysis. The purpose of limiting v' is to prevent adverse and unanticipated loads (positive or negative lift forces) being placed on the helicopters rotors and hence increasing the pilots work load during the crucial stages of landing and take-off (CAA, 2004 & 2008). Turbulence fluctuations need only be assessed at and above the position of the helicopters' rotors, typically around 3 meters above the helideck.

6. Availability

Helideck availability can be determined from the number of cases with different wind speeds and directions which fall within and outside the $v' = 1.75$ m/s limit stated in CAP437. If the frequency of these winds is taken from the wind rose, the overall availability of the helideck can be obtained. The acceptable frequency will usually be determined by the installation duty holder and may be dependent on a number of operational factors.

Conclusion

Computational fluid dynamics has become the principal means to determine the turbulence environment over offshore helidecks – this has been driven by the requirements of the Civil Aviation Authority CAP437 standard and advances in computing power which allows the CFD calculations to be carried out routinely. However, the level of detail in CFD models used to find the specific turbulence quantities required in the CAA standard is not always applied correctly. This paper has reviewed the CAA standard, turbulence physical effects and turbulence models to provide best practice guidance on applying the standard.

The consequence of not applying appropriate detail in the CFD models will have some dependence on the helideck design itself. For a well-designed helideck, with good air gap and well-remote from other weather deck structures, the wind flow over it should be relatively undisturbed, exhibit simple shear and may indeed be reasonably calculated with basic turbulence models (EVMs). However, for a poorly designed helideck this will not be the case and a properly detailed turbulence modeling approach is required. As it is generally the case that the aim of the analysis is to determine whether a helideck is "well-designed" or not, then it will generally be necessary to adopt the more detailed turbulence modeling approach.

Similarly, as the availability of the helideck is usually the output of a CAP437 study, using an inappropriate CFD modeling approach may reduce the accuracy of this calculation. The example calculation provided in this paper demonstrates that even with a well-designed helideck, if an EVM model for turbulence is used rather than a DSM method, it will likely over-predict the level of turbulence over the helideck by a large degree. The consequence for the duty holder in this event is that the availability may be calculated much lower than is actually the case.

References

Civil Aviation Authority, 2013. Standards for Offshore Helicopter Landing Area. Safety Regulation Group Report CAP437, Seventh edition incorporating Amendment 01/2013

Civil Aviation Authority, 2004. Helicopter Turbulence Criteria for Operations to Offshore Platforms. Safety Regulation Group Paper 2008/02

Civil Aviation Authority, 2009. Offshore Helideck Research. Safety Regulation Group Paper 2008/02

Civil Aviation Authority, 2009. Helideck Design Considerations – Environmental Effects. Safety Regulation Group Paper 2008/03

Cheah S.C. Cheng L. Cooper D. & Launder B. E. 1993. On the structure of turbulent flow in spirally fluted tubes. Proc 5th Int. Symp Refined Flow Modeling and Turbulence Measurements.

Davidson P. A. 2004. Turbulence – An introduction for Scientists and Engineers. Oxford University Press. First published 2004.

Ellis L. B. & Joubert P.N. 1974. Turbulent shear flow in a curved duct. J. Fluid Mech (56) pp 337-351

International Standards Organisation, 2007 Petroleum and natural gas industries - Offshore production installations - Heating, ventilation and air-conditioning. ISO 15138. Second Edition 2007

Launder B.E & Spalding D.B. 1974 The numerical computation of turbulent flows. Computer Methods in Applied Mechanics and Engineering (3) pp 269-289

Menter, F.R., 1994. Two-equation eddy-viscosity turbulence models for engineering applications, AIAA Journal, Vol. 32, No. 8, pp. 1598-1605

Menter, F. 2012. Best Practice: Scale-Resolving Simulations in ANSYS CFD, ANSYS Best Practice Guide v. 1.0

Robinson C. M.E. 2001 Advanced CFD Modeling of Road-Vehicle Aerodynamics, PhD These, UMIST

Speziale C.G., Sarkar, S. and Gatski T. B. 1991. Modeling the pressure-strain correlation of turbulence: an invariant dynamical systems approach. J Fluid Mech 227 pp245-272.

Tritton D. J. 1988. Physical Fluid Dynamics. Oxford Science Publications, Second Edition 1988

Versteeg H.K. & Malalasekera W. 1995An Introduction to Computational Fluid Dynamics Longman Scientific and Technical First published 1995

Introducing a Dynamic Calculation Method for Pressure Relief Load and Orifice Sizing

Steve Streblow
Contek Solutions LLC
6221 Chapel Hill Boulevard, Suite 300
Plano, TX, 75093
sstreblow@contekllc.com

Jay Riha
Contek Solutions LLC
6221 Chapel Hill Boulevard, Suite 300
Plano, TX, 75093
jriha@contekllc.com

Prepared for Presentation at
American Institute of Chemical Engineers
2015 Spring Meeting
11th Global Congress on Process Safety
Austin, Texas
April 27-29, 2015

UNPUBLISHED

Introducing a Dynamic Calculation Method for Pressure Relief Load and Orifice Sizing

Steve Streblow
Contek Solutions LLC
6221 Chapel Hill Boulevard, Suite 300
Plano, TX, 75093
sstreblow@contekllc.com

Jay Riha
Contek Solutions LLC
6221 Chapel Hill Boulevard, Suite 300
Plano, TX, 75093
jriha@contekllc.com

Keywords: dynamic pressure relief

Abstract

This paper presents a low cost, dynamic calculation method for pressure relief load and orifice sizing, as well as several case studies detailing the additional information that can be found with using this more detailed method and the more realistic information that can be determined versus the method typically used by many. Using this more detailed method will allow for greater insight into the overpressure event, resulting in safer emergency response and pressure relief system design.

1. Introduction

For many simple systems, such as single component mixtures operating close to the relief valve set pressure, the well-established simplifying assumptions and equations in API 521, such as constant level, heat input, heat of vaporization, etc. are adequate. However, using these well-established methods for steady-state relief does not give one much insight into the time aspects, such as how long it takes to get to relieving conditions, how long it remains there, time the boil the vessel dry or all flow regimes.

For more complex systems, such as multiple components, multiple phases, operating conditions much different than relieving conditions, using the above mentioned methods may result in less safe designs due to failure to understand the dynamic relief event.

This paper presents a more detailed, dynamic method for overpressure work. Several real life examples for a pool fire scenario will be discussed, showing some of the significantly different results obtained with the more detailed method and the profound insights gained. Use of this method will bring many of the benefits of dynamic simulation to the majority of engineers, without the associated cost, resulting in safer relief system design.

2. Background

When complex systems relieve pressure due to excessive heat input such occurs during an external pool fire, engineers have made several simplifying assumptions to allow for rapid determination of relief load, orifice sizing and subsequent activities such as pipe sizing in the over-pressure protection field. For example, with a partially liquid-full vessel, assumptions of constant composition, heat input, heat of vaporization, level, molecular weight, phase, properties, and temperature are common. How good are these assumptions? Oftentimes, these are not good and result in systems which are undersized based on calculations which provide little insight into what is physically going on during the transient event.

For example, from API 521 [1], pool fire heat input is calculated from:

Q= C2 x F x Aws^0.82 [Eq. 1]

Where

C2	= 34,500(drainage/firefighting are inadequate)
F	= 1 (environmental factor for uninsulated vessel)
Aws	= Wetted Surface Area

For a single component boiling fluid, energy output is calculated from Q= mass flowrate x latent heat of vaporization at relief conditions.

In real life, such as an initial oil-water-gas separator, it is much more complicated, with multiple phases, such as heavy liquid-light liquid-vapor, multiple components, liquid expansion upon heating, and increasing liquid level and pressure. This problem then becomes like a batch reactive distillation problem, where heavy components crack, light components are relieved first, and the relieving molecular weight and latent heat of vaporization are changing.

This is important because an undersized over-pressure protection system can result in catastrophic issues such as loss of containment resulting in fire, explosion or toxic effects. Also, lack of insight of the transient event does not allow one to know how long it will take before loss of containment is expected to occur. Gaining knowledge of the dynamics allows for effective emergency response to better mitigate impact of the event.

This problem is well-suited to dynamic simulation. Unfortunately, many engineers do not have access to dynamic process simulation software, which is generally more expensive than the more commonly available steady-state process simulation software.

This method is suitable for complex systems such as multi-phase, multi-component mixtures in pool fire scenarios such as initial oil-water-gas separation in the upstream area, but can also be extended to other types of problems, such as different types of equipment (i.e. process and storage tanks), and different types of scenarios, such as maximum heat input from a heater.

Contek has seen many companies with pressure vessels that were not properly engineered. Many companies simply order a vessel, with standard openings, and ask the vessel manufacturer to also supply a relief valve, but they do not provide all, most, or even any of the information needed to make sure that the relief system is adequate. We encourage you to review this paper and if you notice or suspect that the vessels at your company are not properly engineered, contact a qualified company to study them and provide you with the best path forward.

3. Method and Benefits

This method is one in which a base spreadsheet is used to determine the constant volume parameters, initial conditions and boundary conditions. An appropriate time step is then chosen, and through transformation of variables, the problem is converted to a constant flow rate process which then allows for solution of thermodynamic properties in a steady-state process simulator. The process simulation results are then placed back in the spreadsheet, and the calculation is complete for that time step, and the starting conditions for the next time step are calculated. After the pressure relief event occurs, the orifice size is automatically calculated at the end of each time step. This process is repeated until enough results are available to have confidence that the maximum orifice size has been calculated and is the one which should be specified.

By using this method, one can determine the time it takes until the pressure relief event starts, the fluid being relieved during the event, the liquid level in the vessel, and the time versus temperature profile, which is critical to the evaluation of vessel integrity during the relief event. Sometimes the liquid in the vessel may initially swell and compress the vapor. Therefore, the initial relief may be vapor, but if the liquid then completely fills the vessel, the relieving fluid will be liquid. As boiling becomes more vigorous, it may next relieve a two-phase mixture, with a final period of vapor release. There may also be a supercritical relief phase. This sequence will not be understood and properly designed for unless a dynamic analysis is conducted.
Many times when pressure relief work by others is reviewed, the documentation is lacking or the underlying thoughts are not stated. This can be problematic when the facility is expanded, since the basis needs to be recreated. This method is thoroughly documented, starting at the beginning with the objective, then proceeding through the references, assumptions, basis, constants, formulas, scenarios, calculations, conclusions and finishing with the recommendations. It includes an extensive section on scenario selection which has about 100 scenarios identified and requires a written response to each one.

4. Case Studies

Thirteen case studies are presented. They cover different geographical areas, different operating companies, several different fluids, pressure ranges, and vessel orientations.

4.1 Case Study 1 – Three Phase Separator

The first case study is for a pool fire impacting a horizontal three phase oil, water, and gas separator, which operates at 80% liquid full at 120 degrees F and 500 psig. Since the operating pressure is much lower than the relieving pressure of 1718.2 psig, the temperature of the fluid will need to be increased substantially before the relief starts. Due to the large temperature increase to 388 degrees F, the fluid expands, compressing the vapor. Due to the large pressure increase, the vapor condenses to a liquid. Therefore, at the start of the relief event, the fluid relieving will be liquid. Note that taking into account the change in liquid level in the vessel means that the heat input from the fire increases from the initial conditions. By using the dynamic method, it is determined that it takes twenty seven minutes before the relief event starts. At seventy two minutes, the maximum relieving vapor flowrate is reached, at a temperature of 606 degrees F. This data can be used to determine if the vessel will survive this event, since it experiences an internal temperature well above the design temperature for more than one hour. Using a simpler approach, as is commonly done by many other companies in the oil and gas industry will underestimate the relief load, not identify the fact there will be liquid and two-phase relief in addition to vapor relief, and does not evaluate how long it will take before the relief starts or how long the vessel will remain in the over-temperature condition. See Tables 1 and 2 below for further details.

Table 1. Case Study 1 Parameters

Vessel	Production Unit
Drainage	Inadequate
Fire Fighting	Inadequate
Insulation	Bare
Case	External Pool Fire
Orientation	Horizontal
Seam to Seam [Feet]	12
Diameter [Feet]	3
Normal Liquid Level [Feet]	2.4
Liquid Volume	50% water, 50% hydrocarbon
Operating Temperature [Degrees F]	120
Design Temperature [Degrees F]	130
Operating Pressure [PSIG]	500
Design Pressure [PSIG]	1420
Relief Valve Set Pressure [PSIG]	1420
Relieving Pressure [PSIG]	1718.2

Table 2. Case Study 1 Details

Time [min]	Temperature [Degrees F]	Pressure [PSIG]	Vapor Flow [Ft^3/ Hr]	Oil Flow [Ft^3/ Hr]	Water Flow [Ft^3/ Hr]	Relief Flow [Ft^3/ Hr]	Relief Phase	Vessel Liquid Level, Ft
0.0	120	500	12.4	38.4	38.5			2.40
26.8	388	1420	0.0	45.7	43.7			3.00
33.7	448	1420	0.0	47.8	45.7	35.8	Liquid	3.00
40.7	505	1420	0.0	45.9	48.1	40.1	Liquid	3.00
47.6	554	1420	11.3	42.0	50.2	122.0	2 Phase	3.00
54.5	587	1420	42.6	36.8	49.5	342.0	Vapor	2.77
62.5	601	1420	86.9	31.3	43.9	546.8	Vapor	2.35
71.7	606	1420	122.8	24.6	34.8	607.1	Vapor	1.89
82.4	607	1420	146.6	18.2	2.2	437.9	Vapor	

4.2 Case Study 2– Two Phase Separator

The second case study is for a pool fire impacting a vertical two phase oil-gas separator, which operates at 80% liquid full at 120 degrees F and 15 psig. Since the operating pressure is much lower than the relieving pressure of 151.3 psig, the temperature of the fluid will need to be increased substantially before the relief starts. Due to the large temperature increase to 448 degrees F, the fluid expands, compressing the vapor. Due to the large pressure increase, the vapor condenses to a liquid. At the start of the relief event, the fluid relieving will be liquid. By using the dynamic method, it is determined that it takes 3 minutes before the relief event starts. At four minutes, the relief phase changes from liquid to two-phase vapor-liquid. At five minutes, the relief phase changes from two-phase vapor-liquid to vapor. At this point the heat input starts decreasing. After this time, the relief will be vapor. At six minutes, the maximum relieving vapor flowrate is reached, at a temperature of 793 degrees F. This data can be used to determine if the vessel will survive this event, which experiences an internal temperature well above the design temperature for several minutes. See Tables 3 and 4 below for further details.

Table 3. Case Study 2 Parameters

Vessel	Fuel Gas Scrubber
Drainage	Inadequate
Fire Fighting	Inadequate
Insulation	Base
Case	External Pool Fire
Orientation	Vertical
Seam to Seam [Feet]	2
Diameter [Feet]	0.67
Normal Liquid Level [Feet]	1.6
Liquid Volume	100% hydrocarbon
Operating Temperature [Degrees F]	120
Design Temperature [Degrees F]	150
Operating Pressure [PSIG]	15
Design Pressure [PSIG]	125
Relief Valve Set Pressure [PSIG]	125
Relieving Pressure [PSIG]	151.3

Table 4. Case Study 2 Details

Time [min]	Temperature [Degrees F]	Pressure [PSIG]	Vapor Flow [Ft^3/ Hr]	Oil Flow [Ft^3 /Hr]	Water Flow [Ft^3/ Hr]	Relief Flow [Ft^3/Hr]	Relief Phase	Vessel Liquid Level, Ft
0.0	112	15	0.14	0.56				1.60
2.9	448	151	0.00	0.70				2.00
3.9	551	151	0.18	0.76		15.47	2 Phase	2.00
4.8	654	151	0.44	0.76		32.23	2 Phase	2.00
5.3	707	151	0.38	0.72		51.57	2 Phase	2.00
5.7	755	151	0.80	0.66		97.73	Vapor	1.89
6.2	793	151	2.04	0.36		208.77	Vapor	1.02
7.0	874	151	2.59	0.00		148.46	Vapor	0.00

4.3 Case Study 3 - Two Phase Separator

The third case study is for a pool fire impacting a vertical two phase oil-gas separator, which operates at 80% liquid full at 120 degrees F and 0 psig. Since the operating pressure is much lower than the relieving pressure of 121 psig, the temperature of the fluid will need to be increased substantially before the relief starts. Due to the large temperature increase to 353 degrees F, the fluid expands, compressing the vapor. Due to the large pressure increase, the vapor condenses to a liquid. At the start of the relief event, the fluid relieving will be liquid. By using the dynamic method, it is determined that it takes three minutes before the relief event starts. At six minutes, the relief phase changes from liquid to two-phase vapor-liquid. At seven minutes, the heat input starts to decrease. At eight minutes, the relief phase changes from two-phase vapor-liquid to vapor. After this time, the relief will be vapor. At nine minutes, the maximum relieving vapor flowrate is reached, at a temperature of 788 degrees F. This data can be used to determine if the vessel will survive this event, which experiences an internal temperature well above the design temperature for several minutes. See Tables 5 and 6 below for further details.

Table 5. Case Study 3 Parameters

Vessel	Blowcase
Drainage	Inadequate
Fire Fighting	Inadequate
Insulation	Bare
Case	External Pool Fire
Orientation	Horizontal
Seam to Seam [Feet]	1.92
Diameter [Feet]	1
Normal Liquid Level [Feet]	0.8
Liquid Volume	100% Hydrocarbon
Operating Temperature [Degrees F]	120
Design Temperature [Degrees F]	650
Operating Pressure [PSIG]	0
Design Pressure [PSIG]	150
Relief Valve Set Pressure [PSIG]	100
Relieving Pressure [PSIG]	121

Table 6. Case Study 3 Details

Time [min]	Temperature [Degrees F]	Pressure [PSIG]	Vapor Flow [Ft^3/ Hr]	Oil Flow [Ft^3/ Hr]	Water Flow [Ft^3/ Hr]	Relief Flow [Ft^3/Hr]	Relief Phase	Vessel Liquid Level, Ft
0.0	120	0	0.23	1.45				0.80
3.2	353	121	0.00	1.68				1.00
4.5	461	121	0.00	1.83		6.79	Liquid	1.00
5.8	569	121	0.62	1.83		35.38	2 Phase	1.00
7.1	673	121	2.05	1.76		97.55	2 Phase	1.00
7.8	724	121	2.00	1.62		178.47	Vapor	0.92
8.5	764	121	4.93	1.13		352.54	Vapor	0.64
8.9	778	121	3.80	0.70		430.07	Vapor	0.44
9.4	788	121	5.29	0.08		443.33	Vapor	

4.4 Case Study 4- Three Phase Separator

The fourth case study is for a pool fire impacting a horizontal three phase oil, water, gas separator, which operates at 80% liquid full at 82 degrees F and 102 psig. Since the operating pressure is much lower than the relieving pressure of 1742.4 psig, the temperature of the fluid will need to be increased substantially before the relief starts. As is quite common for these vessels in upstream and midstream oil and gas, the design temperature of 120 degrees F was only slightly above maximum ambient temperature. The operator wanted the relief valve set high, so that it would minimize probability of it ever flowing. Due to the large temperature increase to 415 degrees F, most of the liquid water vaporizes. Due to the large pressure increase, most of the hydrocarbon gas condenses. Overall, the liquid level decreases substantially. At the start of the relief event, the fluid relieving will be vapor. By using the dynamic method, it is determined that it takes thirty five minutes before the relief event starts. For the next several minutes, the liquid level decreases. Then it increases, before reaching a maximum and decreasing again. At sixty five minutes, the maximum relieving vapor flowrate is reached, at a temperature of 585 degrees F. This data can be used to determine if the vessel will survive this event, which experiences an internal temperature well above the design temperature for thirty minutes. See Tables 7 and 8 below for further details.

Table 7. Case Study 4 Parameters

Vessel	Production Unit
Drainage	Inadequate
Fire Fighting	Inadequate
Insulation	Bare
Case	External Pool Fire
Orientation	Horizontal
Seam to Seam [Feet]	10
Diameter [Feet]	2.5
Normal Liquid Level [Feet]	2
Liquid Volume	50% water, 50% oil
Operating Temperature [Degrees F]	82
Design Temperature [Degrees F]	130
Operating Pressure [PSIG]	102
Design Pressure [PSIG]	1440
Relief Valve Set Pressure [PSIG]	1440
Relieving Pressure [PSIG]	1742.4

Table 8. Case Study 4 Details

Time [min]	Temperature [Degrees F]	Pressure [PSIG]	Vapor Flow [Ft^3/ Hr]	Oil Flow [Ft^3/ Hr]	Water Flow [Ft^3/ Hr]	Relief Flow [Ft^3/Hr]	Relief Phase	Vessel Liquid Level, Ft
0.0	82	102	31.60	95.21	95.61			2.00
34.6	415	1742	112.39	109.19	1.06			1.24
46.6	501	1742	116.78	127.07	1.07	26.32	Vapor	1.40
55.6	553	1742	152.22	97.35	1.07	44.01	Vapor	1.14
61.6	576	1742	126.98	96.71	89.55	212.45	Vapor	1.95
65.3	585	1742	154.35	72.25	82.07	324.99	Vapor	1.64
69.5	593	1742	188.70	55.03	72.95	314.91	Vapor	1.40
72.5	597	1742	176.34	46.20	66.14	310.02	Vapor	

4.5 Case Study 5- Three Phase Separator

The fifth case study is for a pool fire impacting a vertical three phase oil, water, gas separator, which operates at 80% liquid full at 200 degrees F and 60 psig. Since the operating pressure is much lower than the relieving pressure of 211.8 psig, the temperature of the fluid will need to be increased substantially before the relief starts. Initially the liquid level increases, and then proceeds to decrease. By using the dynamic method, it is determined that it takes fourteen minutes before the relief event starts. At thirty nine minutes, the maximum relieving vapor flowrate is reached, at a temperature of 366 degrees F. This data can be used to determine if the vessel will survive this event, which experiences an internal temperature somewhat above the design temperature for thirty minutes. See Tables 9 and 10 below for further details.

Table 9. Case Study 5 Parameters

Vessel	Vapor Recovery Tower
Drainage	Inadequate
Fire Fighting	Inadequate
Insulation	None
Case	External Pool Fire
Orientation	Vertical
Seam to Seam [Feet]	25
Diameter [Feet]	4
Normal Liquid Level [Feet]	20
Liquid Volume	10% water, 90% oil
Operating Temperature [Degrees F]	200
Design Temperature [Degrees F]	300
Operating Pressure [PSIG]	60
Design Pressure [PSIG]	175
Relief Valve Set Pressure [PSIG]	175
Relieving Pressure [PSIG]	211.8

Table 10. Case Study 5 Details

Time [min]	Temperature [Degrees F]	Pressure [PSIG]	Vapor Flow [Ft^3/ Hr]	Oil Flow [Ft^3 /Hr]	Water Flow [Ft^3/ Hr]	Relief Flow [Ft^3/ Hr]	Relief Phase	Vessel Liquid Level, Ft
0.0	199	59	214.8	773.5	85.9			20.00
13.6	323	212	159.1	827.2	87.7			21.29
23.5	359	212	4208.3	758.7	63.2	5629.2	Vapor	19.13
31.6	372	212	4258.6	701.4	34.6	6812.3	Vapor	17.13
39.4	366	212	4632.9	653.9	0.9	7524.4	Vapor	15.24
47.0	443	212	690.7	670.9	0.9	534.5	Vapor	15.63

4.6 Case Study 6- Three Phase Separator

The sixth case study is for a pool fire impacting a vertical three phase oil, water, gas separator, which operates at 80% liquid full at 200 degrees F and 60 psig. Since the operating pressure is much lower than the relieving pressure of 211.8 psig, the temperature of the fluid will need to be increased substantially before the relief starts. Initially the liquid level increases, and then proceeds to decrease. By using the dynamic method, it is determined that it takes eight minutes before the relief event starts. At twenty three minutes, the maximum relieving vapor flowrate is reached, at a temperature of 370 degrees F. This data can be used to determine if the vessel will survive this event, which experiences an internal temperature somewhat above the design temperature for fifteen minutes. See Tables 11 and 12 below for further details.

Table 11. Case Study 6 Parameters

Vessel	Vapor Recovery Tower
Drainage	Inadequate
Fire Fighting	Inadequate
Insulation	None
Case	External Pool Fire
Orientation	Vertical
Seam to Seam [Feet]	25
Diameter [Feet]	2.5
Normal Liquid Level [Feet]	20
Liquid Volume	10% water, 90% oil
Operating Temperature [Degrees F]	200
Design Temperature [Degrees F]	300
Operating Pressure [PSIG]	60
Design Pressure [PSIG]	175
Relief Valve Set Pressure [PSIG]	175
Relieving Pressure [PSIG]	211.8

Table 12. Case Study 6 Details

Time [min]	Temperature [Degrees F]	Pressure [PSIG]	Vapor Flow [Ft^3/ Hr]	Oil Flow [Ft^3/ Hr]	Water Flow [Ft^3/ Hr]	Relief Flow [Ft^3/ Hr]	Relief Phase	Vessel Liquid Level, Ft
0.0	200	60	83.93	302.15	33.57			20.00
7.8	323	212	62.332	323.14	34.26			21.29
13.9	360	212	1785.52	294.35	23.73	3911.72	Vapor	18.95
18.5	372	212	1671.19	272.31	12.47	4633.58	Vapor	16.97
22.9	370	212	1695.57	255.02	0.38	4921.05	Vapor	15.21
26.9	441	212	259.607	261.17	0.37	357.98	Vapor	15.58

4.7 Case Study 7- Three Phase Separator

The seventh case study is for a pool fire impacting a horizontal three phase oil, water, gas separator, which operates at 80% liquid full at 200 degrees F and 69.09 psig. Since the operating pressure is much lower than the relieving pressure of 151.3 psig, the temperature of the fluid will need to be increased substantially before the relief starts. Initially the liquid level increases, and then proceeds to decrease. By using the dynamic method, it is determined that it takes twenty four minutes before the relief event starts. At ninety minutes, the maximum relieving vapor flowrate is reached, at a temperature of 358 degrees F. This data can be used to determine if the vessel will survive this event, which experiences an internal temperature somewhat above the design temperature for sixty five minutes. See Tables 13 and 14 below for further details.

Table 13. Case Study 7 Parameters

Vessel	Heater Treater
Drainage	Inadequate
Fire Fighting	Inadequate
Insulation	None
Case	External Pool Fire
Orientation	Horizontal
Seam to Seam [Feet]	20
Diameter [Feet]	6
Normal Liquid Level [Feet]	4.8
Liquid Volume	50% water, 50% oil
Operating Temperature [Degrees F]	200
Design Temperature [Degrees F]	400
Operating Pressure [PSIG]	69.09
Design Pressure [PSIG]	125
Relief Valve Set Pressure [PSIG]	125
Relieving Pressure [PSIG]	151.3

Table 14. Case Study 7 Details

Time [min]	Temperature [Degrees F]	Pressure [PSIG]	Vapor Flow [Ft^3/ Hr]	Oil Flow [Ft^3/ Hr]	Water Flow [Ft^3/ Hr]	Actual Relief [Ft^3/ Hr]	Relief Phase	Vessel Liquid Level, Ft
0.0	200	69	360.8	1108.4	1108.4			4.80
24.3	281	151	280.2	1154.3	1144.6			5.00
62.1	345	151	20940.9	956.0	1078.6	4771.9	Vapor	4.41
89.8	358	151	24066.5	826.1	947.6	5442.0	Vapor	3.89
105.5	366	151	12388.6	828.4	886.7	2696.4	Vapor	3.78

4.8 Case Study 8- Three Phase Separator

The eighth case study is for a pool fire impacting a horizontal three phase oil, water, gas separator, which operates at 80% liquid full at 65 degrees F and 30 psig. Since the operating pressure is much lower than the relieving pressure of 90.8 psig, the temperature of the fluid will need to be increased substantially before the relief starts. Initially the liquid level increases, and then proceeds to decrease. By using the dynamic method, it is determined that it takes forty nine minutes before the relief event starts. At one hundred and nineteen minutes, the maximum relieving vapor flowrate is reached, at a temperature of 332 degrees F, which is below the design temperature. See Tables 15 and 16 below for further details.

Table 15. Case Study 8 Parameters

Vessel	Heater Treater

Drainage	Inadequate
Fire Fighting	Inadequate
Insulation	Bare
Case	External Pool Fire
Orientation	Vertical
Seam to Seam [Feet]	20
Diameter [Feet]	6
Normal Liquid Level [Feet]	16
Liquid Volume	67% water, 33% oil
Operating Temperature [Degrees F]	65
Design Temperature [Degrees F]	400
Operating Pressure [PSIG]	30
Design Pressure [PSIG]	75
Relief Valve Set Pressure [PSIG]	75
Relieving Pressure [PSIG]	90.8

Table 16. Case Study 8 Details

Time [min]	Temperature [Degrees F]	Pressure [PSIG]	Vapor Flow [Ft^3/ Hr]	Oil Flow [Ft^3 /Hr]	Water Flow [Ft^3/ Hr]	Actual Relief [Ft^3/ Hr]	Relief Phase	Vessel Liquid Level, Ft
0.0	65	30	434.9	573.8	1166.1			16.00
48.8	223	91	333.2	622.0	1219.6			16.93
77.2	311	91	2582.1	640.8	1266.7	1142.0	Vapor	17.54
91.0	330	91	12078.1	637.8	1232.9	11957.4	Vapor	17.20
119.1	332	91	40299.1	633.1	1068.6	19903.1	Vapor	15.65

4.9 Case Study 9- Two Phase Separator

The ninth case study is for a pool fire impacting a vertical two phase water gas separator, which operates at 19% liquid full at 70 degrees F and 40. Since the operating pressure is much lower than the relieving pressure of 1742.4 psig, the temperature of the fluid will need to be increased substantially before the relief starts. Initially the liquid level increases, and then proceeds to decrease. By using the dynamic method, it is determined that it takes ninety seven minutes before the relief event starts. At one hundred nineteen minutes, the maximum relieving vapor flowrate is reached, at a temperature of 611 degrees F, which is well above the design temperature. See Tables 17 and 18 below for further details.

Table 17. Case Study 9 Parameters

Vessel	Inlet Separator
Drainage	Inadequate
Fire Fighting	Inadequate
Insulation	None

Case	External Pool Fire
Orientation	Vertical
Seam to Seam [Feet]	10
Diameter [Feet]	2.5
Normal Liquid Level [Feet]	1.92
Liquid Volume	100% water
Operating Temperature [Degrees F]	70
Design Temperature [Degrees F]	125
Operating Pressure [PSIG]	40
Design Pressure [PSIG]	1440
Relief Valve Set Pressure [PSIG]	1440
Relieving Pressure [PSIG]	1742.4

Table 18. Case Study 9 Details

Time [min]	Temperature [Degrees F]	Pressure [PSIG]	Vapor Flow [Ft^3/Hr]	Oil Flow [Ft^3/Hr]	Water Flow [Ft^3/Hr]	Relief Flow [Ft^3/Hr]	Relief Phase	Vessel Liquid Level, Ft
0.0	70	40	187.40	0.00	57.70			1.92
96.9	580	1440	174.83	0.00	70.24			1.92
104.1	606	1742	174.50	0.00	70.58			2.33
118.8	611	1742	353.79	0.00	53.53	154.32	Vapor	2.34
137.3	614	1742	382.50	0.00	34.83	130.61	Vapor	1.78

4.10 Case Study 10- Two Phase Separator

The tenth case study is for pool fire impacting a vertical two phase water gas separator, which operates at 25% liquid full at 70 degrees F and 5 psig. Since the operating pressure is much lower than the relieving pressure of 1210 psig, the temperature of the fluid will need to be increased substantially before the relief starts. Initially the liquid level increases, and then proceeds to decrease. By using the dynamic method, it is determined that it takes fifty nine minutes before the relief event starts. At seventy six minutes, the maximum relieving vapor flowrate is reached, at a temperature of 567 degrees F, which is well above the design temperature. See Tables 19 and 20 below for further details.

Table 19. Case Study 10 Parameters

Vessel	Inlet Separator
Drainage	Inadequate
Fire Fighting	Inadequate
Insulation	None
Case	External Pool Fire

Orientation	Vertical
Seam to Seam [Feet]	11
Diameter [Feet]	2
Normal Liquid Level [Feet]	2.75
Liquid Volume	100% Water
Operating Temperature [Degrees F]	70
Design Temperature [Degrees F]	125
Operating Pressure [PSIG]	5
Design Pressure [PSIG]	1000
Relief Valve Set Pressure [PSIG]	1000
Relieving Pressure [PSIG]	1210

Table 20. Case Study 10 Details

Time [min]	Temperature [Degrees F]	Pressure [PSIG]	Vapor Flow [Ft^3/ Hr]	Oil Flow [Ft^3/ Hr]	Water Flow [Ft^3/ Hr]	Relief Flow [Ft^3/Hr]	Relief Phase	Vessel Liquid Level, Ft
0.0	70	5	119.46	0.00	45.82			2.75
58.7	541	1000	108.79	0.00	56.54			2.75
62.6	565	1210	107.97	0.00	57.36			3.39
75.5	567	1210	353.65	0.00	43.04	251.20	Vapor	3.44
91.9	568	1210	373.75	0.00	28.17	202.98	Vapor	2.58

4.11 Case Study 11- Two Phase Separator

The eleventh case study is for a pool fire impacting a vertical two phase water gas separator, which operates at 18% liquid full at 80 degrees F and 850 psig. Since the operating pressure is much lower than the relieving pressure of 1742.4 psig, the temperature of the fluid will need to be increased substantially before the relief starts. Initially the liquid level increases, and then proceeds to decrease. By using the dynamic method, it is determined that it takes thirteen minutes before the relief event starts. At thirty three minutes, the maximum relieving vapor flowrate is reached, at a temperature of 330 degrees F, which is well above the design temperature.
See Tables 21 and 22 below for further details.

Table 21. Case Study 11 Parameters

Vessel	Inlet Separator
Drainage	Inadequate
Fire Fighting	Inadequate
Insulation	None
Case	External Pool Fire
Orientation	Vertical
Seam to Seam [Feet]	10
Diameter [Feet]	2
Normal Liquid Level [Feet]	1.83
Liquid Volume	100% Water
Operating Temperature [Degrees F]	80
Design Temperature [Degrees F]	125
Operating Pressure [PSIG]	850
Design Pressure [PSIG]	1440
Relief Valve Set Pressure [PSIG]	1440
Relieving Pressure [PSIG]	1742.4

Table 22. Case Study 11 Details

Time [min]	Temperature [Degrees F]	Pressure [PSIG]	Vapor Flow [Ft^3/ Hr]	Oil Flow [Ft^3/ Hr]	Water Flow [Ft^3/ Hr]	Relief Flow [Ft^3/ Hr]	Relief Phase	Vessel Liquid Level, Ft
0.0	80	850	119.04	30.04	3.60			1.83
12.7	256	1440	129.78	19.39	3.51			1.83
18.5	308	1742	149.41	3.28	0.00			1.25
33.1	330	1742	163.25	3.05	0.00	13.05	Vapor	0.18
48.6	352	1742	163.32	2.76	0.00	12.12	Vapor	0.17
65.5	374	1742	163.63	2.40	0.00	11.12	Vapor	0.15
84.4	395	1742	164.11	1.95	0.00	9.92	Vapor	0.13
106.9	415	1742	164.40	1.40	0.00	8.21	Vapor	0.11

4.12 Case Study 12- Glycol Absorber

The twelfth case study is for a pool fire impacting a vertical glycol absorber, which operates at 16% liquid full at 80 degrees F and 850 psig. Since the operating pressure is much lower than the relieving pressure of 1210 psig, the temperature of the fluid will need to be increased substantially before the relief starts. The liquid level increases throughout the event.

By using the dynamic method, it is determined that it takes eight minutes before the relief event starts. At eighty one minutes, the maximum relieving vapor flowrate is reached, at a temperature of 739 degrees F, which is well above the design temperature. See Tables 23 and 24 below for further details.

Table 23. Case Study 12 Parameters

Vessel	Glycol Absorber
Drainage	Inadequate
Fire Fighting	Inadequate
Insulation	None
Case	External Pool Fire
Orientation	Vertical
Seam to Seam [Feet]	19
Diameter [Feet]	2.5
Normal Liquid Level [Feet]	3
Liquid Volume	Mostly TEG
Operating Temperature [Degrees F]	80
Design Temperature [Degrees F]	100
Operating Pressure [PSIG]	850
Design Pressure [PSIG]	1000
Relief Valve Set Pressure [PSIG]	1000
Relieving Pressure [PSIG]	1210

Table 24. Case Study 12 Details

Time [min]	Temperature [Degrees F]	Pressure [PSIG]	Vapor Flow [Ft^3/ Hr]	Oil Flow [Ft^3 /Hr]	Water Flow [Ft^3/ Hr]	Relief Flow [Ft^3/ Hr]	Relief Phase	Vessel Liquid Level, Ft
0.0	92	845	353.46	0.00	80.44			3.00
7.6	160	1000	350.96	0.00	82.95			3.00
18.2	249	1210	346.70	0.00	87.20		Vapor	3.09
40.7	421	1210	483.08	0.00	96.79	90.79	Vapor	3.25
61.5	584	1210	467.68	0.00	106.68	95.17	Vapor	3.61
80.6	739	1210	464.95	0.00	110.99	104.23	Vapor	3.98
99.1	812	1210	415.30	0.00	96.56	59.10	Vapor	4.14

4.13 Case Study 13- NGL (Natural Gas Liquids) Vessel

The thirteenth case study is for a pool fire impacting a horizontal NGL vessel, which operates at 32% liquid full at 80 degrees F and 150 psig. Since the operating pressure is much lower than the relieving pressure of 266.2 psig, the temperature of the fluid will need to be increased substantially before the relief starts. The liquid level increases initially and then decreases. By using the dynamic method, it is determined that it takes fourteen minutes before the relief event starts. At sixty six minutes, the maximum relieving vapor flowrate is reached, at a temperature of 268 degrees F, which is well above the design temperature. See Tables 25 and 26 below for further details.

Table 25. Case Study 13 Parameters

Vessel	NGL vessel
Drainage	Inadequate
Fire Fighting	Inadequate
Insulation	None
Case	External Pool Fire
Orientation	Horizontal
Seam to Seam [Feet]	64
Diameter [Feet]	9.07
Normal Liquid Level [Feet]	2.92
Liquid Volume	Hydrocarbon
Operating Temperature [Degrees F]	60
Design Temperature [Degrees F]	150
Operating Pressure [PSIG]	150
Design Pressure [PSIG]	250
Relief Valve Set Pressure [PSIG]	220
Relieving Pressure [PSIG]	266.2

Table 26. Case Study 13 Details

Time [min]	Temperature [Degrees F]	Pressure [PSIG]	Vapor Flow [Ft^3/ Hr]	Oil Flow [Ft^3 /Hr]	Water Flow [Ft^3/ Hr]	Relief Flow [Ft^3/ Hr]	Relief Phase	Vessel Liquid Level, Ft
0	60	150	14,015	-	5322.03			2.92
14	120	220	13,856	-	5480.35			2.92
21	152	266	13,790	-	5546.18			3.00
35	193	266	25,227	-	4959.90	10,904.33	Vapor	3.04
50	232	266	27,983	-	4022.17	12,058.26	Vapor	2.72
66	268	266	31,376	-	2631.63	12,640.33	Vapor	2.20
86	303	266	35,499	-	655.92	11,926.00	Vapor	1.44
96	315	266	24,256	-	0.00	7,107.82	Vapor	0.36

5. Conclusions

What conclusions can be drawn from comparing the thirteen cases?

When the operating pressure is much less than relieving pressure, the vapor may actually be compressed back into the liquid phase. When the operating pressure is much less than relieving pressure, the relieving temperature may be much higher than the operating temperature and design temperature. By using this method, one can get a temperature versus time curve, which can be used to determine if the vessel is likely to fail due to exceeding the rated temperature.

When the operating level is high in a vessel, the liquid may expand enough at the relieving conditions that there will be liquid relief, followed by two-phase flow, and then finally vapor. Without a dynamic model, one may not realize that there could be relief cases other than vapor. The cases other than vapor may mean that the outlet should go to a closed system to prevent personnel exposure to hot liquids and potential sources for pool fires and may result in larger orifice and pipe sizes.

With the liquid expanding due to the increase in temperature, the heat input actually increases, due to a larger wetted area. Without using this method, one may undersize the relief valve. Note that in NFPA 30, there is no credit given for normal liquid level. Steady-state external fire relief evaluated using NFPA 30 is based on full straight-side height.

By using this method, one can determine how long it takes before the relief occurs, thus allowing for proper emergency response by determining time available for evacuation or attempting to extinguish the fire. By using this method, one can determine how long it takes before the vessel is no longer wetted by liquid and therefore at increased potential for failure.

By using the dynamic method, one can determine when during the event the maximum relief load occurs. Sometimes it is right at the beginning, sometimes at the end, and sometimes at an intermediate point.

6. References

[1] API Standard 521, Pressure-relieving and De-pressuring Systems, Sixth Edition, January 2014

Quantifying Human Error Probabilities through a Bayesian Network using Expert Opinion and an Auditing Tool: An Application to the Tokai-Mura Accident

J. P. Duarte
Department of Telecommunications and Control Engineering
Polytechnic School, University of São Paulo
Av. Prof. Luciano Gualberto, Travessa 3, 158, 05508-900 São Paulo, SP, Brazil
julianapduarte@usp.br

A. C. O. Ribeiro[1], A. L. Sousa[2], P.F. Frutuoso e Melo[3]
[1]Bayer Crop Science Brazil S.A.
Estrada Boa Esperança 650, 26110-100 Belford Roxo, RJ, Brazil
antonio.ribeiro@bayer.com

[2]National Commission of Nuclear Energy (CNEN)
Rua General Severiano 90, 22290-901 Rio de Janeiro, RJ, Brazil
alsousa@cnen.gov.br

[3]Graduate Program of Nuclear Engineering, COPPE, Federal University of Rio de Janeiro
Av. Horácio Macedo 2030, Bloco G, sala 206, 21941-914 Rio de Janeiro, RJ, Brazil
frutuoso@nuclear.ufrj.br

Prepared for Presentation at
American Institute of Chemical Engineers
2015 Spring Meeting
11th Global Congress on Process Safety
Austin, Texas
April 27-29, 2015

UNPUBLISHED

Quantifying Human Error Probabilities through a Bayesian Network using Expert Opinion and an Auditing Tool: An Application to the Tokai-Mura Accident

J. P. Duarte
Department of Telecommunications and Control
Polytechnic School, University of São Paulo
Av. Prof. Luciano Gualberto, Travessa 3, 158, 05508-900 São Paulo, SP, Brazil
julianapduarte@usp.br

A. C. O. Ribeiro[1], A. L. Sousa[2], P.F. Frutuoso e Melo[3]
[1]Bayer Crop Science Brasil S.A.

[2]National Commission of Nuclear Energy (CNEN)

[3]Graduate Program of Nuclear Engineering, COPPE, Federal University of Rio de Janeiro

Keywords: Tokai-Mura accident, Human Reliability Analysis, THERP, CREAM, Expert opinion elicitation

Abstract

This paper presents a human reliability analysis (HRA) model for allowing the incorporation of elements that represent the facility conditions in determining human error probabilities (HEP) used in probabilistic safety analyses of process plants. It is based on an approach presented in an earlier paper by the same authors and the difference is that the model has been implemented in a Bayesian network and run by means of the Netica™ tool. The results in terms of HEPs are the same as those of the earlier paper and a sensitivity analysis was performed by means of the network developed.

1. Introduction

This paper is based on the model presented in [1]. The idea here is to take the same model and develop it by means of a Bayesian network. The reasons for doing so is to take advantage of the tools Bayesian networks [2], like easiness for modeling dependent events, which is the case when one models human failure analyses for risk analysis purposes of process plants.

In a previous work [1] a model for assessing human error probabilities for process plants

Note: Do not add page numbers. Do not refer to page numbers when referencing different portions of the paper

was developed by considering, as a starting point, human failure probabilities taken from THERP [3] and CREAM [4].

Those human error probabilities do not take into account elements that represent the facility conditions in determining human error probabilities (HEP) used in probabilistic safety analyses of process plants.

An approach to show the predominance of human factors as accident cause was presented, as well as existing methodologies for HEP determination and their deficiencies in incorporating socio-technical elements that influence them. Such elements are: control center design, remote operations, human-machine interface, training, communications, environmental factors, workloads and staffing levels, safety culture, procedures, maintenance, management of change, and incident investigation [5].

A mathematical model, based on Bayesian networks, is now proposed to incorporate these elements in an easier way.

This paper is organized as follows: section 2 briefly discusses Bayesian networks. Section 3 presents the application of a Bayesian network for the Tokai-Mura failure event, as had been in the previous paper [1] and section 4 presents the conclusion reached.

2. Bayesian Networks

Given the complexity involved in Bayesian inference when it comes to systems with more than two variables, Bayesian networks are recommended [2]. Bayesian networks (BN) are directed acyclic graphs that, in a probabilistic way, represent dependencies between variables. Network nodes represent random variables (discrete or continuous) and directed arcs illustrate the dependency relationships among variables [6]. The relationship between cause and effect is expressed by conditional probabilities.

BNs are useful for aggregating expert opinions. The complexity of a Bayesian network depends on the level of information that can be obtained and the importance that the analyst gives to such information.

Each node has an associated conditional probability table (CPT) that quantifies the effects that parents exert on a node, i.e., the probability of the node being in a specific state, given its parent states. For each variable A that has as parents $X_1, \ldots, X_i$, there is a table of conditional probabilities $P(A \mid X_1, \ldots, X_i)$.

3. Tokai-Mura Human Failure Analysis by a Bayesian Network

It should be recalled from [1] that the purpose of our model is to modify basic human failure probabilities by considering plant characteristics.

The purpose it to take a ground human failure probability (HEP) and modify it by considering a proposal originally set in [7]. Notice that the difference between our model and that of [7] is the way we have modified the ground human error probabilities: we propose this modification by means of three grades (1, 2, and 3) and an auditing factor (r_i).

Grade # 1 measures the relative importance of the factor influences and is achieved through an array of relationship factors by which it is possible to analyze the distribution and intensity of factor relationships.

Grade # 2 measures the weight of each factor or element influence through the elicitation of expert opinion [8]. Twelve experts attributed a degree 1-5, where 1 is the lowest contribution degree and 5 is the largest one to the importance of each factor. The grade of each element was calculated by the average of the grades assigned by the experts.

As seen, the fourth factor (training) was considered the most relevant (grade equal to 4.7). It should be noted that these weights were attributed by the experts prior to the model application to the Tokai-Mura event (as will be seen, factors 2 and 10 have no influence on the analysis). The participants were senior engineers working in the nuclear, chemical, and petrochemical industries and Brazilian regulatory agencies.

Grade 3 represents the incidence weights of factors or elements as root causes or contributors, being established from an analysis of abnormal events (incidents, accidents and near misses, [9], [10]) in the plant, which shows the number of times that each element contributed as a root cause. The incidence of the factors or elements shown in Fig. 1 (boxes at the left) appears as a root cause of the abnormal events analyzed and this normalized figure was used as a weight for Grade 3. If an event history is unavailable, either data from a similar plant is used or Grade 3 is set equal to 1.

The r_i factor measures the degree of implementation of each factor assessed by the plant auditing process. To measure the degree of implementation of the elements a questionnaire was used to assess compliance of each factor. Each verification item is scored from 1 – 5, where 1 means noncompliance and 5 means full compliance. No scoring for elements 2 and 10 were considered because they did not apply to the Tokai-Mura plant. Modifications and grouping of elements in the original OGP questionnaire have been made to simplify the scoring and model application.

In order to correct the original HEP values of existing HRA techniques we proposed a quantitative model that is taken from [7] and adapted to our purposes. The correction to the ground human error probabilities are performed by means of the discussed Grades 1, 2, and 3 and also the auditing factor r_i. Notice that the ground probabilities are taken from THERP [3] and CREAM [4], as already done in our earlier paper [1].

Figure 1 displays the Bayesian network developed for analyzing the Tokai-Mura event. The boxes on the first column (left) display the OGP elements that are central to estimate Grade # 1. As discussed in [1], elements 2 and 10 do not apply to the Tokai-Mura plant. The second column displays the failure events that appear in the human failure event tree of [2] and finally the box in the third column displays the Tokai-Mura human failure probability. It should be noticed that the results of both analyses match. This means that Bayesian networks may be a useful tool for performing human reliability analyses in a systematic way and also perform a sensitivity analysis of the results in order to shed light on the most important contributors to the final result.

The data displayed in the first column of Fig. 1 is related to the results of the auditing procedure that indicates the degree of implementation of each element (that is, r_i). Next, for each failure event of Fig 1 we connect to it all elements that impact it and in the box for each failure event we can write down an equation for calculating the failure probability for each event. Again, failure probabilities are displayed in percentages. The first percentage is the failure probability and its complement is the success probability. For example, for event A (Failure of correct directing and/or consideration of mass, volume and geometry in the safe preparation of batch (P2 inadequate planning) the failure probability is 2.51×10^{-2}. The equations in each box of a failure event are exactly the ones developed in the earlier paper for each case. Table 1 describes all failure events for the analysis. Events B, C, E, and G are not displayed here because their contribution is negligible [1].

Table 1. Description of failure events

Failure Event	Event Description
A	Failure of correct directing and/or consideration of mass, volume and geometry in the safe preparation of batch (P2 inadequate planning)
D	Failure to conduct a thorough briefing in batch preparation (I2 - decision error with P2 - planning error)
F	Failure of the supervisor and the production head to detect and correct the operator group decision on the tank mode use and number of batches
H	Supervisor fails to stop production

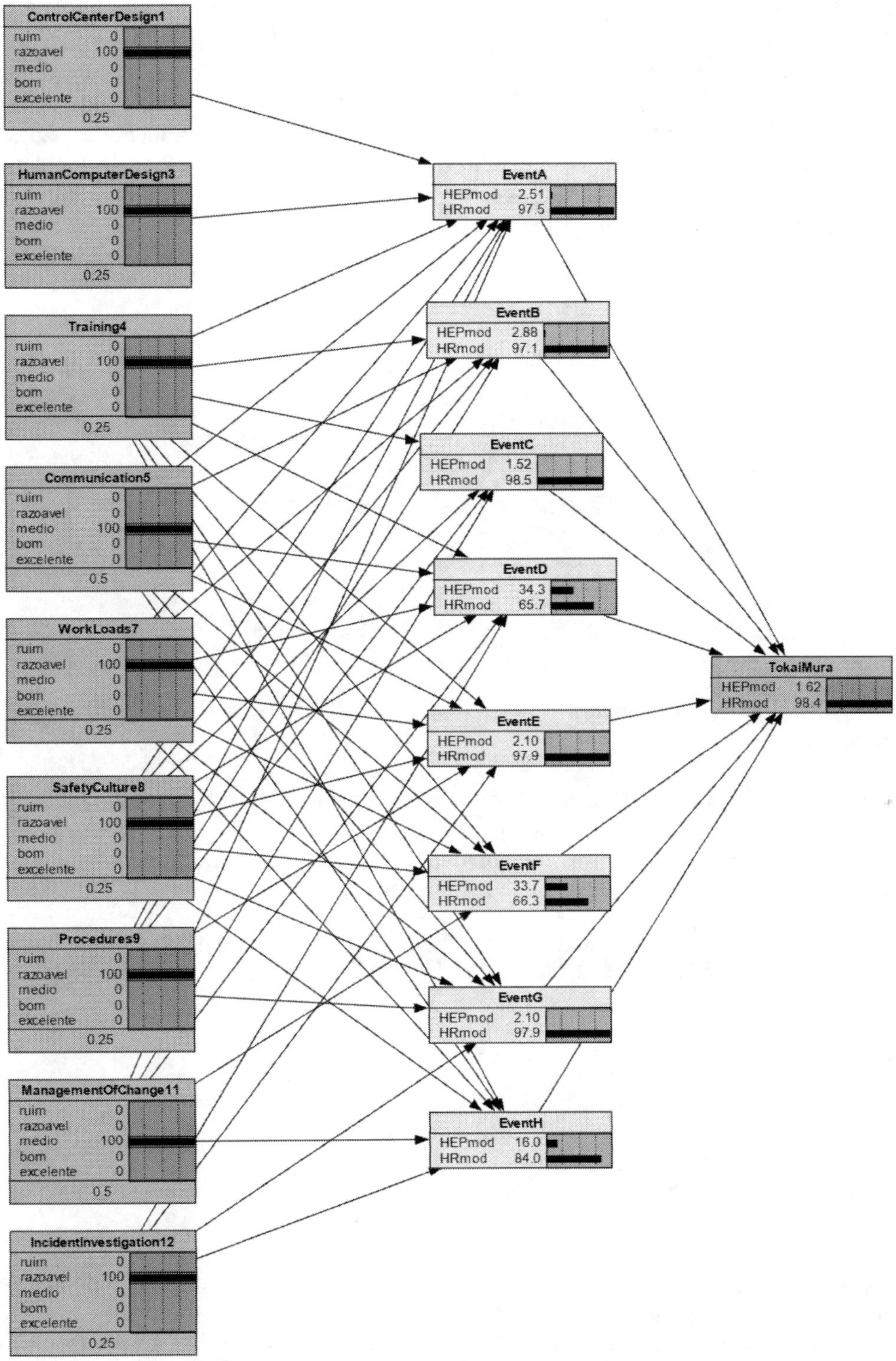

Figure 1. Bayesian Network for modeling the Tokai-Mura event

Table 2 displays an initial sensitivity analysis related to the r_i factor. Notice that the first column displays the OGP elements, the second one displays the human failure probability range and the last one displays the relative percentage variation. The range of column 2 was obtained as follows: we varied the degree of implementation from 'Bad' (which means a degree of implementation equal to zero) to 'Excellent' (which means a degree of implementation equal to 100%). The first figure in the second column is the failure probability for 'Bad' implementation degree and the second one is the failure probability for the 'Excellent' degree of implementation.

Table 2. Sensitivity Analysis for the r_i Factor

Element	Range for HEP (x 10^{-2})	Relative variation (%)
1	1.65 – 1.58	4.3
3	1.62 – 1.62	0.0
4	3.45 – 0.19	201.2
5	2.06 – 1.29	47.5
6	1.62 – 1.62	0.0
7	1.87 – 1.08	48.8
8	1.75 – 1.29	28.4
9	2.03 – 0.84	73.5
11	2.85 – 0.94	117.9
12	1.74 – 1.32	25.9

It can be seen from Table 1 that elements 4, 11, 9, 7, and 5 are the most important for the Tokai-Mura.

Table 3 shows the sensitivity analysis for the human failure event *A*. A similar sensitivity analysis was performed for human failure events D, F, and H. Notice that human failure events B, C, E, and G were not considered here since their contribution to the global human failure event is negligible, as can be seen from Fig 1.

Table 3. Sensitivity Analysis for the Human Failure Event *A*

Element	Range for HEP (x 10^{-2})	Relative variation (%)
1	3.38 – 2.04	53.4
3	2.54 – 2.42	4.8
4	3.48 – 0.95	100.8
5	2.80 – 2.25	21.9
7	2.68 – 2.07	24.3
8	2.59 – 2.28	12.4
9	3.04 – 1.41	64.9
11	3.57 – 1.77	71.7

It may be seen from Table 3 that Element # 4 is the most important since it doubles the event human failure probability. Element # 9 is the second most important. A similar behavior may be found in Table 4, 5, and 6. Notice that these results may help making decisions as to where to implement modification when resources are limited.

Table 4. Sensitivity Analysis for the Human Failure Event *D*

Element	Range for HEP (x 10^{-2})	Relative variation (%)
4	49.9 – 11.2	112.8
5	38.9 – 30.3	25.1
7	37.0 – 27.5	27.7
8	35.6 – 30.7	14.3
11	51.4 – 22.9	83.1
12	36.3 – 29.1	21.0

Table 5. Sensitivity Analysis for the Human Failure Event *F*

Element	Range for HEP (x 10^{-2})	Relative variation (%)
4	50.3 – 10.1	119.3
5	38.5 – 29.4	27.0
7	36.4 – 26.5	29.4
8	35.0 – 29.9	15.1
11	51.9 – 21.8	89.3

Table 6. Sensitivity Analysis for the Human Failure Event *H*

Element	Range for HEP (x 10^{-2})	Relative variation (%)
4	30.3 – 2.34	174.8
5	19.8 – 12.9	43.1
7	18.1 – 10.9	45.0
8	17.0 – 13.2	23.8
11	31.9 – 7.98	149.5
12	17.6 – 12.0	35.0

4. Conclusions

It has become clear at safety conferences and congresses in the nuclear and chemical and petrochemical fields that existing laws and regulations, especially some requirements of international regulatory bodies such as the CSB (Chemical Safety Board, USA), are more and more explicit in regarding the implementation of human reliability analysis (HRA) as a way of risk reduction. However, most organizations still do not have efficient mechanisms to understand and implement policies for human factors analyses. This work offers a contribution to include in a comprehensive manner the elements that influence human error. Also, improvements on plant management that can be taken into account by the 12 factors considered can be easily evaluated by the Bayesian network, thus allowing for estimating human reliability improvements.

A benefit of the Bayesian network model is the fact that a sensitivity analysis can be easily performed to analyze the impact of each of the 12 factors mentioned and thus allowing for a more realistic plant behavior modeling in face of abnormal events.

A contribution of the proposed model is to allow seeing how elements relate and how they influence HEP quantification, which allows directing efforts in the short and long term to reduce HEPs or even review the effectiveness of the efforts being made to reduce them.

5. References

[1] A. C. Ribeiro, A. L. Sousa, J. P. Duarte, and P. F. Frutuoso e Melo, Human Reliability Analysis of the Tokai-Mura Accident through a THERP-CREAM and Expert Opinion Auditing Approach, submitted to the Nuclear Engineering and Design journal, 2014.

[2] F. V. Jensen, An Introduction to Bayesian Networks, UCL Press, University College, London, 1996.

[3] A. D. Swain and H. E. Guttman, Handbook of Human Reliability Analysis with Emphasis on Nuclear Power Plant Applications, NUREG/CR-1278, US Nuclear Regulatory Commission, Washington, DC, 1983.

[4] E. Hollnagel, Cognitive Reliability and Error Analysis Method (CREAM), Elsevier, Oxford, 1998.

[5] CCPS, Human Factors Methods for improving Performance in the Process Industries, Center for Chemical Process Safety, American Institute of Chemical Engineers, Wiley, New York, 2007.

[6] J . Pearl, Probabilistic Reasoning in Intelligent Systems. Morgan Kaufmann Publishers, San Francisco, CA, 1988.

[7] S. G. Kariuki, Integrating Human Factors into Chemical Process Quantitative Risk Analysis, PhD dissertation, Berlin Technical University, Berlin, 2007.

[8] B. M. Ayyub, Elicitation of Expert Opinions for Uncertainty and Risks. CRC Press, Boca Ratton, FL, 2001

[9] CCPS, Guidelines for Chemical Process Quantitative Risk Analysis, Center for Chemical Process Safety, American Institute of Chemical Engineers, Wiley, New York, NY, 2000.

[10] IAEA, IAEA Safety Glossary, Terminology Used in Nuclear Safety and Radiation Protection, International Atomic Energy Agency, Vienna, Austria, 2007.

396889 Numerical Modeling of Pool Fire inside an Offshore Methanol Storage Tank

Abstract:

Methanol is widely used in the oil and gas industry as a reactant for the production of methyl tertiary butyl ether (MTBE) and as down-hole fluid additive to reduce hydrate formation and prevent freezing of solutions during injection into gas wells. Methanol is classified as a flammable liquid with a flammable range from 6% to 36% bounding the lower and upper flammable limits, respectively.

There are many experimental and numerical studies carried out for the methanol liquid pool fires in an open air environment, or a wind tunnel, with multiple pool fire sizes. However, few numerical studies have been performed for methanol pool fires inside or emanating from a storage container (typical storage units containing man-way entrances). The inadequate oxygen inside the container will limit the duration of the fire, and extinction and re-ignition of sectional small fires propose a challenge for numerical simulation of such kind of fire. In this paper, the Fire Dynamics Simulator (FDS) tool is used in the analysis of methanol pool fires. FDS is a computational fluid dynamics (CFD) fire model using large eddy simulation (LES) turbulence model. The basis for FDS has been developed from the mathematical background common to many CFD models with an emphasis on the slow moving flow and heat transfer caused by fires.

Methanol tanks typically have a pressure vacuum safety valve (PVSV) installed to relieve static pressure inside the tank when the pressure difference between the vessel pressure and the ambient pressure is small. If the methanol pumps continue to run and the tank PVSV or man-way is opened, air (e.g. oxygen) may ingress and begins mixing with the denser methanol vapors to a flammable range. Assuming the vapors find an ignition source, a pool fire is possible with the flammable vapors burning above the methanol liquid surface. This paper considers two scenarios for air ingress that lead to pool fires, a PVSV opening during normal venting operations and a man-way opening during maintenance to the inside of the tank. The relationship between the fire duration and tank fill levels are obtained and discussed.

Credibility of Designing Refrigerated Atmospheric Storage Tank Pressure Relief Systems for a Tornado

Syed Usman Ahmed
Agrium Wholesale – Engineering & Risk Management
11751 River Road, Fort Saskatchewan, Alberta, Canada, T8L-4J1
Syed.Ahmed@agrium.com

Keywords: Plant Operations, Pressure Relief Devices, Pressure Relief Systems, Process Safety and Storage Tanks

Abstract

Atmospheric pressure changes occur due to varying weather patterns. As per API 2000, a change in barometric pressure is a possible cause of overpressure or vacuum in refrigerated storage tanks due to the potential for a substantial amount of vaporization or condensation. To serve as a valid layer of protection, relief systems are required to be sized to handle the worst case scenario with the largest required relief area. An extreme case of a drastic change (sudden decrease followed by a sudden increase) in barometric pressure is the passing of a severe electrical storm front (i.e. tornado). This causes a sudden atmospheric pressure drop and results in a loss of power to the site (all electrically driven equipment would cease operation) as the system tries to re-establish vapor-liquid equilibrium within the storage tank.

This paper will analyze Agrium's anhydrous ammonia storage tank, 100-T at Leal Distribution Terminal in North Dakota and delve into the reasons a tornado would not cause a realistic overpressure scenario for the storage tank's pressure relief devices.

1. Introduction

Severe weather changes can result in the following:

- High winds
- Precipitation
- Rapid temperature change
- Sudden change in atmospheric pressure due to storm front
- Loss of site-wide electrical power due to storm

If the material being stored is refrigerated liquid at or near its boiling point, low atmospheric pressure (decrease in barometric pressure) during a severe storm will cause the stored material to re-establish vapor-liquid equilibrium. Liquids that are not at or near their boiling point are not affected by weather changes.

API 2000 Sections 4.2.5.11 and 5.2.1.2 state: *A rise or drop in barometric pressure is a possible cause of vacuum or overpressure in a tank. This should be considered for refrigerated storage tanks. A change in atmospheric pressure can result in a substantial amount of vaporization or condensation* [1]. The worst credible case is the basis for relief system design. As such, the design basis scenario is the passing of a severe electrical storm front (i.e. tornado) causing a sudden atmospheric pressure drop and resulting in a loss of power to the site (all electrically driven equipment would cease operation); this will result in an overpressure scenario caused by:

- Flash vapor generated from liquid inventory
- Expansion of vapor in the enclosed vapor space (head of tank)
- Normal boil-off

2. Designing Storage Tank Pressure Relief Systems for a Tornado

2.1 Background Information

Determining accurate barometric pressure measurements in tornado cores is difficult due to the harsh physiology of the tornado itself. API 2000 Section 5.2.1.2 suggests using 2000 Pascals/hr as the absolute rate of pressure change (d_{patm}/dt) if local atmospheric pressure change data is not available [1]; however, this typically is not a concern as weather networks have historical weather data that is monitored and collected by local airports through the use of pressure probes. For a pressure probe to remain stationary during a tornado passage, special considerations are applied to the physical geometry of the device in order to increase the accuracy of the measurements [2]. From Figures 1 and 2 below, it is evident that the pressure plunge that takes place as a tornado passes is very transitory (i.e. a few seconds).

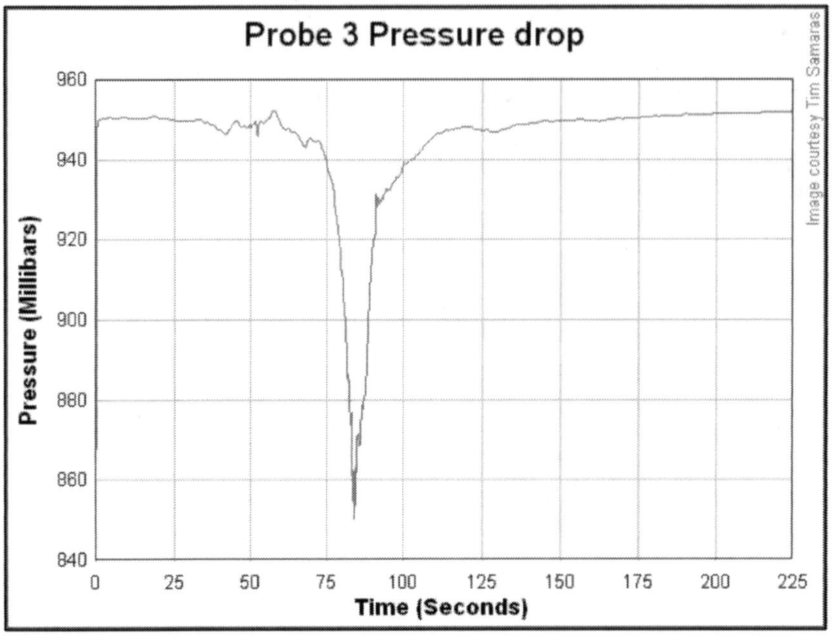

Figure 1: Tornado on June 24, 2003 (Manchester, SD) [3].

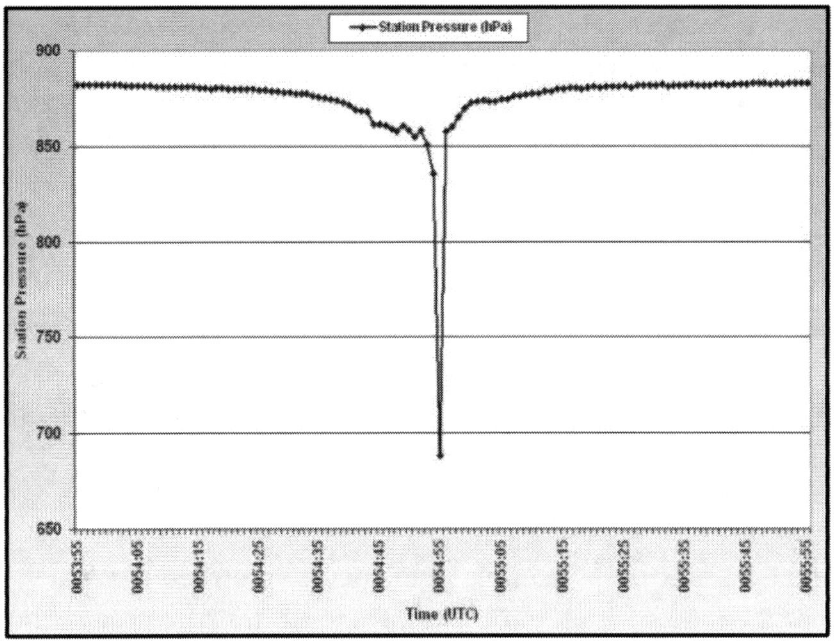

Figure 2: Tornado on April 21, 2007 (Tulia, TX) [3].

2.2 Design Basis

The storage tank must be designed so that it will remain in a safe condition under severe meteorological events, including those that could result in the most severe tornado that could to occur at the site. Referring to Figure 3, the U.S. Nuclear Regulatory Commission divides the Contiguous United States into three tornado intensity regions at the 10^{-7} probability level and each of these regions is characterized by a pressure drop, which is related to the weather conditions during the tornado as provided in Table 1. It is evident from Table 1 that the pressure drop is not large, but occurs very rapidly (2 – 3 seconds).

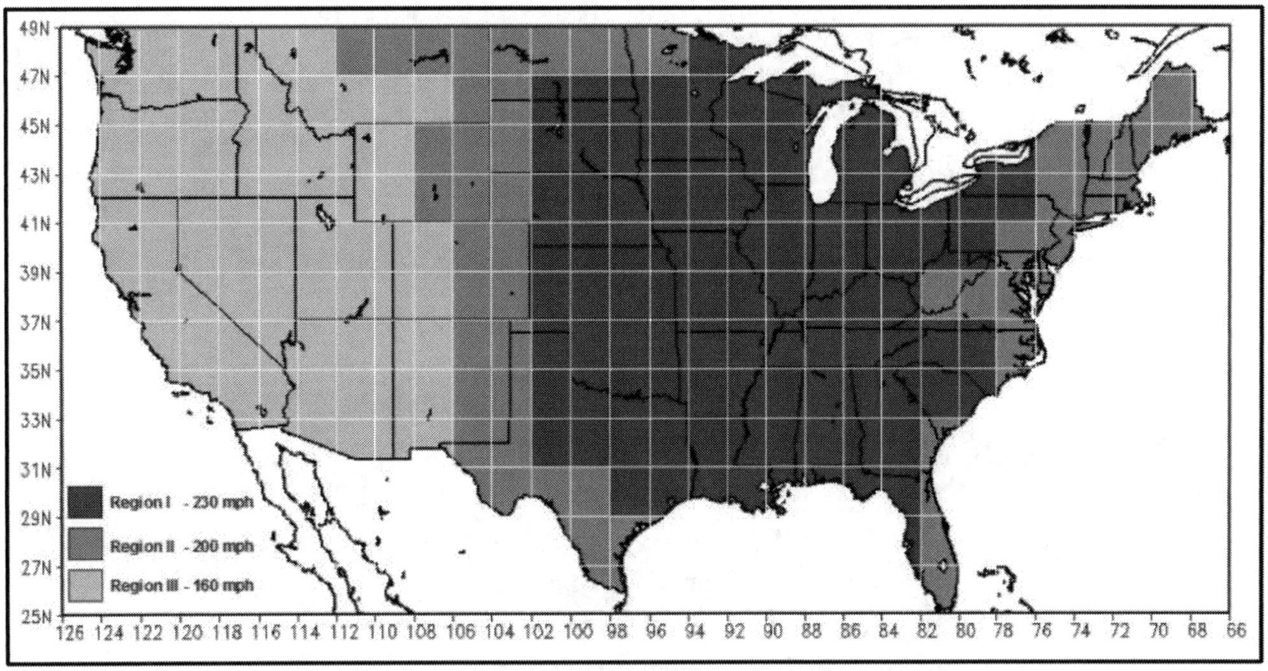

Figure 3: Tornado Intensity Regions for the Contiguous United States for Exceedance Probabilities of 10^{-7} per year [4]

Table 1: Design-Basis Tornado Characteristics [4]

Region	Maximum wind speed m/s (mph)	Translational speed m/s (mph)	Maximum rotational speed m/s (mph)	Radius of maximum rotational speed m (ft)	Pressure drop mb (psi)	Rate of pressure drop mb/s (psi/s)
I	103 (230)	21 (46)	82 (184)	45.7 (150)	83 (1.2)	37 (0.5)
II	89 (200)	18 (40)	72 (160)	45.7 (150)	63 (0.9)	25 (0.4)
III	72 (160)	14 (32)	57 (128)	45.7 (150)	40 (0.6)	13 (0.2)

2.3 Required Relief Rate

As per API Standard 2000, the total required relief rate is the sum of that caused by vapor expansion and that caused by flashing. The description of the calculation method holds true for only certain cases (i.e. does not take into consideration the required relief due to normal boil-off). Further, the actual equations in API 2000 shall not be used as it is not valid for a refrigerated liquid at or near its boiling point. The total required relief rate is described in Equation 1 as:

$$\dot{W} = \dot{W}_{AL} + \dot{W}_{AG} + \dot{W}_{AB} \quad (1)$$

where,

$\dot{W}_{AL}$ = Flash vapor generation due to atmospheric pressure drop, lb/hr

$\dot{W}_{AG}$ = Vapor expansion due to atmospheric pressure drop, lb/hr

$\dot{W}_{AB}$ = Normal boil-off (must be taken into account due to loss of power), lb/hr

In order to calculate the required relieving capacity the following parameters shall be used:

- The tank is filled to 100% of its design liquid capacity
- No inlet flow[1]
- The absolute rate of pressure change of atmospheric pressure (d_{patm}/dt) for the site as determined from Table 1

Refer to Section 6.1 for sample calculation.

2.3.1 Flash Vapor Generation

Apply the conditions below to 100% liquid inventory in the storage tank to determine the flash vapor generation over a one hour period (using an adiabatic flash):

- The initial conditions in the tank are as follows:
  - Site design atmospheric pressure
  - Storage tank design pressure
  - Liquid inventory at bubble point

- The final conditions are as follows:
  - Site design atmospheric pressure less the absolute rate of pressure change of atmospheric pressure
  - Storage tank design pressure
  - Bubble point

2.3.2 Vapor Expansion

The drop in atmospheric pressure results in a change in density in the vapor space. Hence, the change in density results in vapor expansion which requires relief and is described in Equation 2 as:

---

[1] There is little difference in the overall relief requirements if inlet flow continues compared to assuming no inlet flow. It has been determined that the overall relief requirements for the storage tank contents are primarily dictated by the liquid inventory in the tank and not the inlet flow stream. Between 0.25% and 0.5% of the total vapor generated is from the liquid inflow (including flash), compared to that which flashes from the liquid in the tank. Therefore, 100% liquid inventory with no inlet flow shall be used as the relief system design basis.

$$\dot{W}_{AG} = V\frac{(\rho_1 - \rho_2)}{1\ hr} \quad (2)$$

where,

$\dot{W}_{AG}$ = Mass flow rate requiring venting due to vapor expansion, lb/hr

V = Volume of vapor space above maximum liquid level, ft^3

ρ_1= Density of vapor at P_1 and its dew point temperature, lb/ft^3

ρ_2= Density of vapor at P_2 and its dew point temperature, lb/ft^3

P_1 = Design pressure of storage tank

P_2 = Design pressure of storage tank less the absolute rate of pressure change (d_{patm}/dt)

2.3.3 Normal Boil-Off (Evaporation due to Ambient Heat Input)

The normal boil-off rate is equal to the vapor generation rate as a result of the normal heat input to the tank. The vapor generation rate is found using the normal ambient heat input and the heat of vaporization of the stored liquid at relief conditions (i.e. pressure and temperature)[2] .

$$\dot{W}_{AB} = \frac{Q}{\Delta H_{vap}} \quad (3)$$

where,

Q = Heat input (as per Heat & Material Balance), Btu/hr

ΔH_{vap} = Heat of vaporization of the stored liquid at the reliving pressure and temperature, Btu/lb

2.4 Pressure Relief Devices

There are various types of pressure relief devices that can be used for pressure and vacuum relief: open vents, direct-acting vent valves, and pilot-operated vent valves.

Open vents are always open and provide overpressure or vacuum protection for atmospheric storage tanks by allowing the tank to in-breath and out-breath at any pressure differential. Open vents are often equipped with a weather hood or turndown (see Figure 4) to prevent rain or snow from entering the tank.

---

[2] The heat of vaporization is obtained from simulation software or thermodynamic and physical properties charts/tables for pure liquids.

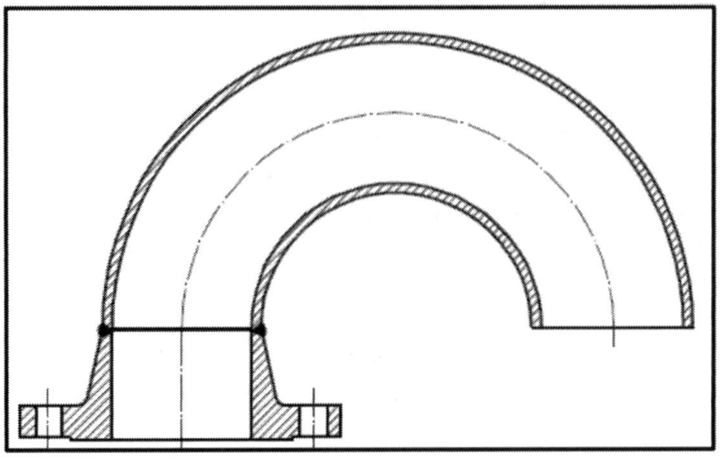

Figure 4: Open vent [1].

Direct-acting vent valves are available to provide pressure relief, vacuum relief, or a combination of pressure and vacuum relief. Combination vent valves may be of a side-by-side configuration. Side-by-side vent valves or pressure-relief vent valves are available with flanged outlets for pressure discharge when it is necessary to vent the pressure-relief vapors away.

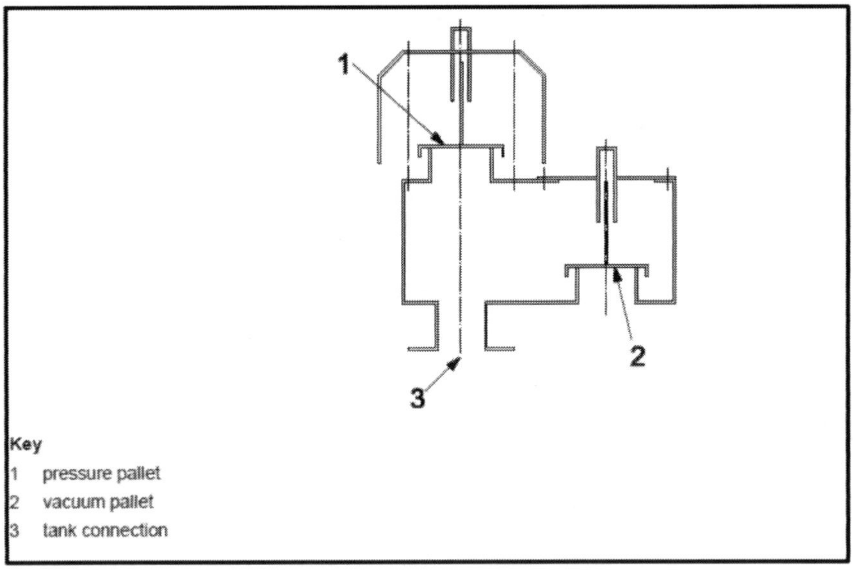

Figure 5: Side-by-side pressure/vacuum vent [1].

Pilot-operated vent valves are available to provide pressure relief, vacuum relief or a combination of pressure and vacuum relief. Unlike side-by-side direct-acting vent valves, pilot-operated vent valves relieve pressure or vacuum through the same opening to atmosphere.

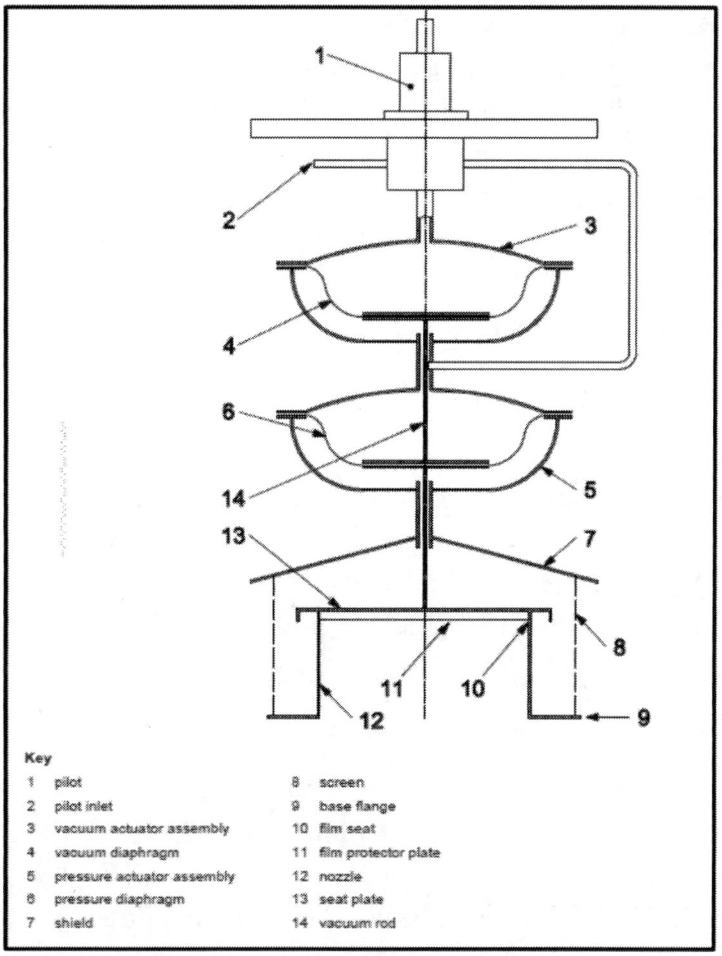

Figure 6: Pilot-operated pressure/vacuum vent [1].

2.4.1 Orifice Area – Anderson Greenwood Low Pressure POPRVs

To size a pressure relief valve, the required relief rate for the controlling contingency must be known, along with the fluid properties at relief conditions. The required orifice area is calculated from vendor supplied formulae to determine the size of the relief device.

Referring to Formula 11 on page 33 of the Anderson Greenwood Low Pressure Pilot Operated Pressure Relief Valves (POPRV) Catalog [5], the required orifice area can be calculated based on the mass flow rate at actual conditions or volumetric flow rate at standard conditions as shown in Equation 4:

$$A_{(in^2)} = \frac{W\sqrt{TZ}}{735K_dP_1F\sqrt{M}} \tag{4}$$

where,

W = Required relief rate, lb/hr

T = Absolute relieving temperature[3], °R
P_1= Pressure at valve inlet during flow, psia
M = Molecular weight of the flowing gas, MW
Z = Compressibility factor of the flowing gas
K_d = Coefficient of discharge
F = Subsonic flow factor

Refer to Section 6.2 for sample calculation.

3. Results & Conclusion

Assuming that vapor-liquid equilibrium could be instantly established, 3.95 x 10^6 lb/hr of NH_3 vapor would have to be relieved if a tornado came in direct contact with 100-T, Ammonia Storage Tank, causing a rapid decrease in atmospheric pressure. The corresponding required orifice area is 10,396 in^2. Therefore, the system would require 92 POPRVs as each POPRV has an orifice area of 113 in^2.

4. Conclusion

Based on the analysis conducted at Leal Distribution Terminal, it is evident that a tornado which results in a sudden decrease in barometric would not cause a realistic overpressure scenario for the storage tank's pressure relief devices to be sized due to the following:

- The unlikely probability (10^{-7} chances per year) of encountering such a tornado that exceeds the intensity of Region II (200 mph winds)
- The near impossibility of the system (ammonia within the storage tank) to re-establish vapor-liquid equilibrium in a few seconds (i.e. time span of pressure drop) thus resulting in an overpressure scenario
- The more likely damage mechanism of very high wind velocities, heavy rain, and wind-driven debris renders overpressure protection a moot point.

Therefore, when determining the required relief rate for the Weather Changes / Atmospheric Pressure Drop contingency the absolute rate of pressure change of atmospheric pressure (dp_{atm}/dt) used should not be that for a tornado in the region, rather the greater dp_{atm}/dt for the site over at least 50 years, or that provided in API Standard 2000 Section 5.2.1.2 (2000 pascals/hr), whichever is greater.

---

[3] Storage tank relief pressure for overpressure and normal maximum temperature at site for vacuum

5. References

[1] *ANSI/API STANDARD 2000: Venting Atmospheric and Low-pressure Storage Tanks*. 6th ed. N.p.: American Petroleum Institute, 2009. Print.

[2] Samaras, Timothy M., and Julian J. Lee. *Pressure Measurements Within a Large Tornado*. Proc. of Eighth Symposium on Integrated Observing and Assimilation Systems for Atmosphere, Oceans, and Land Surface, Washington State Convention and Trade Center, Seattle. Applied Research Associates, Inc., n.d. Web.

[3] Edwards, Roger. "The Online Tornado FAQ." *Storm Prediction Center*. NOAA / National Weather Service, 07 Mar. 2014. Web. 24 July 2014. <http://www.spc.noaa.gov/faq/tornado/index.html#pressure1>.

[4] Regulatory Guide 1.76, "Design-Basis Tornado and Tornado Missiles for Nuclear Power Plants," *U.S. Nuclear Regulatory Commission*, Washington, DC, March 2007.

[5] Pentair. *Anderson Greenwood Low Pressure POPRV Catalog*. N.p.: Pentair Valves & Controls, 2013. Print.

6. Appendix – Sample Calculation

6.1 Required Relief Rate

Data and Constants:

		Reference
Tank MAWP	1 psig	Site General Arrangement
Tank Design Temperature	-39°F	Site General Arrangement
P_{atm}, Atmospheric Pressure	13.8 psia	Atmospheric Pressure at site
P_{set}, Set pressure of PSV	1 psig	P_{set} = Current set pressure of PSVs
P_1, Relief Pressure	14.9 psia	$P_{relief} = (1.10)\ P_{set} + P_{atm}$
T_{relief}, Relief Temperature	-26.27°F	Simulation Software
P_2, Absolute Operating Pressure of Tank (includes 10% overpressure) - Absolute Rate of Pressure Change	14.5 psia	$P_{relief} - d_{patm}/dt$
ρ_1, Density of vapor at P_1 and dew point Temperature	0.0554 lb/ft^3	Simulation Software
ρ_2, Density of vapor at P_2 and dew point Temperature	0.0540 lb/ft^3	Simulation Software
Q, Ambient Heat Input	680000 Btu/hr	Site General Arrangement
ΔH_{vap}, Heat of Vaporization	607.9675 Btu/lb	Simulation Software
r, Radius of head	128 ft	Site General Arrangement
h, Height of head	28.72 ft	Site General Arrangement
d_{patm}/dt, Absolute Rate of Pressure Change	0.4 psi/sec	[4]
$\dot{W}$, Total relief rate due to vapor expansion, flashing & normal boil-off	Calculate	
$\dot{W}_{AB}$, Relief rate due to normal boil off	Calculate	
$\dot{W}_{AL}$, Relief rate due to flashing	Calculate	Simulation Software
$\dot{W}_{AG}$, Relief rate due to vapor expansion	Calculate	
V_{head}, Volume of Head/Dome section	Calculate	

Calculation:

- The volume of the head/dome section is:

$$V_{head} = \frac{\pi}{6}(3r^2 + h^2)h = \frac{\pi}{6}(3(128ft)^2 + (28.72ft)^2)28.72ft$$

$$V_{head} = 751539\ ft^3$$

- The drop in atmospheric pressure results in a change in density in the vapor space. Hence, the change in density caused by an atmospheric pressure drop results in vapor expansion which requires relief and is described in Equation 3 as:

$$\dot{W}_{AG} = V\frac{(\rho_1 - \rho_2)}{1\ sec} = 751539\ ft^3 \frac{\left(0.0554\frac{lb}{ft^3} - 0.0540\frac{lb}{ft^3}\right)}{1\ sec}$$

$$\dot{W}_{AG} = 1052\frac{lb}{sec}\ \text{of NH3 Vapor at} - 26.27°F\ \text{and}\ 14.9\ psia$$

- A flash simulation was performed to determine the amount of vapor generated due to the flashing of liquid in the storage tank caused by an atmospheric pressure drop. The inlet stream was evaluated at the relieving pressure of the tank (14.9 psia) and the corresponding saturation temperature (-26.27 °F). The flash unit was set to a pressure drop of 0.4 psia and was operating adiabatically (Duty = 0 Btu/hr).

$$\dot{W}_{AL} = 804\ \frac{ft^3}{sec}\ \text{at} - 27.70°F\ \text{and}\ 14.5\ psia$$

The equivalent mass flow rate at relieving conditions was obtained from simulation software:

$$\dot{W}_{AL} = 804\ \frac{ft^3}{sec}\Bigg|0.0554\frac{lb}{ft^2}$$

$$\dot{W}_{AL} = 45\ \frac{lb}{sec}\ \text{of NH3 Vapor at} - 26.27°F\ \text{and}\ 14.9\ psia$$

- The normal boil-off rate is calculated dividing the ambient heat input by the heat of vaporization.

$$\dot{W}_{AB} = \frac{Q}{\Delta H_{vap}} = \frac{680000\frac{Btu}{hr}}{607.9675\frac{Btu}{lb}} = 1119\frac{lb}{hr}$$

$$\dot{W}_{AB} = 0.31\ \frac{lb}{sec}\ \text{of NH3 Vapor at} - 26.27°F\ \text{and}\ 14.9\ psia$$

- Total flow rate is the sum of that caused by vapor expansion, flashing and normal boil-off:

$$\dot{W} = \dot{W}_{AG} + \dot{W}_{AL} + \dot{W}_{AB} = 1052\frac{lb}{sec} + 45\frac{lb}{sec} + 0.31\frac{lb}{sec}$$

$$\mathbf{\dot{W} = 3.95 \times 10^6 \frac{lb}{hr} \text{ of NH3 Vapor at } - 26.27°F \text{ and } 14.9 \text{ psia}}$$

6.2 Required Orifice Area

Data and Constants:

		Reference
$A_{(in2)}$, Orifice Area	Calculate	[5]
F, Subsonic Flow Factor	Calculate	[5]
$k = C_p/C_v$	Calculate	Simulation software
K_d, Subsonic Valve Coefficient	Calculate	[5]
M, NH3 Molecular Weight	17.0306	Simulation software
Z, Compressibility Factor	0.9853	Simulation software

Calculations:

$$A_{(in^2)} = \frac{W\sqrt{TZ}}{735K_dP_1F\sqrt{M}} \rightarrow T = T_{relief}$$

where

$$K_d = 0.650\left(\frac{P_2}{P_1}\right)^{-0.349} \rightarrow P_1 = P_{relief} \text{ and } P_2 = P_{atm}$$

$$K_d = 0.650\left(\frac{13.8}{14.9}\right)^{-0.349} = 0.6676$$

$$k = \frac{C_P}{C_V} = \frac{0.4868}{0.3643}$$

$$k = 1.3363$$

$$F = \sqrt{\frac{k}{k-1}\left[\left(\frac{P_2}{P_1}\right)^{\frac{2}{k}} - \left(\frac{P_2}{P_1}\right)^{\frac{k+1}{k}}\right]}$$

$$F = \sqrt{\frac{1.3363}{1.3363-1}\left[\left(\frac{13.8\ \text{psia}}{14.9\ \text{psia}}\right)^{\frac{2}{1.3363}} - \left(\frac{13.8\ \text{psia}}{14.9\ \text{psia}}\right)^{\frac{1.3363+1}{1.3363}}\right]} = 0.2602$$

$$A_{(\text{in}^2)} = \frac{3.95 \times 10^6 \frac{\text{lb}}{\text{hr}} \sqrt{433.40°\text{R} \cdot 0.9853}}{735 \cdot 0.6676 \cdot 14.9\ \text{psia} \cdot 0.2602\sqrt{17.0306}}$$

$$\mathbf{A_{(in^2)} = 10396\ in^2}$$

Results:

Assuming that vapor-liquid equilibrium could be instantly established, 3.95×10^6 lb/hr of NH3 Vapor would have to be relieved if a tornado came in direct contact with 100-T, Ammonia Storage Tank, causing a rapid decrease in atmospheric pressure. The corresponding required orifice area is 10,396 in^2. Therefore, the system would require 92 POPRVs as each POPRV has an orifice area of 113 in^2.

397003 Cultura De Liderazgo y Seguridad – Independientemente Del Idioma

Abstract:

La seguridad en el ámbito de trabajo es vital para el bienestar y sostenibilidad de cualquier negocio. No se trata de un requerimiento, o meramente un lineamiento corporativo; sino es algo que debe ser abrazado por el personal y su liderazgo en cualquier sector productivo. Una vez dicho esto, prevalece una gran diversidad en regiones y ambientes de trabajo, lo cual puede complicar el proceso de crear e implementar una cultura de seguridad común. Dado que los empleados son originarios de diferentes geografías y culturas traídas de sus empleos anteriores, la definición del término "seguridad en el lugar de trabajo" se hace subjetiva. En adición a la variedad de culturas; pueden confluir lenguajes y dialectos en grupos de trabajo por lo que la barrera lingüística frecuentemente juega un importante rol.

Este papel discutirá como un liderazgo de cultura segura puede considerar las diferencias lingüísticas para promover una cultura de seguridad común; al momento de implementar un plan de seguridad para una fuerza laboral diversa. Las discusiones enfatizaran aprendizajes sobre la curva Bradley y demostrarán la importancia de seguir prácticas de trabajo seguras para alcanzar aproximarse a un área 100% libre de accidentes. Ejemplos reales de regiones de trabajo internacionales y equipos de trabajo diversos se presentaran para subrayar esta necesidad y compartir lecciones aprendidas.

398031 Ensure Compliance to NFPA Standards through Corporate Level Combustible Dust Audits

Abstract:

The Occupational Safety and Health Administration (OSHA) CPL 03-00-008 Combustible Dust National Emphasis Program (reissued) has been in effect since March 11, 2008. The traditional strategy of regulating only major industry sectors has long passed. Currently industries handling hazardous combustible solid particulates/fibers, including small manufacturing businesses, are subject to Instruction compliance. Industries covered by this Instruction must also comply with applicable OSHA and related National Fire Protection Association (NFPA) Standards. Managing compliance is a necessity but can sometimes be perceived as an overwhelming task. After conducting a number of corporate audits for compliance to the combustible Dust NEP, some common technical and administrative themes will be shared that could help an EH&S audit group and initiate a revision of their current protocols especially for legacy facilities and new design/construction.

Using NFPA 654 as the reference standard, audit protocol items related to facility dust hazard assessments will be presented. These include but are not limited to: dust control, ignition control, damage control, and administrative issues such as training for new and veteran affected employees and contractors, Management of Change, Process Hazard Analysis (PHA), Inspections and Preventive Maintenance, and Safety Data Sheets. Reports from individual facility findings and recommendations should follow a standard format, and allow for local management to obtain a clear understanding of the issues related to their specific site and compare findings with plants of similar technology. In legacy facilities, which were built or acquired prior to current combustible dust regulations, there are additional unique findings that are specific to industry type, age of the facility, and vendor installed equipment.

Corporate audits can be planned to identify common strengths and opportunities across the organization to allow senior management to understand the prioritized recommendations in a systematic action plan based on safety needs. A "report card" has been developed to provide an overview of a corporation's combustible dust current program, which compares actual compliance to Instruction guidance and NFPA standards. Common field trends will be presented as well as guidance for establishing and properly maintaining engineering and administrative systems for a safer combustible dust program.

Combining Traditional UOP HAZOP Analysis With Dynamic Simulation A New Process Safety / Risk Assessment Tool

Scott M. Wozniak
UOP, A Honeywell Company
25 E. Algonquin Road, Des Plaines, IL. 60017-5017
Scott.Wozniak@Honeywell.com

Bill Wiede
UOP, A Honeywell Company
25 E. Algonquin Road, Des Plaines, IL. 60017-5017
Bill.Wiede@Honeywell.com

Prepared for Presentation at
American Institute of Chemical Engineers
2015 Spring Meeting
11th Global Congress on Process Safety
Austin, Texas
April 27-29, 2015

UNPUBLISHED

Combining Traditional UOP HAZOP Analysis With Dynamic Simulation A New Process Safety / Risk Assessment Tool

Scott M. Wozniak
UOP, A Honeywell Company
25 E. Algonquin Road, Des Plaines, IL. 60017-5017
Scott.Wozniak@Honeywel.com

Bill Wiede
UOP, A Honeywell Company
25 E. Algonquin Road, Des Plaines, IL. 60017-5017
Bill.Wiede@Honeywell.com

Keywords: HAZOP, Dynamic Simulation, Process Safety, Risk Assessment and Flowsheets

Abstract

There is a clear link between safety and reliability in process design and operation. The traditional Hazard and Operability (HAZOP) analysis identifies potential operational failures and weak points in a process design; evaluates risks introduced by them and develops effective risk reduction strategies before a customer takes delivery of a process design.

This paper shows how our current HAZOP methodology is supported by dynamic simulation and modeling techniques. By leveraging this process technology, more robust HAZOP studies result by integrating our current UOP Process Technology library of **UNISIM®** flowsheet templates in all phases of the HAZOP work process.

We will show how dynamic simulation can determine consequences of deviations from normal design exactly and how this tool can potentially identify new unknown hazards and safety issues.

1. Introduction

Process Safety can help provide better assurance for communities and for the refinery, petrochemical and chemical plants through a HAZOP study, a method of identifying both deviations to process design that could cause potential hazards, and safeguards required to minimize potential hazards.

The HAZOP study is traditionally a qualitative study. To give it more quantitative character dynamic simulation is sometimes used to complement the HAZOP study. Combining HAZOP with dynamic simulation provides the means for investigating (and demonstrating) the consequences of deviations from normal operating conditions. This paper shows a few examples using dynamic simulation in a HAZOP.

2. HAZOP

Companies and the public are increasingly asking for better assurance that sites and communities will not be adversely impacted by fires, explosions and releases from oil refineries and chemical plants. The banking and the insurance industries are also increasingly asking for this assurance. Process Safety can help provide that assurance through a HAZOP study, a method of identifying both deviations to process designs that could cause potential hazards, and the safeguards required to minimize potential hazards. The HAZOP study is recognized worldwide as a primary methodology for hazard identification conducted for the oil, petrochemical and chemical industries for satisfying two basic requirements. The first is checking a new design for safety and operability problems whilst it is at the process instrumentation diagram stage. The second is the identification of latent safety and operability problems in existing plant which have yet to be revealed during operations.

A HAZOP study provides benefits to the owner and contractor as it:

- Identifies improvements for the safe operation of the process unit at an earlier stage in the project making it easier and usually significantly less expensive to make those changes.
- Reduces the chances of unplanned shutdowns.
- Significantly reduces time and costs for future HAZOP studies due to changes made to the process unit during construction or later revalidations (a government requirement in some parts of the world) with the use of the electronic version of the HAZOP study report.
- Helps answer questions during the training of operators and maintenance personnel about deviations or unusual scenarios that may occur in the operation of the process unit.
- Provides information for developing process-unit-specific operating and maintenance procedures.
- Identifies links to process equipment "outside battery limits"
- Provides guidance for developing mechanical integrity programs, including information required by the ANSI/ISA (USA) or IEC 1508 (Europe) instrumentation standards.
- Contributes toward demonstrating to the communities that potential hazards have been identified.

3. HAZOP Study

3.1 HAZOP Methodologies

UOP HAZOP studies are well documented and include all of the supporting information used during the HAZOP sessions in a way that is useful to the owner and the contractor.
HAZOP studies of UOP process units are conducted with a team of experts including a Chairman/HAZOP Team Leader, Process Design Engineer, Process Engineer, Technical Service Engineer, Control Engineer, and the Project Engineer. Specialists to be consulted on specific points may include: Electrical Engineer, Piping Engineer, Research Scientist, Civil Engineer, Chemist, and the Flowsheeting Specialist, as necessary (Figure 1).

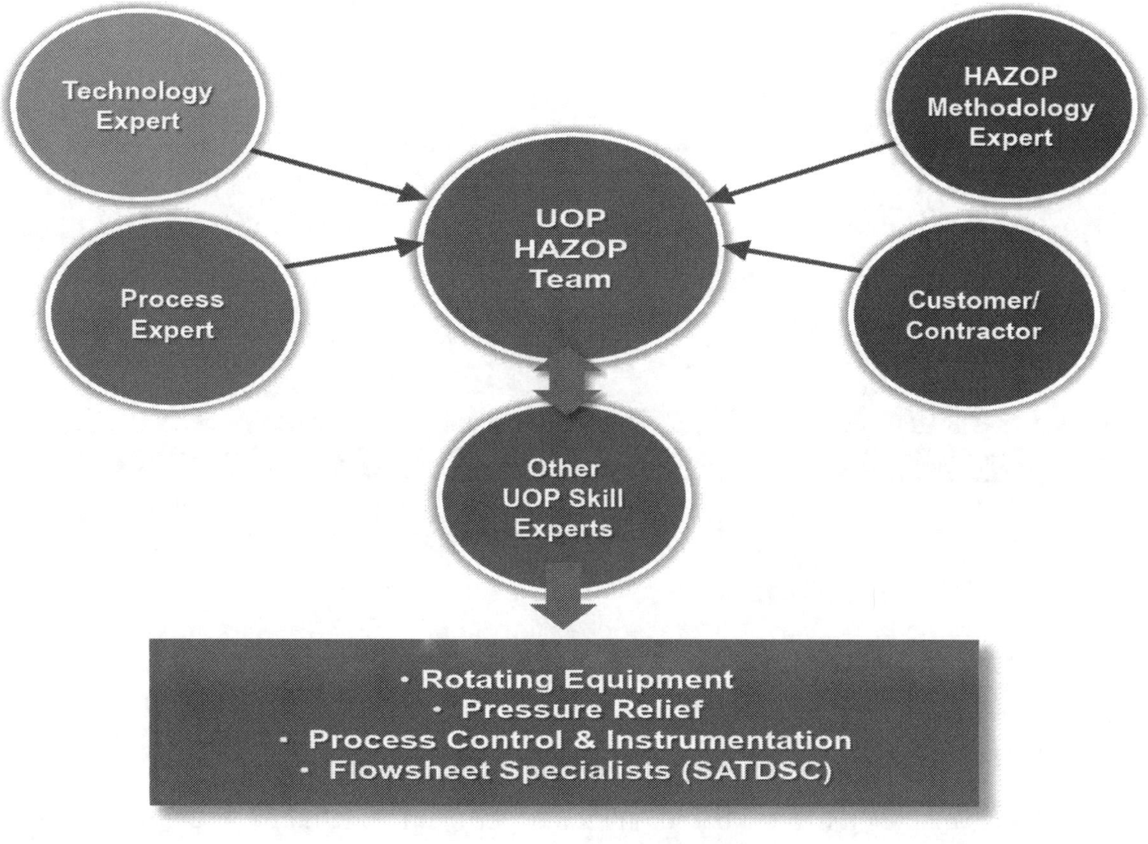

Figure 1. HAZOP Team

HAZOP is carried out as a structured study. The method relies on using guidewords (such as, no, more, less) combined with process parameters (e.g., temperature, flow, pressure) that aim to reveal deviations (such as less flow, more temperature and so on) of the process intention or normal operation. HAZOPs are best to be done early in the project. Figure 2 shows the cost of changes vs. project life cycle and one can see the ability to influence changes decreases over time while the cost of changes increases over time as the project progresses. The potential for adding value to the products of a project are obviously highest during the conceptual phase of the project and lowest during the finishing phase. Between these two extremes, the curve tends to follow a reverse "S" curve as shown in the figure. Conversely, the cost of making changes is lowest in the planning phases, but rises more and more steeply as the project progresses through the two production phases. In construction, for example, it has been suggested that the cost to make a change, or fix a safety issue, increases by ten times through each succeeding major phase.

If the Cost-to-Change curve is considered in conjunction with that of Adding-Value, the implications to management decision making become readily apparent. The intersection of the two curves probably represents the point at which a change in scope changes from a constructive opportunity into a destructive intervention.

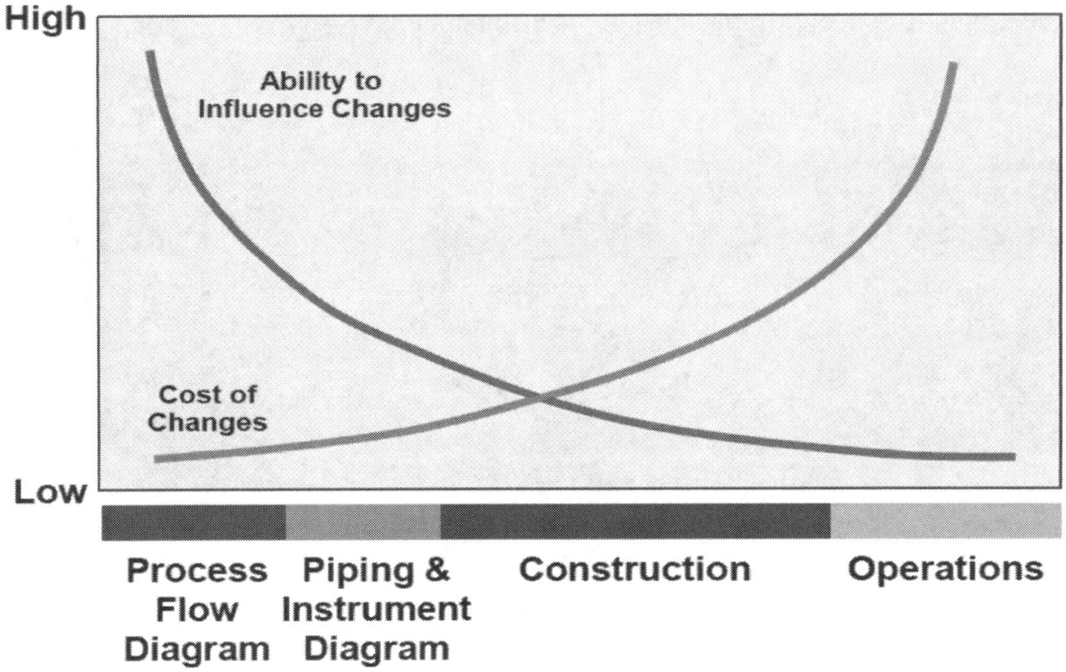

Figure 2. Cost of Changes vs. Project Life Cycle

3.2 HAZOP Consequences

The consequences are then risk rated based on their severity and likelihood. Figure 3 shows a Risk Ranking Matrix that is used in all of our HAZOP reviews. The Risk Ranking Matrix has four levels of risk which are the following:

1. Risk Ranking D: Risk not acceptable
2. Risk Ranking C: Risk acceptable with additional administrative or engineering controls which lower the risk ranking to "A" or "B"
3. Risk Ranking B: Risk acceptable with additional administrative or engineering controls
4. Risk Ranking A: Risk acceptable even if no further action taken.

If during the HAZOP a Risk Ranking of "D" or "C" is determined by the HAZOP team the HAZOP leader may engage the Flowsheeting Specialist to see if a dynamic simulation could determine a better understanding of the issue and the severity. The dynamic simulation may determine that the severity may not be as severe as the HAZOP team determined by qualitatively assessing the severity quantitatively and potentially move the Risk Ranking to an "A" or "B".

Severity:

1. Catastrophic (Death or sever Occupational illness; Major environmental impact; Production loss of more than six months; Facility/equipment damage of more than $10MM)
2. Severe (Injury or moderate occupational illness; Serious environmental impact or permit violation; Production loss of between one month and six months; Facility/equipment damage between $1MM and $10MM)
3. Moderate (Minor injury or occupational illness; Minor environmental impact; Production loss of between one week and one month; Facility/equipment damage between $100M and $1MM)
4. Low (No injury or occupational illness; No environmental impact; Production loss of less than one week; Facility/equipment damage of less than $100M)

Likelihood:

1. Frequent (Potential to occur frequently, many times per year)
2. Occasional (Potential to occur occasionally, once per year)
3. Moderate (Potential to occur under unusual circumstances, once or twice within the facility life time)
4. Unlikely (Could possibly occur or known to have occurred within the petroleum/petrochemical industry, but not likely over the facility lifetime)

Ranking:

Likelihood \ Severity	1	2	3	4
1	D	D	C	A
2	D	C	B	A
3	C	B	A	A
4	B	A	A	A

Risk Ranking	Definition
D	Risk not acceptable
C	Risk acceptable with additional administrative or engineering controls which lower the risk ranking to "A" or "B"
B	Risk acceptable with additional administrative or engineering controls
A	Risk acceptable even if no further action taken
NR	Not Ranked (insufficient information)

Figure 3. Risk Ranking HAZOP

The application of the HAZOP technique to a detailed process unit design is a complex and time consuming task. UOP has created a model where the HAZOP worksheets are pre-populated by an experienced HAZOP leader who during consultations with our skill experts will allow us to make the HAZOP reviews more efficient and less time consuming during the HAZOP meetings.

3.3 Using Dynamic Simulation in UOP HAZOP Reviews

In the past 30 years, much research effort has been dedicated to the development of the dynamic analysis. The dynamic analysis approach for a HAZOP of a process unit seems to be the most straightforward procedure which gives quantitative results.

Usually, the HAZOP analysis does not consider the duration and magnitude of the deviations generated during the operation. However, what exactly does the deviation 'less flow' mean: 80% or 10% of the usual operation value? Does the deviation occur as an immediate (step) decrease of the flow lasting five minutes or more, or is it only an impulse? Is this decrease continuous at some rate? Answers to these questions can be obtained using dynamic simulation which is one of the newer features. Consider the example where a valve is left open on a line going to a storage tank. Dynamic simulation will assist on the duration of how long it will take to overfill that tank even if there are multiple inlets to the storage tank. In dynamic simulation the extent of the deviations can be easily incorporated and possible consequences investigated.

The dynamic analysis serves for analyzing the time and direction of shifting from one steady state to another one due to a failure deviation. It is also very useful for the investigation of dynamic behavior of the system with respect to the time and length of the failure. The results of safety analysis can be used in HAZOP studies. HAZOP, with dynamic simulation, has the potential to become a very practical and robust tool. We will illustrate this by presenting examples of dynamic simulation for conventional column on loss of reflux, for UOP's Phenol™ Process operating pressure evaluation.

However, it is important to point out that the implementation of dynamic simulation for any chosen system is dependent on an adequate steady state model that is dependent on physical properties, kinetics, mass and heat transfer behaviors. During a HAZOP study, possible deviations are generated by rigorous questioning, prompted by a series of standard "guidewords" applied to the intended design. After the guideword is matched with a parameter, a deviation is generated. The next step in the HAZOP study is to look for potential consequences. At this moment, the application of dynamic analysis is useful in order to find quantitative consequences.

The HAZOP study is traditionally a qualitative study. By definition, a dynamic simulation is the imitation of the operation of a real-world process or system over time, which means that in principle it should be a realistic way of representing an actual process. Combining HAZOP with dynamic simulation does provide the means for investigating and demonstrating the consequences of deviations from normal operating conditions. Dynamic simulation now provides the UOP HAZOP team the ability to quickly investigate and test the effectiveness of various suggested strategies dealing with high risk situations (intolerable levels of risk). Figure 4 shows the magnitude of risk from intolerable risk to ALARP "as low as reasonably practicable" to tolerable risk. To move from an intolerable risk to ALARP decreases the risk with increasing cost. In process safety we cannot accept intolerable risk so one must move from the intolerable level of risk to a risk level of ALARP or tolerable risk. In the past this would involve increasing cost, but we will demonstrate that by using dynamic simulation we may be able to move from an intolerable risk to ALARP by showing a less severe consequence with no significant additional cost.

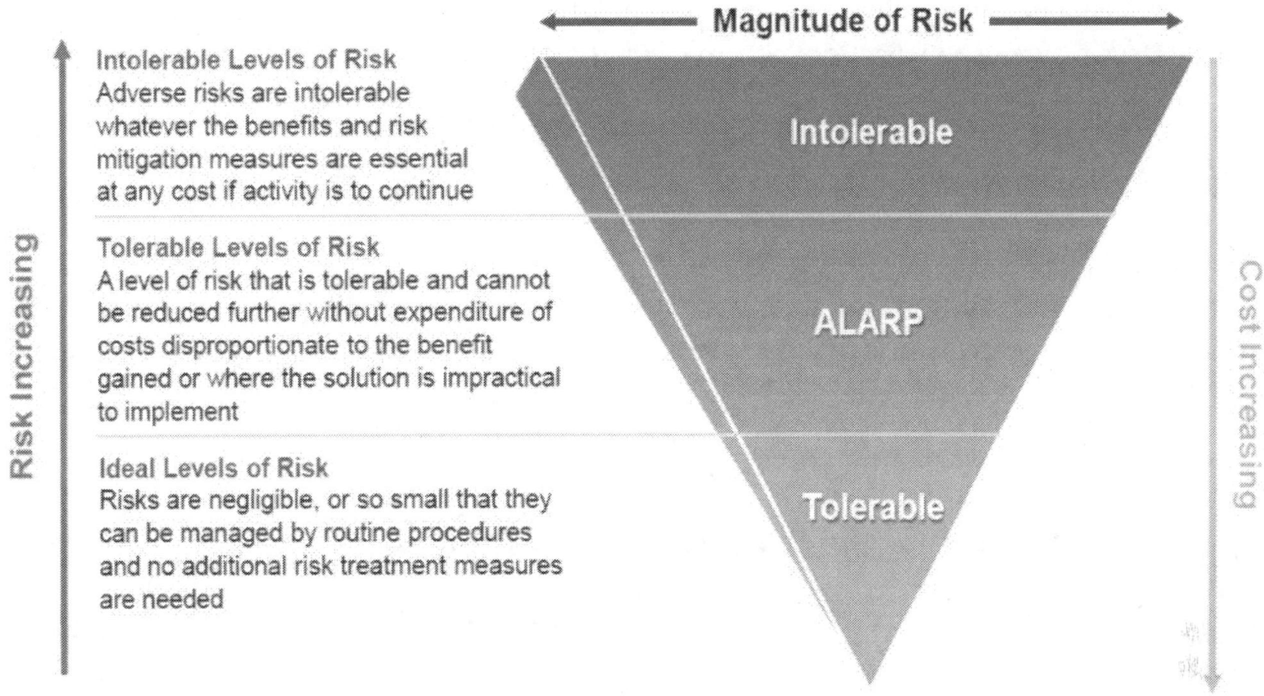

Figure 4. ALARP

By having a dynamic simulation of the process as a support tool, an extensive, easier and more complete study can be accomplished. That provides a systematic screening of process deviations associated with possible high risk events, determining the threshold values that may lead to such events and enabling the examination of a particular design for the adequate safe range of operation. Dynamic simulation should be seen as a tool that complements the traditional HAZOP procedure, it does not replace it. There are still many processes that cannot be modelled accurately enough due to a lack of enough quantitative information, particularly in emergency situations.

Process safety is paramount in the secure and viable operation of any refinery, petrochemical and chemical process plant. Many techniques and methodologies have been defined over the years to undertake steps that will ensure a safe operating environment at a chemical facility. However, until now, process simulation has found very little or no use in safety-related studies such as HAZOP. In this paper, we introduce a systematic framework based on a quantitative HAZOP (that is, HAZOP supported by dynamic simulation related to process malfunctions.) The quantitative HAZOP differs from standard HAZOP in documenting results, classification of frequency and consequences of process deviations, and application of a risk potential matrix (Risk Ranking Matrix).

Dynamic simulations can be used to support the HAZOP in the following ways:

- An important part of the HAZOP procedure is to think of the consequences of a certain deviation

- For complex and nonlinear systems, it is not straightforward to assess the consequences. Hence, the use of dynamic simulation with the deterministic models can be helpful in assessing the effect of faults on the operations and dynamics of the process.

- The purpose of the methodology is to determine the effect of operational disturbances on the safety of the plant and devise ways to reduce the risk of the consequences.

HAZOP can provide qualitative answers regarding the magnitude of the deviations that will lead to severe consequences. It cannot provide the time it takes to reach a "no return" stage of an accident after the deviation has occurred, or the action that can be taken in order to prevent the accident. Dynamic simulation can provide this information that a traditional HAZOP could not provide. Dynamic simulation in HAZOP can provide quantitative assessment of the consequences of abnormal operating conditions.

4. Dynamic Simulation and HAZOP Analysis

4.1 Background

Steady-state process templates (flowsheets) are used extensively in the engineering design workflow for all UOP process technologies. UOP engineers utilize the Honeywell **UNISIM®** Design Suite of process modeling tools (See References [1-3]) in the preparation of a process design package to our customers called Schedule A. This suite of tools allows for very robust development of steady state and dynamic process models.

In the process phase of the Schedule A workflow we have developed a large database of validated steady-state process templates for most product offerings in the refining and downstream petrochemical areas (see Figures 5 and 6). Modifications to these templates are made based on customer specific requirements. All the equipment and piping associated with a project are sized based on flow rates calculated from steady-state heat and mass balances which interface all of our engineering sizing tools. Simulation results are downloaded to our standalone hydraulic tools to determine line sizes and generate tabulation reports. If needed, various parts of the steady-state template can be checked using the equivalent dynamic simulation template.

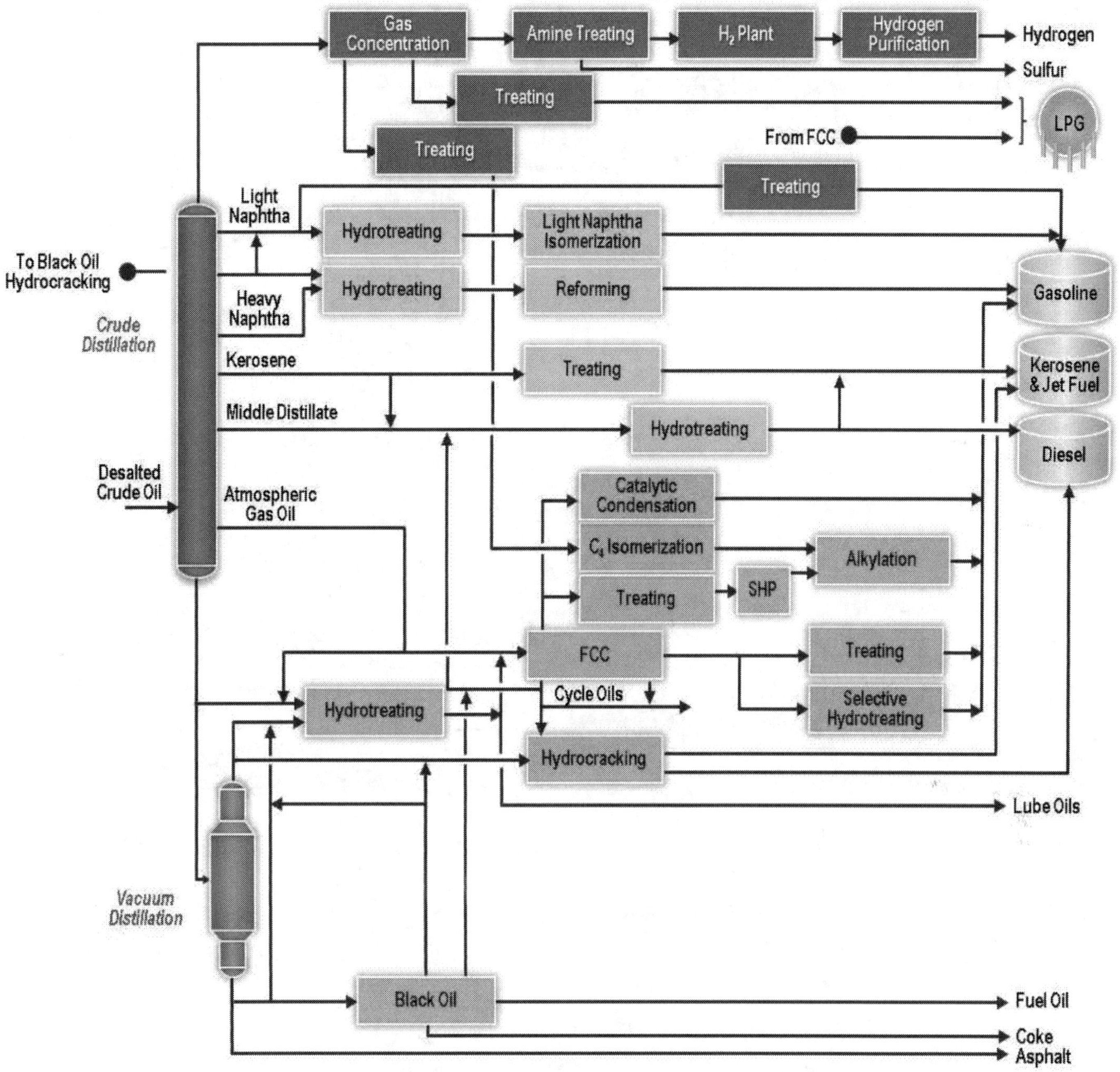

Figure 5. UOP Refining Processes

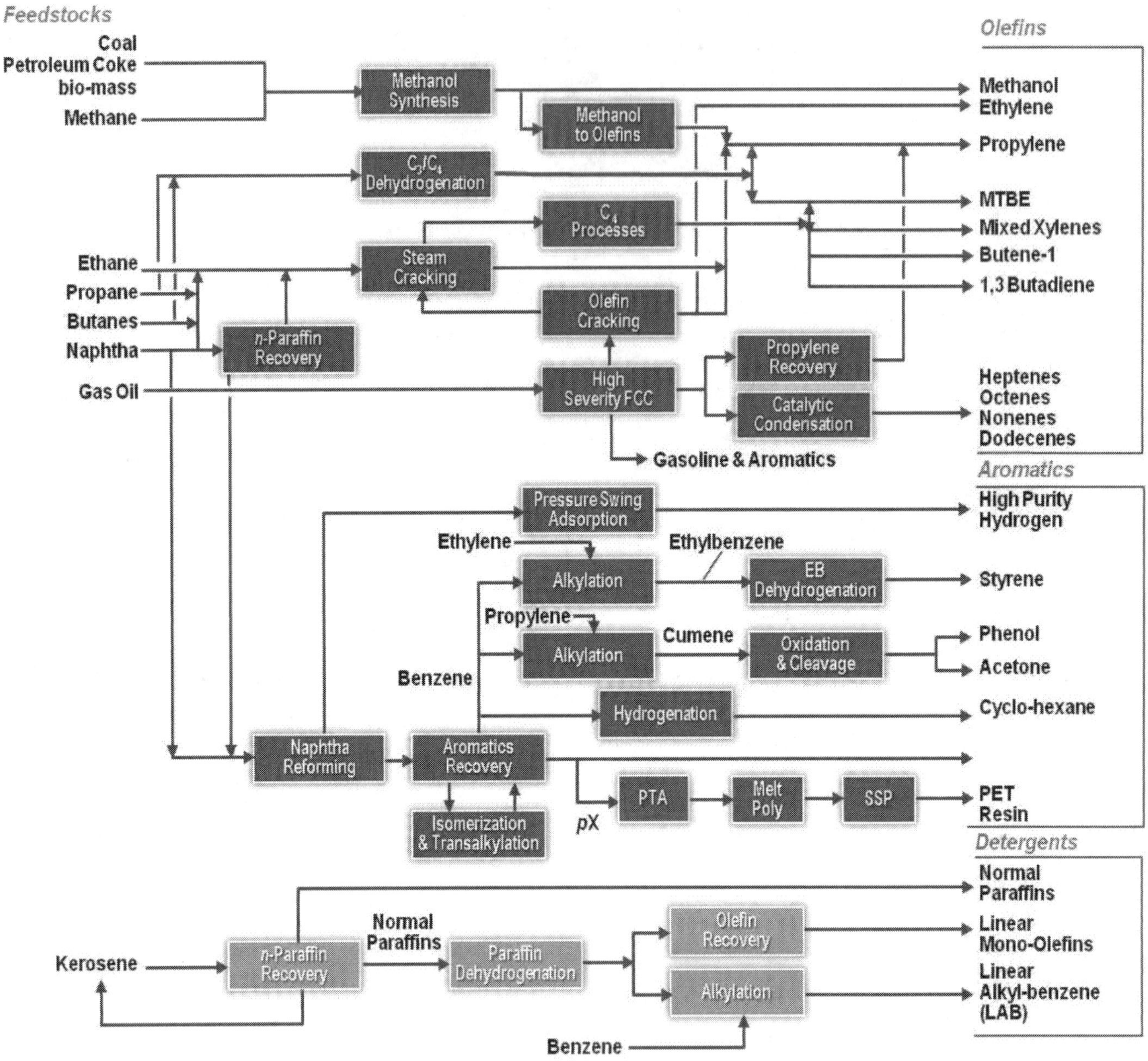

Figure 6. UOP Petrochemical Processes

4.2 Dynamic Simulation Templates

Transitioning the converged steady template to a dynamic simulation template requires a different approach towards specifying unit operations. In our implementation of dynamic templates a sequence of important actions are required to insure a successful high fidelity dynamic analysis:

1) Converge a realistic steady-state simulation of the problem to be studied.
2) Add valves and pipes to ensure pressure-driven flow between unit operations.
3) Set terminal feed and product pressures.

4) Determine approximate or include exact dimensions of all process equipment.
5) Specify sizing data for all pipe runs (lengths, diameters, etc.). Size the CVs.
6) Include nozzle elevations for all valves, piping and process equipment.
7) Add all necessary controllers (FICs, TICs, etc) to all control loops.
8) Run the dynamic simulation: tune the controllers, check pressure drops, and line out the model to match the original steady-state performance.
9) Add events to the Event Scheduler to model upsets and other conditions.

The above steps need to be completed only once and serve as the starting point for dynamically evaluating all normal and upset operating scenarios.

Each of our dynamic models is preconfigured to include upset conditions, strip charts and tabulate historical data. In **UNISIM®**, the event scheduler is used to model the preconfigured dynamic upset scenarios. Additional upsets are also manually activated as a dynamic run is proceeding during time integration.

Over the years we have established a strategy and timetable to implement dynamic simulation methods into our current UOP process work flow. We use dynamic process templates to supplement the work as needed during the Schedule A projects. Various sections of these dynamic templates are used to identify specific design issues for each project (e.g. a project HAZOP review, emergency shutdown events, customer inquiries, etc.) that normally would not be recognized through the use of steady state simulators.

Sections of the dynamic templates are utilized to explore transient behavior around individual equipment, e.g. reactors, columns, etc. There are many different levels of fidelity possible when creating dynamic flow sheets. In UOP engineering we use dynamic templates that implement the basic regulatory control schemes shown on all our process flow diagrams for all UOP process units.

From a full dynamic template of a UOP process unit, the process engineer is able to break down the process flows and create smaller process snapshots for analysis. To consider and evaluate time dependent behavior in our Schedule A work process, a separate group of flowsheeting skill specialists are utilized on Schedule A projects to assist in preparing and designing these dynamic flowsheets.

We have realized substantial benefits in leveraging our expertise in steady-state design modeling to include dynamic design modeling as well. We have recently included our flowsheeting specialists to support our internal HAZOP team reviews. This support includes the preparation and evaluation of dynamic models as needed from the working HAZOP worksheets. Supporting data (e.g. strip charts, tabular historical data of key events / new hazards / safety issues) from these dynamic runs are included in the reports. Additional dynamic runs are made to address any customer concerns after the reports are issued.

With the use of our dynamic templates from the steady-state design templates we can now quantitatively test the effectiveness of proposed strategies for emergency situations. Furthermore, dynamic analysis has helped us identify new unknown hazards and safety issues.

We can now accurately determine the consequences of any deviations from normal design. We have also included dynamic analyses in our HAZOP reviews of new technologies proposed by the R&D Process Design and Development groups.

4.3 Dynamic HAZOP Analysis

Over the years we have performed many dynamic analyses to support our internal HAZOP reviews. Most of these are very detailed and require specific customer information such as "as built" equipment and line size data as well as detailed valve performance data. Other flowsheets are more simply configured and are used occasionally in our HAZOP analysis to confirm and/or support the qualitative conclusions normally encountered during the reviews. The following examples illustrate some process scenarios typically encountered during HAZOP reviews.

4.3.1 Conventional Column – Loss of Reflux

The majority of process units utilize two and three phase conventional separation columns. Figure 7 shows a conventional two phase column with a full reflux condensing circuit with vapor overhead product and a vertical thermosiphon reboiler.

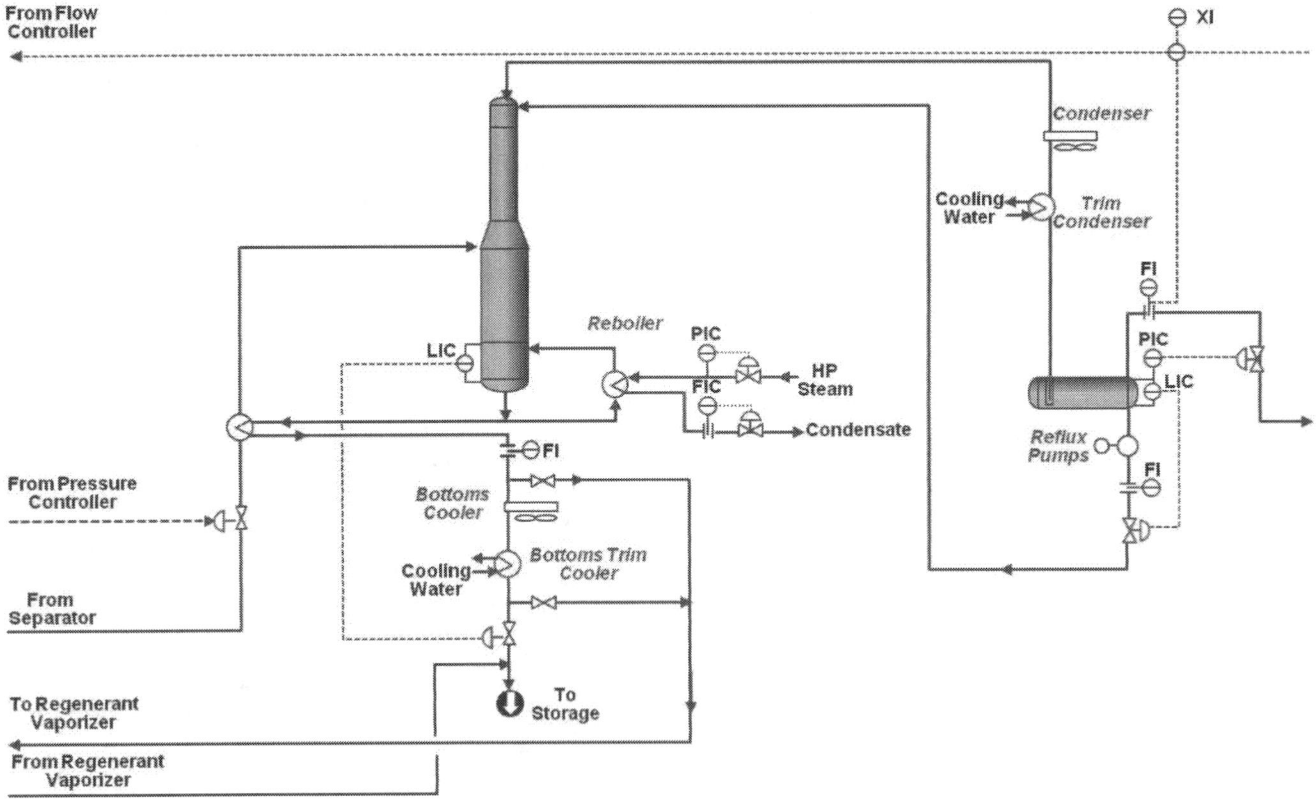

Figure 7. Conventional Column

In a typical HAZOP analysis, there could be potential hazards associated with the installation of engineer controlled emergency isolation valves. In those situations, an evaluation will be conducted to determine whether the installation of the engineer controlled emergency isolation valves provides the safer option for personnel, the environment or surrounding equipment than if these valves were not installed. As shown in Figure 8, these isolation valves can be local or remote manually operated and are typically not interlocked to other instrumented devices.

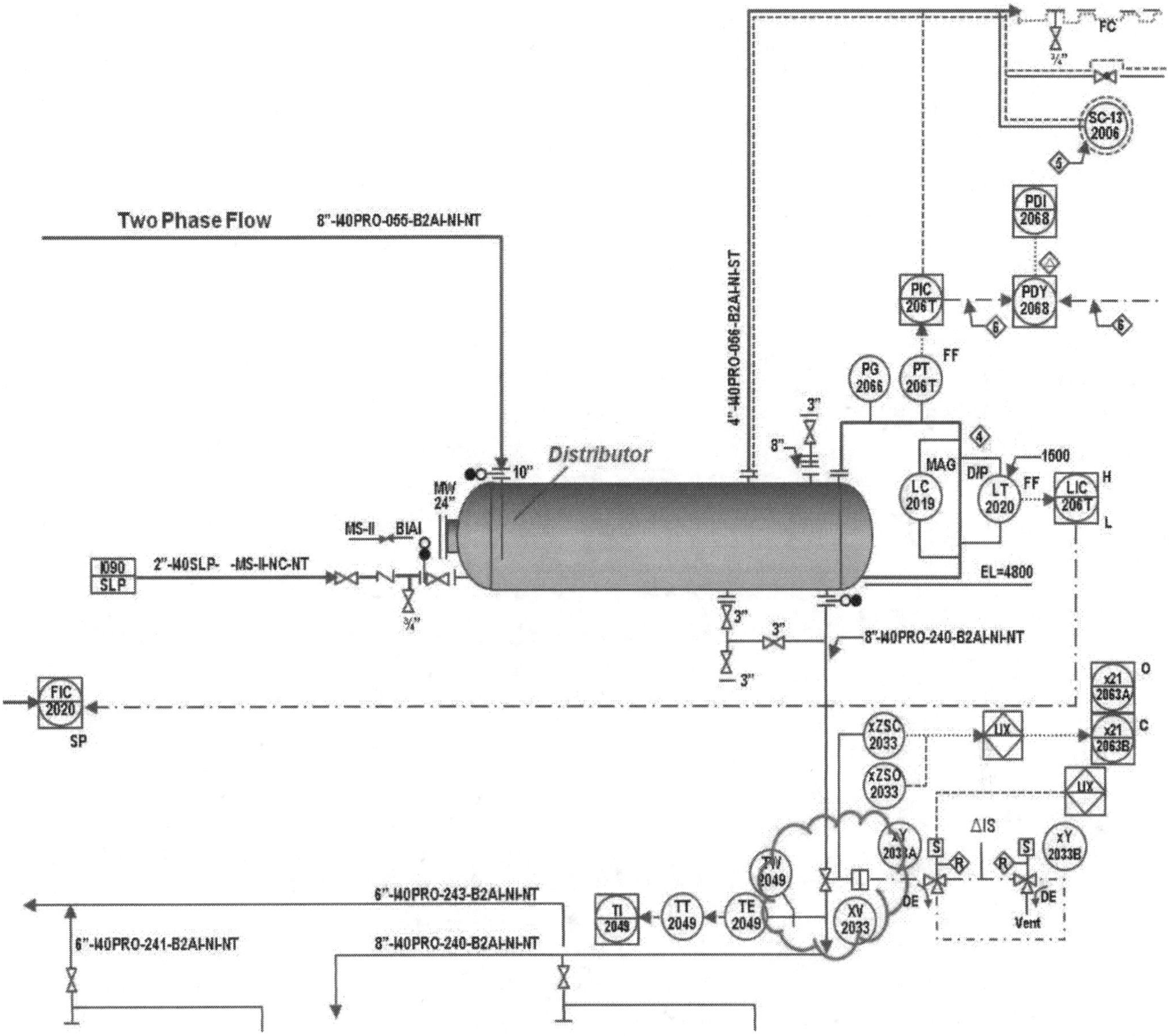

Figure 8. Expanded Piping and Instrument Diagram View – Column Receiver

From our database of **UniSim®** templates, a dynamic model in of the conventional column shown in Figure 9 follows our standard process flow diagram view of equipment, piping and basic piping and instrument diagram control designations. This model has event schedules preconfigured with a number of upset scenarios. If needed, adhoc scenarios can easily be added to the model during the HAZOP team analysis.

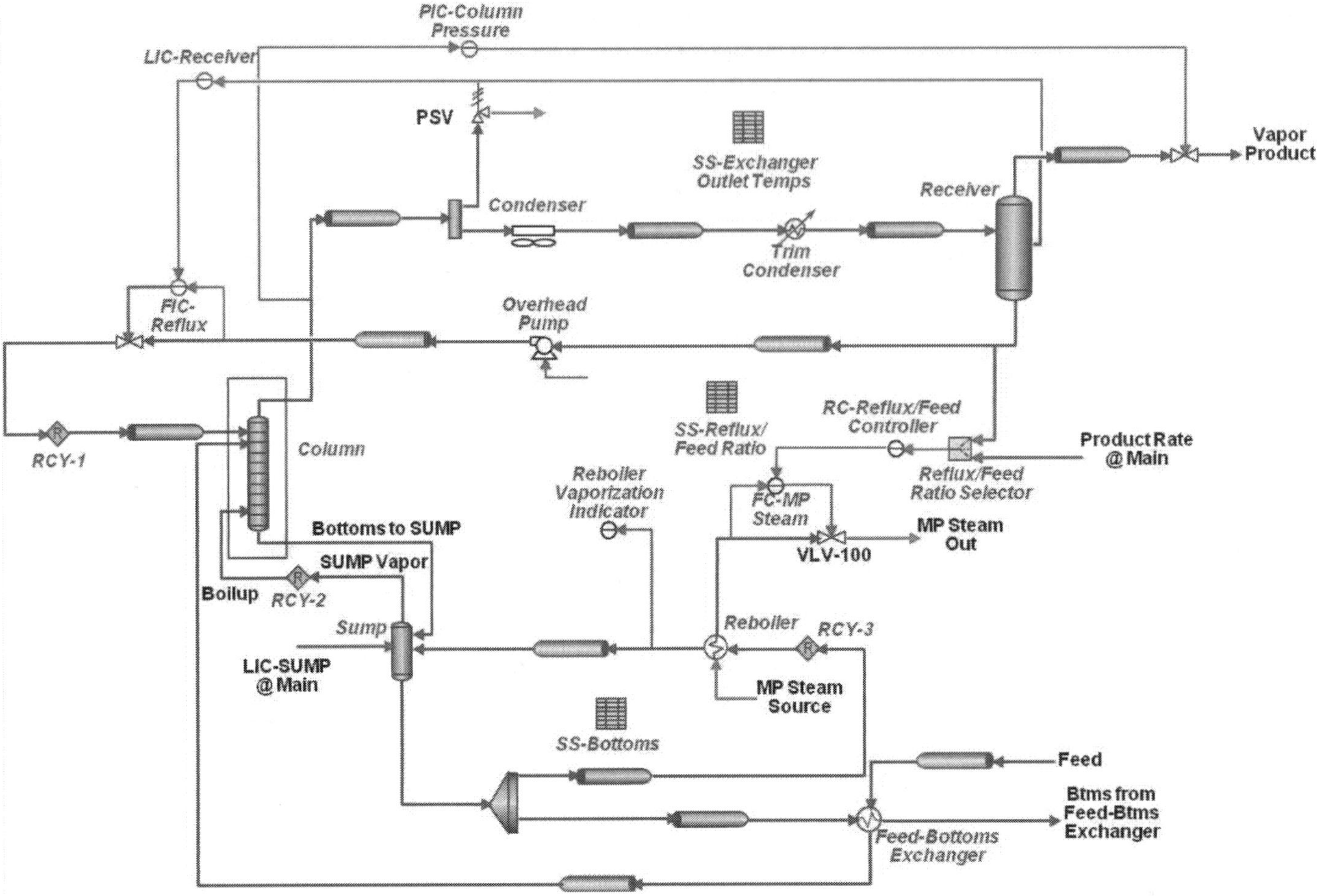

Figure 9. UNISIM® Dynamic Conventional Column Model

Figures 10 through 12 illustrate via strip charts, a time sequence of events when a manually operated isolation valve is inadvertently closed during normal operation. First the column is lined out to steady state conditions; then the isolation valve is triggered to fully close. As Figure 10 indicates, the overhead vapor from the column immediately spikes upward then moves downward rapidly as it is being displaced by the increasing liquid level in the receiver. After 3.5 minutes the liquid level completely fills the receiver and overflows through the vapor line at a large rate (see Figure 11). Also, with no reflux flow returning to the column, the sump liquid level in the column starts to decrease. As indicated in Figure 12, the liquid level in the sump ultimately drops to zero after another 7.5 minutes. During this upset event the column operating pressure increased close to the pressure safety valve set pressure but did not open.

Obviously during normal operation, the appropriate trip activation or operation actions in response to safety alarms would have been in place to mitigate any severe operating scenarios such as the one performed in this simple example. As this example shows, the continued use of dynamic simulation in our HAZOP reviews has allowed us to quantitatively evaluate the effect of time on process safety issues particularly as they relate to equipment failures, startup and shutdown scenarios. We can also readily incorporate vendor data and validate their performance. For instance we use actual valve flow coefficients C_V and stroke time data for all control and isolation valves whenever dynamic simulation work is being implemented in designs.

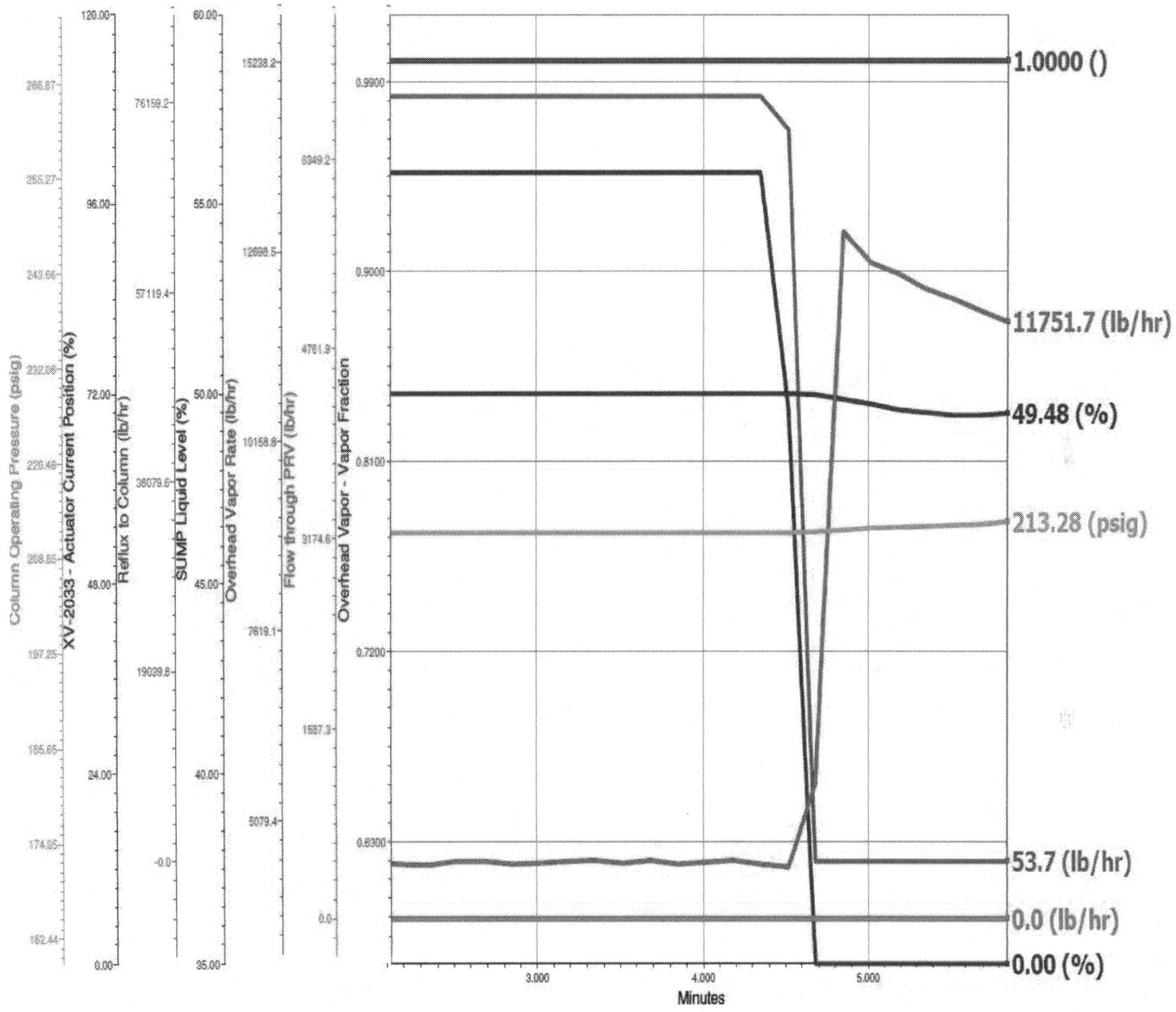

Figure 10. Dynamic Upset – Loss of Reflux: Isolation Valve Closes

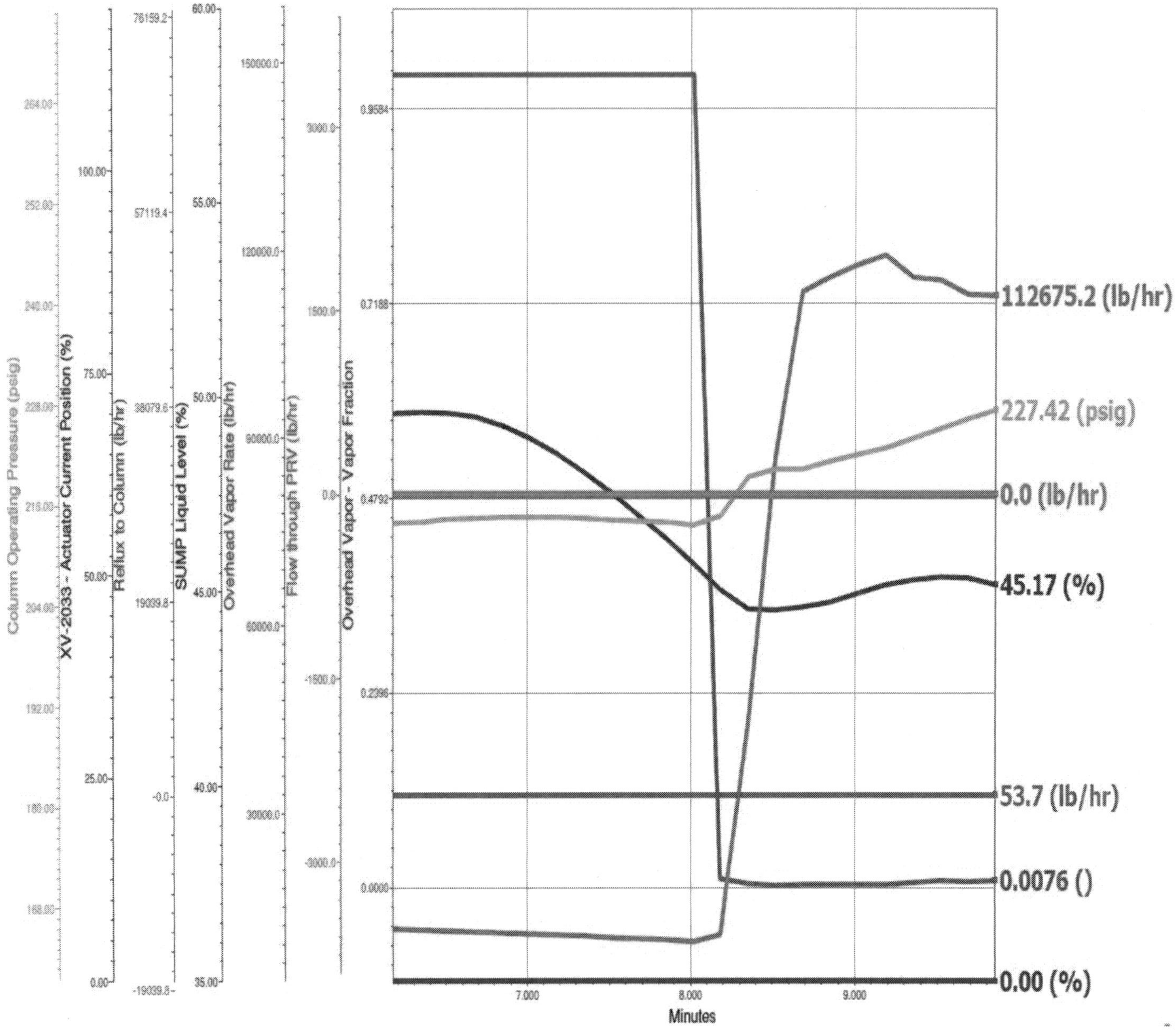

Figure 11. Dynamic Upset – Loss of Reflux→Receiver Liquid Full > 3.5 minutes

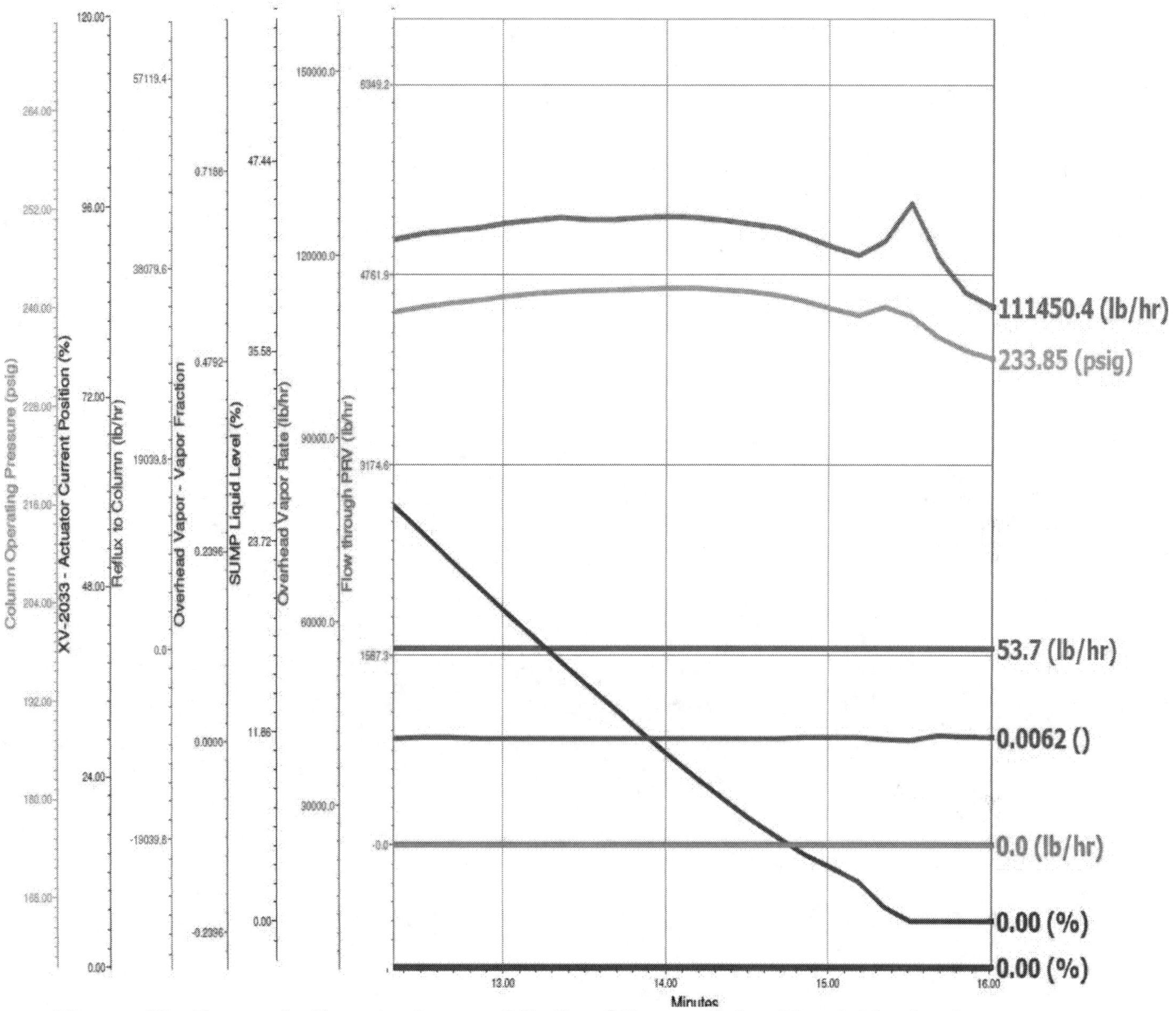

Figure 12. Dynamic Upset – Loss of Reflux→Sump Dries Up at 11 minutes

4.3.2 UOP Phenol Process - Operating Pressure Evaluation

Dynamic studies are also performed for customers based on their own internal HAZOP reviews. One of our customers requested that we evaluate the possibility of running oxidizer tanks at higher pressures so as to increase product flow.

The process flow diagrams for the oxidizer section were taken from our library of design templates and modified to include all customer specific information. Figure 13 shows a dynamic model representation of a section of the oxidizer feed train. Piping and instrument diagrams were also used to accurately define the placement of all control and isolation valves as well as control logic systems in the dynamic simulation. Some elevations and line sizing information were also extracted from these drawings. The remaining data was obtained from previous hydraulic reports. Equipment sizes were also incorporated into the model.

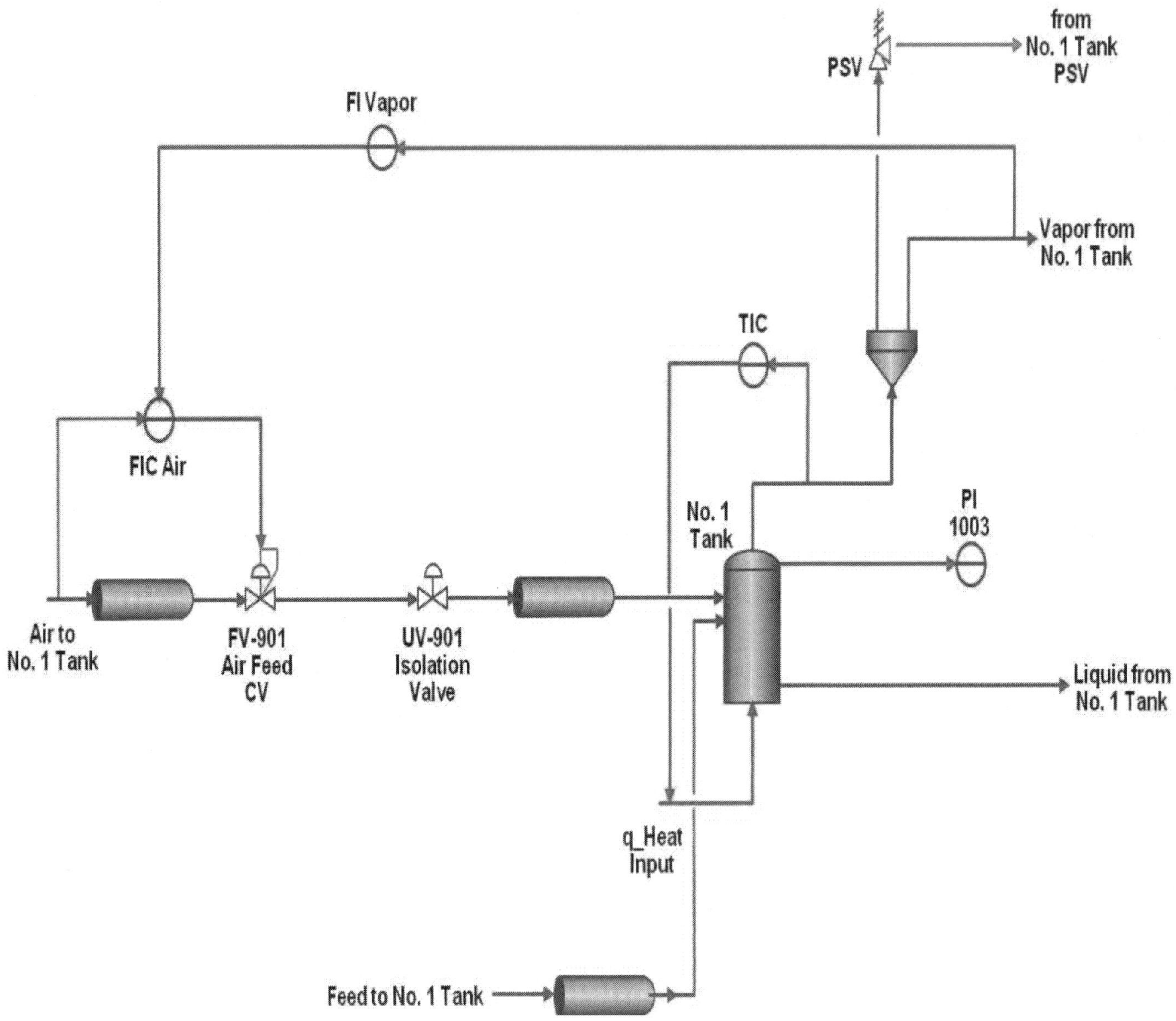

Figure 13. UNISIM® Dynamic Oxidizer Model

Figure 14 shows a typical air feed valve / isolation valve configuration. These valves utilized the "as built" detailed valve performance data provided by the customer.

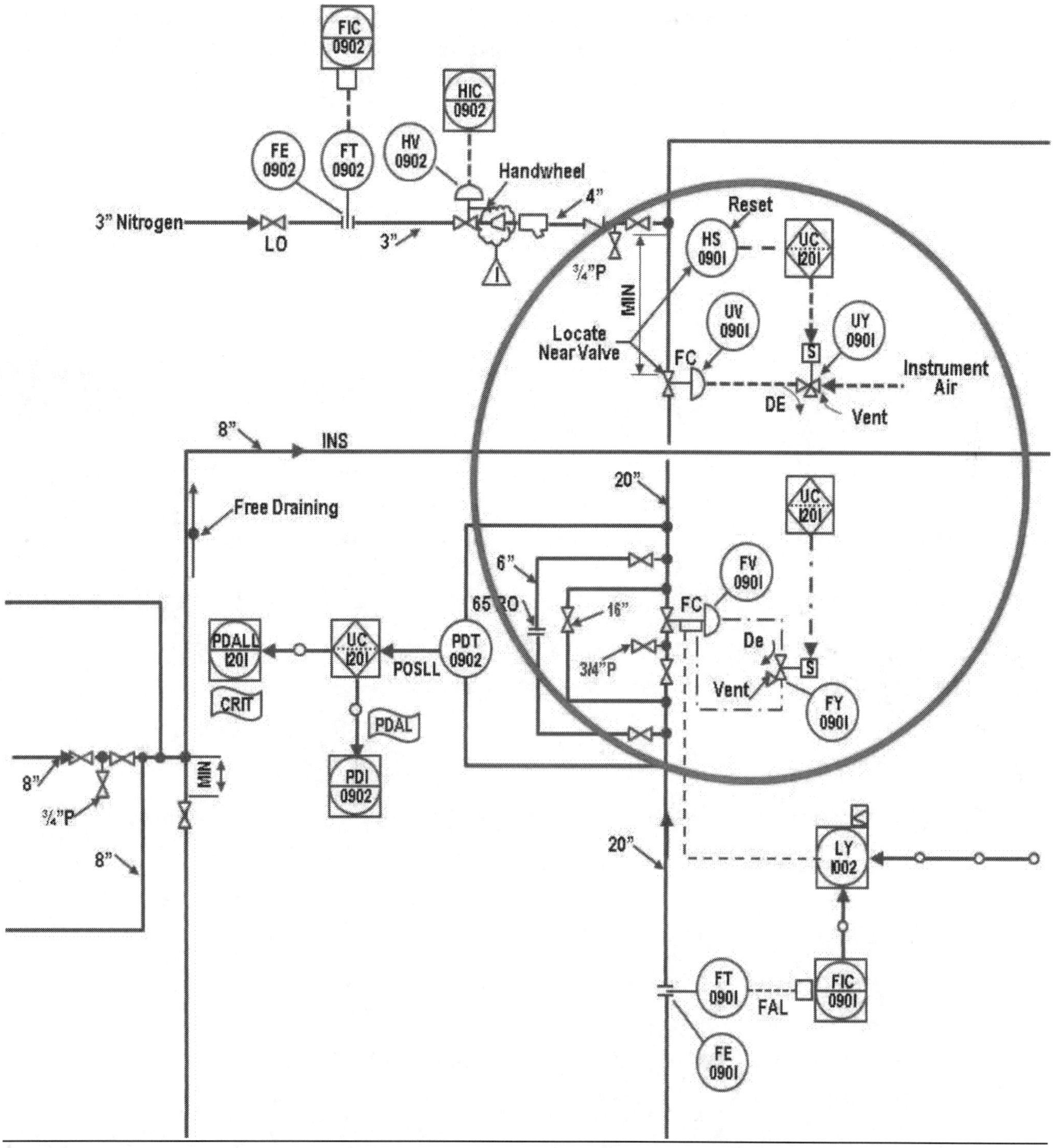

Figure 14. Piping and Instrument Diagram View – Air Feed / Isolation Valves

The results of the dynamic analysis showed that the customer can indeed run their unit at higher operating pressures. A higher pressure trip point was recommended for the oxidizer shutdown system which would still prevent overpressure of the oxidizer tanks from reaching their set relief pressures. Based on the "as built" stroke time valve data obtained from the customer, the results shown in Figure 15 indicate that the valves can perform quickly in avoiding pressure relief of the oxidizer tanks at the highest possible operating pressure.

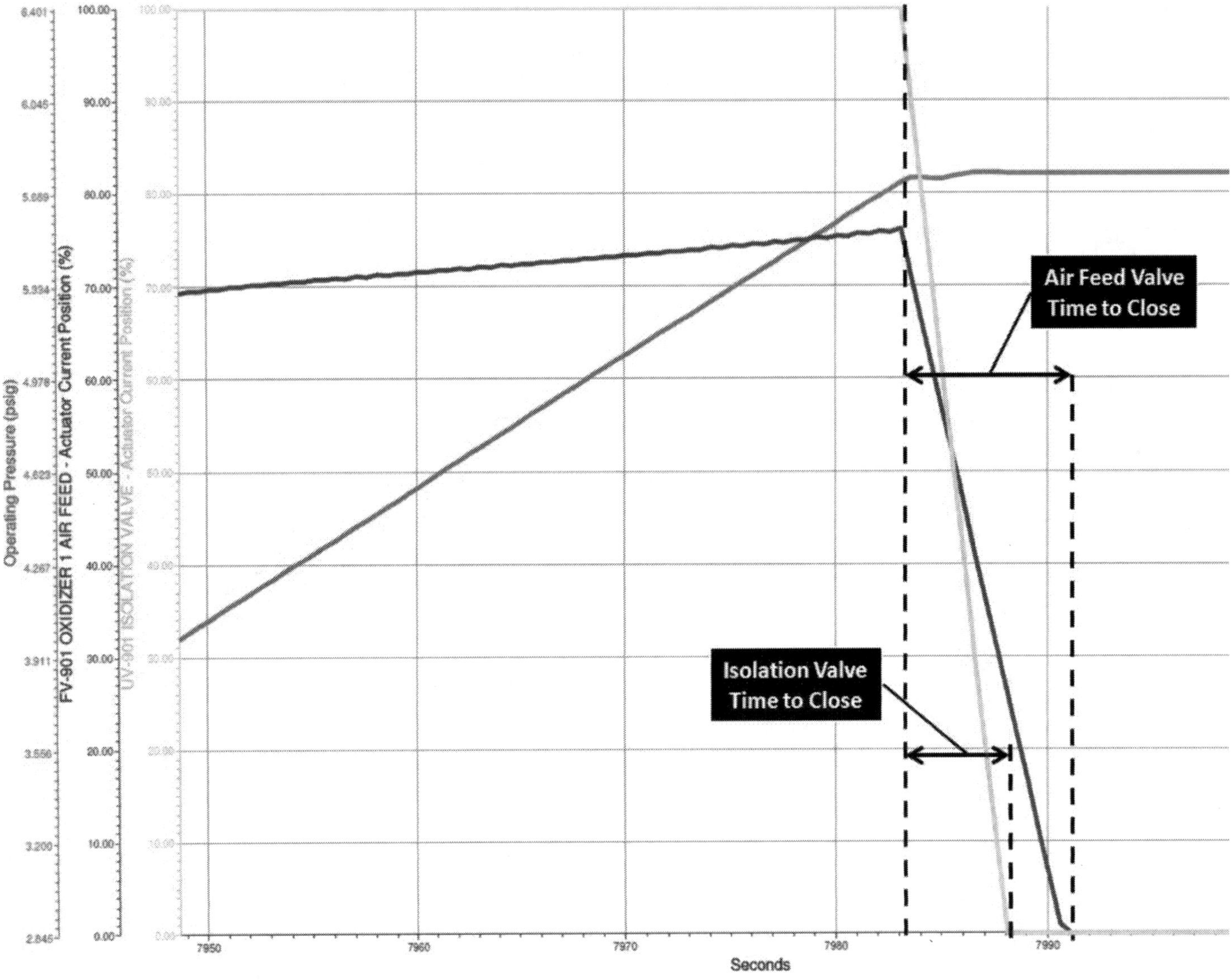

Figure 15. Dynamic Upset – Valve Stroke Time

5. Conclusion

The combination of traditional HAZOP analysis with dynamic simulation facilitates accurate determination of the consequences of deviations from normal design. Potential hazards and safety issues can be better identified. By dividing the problem model into smaller, independent flowsheet sections, the level of complexity associated with the dynamic simulations can be reduced.

HAZOP teams can now test the effectiveness of mitigation strategies for many emergency situations. The use of dynamic simulation as a process safety / risk assessment tool provides significant added value to HAZOP reviews and reports.

6. References

[1] Process Design Simulation: Fundamentals – Dynamic Modeling Using **UNISIM®** Design Suite Implementation, PDS-4528 R380 Student Guide, Rev 01.0, 07/2010.

[2] **UNISIM®** Design Dynamic Modeling-Reference Guide, May 2012 R410 Release.

[3] **UNISIM®** Operations Guide, May 2012 R410 Release.

11TH GLOBAL CONGRESS ON PROCESS SAFETY

How close is my operation to a Major Accident?

Abouzar Yousefi
Repsol
Mendez Alvaro, 44, 28045 Madrid, Spain
abouzar.yousefi@repsol.com

Manuel Rodriguez Hernández
Universidad Politécnica de Madrid
José Gutiérrez Abascal, 2 28006 Madrid, Spain
manuel.rodriguezh@upm.es

Jaime Martin Juez
Repsol
Mendez Alvaro, 44, 28045 Madrid, Spain
jmartinj@repsol.com

Prepared for Presentation at
American Institute of Chemical Engineers
2015 Spring Meeting
11th Global Congress on Process Safety
Austin, Texas
April 27-29, 2015

UNPUBLISHED

How close is my operation to a Major Accident?

Abouzar Yousefi
Repsol
Mendez Alvaro, 44, 28045 Madrid, Spain
abouzar.yousefi@repsol.com

Manuel Rodriguez Hernández
Universidad Politécnica de Madrid

Jaime Martin Juez
Repsol

Keywords: Major Accident, Process Safety, Safety Indicator, Accident Model,

Abstract

Despite many efforts made to prevent major accidents in upstream industry, these accidents continue to occur. It seems that our thinking about how accidents happen and all the management systems based on that approach, while reasonably successful, does not enable us to achieve the goal of zero accidents.

Different accident models have been introduced to explain how accidents happen. Accident models have evolved over time from being linear and deterministic to be non-linear and non-deterministic. These models not only describe how potentially hazardous situations develop and what processes form the basis for this development, but also serve the goal of helping to define measures to prevent and control them.

To understand how effective are these preventive and control measures, safety performance indicators can be used. These indicators are of a reactive, or a proactive nature, depending on the characteristics of the indicators. Reactive indicators or lagging indicators are used more commonly in upstream operations. These indicators are post-event types and generally show the safety level of the past time.

Safety level is not static and depends on many risk factors that change in presence and intensity over location and time. On the other hand major accidents have a very low frequency. Therefore reactive indicators based on these events cannot neither indicate how safe the operation is in real time nor warn how close the operation is to a major accident. For this reason proactive indicators or leading indicators are necessary to provide enough information at the right time to decision makers before a major accident happens.

Considering various accident models, it is recommended to use an adequate model for upstream operation and develop methods for the establishment of indicators, which can unveil early warnings of major accidents by measurements of underlying causes and contributing factors to major accidents.

1. Introduction

Major accident is defined in different sources. As an example International Association of Oil & Gas Producers (IOGP) has defined a major incident as an unplanned event with escalation potential for multiple fatalities and/or serious damage, possibly beyond the asset itself. Typically there are hazardous releases, but also include major structural failure or loss of stability that could put the whole asset at risk [1]. Also, it has been defined by governing regulations in UK [2], Norway [3], Australia [4] and others. While each definition differs slightly, a major accident is generally described as a fire, explosion, or release of a hazardous chemical substance that involves the death or serious injury to multiple individuals, severe environmental damage and/or loss of major financial assets [5].

Major accidents continue to occur in the process industry and often have serious consequences. Over the last generation, successive major accidents, from the explosion at Flixborough (1974), the deadly toxic gas release in Bhopal (1984), the explosion at Piper Alpha (1988) to more recent examples including the explosions at BP Texas City (2005), Buncefield (2005) and the blow-out and subsequent environmental disaster in the Gulf of Mexico (2010) have raised concerns with the public, stakeholders and regulators [6]. The Gulf of Mexico accident (2010) has highlighted a number of serious problems in scientific thinking about safety. One of these is that our current thinking about how accidents happen and all the management systems based on that approach, while reasonably successful, does not appear to enable us to achieve the goal of zero accidents [7].

Undoubtedly the main reason we want to understand how accidents happen is because we wish to prevent future ones [8].The best way to do this is to have a solid theoretical basis that can be shown both to describe what actually happens and to serve the goal of helping define effective preventive measures; the categorization of causal and contributing factors and the understanding of the mechanisms should be explicitly related to opportunities to prevent, remedy and improve. Theories about how accidents happen have evolved over time and different accident models have been introduced from being linear and deterministic to be non-linear and non-deterministic. In section 2 a summary of accidents models is provided.

On one hand, to know how close an operation is to a major accident, it is necessary to recognize how developed the path to a major accident is in that operation considering accident models. On the other hand, safety is often defined as the absence of unwanted events [9]. In other words, when safety is missing unwanted events occur. Such definition of safety is, however, indirect rather than direct since safety is defined by what happens when it is absent or missing. Properly speaking, it is therefore definition of lack of safety (or unsafety) rather than of safety. One consequence of this is that safety management relies on measurements that refer to the absence of safety rather than to the presence of safety [10]. Counting accidents or calculating frequency rates of accidents are examples of measurement based on the absence of safety. In section 3 a summary of safety measurements is provided.

Although these measurements are useful, they are not able to indicate how close an operation is to an accident in general or to a major accident in particular. Therefore considering various accident models, in section 4 it is recommended in detail how to use an adequate model for upstream operation and develop methods for the establishment of indicators, which can unveil early warnings of major accidents by measurements of underlying causes and contributing factors to major accidents.

2. Accident models

The historical development of accident models and various approaches for accident analysis have been discussed by engineers, scientists, cognitive psychologists, and sociologists. One of the earliest accident causation models is the Domino theory proposed by Heinrich in the 1940s [11], which describes an accident as a chain of discrete events which occur in a particular temporal order. This theory belongs to the class of sequential accident models or event-based accident models, which underlie most accident models such as Failure Modes and Effects Analysis (FMEA), Fault Tree Analysis (FTA), Event Tree Analysis, and Cause-Consequence Analysis [12]. These models work well for losses caused by failures of physical components or human errors in relatively simple systems. Typically, in these models, causal factors in an accident which are not linked to component failures are classified as human error as a kind of catchall or "garbage can" [13]. These models are limited in their capability to explain accident causation in the more complex systems that were developed in the last half of the 20th century.

In the 1980s, a new class of epidemiological accident models tried to explain accident causation in complex systems. Epidemiological models regard events leading to accidents as analogous to the spreading of a disease, i.e., as the outcome of a combination of factors, some manifest and some latent, that happen to exist together in space and time. Reason's Swiss cheese model [14, 15] is a major contribution to this class of models, and has greatly influenced the understanding of accidents by highlighting the relationship between latent and immediate causes of accidents.

Sequential and epidemiological accident models are inadequate to capture the dynamics and nonlinear interactions between system components in complex sociotechnical systems. New accident models, based on systems theory, classified as systemic accident models, endeavor to describe the characteristic performance on the level of the system as a whole, rather than on the level of specific cause-effect "mechanisms" or even epidemiological factors [16]. Systemic safety models have their roots in systems theory and cybernetics. Systems theory includes the principles, models, and laws necessary to understand complex interrelationships and interdependencies between components (technical, human, organisational and management) of a complex system. Safety models based on systems theory view accidents as emergent phenomena that arise from interactions among system components, where the interactions may be nonlinear and involve multiple feedback loops [17].

A major difference between systemic accident models and sequential/epidemiological accident models is that systemic accident models describe an accident process as a complex and interconnected network of events while the latter describes it as a simple cause-effect chain of events. Two notable systemic modelling approaches, Rasmussen's hierarchical sociotechnical framework (AcciMap) [18] and Leveson's Systems-Theoretic Accident Model and Processes (STAMP) model [19], endeavor to model the dynamics of complex sociotechnical systems.

Modern technology and automation has significantly changed the nature of human work from mainly manual tasks to predominantly knowledge intensive activities and cognitive tasks. This has created new problems for human operator performance (such as cognitive load) and new kinds of failure modes in the overall human-machine systems. Cognitive systems engineering has emerged as a framework to model the behavior of human-machine systems in the context of the environment in which work takes place [20]. Human factors should be evaluated in three hierarchical levels [21]: The first level should concern the cognitive behavior of human beings during the control of processes that occur through the man-system interface. At this level human errors are evaluated through human reliability techniques of first and second generation. In the second level, the focus is on the cognitive behavior of human beings when they work in groups, as in nuclear power plants. At this level the focus is on the anthropological aspects that rule the interaction among human beings. In the third level, the interest is in the influence of organizational culture on human beings as well as on the tasks being performed. At this level the economical and political aspects that shape the company organizational culture are added to the factors of the second level. Based on the abovementioned principles, various tools, techniques, models and methods have been introduced: THERP - Technique for Human Error Rate Prediction [22], HCR - Human Cognitive Reliability [23], ASEP - Accident Sequence Evaluation Program [24], HEART - Human Error Assessment and Reduction Technique [25], CREAM - Cognitive Reliability and Error Analysis Method [26], HFACS- Human Factors Analysis and Classification System [27], FRAM- Functional Resonance Accident Model [16], ATHEANA - A Technique for Human Error Analysis [28], etc.

3. Safety related indicators

As mentioned already safety is a concept often understood on the absence of undesired events and not easy to define directly. It is even more difficult to define what constitutes an adequate level of safety. In other words it is difficult to assess safety directly, because its appearance is indicated by absence of accidents. Then the only direct measure of achieved performance is in terms of absence of harm or loss, and reducing the cost of losses provides direct evidence of performance improvements.

The term Indicator in the safety field is rather new, but safety measurements were performed in the 1980s and before with terms like Index, Rate and Measurements [29]. In addition, there are terms such as Performance Measure, Performance Indicator and Metric, but these terms are not always used in a consistent way [30]. An indicator is a measurable representation of an aspect of reality [29]. According to Hale [31] safety indicators are used to: (1) monitor the level of safety in a system; (2) decide where and how to take action; and (3) motivate those in position to take the necessary actions to take them.

Until recently, the primary safety indicators have been things that go wrong like the number of injuries, the cost of accidents and claims, etc. Lost time injury frequency (LTIF) and total recordable incident rates (TRIR) are more common used safety indicators. Over the last decades, substantial improvements in the areas of LTIF and TRIR are reported [32]. It was for a long time assumed that personal injuries reflected a facility's major hazard risk, but since, personal (occupational) safety is about avoiding cuttings, trips and falls type hazards among employees, it does not represent management of major hazards that have the potential to cause major accidents [33]. Therefore, good personal safety performance does not ensure the satisfactory condition of an asset and its operation [34]. The recent undesired incidents in oil and gas facilities are clear signs of this, for example the explosion at a Shell Chemical Company plant in Deer Park, Texas in 1997 [35], the BP Texas City refinery disaster in 2005 [36] and Gulf of Mexico blowout in 2010 [5]. Relying on injury rates as an indicator for safety level significantly hindered BP's perception of process risk. As a result, BP's corporate safety management system for its US refineries did not effectively measure and monitor process safety performance [36]. There are different opinions to answer why some people do not properly distinguish between process safety and personal safety indicators. Hale [37] argues that the tendency to conflate process safety and personal safety, which is in relation to the major and minor accidents, goes back to a misinterpretation of the Heinrich triangle.

3.1 Personal Safety Indicator versus Process Safety Indicator

Hopkins makes a clear distinction between personal and process safety [38]: It is really a distinction between different types of hazards. Process safety hazards are those arising from the processing activity in which a plant may be engaged. Typical process safety incidents involve the escape of toxic substances and the release of flammable material which may or may not result in fires or explosions. Many process safety incidents either damage the plant or have the potential to damage the plant. Moreover, they have the potential to generate multiple fatalities. The term process safety originates in the US; in some other parts of the world process safety is referred to as asset integrity or technical integrity. Personal safety hazards, on the other hand, affect individuals but may have little to do with the processing activity of the plant. Typically they give rise to incidents such as falls, trips, crushings, electrocutions and vehicle accidents. It turns out that most injuries and fatalities are a result of personal safety hazards rather than process hazards and, as a result, injury and fatality statistics tend to reflect how well an organisation is managing personal safety hazards rather than process safety hazards. Any organisation that seeks to assess how well it is managing process safety hazards cannot therefore rely on injury and fatality data; it must develop indicators that relate specifically to process hazards. That is one of the primary conclusions of recent major accident reports [39].

In recent years, many organizations have recognized the importance of measuring process safety performance in order to improve process safety management. In 2003, the Organisation for Economic Cooperation and Development (OECD) developed and published the initial version of the Guidance on Developing Safety Performance Indicators related to Chemical Accident Prevention Preparedness and Response which has been revised in 2008 [40]. In 2006, the UK Health and Safety Executive (HSE) published a guide for development of leading and lagging process safety metrics: Developing Process Safety Indicators, A Step-By-Step Guide for Chemical and Major Hazard Industries [41]. In 2007, the Center for Chemical Process Safety (CCPS) released a document on Process Safety Leading and Lagging Metrics [42] which has been updated later. In 2010, American Petroleum Institute (API) published a recommended practice called Process Safety Performance Indicators for the Refining and Petrochemical Industries [43] which is under revision. In 2011, International Association of Oil & Gas

Producers (IOGP) published the Report number 456 called Process Safety - Recommended Practice on Key Performance Indicators [44]. In 2011, the European Chemical Industry Council (Cefic) published a document called Guidance on Process Safety Performance Indicators with a focus on measuring process safety incidents as a lagging indicator for process safety performance [45].

3.2 Lagging Indicator versus Leading Indicator

Reactive indicators ('lagging indicators') are after-the-event type of indicators, while proactive indicators ('leading indicators') assume measurements of underlying causes and contributing factors to accidental events, such as (inadequate) training and supervision, and thus providing early warnings [46].

OECD Guidance on Developing Safety Performance Indicators [40] divides safety performance indicators into two types: "outcome indicators" and "activities indicators."

- Outcome indicators are designed to help assess whether safety-related actions (policies, procedures and practices) are achieving their desired results and whether such actions are leading to less likelihood of an accident occurring and/or less adverse impact on human health, the environment and/or property from an accident. They are reactive, intended to measure the impact of actions that were taken to manage safety and are similar to what are called "lagging indicators" in other documents. Outcome indicators often measure change in safety performance over time, or failure of performance. Thus, outcome indicators tell you whether you have achieved a desired result (or when a desired safety result has failed). But, unlike activities indicators, they do not tell you why the result was achieved or why it was not.
- Activities indicators are designed to help identify whether enterprises/organisations are taking actions believed necessary to lower risks (e.g., the types of policies, procedures and practices described in the Guiding Principles). Activities indicators are proactive measures, and are similar to what are called "leading indicators" in other documents. They often measure safety performance against a tolerance level that shows deviations from safety expectations at a specific point in time. When used in this way, activities indicators highlight the need for action when a tolerance level is exceeded. Thus, activities indicators provide enterprises with a means of checking, on a regular and systematic basis, whether they are implementing their priority actions in the way they were intended. Activities indicators can help explain why a result (e.g., measured by an outcome indicator) has been achieved or not.

Traditionally, the focus of safety indicators for major hazards has been on incident indicators or lagging indicators, which means recording the number of accidents or near-misses. Major accidents are rare and occur so infrequently specially at a site level. Many companies have not seen one at all and some have seen only just. The frequency of major accidents therefore is not a good indicator for the state of safety of a particular plant [47]. Also it has been claimed that these indicators may not be useful as early warnings [36], and the focus on developing indicators that provide feedback before an accident occurs (so-called leading indicators) has increased [31]. Leading indicators are generally intended to help in identifying early signals of deteriorating performance and thereby to provide means for warning of impending problems before an actual incident or accident occurs. Early warnings from these indicators should initiate a further investigation of the causes of the symptoms to ensure that corrective actions will be timely initiated and rightly directed [30].

After distinguishing two dimensions of safety indicators: personal safety versus process safety indicators and lead versus lag indicators it is necessary to highlight another issue about safety indicators. Maybe it is not relevant whether the indicator is current or after the fact. The issue is whether, in the relevant time period, there are sufficient instances of the events being counted to be able to talk meaningfully about a rate. If there are, then charting this indicator over time will indeed provide evidence of whether the safety management system is improving. On the other hand, if years can go by without the occurrence of a single countable event, it is not possible to compute a meaningful annual rate, nor is it possible to conclude from one occurrence that safety is deteriorating. This is what Hopkin called a zoom effect at work [38]. Fatality rate can be sensibly computed for an industry, and possibly even for a very large company. Regulators can therefore treat the industry fatality rate as an indicator of how well safety is being managed in that industry and seek to drive it down by encouraging various industry-wide safety initiatives. The same may be true for a very large company. However, once we zoom in on a particular company site we may find that it goes for many years without a fatality. Under these circumstances it will not make sense to treat the fatality rate at the site as an indicator of safety. Management at such a site will need to identify more frequently occurring events, such as injuries, in order to be able to judge how well they are managing safety over time. Similarly, catastrophic accidents occur so infrequently at any one major hazard site that they cannot be used as an indicator of how well process safety is being managed at that site, whereas loss of containment events may be frequent enough to serve the purpose. In short, where harmful events are occurring frequently enough to be able to talk about a rate, this rate can be used to measure safety; where harmful events are rare, we must look to more frequently occurring precursor events in order to be able to measure safety. Putting this another way, there are circumstances in which so-called lag indicators are perfectly good indicators of how well safety is being managed, and circumstances in which they are not. It depends, in part, on the zoom's level. As mentioned by Hopkins [38], if one is interested in how well safety is being managed, the distinction between lead and lag indicators becomes largely irrelevant.

4. Discussion and Conclusion

Different approaches have been followed to evaluate operational safety at process plants and consequently different process safety indicators have been introduced. Although process safety indicators provide important information on the effectiveness of safety systems and try to provide early warnings of impending catastrophic failures, most of the existing indicators are considered as "lagging" indicators and they are expected to show an impact only when a downward trend has already started. Therefore these indicators are not able to declare how close an operation is to a major accident and development of new indicators can help warning when an operation is out of safe zone.

For early warnings, we need to look further back in the causes and the underlying conditions and factors that lead to a major accident in process industry. The acknowledgement of industrial installations as complex systems points to systems thinking theory as the most suitable tool to deal with major accident in process industry. Hence it is recommended to select a system theory based accident model like STAMP which best fits the process and accordingly map the process into the model in order to define a set of indicators which could demonstrate the progress of a major accident through its modeled path. Required information for indicators has to be obtained from the operation and processed accordingly in order to indicate the progress toward major accident. This can answer questions such as: How close is an operation to a major accident? And meanwhile it will support management decision making about required actions to restore safety.

No doubt, realization will be accompanied by numerous problems and unforeseen difficulties. Essential will be sensible choosing and observing a set of indicators based on the selected model as no single indicator can be a sure sign of imminent major accident.

5. References

Reference in order of first citation

[1] "Report No. 415, Asset integrity - the key to managing major incident risks", International association of Oil and Gas Producers, Dec 2008

[2] "Offshore Installations, the offshore installations (Safety Case) regulations", UK HSE, 2005

[3] "Guidelines Regarding the Management Regulations", Norway PSA, December 16, 2014

[4] "Offshore Petroleum and Greenhouse Gas Storage (Safety) Regulations", 2009 (Select Legislative Instrument 2009 No. 382 as amended), Australia, prepared on 1 January 2012

[5] "Investigation report overview, explosion and fire at the Macondo well", US Chemical and Safety and Hazard Investigation Board (CSB), June 5, 2014

[6] ''Corporate governance for process safety, Guidance for senior leaders in high hazard industries'', OECD Environment, Health and Safety Chemical Accidents Programme, June 2012

[7] B. Ale et al, Towards BBN based risk modelling of process plants, Safety Science, 69 (2014) 48–56

[8] Wagenaar, W.A. & Hudson, P.T.W., The analysis of accidents with a view to prevention. Report for Shell International SIPM, The Hague. Department of Experimental Psychology, Leiden University, The Netherlands (1987)

[9] Hollnagel, Risk + barriers = safety?, Safety Science 46 (2008) 221–229

[10] Hollnagel, Is safety a subject for science, Safety Science 67 (2014) 21–24

[11] Heinrich, H. W., Peterson, D., and Roos, N. (1980). Industrial accident prevention. McGraw-Hill, New York.

[12] Leveson, N.G., 1995. Safeware: System Safety and Computers. Addison Wesley, Reading, MA.

[13] Hollnagel, E. (2001). Anticipating Failures: What should Predictions be About? In The Human Factor in System Reliability – Is Human Performance Predictable? RTO Meeting Proceedings 32, RTO-MP-32, January, Research and Technology Organization, North Atlantic Treaty Organization, Cedex, France: RTO/NATO.

[14] Reason, J. (1990). Human Error. Cambridge: Cambridge University Press.

[15] Reason, J. (1997). Managing the Risks of Organizational Accidents. England: Ashgate.

[16] Hollnagel, E., 2004. Barriers and Accident Prevention. Ashgate Publishing Limited, Aldershot, UK.

[17] Perrow, C. Normal Accidents: Living with High-Risk Technologies. 1984, New York: Basic Books

[18] Rasmussen, J. Risk Management in a Dynamic Society: A Modelling Problem, Safety Science, 27 (1997), 2, 183-213.

[19] Leveson, N. A New Accident Model for Engineering Safer Systems, Safety Science, 42 (2004), 237-270.

[20] Hollnagel, E. & Woods, D. D. Cognitive Systems Engineering: New wine in new bottles. International Journal of Man-Machine Studies, 18 (1983), 583-600.

[21] Alvarenga et al, A critical review of methods and models for evaluating organizational factors in HRA - Human Reliability Analysis - Progress in Nuclear Energy 75 (2014) 25-41

[22] Swain, A.D., Guttman, H.E., 1983. Handbook of Human Reliability Analysis with Emphasis on Nuclear Power Plant Applications. NUREG/CR-1278. U.S. Nuclear Regulatory Commission, Washington, D.C.

[23] Hannaman, G.W., Spurgin, A.J., Lukic, Y., 1984. Human Cognitive Reliability Model for PRA Analysis, NUS-4531. Draft EPRI Document. Electric Power Research Institute, Palo Alto, CA.

[24] Swain, A.D., 1987. Accident Sequence Evaluation Program Human Reliability Analysis Procedure. NUREG/CR-4772/SAND86-1996. Sandia National Laboratories, U.S. Nuclear Regulatory Commission, Washington, DC.

[25] Williams, J., 1988. A data-based method for assessing and reducing human error to improve operational performance. In: IEEE Conference on Human Factors in Power Plants. Monterey California.

[26] Hollnagel, E., Cognitive Reliability and Error Analysis Method (CREAM). Elsevier Science (1998) New York.

[27] Wiegmann, D.A., Shappell, S.A., 2003. A Human Error Approach to Aviation Accident Analysis: The Human Factors Analysis and Classification System. Ashgate Publishing Company, Burlington, VT.

[28] NRC, 2007. ATHEANA's User Guide Final Report. NUREG-1880. US Nuclear Regulatory Commission, Washington, D.C.

[29] Øien et al., Building Safety indicators: Part 1 – Theoretical foundation, Safety Science 49 (2011) 148–161

[30] Lehtinen, E. (1995). "A Concept of Safety Indicator System for Nuclear Power Plants", Technical Research Centre of Finland, Espoo, Research Notes 1646, 26 p. + app. 106 p.

[31] Hale Andrew, Why safety performance indicators - Safety Science 47 (2009) 479–480

[32] ''Safety performance indicators – 2010 data'' IOGP, Report No. 455, 2011

[33] Eskogdalen et al, Developing safety indicators for preventing offshore oil and gas deepwater drilling blowouts - Safety Science 49 (2011) 1187–1199

[34] Hassan and Khan, Risk-based asset integrity indicators, Journal of Loss Prevention in the Process Industries 25 (2012) 544-554

[35] EPA/OSHA Joint Chemical Accident Investigation Report, Shell Chemical Company plant in Deer Park, Texas, 1998

[36] Baker et al, The report of the BP U.S. refineries independent safety review panel, 2007

[37] Hale, A., 2001, Conditions of occurrence of major and minor accidents. Institution of Occupational Safety and Heath Journal 5(1), 7–21.

[38] Hopkin Andrew, Thinking about process safety indicators, Safety Science 47 (2009) 460–465

[39] CSB, 2007. Investigation Report, Refinery Explosion and Fire. US Chemical Safety and Hazard Investigation Board.

[40] ''Guidance on developing safety performance indicators related to chemical accident prevention, preparedness and response'' OECD - 2nd edition 2008.

[41] HSE, 2006. Developing process safety indicators: a step-by-step guide for chemical and major hazard industries, UK Health and Safety Executive.

[42] ''Process Safety – Leading and lagging Metrics'', CCPS 2007

[43] ''Process Safety Performance Indicators for the Refining and Petrochemical Industries'', API-2010

[44] ''Process Safety - Recommended Practice on Key Performance Indicators'', Report number 456, IOGP – 2011

[45] ''Guidance on Process Safety Performance Indicators'', European Chemical Industry Council (Cefic) - 2011

[46] Øien et al., Building Safety indicators: Part 2 – Application, practices and results, Safety Science 49 (2011) 162-171

[47] Ale, Ben, More thinking about process safety indicators, Safety Science 47 (2009) 470–471

393997 An Approximate Model for Discharge Density Associated with Critical Flow

Abstract:

The purpose of this presentation (paper) is to present an approximate model derived from the entropy balance associated with isentropic, choked flow utilizing the van der Waal equation of state and an approximate form of the van der Waal equation for high pressures that will provide the stagnation density. This will provide the form of the model where a non-dimensional parameter related to the deviation from isothermal behavior will be estimated via three non-dimensional numbers related to pressure, temperature and ratio of specific heats of the system. The model will be validated against three gases which are ideal and thermodynamically perfect, ideal but not thermodynamically perfect, and a real gas. Pressures will range up to 4 MPa. The relationship of this model against other models in the literature will be explored and its relationship to the Omega method will be discussed.

396653 A Multilevel Safety EARLY Warning Method Based on Bayesian Networks with Applications for Petrochemical Process

Abstract:

Process systems in petrochemical industries have becoming increasingly large and automatic. There exist strong interdependencies between various facilities and components. Once any unit in the system fails, it often triggers cascade failures as a chain reaction, resulting in significant production losses, and even leads to catastrophic accidents. They are the result of a number of failures at different technical and organizational levels.

Proactive methodologies for the development of early warning methods can unveil early deviations in the causal chain. Two main issues exist that are how to describe the interdependency in the complex process system, and how to ratiocinate and demonstrate the reasons at each level of the current abnormal event and also foresee the possible consequence. In this way, a multilevel safety early warning method is proposed for petrochemical process to overcome above problems. The approach combines the advantages of multilevel flow modeling (MFM), HAZOP analysis and dynamic Bayesian network (DBN) as well, which become a special and effective mechanism of safety early warning.

Firstly MFM is used to represent a system in terms of goals, objectives, functions and components each of which can be described at different levels of part-whole decomposition. Secondly HAZOP study is carried out based on the MFM models, by which all the possible deviations and their corresponding potential fault causes and consequences are analyzed carefully. Thirdly dynamic Bayesian Network (DBN) is used to build the fault causal relationships which represent the fault interdependencies in the complex process system. Finally by the inference mechanism of DBN, the most possible initial reason(s) happened in the fault interdependency network with multi reasons and consequences can be found out accurately and also the future possible consequences can be predict for proactive maintenance or emergency decision making.

The analysis shows that the proposed model fits its purpose enabling causal reasoning that explains causes and consequences derived from deviations. Examples from case studies with respect to the Fluidized Catalytic Cracking unit (FCCU) in petrochemical process system are included to illustrate the effectiveness and accuracy of the proposed approach. Based on the proposed multilevel safety early warning method, an expert system is developed and applied to practice in the petroleum industry. Its application is also illustrated to show the effect of accident prevention in the petrochemical enterprises. The salient features of this expert system and its performance on practical petrochemical systems is presented.

11TH GLOBAL CONGRESS ON PROCESS SAFETY

Summary of Macondo Inquiries and Recommendations

Dr Robin Pitblado
DNV GL
1400 Ravello Dr, Katy TX 77441
robin.pitblado@dnvgl.com

Prepared for Presentation at
American Institute of Chemical Engineers
2015 Spring Meeting
11th Global Congress on Process Safety
Austin, Texas
April 27-29, 2015

UNPUBLISHED

Summary of Macondo Inquiries and Recommendations

Dr Robin Pitblado
DNV GL
1400 Ravello Dr, Katy TX 77441
robin.pitblado@dnvgl.com

Keywords: Inquiries, Lessons Learned, Macondo, Process Safety

Abstract

The Macondo event, which commenced on April 20 2010 was the largest oil spill event in history and caused 11 fatalities, 17 injuries and the total loss of the Deepwater Horizon drilling rig. Not surprisingly, it is also amongst the most studied of events, with close to 20 major inquiries having been undertaken in the USA and elsewhere. The purpose of this paper is to summarize the main inquiries findings and recommendations that are publicly available. While this involves some selection, the aim of the paper is to be neutral and to simply report the facts as stated. The aim is that readers will be able to identify what body recommended what and to match this to changes to regulations and good practices post the accident.

1. Introduction

The Macondo accident on April 20 2010 was the largest oil spill event in history and caused 11 fatalities, 17 injuries and the total loss of the Deepwater Horizon drilling rig. It is important that the industry learns key lessons from this accident. While many lessons have been incorporated into new regulations, updated standards, and revised company practices, many engineers are unaware of the full range of findings from the over 209 inquiries that have been undertaken. Also many good recommendations have not been adopted and these deserve fresh consideration.

A list of the most important inquiries since Macondo is provided in Table 1, and there are further reports expected. The CSB has noted it plans two further reports addressing human factors and organizational issues, and we expect more detailed environmental findings as the environmental impacts become more fully understood.

Table 1. Major Macondo Investigation reports summarized in this study

No.	Author	Title	Release Date
1	DWH Commission	National Commission on the BP Deepwater Horizon Oil Spill and Offshore Drilling – Main Report and multiple topic papers	11-Jan 2011
2	DWH Commission	Chief Counsel's Report	Feb 2011
3	BP	DWH Accident Investigation Report	8-Sep-2010
4	Transocean	Macondo Well Incident: Transocean Investigation Reports, Vol. I and II	June 2011
5	USCG	Report of Investigation into the Circumstances Surrounding the Explosion, Fire, Sinking and Loss of Eleven Crew Members Aboard the Mobile Offshore Drilling Unit DWH Vol.1 (MISLE 3721503)	22-Apr-11

6	BOEMRE	JIT - BOEMRE Final Report regarding Macondo Well Blowout	14-Sep-11
7	Republic of the Marshall Islands	Deepwater Horizon Marine Casualty Investigation Report	17-Aug-11
8	Adm Thad Allen	Incident Commanders Report	Oct 2010
9	DNV GL	Forensic Examination of Deepwater Horizon Blowout Preventer, Vol. I and II (Appendices).	March 2011
10	Chemical Safety Board	Investigation Report (Vols 1-2)	5-Jun-14
11	Center for Catastrophic Risk Management (UC Berkeley)	Final Report on the Investigation of the Macondo Well Blowout	March 2011
12	National Academy of Engineering	Macondo Well-Deepwater Horizon Blowout: Lessons for Improving Offshore Drilling Safety	2012
13	National Research Council	Ecosystem Services Approach to DWH	2013
14	US District Court Eastern Louisiana	Court Judgment	4-Sep-2014
15	Norway Petroleum Safety Authority	Interim report – DWH Accident	June 2011
16	Norway Petroleum Safety Authority	PSA's Concluding report on its follow-up for the DWH accident	5-Feb-2014
17	OGP	Getting it Right	Jan 2013
18	OLF	DWH Lessons Learned and Follow-up	June 2012
19	SINTEF	Report on DWH (A19148)	May 2011
20	UK HSL	DWH Fire & Explosion Issues	Apr 2014
21	US Transportation Research Board	Evaluating the Effectiveness of Offshore Safety and Environmental Management Systems (Special Report 309)	2012

DNV GL (2015) has published a fuller review of these inquiries, and this paper provides a shorter summary of key findings and recommendations.

2. National Commission – BP Deepwater Horizon Oil Spill

The National Commission was established by President Obama soon after the event with a mandate to report within 6 months. The Commission was chaired by Bob Graham and William K. Reilly, along with 5 other members and supported by a chief counsel and technical team. The main report was issued in January 2011 and was structured into 3 main sections: I) Path to the Tragedy, II) Explosion and Aftermath, and III) Lessons Learned for Industry, Government, and Energy Policy.

In Part I, the history of development of drilling in the Gulf of Mexico (GoM) and particularly in Deepwater was summarized, along with the technical challenges.

In Part II, the causes and consequences of the disaster were reviewed. Part III addressed lessons learned for industry, government and for energy policy and contained the Commission recommendations. The report emphasized safety culture as a key factor for enhancing safety and discussed culture issues affecting BP, its contractors, and the GoM industry generally. The Commission also identified a conflict of interest in API, which serves as coordinator for industry expertise and standards, but also acts as the industry chief lobbyist. API standards were characterized as "lowest common denominator" rather than "best practice" standards, and since these standards underpin the Department of Interior regulations, the whole regulatory system was compromised. Commission recommendations were grouped as follows:

A. Improving the Safety of Offshore Operations (A1-A5)
B. Safeguarding the Environment (B1-B3)
C. Strengthening Oil Spill Response, Planning, and Capacity (C1-C6)

D. Advancing Well-Containment Capabilities (D1-D3)
E. Overcoming the Impacts of the Deepwater Horizon Spill and Restoring the Gulf (E1-E7)
F. Ensuring Financial Responsibility (F1-F4)
G. Promoting Congressional Engagement to Ensure Responsible Offshore Drilling (G1-G2)

3. National Commission – Chief Counsel's report

The Chief Counsel's report complements the Commission's main report and contains significantly more technical details. This report focuses on the direct technical and managerial causes of the accident and does not address response activities or environmental, social or economic impacts. Main chapters deal with deepwater drilling topics and the Macondo well, deepwater well design, cementing, the temporary abandonment, the negative pressure test, kick detection and response, and the blowout preventer and its maintenance. Many findings are made on technical and management issues, but the report does not make recommendations – these are presented in the Commission main report (1). But the findings do highlight important issues and lessons learned.

It is not possible to list all the findings in this lengthy report, but some main findings are presented. The use of a long production casing made the cementing job more difficult and the use of rupture disks to prevent annular pressure build-up compromised later containment efforts as these were believed to have failed inwards. BP selected the long-term benefits of using the long string without adequate assessment of the short-term risks. The report is clear that the cement job failed but there were several possible reasons and it was not possible to select amongst these. Management findings focused on BP's systems that did not force the Macondo team to identify and evaluate all cementing risks; and changes to the well plan that were not adequately assessed. The BP team did not assess contractor data in sufficient detail and made too simple analysis of well data as an indicator of cementing success.

A final chapter dealt with overarching failures of management. The Chief Counsel's team observed at least the following management failures: (1) ineffective leadership at critical times; (2) ineffective communication and silo'ing of information; (3) failure to provide timely procedures; (4) poor training and supervision of employees; (5) ineffective management and oversight of contractors; (6) inadequate use of technology; and (7) failure to appropriately analyze and appreciate risk. Ultimately, the companies placed undue reliance on timely intervention and human judgment, in light of their failure to provide individuals with the information, tools, and training necessary to be effective.

The Chief Counsel's report concluded: "the blowout occurred in large part because the companies diffused knowledge, responsibility for, and ownership of safety among themselves and among groups of people. The people onshore and on the rig had a false sense of security. They did not recognize the need for individual leadership in addressing the multiple anomalies and uncertainties that they observed. Instead, they relied on many ambiguous "dotted line" relationships within and between the companies and personnel involved."

4. BP Report: Deepwater Horizon Accident Investigation Report

The BP team commenced their investigation very soon after the blowout event commenced. They focused on the direct technical causes of the incident and did not address management related issues or alternative risk management strategies that might have been employed nor the sequence of emergency actions immediately after the explosion. They identified a sequence of eight key barriers that failed and these permitted the accident to occur. The BP team made recommendations in eight areas that would apply to BP and to its contractors:

1) Procedures and technical practices - cementing, well control, well design, negative-pressure testing
2) Capability and competency - strengthen the Technical Authority role and enhance competency
3) Strengthen audit and verification program
4) Process safety performance management - Establish leading and lagging indicators for drilling
5) Develop cementing services assurance
6) Confirm that essential well control and monitoring practices are clearly defined and applied
7) Rig process safety – require HAZOP studies on key parts of drilling systems
8) Establish minimum levels of reliability for BOP systems, develop ROV intervention strategies

5. Transocean: Macondo Well Incident

The Transocean investigation had three main objectives – to establish 1) how the reservoir fluids reached the rig floor, 2) why the BOP failed to stop the flow, and 3) how did the fluids ignite. The Transocean report did not make specific recommendations. Their main conclusion was: "The Macondo incident was the result of a succession of interrelated well design, construction, and temporary abandonment decisions that compromised the integrity of the well and compounded the risk of its failure."

6. USCG JIT - Vol.1 Report into the Deepwater Horizon (Ref: 3721503)

This investigation was conducted under a Memorandum of Agreement with BOEMRE. The USCG made a large number of detailed recommendations related to its areas of responsibilities (these are listed in Table 2).

Table 2. Summary of USCG Recommendations – related to fire & explosion

Ref	Recommendation summary
1A	Labelling of rated electrical equipment in hazardous areas.
1B	Guidance on design of detection and alarm systems.
1C	Guidance on fire and explosion strategies for units with DP.
1D	Specification of minimum explosion design loads.
1E	Explosion risk analysis for each existing facility.
1F/G	Control of location of ventilation inlets. Shut down of ventilation in the event of gas detection.
2A	Fire pumps to be self-contained.
2B	H-60 rated separation between drill floor and adjacent accommodation areas.
2C	Guidelines on engineering evaluations of fire partitions.
2D	Fixed deluge or monitors provided for drill floor.
2E	Fire risk analysis to supplement the prescriptive requirements of the MODU code.
3A	Standards for maximum radiant heat allowable at muster stations and lifeboat stations.

Related to the recommendations related to fire and explosion the USCG (2014) issued guidance for offshore floating or fixed installations recommending modeling of fire and explosion scenarios consistent with API RP 2FB or equivalent.

7. BOEMRE – Report into Causes of Macondo Well Blowout

The BOEMRE investigation was carried out as part of the Joint Investigation Team (JIT) with USCG. It focused on BOEMRE responsibilities, especially the drilling activity and the cause of the blowout. A final section provided 41 recommendations and was divided into six sections

(with recommendations counts in each): wells (6), kick detection and response (8), ignition source (4), BOP (6), regulatory agency (11), and OSC Companies (6).
Some of the more significant recommendations included:

- require the negative pressure testing of wells where appropriate
- require at least two barriers (one mechanical and one cement barrier) for a well that is undergoing temporary abandonment procedures, and to clarify that a float collar/valve is not to be considered to be a "mechanical barrier."
- revise the incident reporting rule at 30 CFR§ 250.188 to capture well kick incidents
- several specific recommendations to enhance BOP operation and intervention
- expand 30 CFR § 250.446 to include documentation and record keeping requirements for major (3-5 year) inspections as required by BOEMRE's adoption of API RP 53

Note, there was no recommendation regarding safety and environmental management systems as BOEMRE had already decided by this time to require adoption of SEMS based on API RP 75.

8. Marshall Islands Registry – DWH Marine Casualty Investigation Report

REPUBLIC OF THE MARSHALL ISLANDS
Office of the Maritime Administrator

The Flag State for the DWH was the Republic of the Marshall Islands. Under international conventions, the Flag State must report the causal factors of all serious and very serious marine casualties.
The report concludes that the causal factors were: deviations from standards of well control engineering; deviation from the well abandonment plans submitted to and approved by the Minerals Management Service; and failure to react to multiple indications that a well control event was in progress.
The report contains 17 recommendations, but the Marshall Islands highlights four: enhanced communications between the Flag State and the Coastal State; IMO should review the MODU code related to aspects of this incident; there should be better definition of responsibilities of the Master and OIM especially during emergencies; and better procedures for actuating the Emergency Disconnect System.

9. Adm Thad Allen: Incident Commander's Report

Once the scale of the incident was established, the event was declared to be a Spill of National Significance and Thad Allen, then Commandant of the USCG, was declared the National Incident Commander. These were the first applications of these terms and the event tested existing laws and regulations very hard. The Oil Pollution Act (OPA 90) and the National Contingency Plan (NCP – 40 CFR Part 300) provided the guiding regulations but there was overlap with a Homeland Presidential Directive for domestic incidents. The NCP had been deployed successfully for thousands of spill events over the past 20 years.
It started to break down during the Deepwater Horizon spill as some aspects were not accepted by the public or by state officials – notably the role of the *responsible party* (here BP) was not understood and some states/local authorities did not accept federal authority.
The National Incident Commander (NIC) concluded that the Oil Pollution Act of 1990 and the NCP provided a sound basis for response and any changes should not change its fundamental governance structure. But many changes are required. There must be better integration of all necessary government agencies before there is an emergency, actions are needed to de-conflict and reconcile the role of the NIC, the NIC must have adequate authorities to execute the role

properly and work with the National Response Team (15 relevant federal agencies), and the private sector must be encouraged to develop modern oil spill response capabilities, matching the challenges of deepwater activities.

10. DNV Forensic Examination of Deepwater Horizon BOP

The Justice Department commissioned DNV (now DNV GL) to carry out a forensic examination of the Transocean BOP recovered from the sea floor once the well had been killed. The BOP was transported to NASA's Michoud facilities in Louisiana where tests were conducted. The report was designed to support legal proceedings, if required, and thus contains a very detailed set of information with 149 figures and 30 tables.

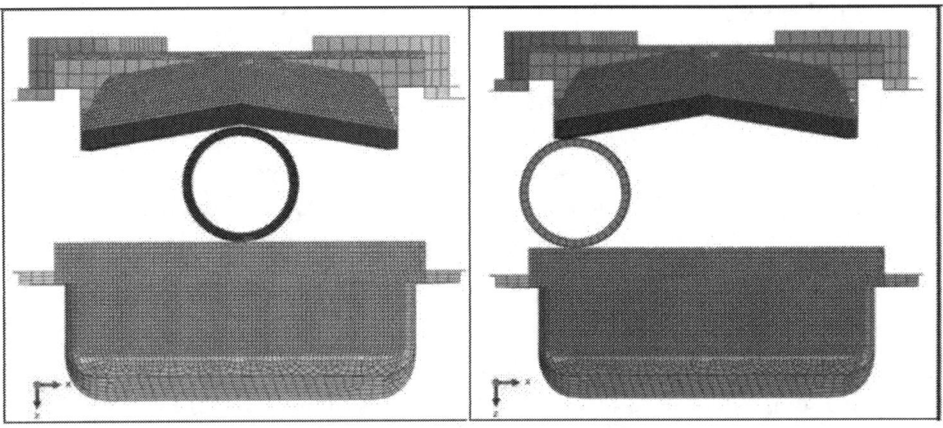

Drill pipe properly centered in BSR | Drill pipe buckled beyond BSR cutting zone

Figure 1. Location of drill pipe inside BSR and its effect on BSR cutting ability

The DNV report made a number of detailed recommendations to industry regarding BOP design and operation issues. These included:

- Study of elastic buckling if drill pipe – as this is what prevented the BSR from functioning properly
- Study of the shear blade surfaces of shear rams – as they could not shear the off-center drill pipe
- Study of well control procedures or practices – as the timing of the closure of the Upper Annular and the Upper VBR's contributed to the drill pipe buckling and moving off center
- Better testing of back-up control systems – as some system components did not perform as intended.
- Better processes to prevent common mode failure of back-up control systems
- Provide better means to verify the operation or state of various components of BOPs in an emergency
- Study of the effectiveness of Remotely Operated Vehicle interventions
- Stipulating requirements for back-up control system Performance – as now for primary control functions

11. CSB Investigations (Vols 1-2)

The US Chemical Safety Board, which is well known for its onshore investigations, undertook an investigation into the Macondo offshore event at the request of Congress. They issued two reports in June 2014. Vol 1 addresses the relationship between the parties involved and describes the many technical challenges associated with the drilling at Macondo. Vol 2 is much more detailed and this focuses on the CSB technical findings on the Deepwater Horizon BOP and its ongoing management as a safety critical element (SCE).

The CSB participated in some of the forensic examinations of the BOP and carried out its own additional tests. The CSB noted several opportunities to improve US regulations. These should include risk based concepts with a focus on major accident events; that safeguards for such accidents be based on a sound philosophy; that there should be an explicit target for risk reduction (e.g. ALARP); and that Safety Critical Elements (SCE's) be identified and managed through life to attain defined performance standards.

The report addressed a number of specific faults for the BOP, but the CSB decided not to make detailed BOP design recommendations, rather to focus on a higher level on issues relating to all Safety Critical Elements. These include that BSEE should develop regulations requiring SCE's to be identified and performance standards defined for each, and that risks be reduced to an ALARP target level. SCE's should be monitored through life; that there should be an independent verification scheme for these; and deficiencies identified be documented and corrected. The CSB recommended that API support this process by developing guidance documents for identifying SCE and reducing risks to ALARP. More specifically also, the API should revise Blowout Preventer Equipment System for Drilling Wells (API Standard-53, 4th edition) to establish additional component level tests to supplement the current system level tests.

12. Center for Catastrophic Risk Management (UC Berkeley)

A group of experienced risk management and organizational specialists resides at the UC Berkeley Center for Catastrophic Risk Management. The final report was issued in March 2011, and this was preceded by 3 progress reports. The study group addressed both technical and organizational causes.

In the three progress reports the Center concluded: 1) this disaster was preventable had existing progressive guidelines and practices been followed; 2) the failures (to contain, control, mitigate, plan, and clean-up) appear to be deeply rooted in a multi-decade history of organizational malfunction and shortsightedness; and 3) thus, as a result of a cascade of deeply flawed failure and signal analysis, decision-making, communication, and organizational - managerial processes, safety was compromised to the point that the blowout occurred with catastrophic effects. At the time of the Macondo blowout, BP's corporate culture remained one that was embedded in risk-taking and cost-cutting.

The Center recommendations included:

- Develop technology systems between government and industry to develop high hazard wells sustainably.
- Develop institutions capable of effective risk assessment and management of high hazard wells.
- Implement such risk based systems in the context of high reliability organizations and capable of responding to major blowout events.
- Utilize validation projects to show such characteristics.

13. National Academy of Engineering

The National Academy of Engineering convened a review panel of 15 experts in response to a request from the Department of Interior. Their report included main sections on well design and construction, the BOP system, MODUs, industry management of offshore drilling, and regulatory oversight; each of these with multiple findings, observations and recommendations. Appendix C in the report groups together all 38 findings, 23 observations, and 68 recommendations. While these are all important, space does not permit listing all of these here, however the Summary chapter highlights the most important and those 13 recommendations are listed here:

1) Develop guidelines for drilling and completion that address credible risks and safety margins
2) Tests should be defined with acceptance criteria for primary cement and mechanical barriers
3) BOP systems need to provide reliable cutting, sealing and separation, with maintenance and training
4) Instrumentation / expert systems should warn of loss of well control with automatic functions
5) Loss of well control should be mitigated by access to well capping and containment capabilities
6) US regulations should be updated to be a hybrid of prescription with proactive goal-setting
7) BSEE should identify and approve safety critical points in well construction and abandonment
8) A single US agency should have responsibility for an integrated approach to system safety
9) The Lessee should have responsibility for well integrity and the drilling contractor for equipment
10) Industry should carry out more R&D on all aspects of drilling safety
11) There needs to be formal education and training of offshore staff to properly implement system safety
12) Develop industry-wide systems for incident reporting and share lessons learned openly
13) Enhance safety culture through training, addressing human factors, system safety, and indicators

14. Norway PSA – Interim Report DWH

The Petroleum Safety Authority is the safety regulator for offshore developments in Norway. It established an internal review team after the Macondo event, but considered also Montara (2009), Snorre A (2004) and Gullfaks C (2010) events and that work generated an interim report in 2011.

The PSA does not believe that the Macondo accident challenged the critical philosophies underpinning the Norwegian regulatory system such as the division of responsibility for regulatory compliance and the demand for systematic and risk-based compliance with functional requirements. But it did challenge whether the PSA is continuously improving the way it seeks to influence safety. An overall observation is that industry needs to significantly improve how it manages major accident risk and that safety culture must be enhanced.

The need for effective barrier management covering all barrier elements is emphasized. There should be an integrated approach to well barriers, including the principle of two independent and tested well barriers as well as barrier monitoring. The DWH accident confirms the importance of defining the ambition level of performance standards for barrier elements and ongoing monitoring of these. In Norway, the effects of cold climate and aging facilities are important in this regard. The BOP may need to have an associated SIL level to ensure its reliability.

The DWH accident has highlighted the need for an effective capping and containment capability. Furthermore there is a need for better training of key staff in dealing with emergencies. Better systems are needed for disconnection from the well, and for enhanced ignition controls especially related to air intakes of power generators. Maintenance of the BOP was an issue and the PSA reemphasized the need to maintain safety critical equipment.

The PSA highlighted that offshore safety needs a systems perspective – complex systems fail in complex ways and simplistic solutions are not likely to be effective. Initiatives are needed to enhance safety culture and the development of a culture of accountability. Companies need to demonstrate their commitment to safety by participating actively in standards development.

The Interim report was issued in 2011 and the PSA decided that a follow-up report was justified in 2014 addressing new information that appeared after that date. The central findings did not change, but extra emphasis was given in several areas, and some recent changes are documented.

The PSA studied well control events on the Norwegian Continental Shelf and identified four main challenges:

- the need for a stronger commitment to technical measures to improve safety
- a bigger commitment to planning, barrier management and better-adapted risk analyses
- paying greater attention to major accident risk – more investigation of incidents
- creating operational parameters for good collaboration in the operator/supplier hierarchy.

In the area of risk assessment, the PSA noted that the DWH accident demonstrated a need for better risk management and processes which lead to more robust solutions and operations. Barrier management has been a PSA focus area for several years, but there is an ongoing need to maintain industry commitment to barrier management and to link this to overall risk management. Finally, with respect to safety culture and organization, the DWH accident highlighted the need for companies to be open to the unknown and paying attention to signals of vulnerability. The Norwegian industry has initiated several projects relating to safety culture, organization and management. The NORSOK D-010 standard for well control has been updated, including key performance indicators and for change management.

PSA agreed with BSEE's four main expectations for the BOP:

- that shear rams have the capacity to cut the equipment being run into the well through the BOP
- that the BOP is maintained as a critical safety system
- that real-time condition monitoring is established by instrumenting the BOP
- that requirements are developed for the competence of BOP maintenance personnel.

15. OLF – DWH Lessons Learned and Follow-up

The Norwegian Oil Industry Association (OLF), now renamed Norwegian Oil and Gas Association, in collaboration with NOFO (the Norwegian Clean Seas Association for Operating Companies) carried a review of the DWH incident, compared regulations in the US and Norway, and made recommendations to enhance Norwegian operations and engineering standards. The group made 45 recommendations for improvements, summarized below.

Prevention: The focus for improvement is to enhance the Norwegian NORSOK drilling standards (D-001: drilling facilities and D-010: well integrity in drilling and well operations). Specific recommendations address critical cementing jobs, lockdown requirements for tubing and casing hangers, negative pressure testing, fluid displacement requirements, well control exercises, BOP backup control systems, and enhanced BOP testing requirements. Enhancement

to well management systems, process safety, and teamwork and communications were also proposed. Action on drill crew expertise is being addressed by OGP (17).
Intervention and Response: Capping and containment is being addressed through the OGP SWRP Project (17). The Macondo unified command system proved to be effective and is now considered a best practice approach and it is recommended for Norway. There are lessons to be learned about protecting workers from chemical exposure during a response through protective equipment and training. OLF commissioned the Norwegian Institute for Water Research to assess the environmental impact to fisheries and of dispersant usage. Fortunately, environmental damage from DWH was less than expected and the use of in-situ burning and underwater use of dispersants was considered beneficial. There remain gaps in knowledge regarding Corexit dispersant and its potential to delay natural degradation of oil. While fisheries have recovered, there were very large costs to the fishing industry due to an extensive closure of the resource. OLF provided a useful summary of recommendations related to well control and response issues (Figure 2).

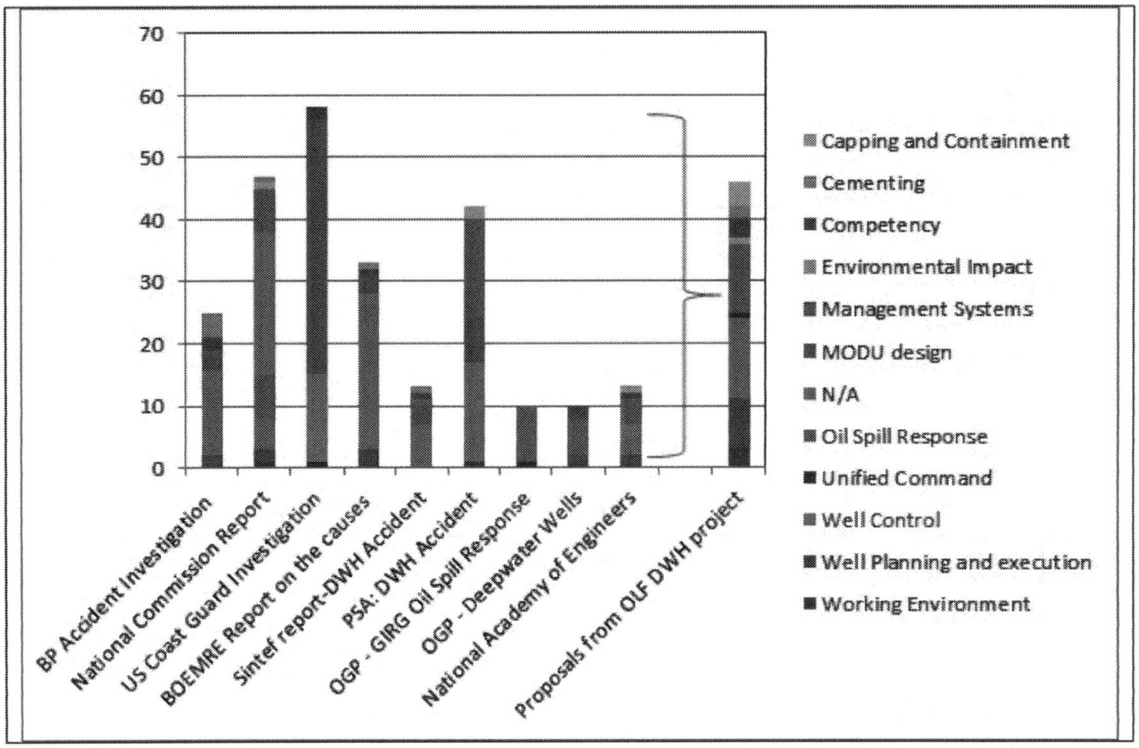

Figure 2. Recommendation count and types from 10 major Macondo inquiries

16. UK Health & Safety Laboratory - DWH Fire & Explosion Issues

The HSE requested the HSL to study fire & explosion issues related to the Macondo event. HSL relied on several key investigations including the BOEMRE, USCG, BP, Transocean, and RMI reports (all summarized above). The HSL focus was on technical issues related to fire and explosion. Based on their analysis, the HSL made several recommendations:

- Control of ignition sources across the rig in the event of blowout should be considered in ALARP demonstrations

- Substantial reductions in risk can be achieved by improving the reliability of detection systems and simplifying or eliminating reliance on human responses
- Preventing the ingestion of gas into large spaces deep within the rig should be a priority
- If there is a residual risk of gas accumulation in internal spaces, the consequences of explosion should be considered. The risks of explosions venting through vulnerable areas of the rig should be minimized; this may involve strengthening some internal partitions and/or providing explosion relief to the outside.

The HSL report summarized fire and explosion related recommendations in other Macondo investigations and found that these generally match UK offshore recommendations. Some later chapters work through examples of mitigations of ignition likelihood, flammable gas dispersion, and explosion damage.

17. Conclusions

As can be seen there have been many investigations into the Macondo accident and these have addressed several different aspects of the accident. Some investigations took a purely technical cause perspective whereas others emphasized more compliance with regulations or addressed organizational or human causes. Early investigations provided very detailed tutorials on deepwater drilling challenges. An indication of the range of recommendations is shown in Table 3 and a more specific focus on technical recommendations appears in the OLF Figure 8.
Many of the key recommendations have been adopted in one form or another. The USA has seen major changes to the MMS regulator, now BSEE and BOEM, with an updated style of operation emphasizing goal-based safety and increased inspector staff to allow greater presence offshore. BSEE has issued new requirements for drilling safety, for BOP recertification, negative pressure tests, and professional engineer sign-offs on casing and cement, and a requirement to estimate worst case blowout events. API and IADC have worked on an interface requirement between lessee and contractor, and several new API standards have been developed. SEMS I and SEMS II have been implemented with the Center for Offshore Safety defining the protocols and approving third party audit service providers. BSEE has provided guidance on safety culture and is working with another federal agency to implement a confidential reporting system. BSEE has established the Ocean Energy Safety Institute to research several longer term problems (e.g. risk, reliability data, BAST). USCG has issued guidance on additional fire and explosion assessments it would like to see. There have been consortia established in the USA and by OGP at 4 international locations to provide emergency response support. There have been standards updates for drilling and well control by API and by NORSOK, and they have made their safety standards freely available. Detailed assessments of fire and explosion lessons have been made and these are starting to find their way into designs.

Table 3. Range of Findings and Recommendations

No.	Title	Prevention	BOP Design & Operation	Contain & Respond	Manageme nt & culture	US Regulatory	International Regulatory
1	DWH Commission Main	X	X	X	X	X	
2	DWH Chief Counsel's report	X	X		X	X	
3	BP Investigation	X	X	X			
4	Transocean Investigation	X	X	X			
5	USCG	X	X	X		X	X
6	BOEMRE	X	X	X		X	
7	Republic of the Marshall	X	X	X		X	X

	Islands						
8	Adm Thad Allen Report			X		X	
9	DNV GL Forensic Investigation		X				
10	Chemical Safety Board	X	X		X	X	
11	Center for Catastrophic Risk Management (UC Berkeley)	X	X	X	X	X	
12	National Academy of Engineering	X	X	X	X	X	
13	National Research Council			X			
14	US District Court Eastern Louisiana	X	X	X			
15	Norway Petroleum Safety Authority – Interim	X	X	X	X		X
16	Norway Petroleum Safety Authority – Final	X	X	X	X		X
17	OGP	X	X	X	X		
18	OLF	X	X	X	X		
19	SINTEF	X	X	X	X		
20	UK HSL			X			
21	US Transportation Research Board				X	X	

Environmental findings of long term damage are too soon to be available, but initial indications are that the damage is less than might have been feared.

There have been changes overseas. The EU has decided to adopt a safety case approach (similar to that in the UK) for all offshore developments in the EU, and Australia expanded coverage of its regulator (NOPSEMA) to address drilling and environmental impacts to achieve close to a single offshore regulator. The UK and Norway both believe their goal-setting approach was not challenged by the Macondo accident, but that many detailed aspects of drilling safety did need enhancement and HSE, PSA and OGP are working on these issues.

There are some important recommendations for the US regulatory system which have not yet been adopted. These include suggestions to adopt a safety case approach with a greater focus on risk assessment with a risk target, and to nominate safety critical items with defined performance standards. It is not yet known if these will be adopted, or perhaps addressed in some other equivalent manner.

References

Macondo Inquiry references – see Table 1 in above text.

DNV GL (2015) The Macondo Event: Summary of Inquiry Findings and Recommendations, DNV GL, Oil & Gas Advisory Services, Houston.

USCG (2014) Outer Continental Shelf Units-Fire and Explosion Analyses, Notice Of Recommended Interim Voluntary Guidelines, Federal Register, May 2, 2014.

Operational Risk Management Aided by Continuous Monitoring of Barriers

11th Global Congress on Process Safety
Apr 27th, 2015

Ed Silva
Sergio Saad

Summary

"In the process industry, post-Macondo (...), it is essential that operators move away from just demonstrating that they understand the level of risk that personnel and the environment are exposed to. We all need to move towards demonstrating that we have adequate and effective barriers and controls in place to manage that risk effectively." (Bower-White, 2013)

Business challenge

Stop working with bow-tie diagrams statically by starting to continuously monitor the barriers' states, in order to

Solution

Use of event-processing technique to automatically update the states of barriers of bow-tie diagrams, based on data from

Results and benefits

Successful prototype implementation and execution;
Possible application to real situations;
Aid in the decision-making in process safety.

Problem statement

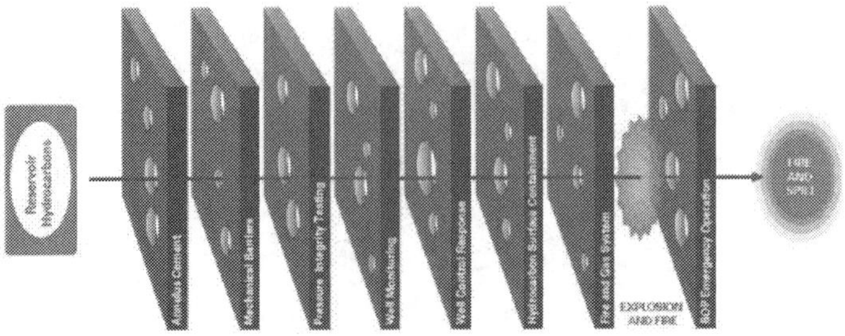

- Continuous monitoring of the states of barriers as a way of as a way to manage operational risk more effectively – no concrete solutions
- Difficulties
  - Information about barriers' states may have origin in several sources
  - Evaluation of the current state of each barrier is done manually ⇒ outdated/inaccurate information
- Alternative
  - Event processing system that provides the automatic updating of the barriers' states to allow their continuous monitoring
  - Constantly updated view of the integrity of barriers ⇒ actual level of risks to which the operations are

Conceptual block diagram

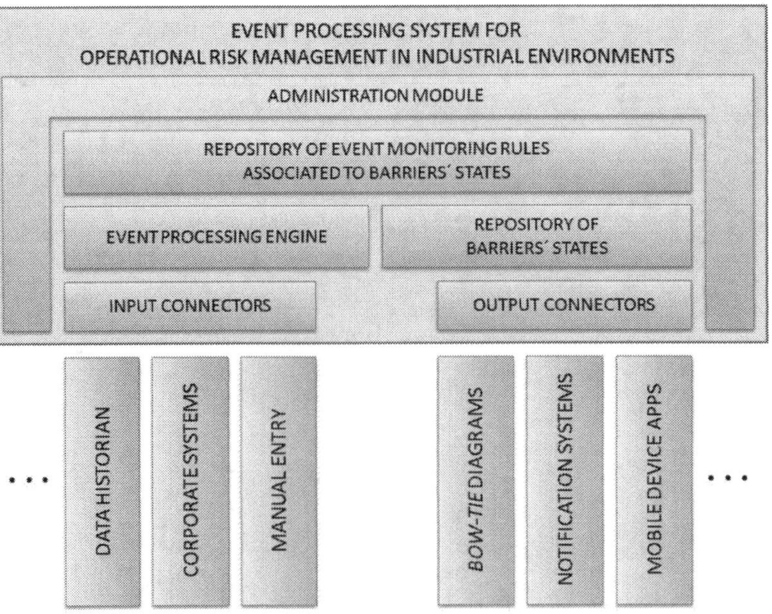

Prototype architecture

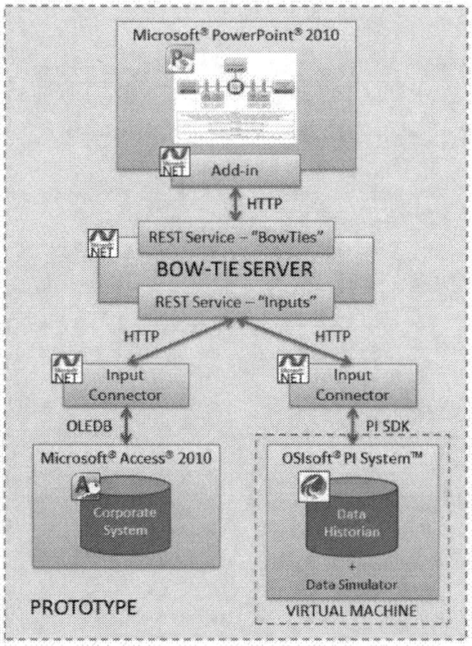

PowerPoint® add-in continuously retrieves bow-tie data from server at user-specified rate and updates barriers' colors

New input values trigger re-evaluation of subscriber barriers' states through re-evaluation of their rules' expressions

Input connectors report value updates to server

Data sources

Business rules and requirements

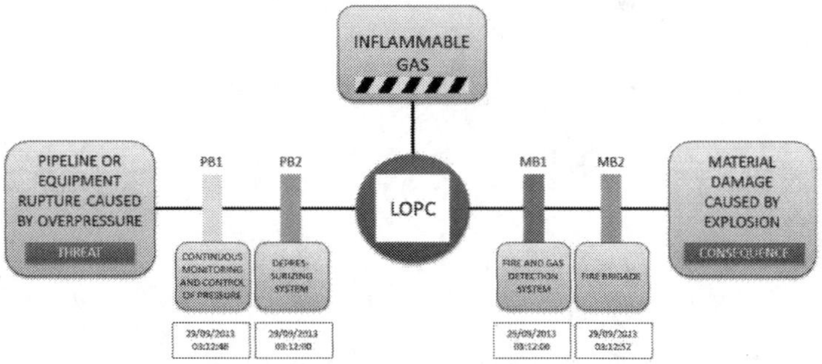

- Barrier states: **Intact**, **Degraded**, **Vulnerable**
- Rules: **boolean expressions + states**
  - Expressions are evaluated according to the order of precedence established by user;
  - Greater precedence true expression = **event** (resulting state);
  - At least one true expression at any time.
- Essential features
  - Rules defined and maintained by end user;
  - Separation between the barriers' states' rules definition/processing and their presentation;
  - Timestamp;
  - Drill-down

Conclusions

- The Event Processing technique is a valid alternative to enable the continuous monitoring of bow-tie diagrams' barriers (successful execution of the prototype system)
- System only is not enough to increase operational safety, it needs to be associated to:
  - A risk management method
  - The engagement of engineers to formulate proper monitoring rules
  - A business process
- Further work
  - Application to other barrier-based risk management methods, either in the oil and gas E&P industry itself or in other industries
  - Security and performance aspects

Thank you!

Ed Silva
Health, Safety and Environment Engineer, MSc
ednevessilva AT gmail.com

Sergio Saad
Computer Engineer, MSc
ssaad33 AT gmail.com

A Spreadsheet Application for Evaluating Two-Phase Emergency Venting Transients Utilizing an Interface to a Thermodynamic Property Package

Subin Hada
Evonik Corporation
4201 Degussa Road, Mobile, AL 36582
subin.hada@evonik.com

Robert D'Alessandro
Evonik Corporation (Retired)
4201 Degussa Road, Mobile, AL 36582
Cheme1978@gmail.com

J. Andrew Jones
Evonik Corporation
4201 Degussa Road, Mobile, AL 36582
andy.jones@evonik.com

Prepared for Presentation at
American Institute of Chemical Engineers
2015 Spring Meeting
11th Global Congress on Process Safety
Austin, Texas
April 27-30, 2015

UNPUBLISHED

A Spreadsheet Application for Evaluating Two-Phase Emergency Venting Transients Utilizing an Interface to a Thermodynamic Property Package

Subin Hada
Evonik Corporation
4201 Degussa Road, Mobile, AL 36582
subin.hada@evonik.com

Robert D'Alessandro
Evonik Corporation (Retired)
4201 Degussa Road, Mobile, AL 36582
Cheme1978@gmail.com

J. Andrew Jones
Evonik Corporation
4201 Degussa Road, Mobile, AL 36582
andy.jones@evonik.com

Keywords: Two-phase relief sizing, bubbly flow, Churn-turbulent flow, process safety

Abstract

This paper presents a numerical method for determining emergency relief requirements for a top-vented cylindrical vessel or batch-type chemical reactor through an ideal nozzle using a spreadsheet application interfaced to a CAPE-OPEN compliant thermodynamic package. A calculation algorithm based on the DIERS recommended homogeneous equilibrium model are presented for single component system. Fluids exhibiting partial vapor-liquid separation with either churn-turbulent or bubbly flow behavior and disengagement parameters greater than unity were considered. A vessel condition subjected to energy input from an uncontrolled external source such as fire was considered with no chemical reaction. Analytical expressions are used; however, more rigorous solution is achieved by relaxing the assumptions of constant physical properties and constant relieving vent rate of the fluid. The use of a spreadsheet significantly simplifies the tedious unsteady state emergency relief calculations while the use of Aspen Properties® provides reliable thermodynamic data throughout the venting transients.

1. Introduction

Emergency pressure relief is often a dynamic event with rapidly changing physical properties. Having reliable physical and thermodynamic data is vital to a successful safety relief valve design. In addition, this property data must be utilized as a function of temperature, pressure, and composition. As a result, solving emergency relief problems that account for changing physical

Note: Do not add page numbers. Do not refer to page numbers when referencing different portions of the paper

properties usually require numerical methods that are typically too complex for spreadsheet applications.

The industrial reality involves the design of safety relief valves for two-phase flow [1, 2, 3]. Calculating two-phase venting requirements using most common analytical solutions like Leung [4, 5] and D'Alessandro [6, 7] assume a constant venting rate and constant physical properties for the fluid over the entire venting process. While one assumption may be more severe than the other, they both put some degree of limitations to the applicability of the method for sizing a relief device.

Several computational codes like SAFIRE™ [8] and SuperChems™ [9] have been available for a number of years. These codes can perform thermophysical properties and phase equilibria calculations numerically on pure component as well as mixtures. However, these packages are relatively expensive and difficult to use. Therefore, a practical and easy to use computational tool that computes the two-phase emergency venting requirement under realistic condition is desired.

The contribution presented in this paper introduces a technique to estimate the relief requirement by utilizing:

- a readily available MS-Excel™ [10] spreadsheet application to perform all the differential mass and energy balance calculations, and
- a CAPE-OPEN compliant interface like Aspen Properties® [11] to generate all the necessary thermophysical properties of the fluid based on different thermodynamic descriptions of the phase equilibrium and thermodynamic and transport properties of each phase.

Using the "socket and plug" concept, thermodynamic properties from Aspen Properties® are imported directly into a MS-Excel™ spreadsheet. Here, Aspen Properties® acts as a thermodynamic plug and MS-Excel™ acts as a thermodynamic socket.

The use of a spreadsheet significantly simplifies the tedious unsteady state emergency relief calculations while the use of Aspen Properties® provides reliable thermodynamic data throughout the venting transient.

The novelty of the presented method is that the more rigorous spreadsheet approach relaxes the assumptions of constant physical properties and constant relieving vent rate of the fluid, thereby, relaxing the limits on the applicability of the available analytical methods.

In addition, in industry, spreadsheet calculations are widely used and when combined with a CAPE-OPEN physical properties package, spreadsheets can be an extremely powerful tool for complex calculations. The technique offered is a good engineering practice that ensures the consistency and reliability of the data used and the tools used are readily available.

2. Methodology

The presented paper focuses on the closed form approximate analytical solutions proposed by D'Alessandro [6, 7] for the two-phase emergency relief requirements of top-vented vertical cylindrical containers when the fluids involved exhibit partial disengagement. D'Alessandro's expressions have an advantage over Leung's expression [4, 5] as they allow values for the disengagement parameter other than unity.

The presented work is limited to tempered (vapor pressure) systems where evaporative cooling is available, i.e., the vessel is partially filled with volatile liquid (pure or mixture) that vaporizes or flashes during the depressurization of the vessel. The methods presented in this paper are based on contributions from the Design Institute for Emergency Relief Systems (DIERS) and the American Institute of Chemical Engineers (AIChE) [1].

2.1 Flow Models

Bubbly and churn-turbulent vessel hydrodynamic flow models are considered in this paper to estimate the degree of vapor-liquid disengagement (liquid swell) as a function of vapor throughput. The vent system was modeled as a simple ideal nozzle. A mathematical model to estimate the vent mass flux and volumetric discharge rate as a function of the vessel pressure and the vent entrance quality was modeled using the homogeneous-equilibrium flow model. Figure 1 shows the schematic diagram of the flow regime and flow models that could be considered in the study. The boxes with solid border lines are the cases considered in this work.

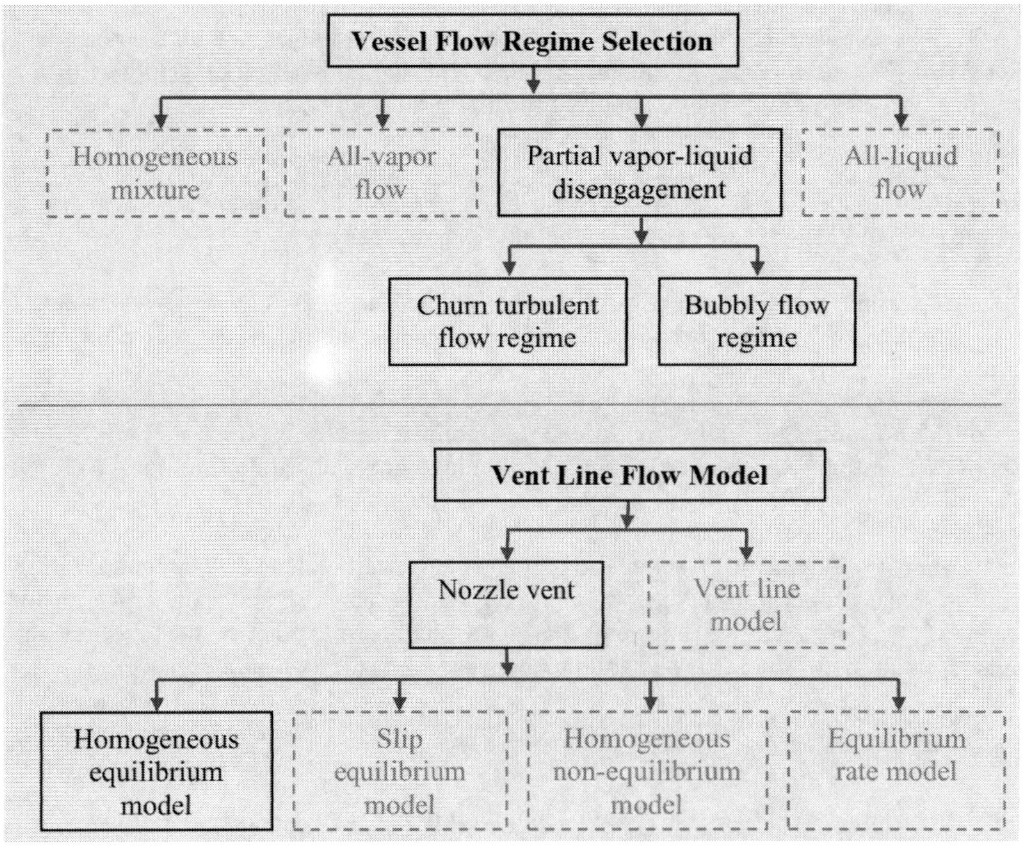

Figure 1: Schematic diagram of flow regime and flow models considered.

Using the DIERS coupling equation, the vessel models and the vent flow models are coupled to provide integrated system models for calculating the volumetric discharge rate from the vessel into the vent [1].

2.1 Spreadsheet Development

The rate of change of temperature, pressure, and composition of the vessel contents with time, as a result of external heat loads from fire, were determined by successive heat and mass balances on the vessel contents at small time increments. The calculations were based on ordinary differential and algebraic equations that were integrated numerically forward in time by the explicit Euler method [12].
Once the spreadsheet was set up to do these calculations, Visual Basic for Application (VBA) [10] was used to automate the calculations for every time step that requires solving a series of finite differential energy and mass balances. For every time step, as the temperature and pressure, the physical and thermodynamic properties were obtained from the thermodynamic modeling package Aspen Properties®. The process is repeated until the relief calculation stop criterion (e.g. complete disengagement condition) is reached.

3. Case Study

Pure propane subjected to external fire without undergoing any chemical reaction was used to illustrate the application for one-component system utilizing the methods shown in this paper. The example is taken directly from D'Alessandro's work [6, 7]. The parameters are summarized in Table 1. The churn-turbulent and bubbly flow fluid behaviors with disengagement parameters greater than unity are considered.

Table 1: Propane storage vessel subjected to fire – given parameters.

Vessel			
Diameter	D_c	m	3.49
Height	h_c	m	10.47
Cross Sectional Area	A_{xc}	m^2	9.55
Volume	V_c	m^3	100.0
Vessel Condition at Relief Device Set Point			
Pressure	P_s	bara	4.5
Temperature	T_s	K	271.38
Mass Content	m_s	kg	50483
Void Fraction	α_s	unitless	0.0523
External Heat Input			
Heat Flux	q_F	kW/m^2	26.41
Heat Input	Q_T	kJ/s	3126
Bubbly Fluid Disengagement Parameter			
Bubble Rise Parameter	k	unitless	1.18
Churn Turbulent Fluid Disengagement Parameter			
Bubble Rise Parameter	k	unitless	1.53

Leung and D'Alessandro assumed constant venting rate and physical properties throughout venting transient whereas this work relaxes this assumption by letting the vent rate and the physical properties vary. It was found that the average vent rate (expressed as a geometric mean of vent rate at set condition and at vent rate at turnaround) calculated using short-cut method was close to vent rate at set condition using more rigorous method.

Table 2: Comparison of required vent rates calculated using short-cut method with more rigorous method at fixed set pressure of 450kPa and 10 % overpressure.

		Disengagement Parameter	Vent Area	Average Vessel Void Fraction		Vent Entrance Quality	Required Vent Rate		
			m^2	At Turnaround	At Disengagement	At Turnaround	At Set Condition	At Turnaround	Average
Bubbly Fluid	Leung	**1.00**	-	0.5639	0.9825	0.0491	-	-	117.90
Bubbly Fluid	D'Alessandro	**1.00**	0.0338	0.5675	1.0000	0.0546	125.06	105.65	114.93
Bubbly Fluid	D'Alessandro	**1.18**	0.0303	0.4772	0.8275	0.0644	111.40	92.10	101.29
Bubbly Fluid	This Work	**1.00**	0.0297	0.5502	0.9995	0.0646	113.64	97.47	-
Bubbly Fluid	This Work	**1.18**	0.0264	0.4696	0.8132	0.0793	100.53	83.55	-
Churn-Turbulent Fluid	Leung	**1.00**	-	0.1898	0.4731	0.2608	-	-	30.00
Churn-Turbulent Fluid	D'Alessandro	**1.00**	0.0117	0.1908	0.4830	0.2592	35.35	25.70	30.14
Churn-Turbulent Fluid	D'Alessandro	**1.53**	0.0107	0.1743	0.3701	0.2932	31.67	22.75	26.84
Churn-Turbulent Fluid	This Work	**1.00**	0.0100	0.1797	0.2751	0.3949	31.15	20.05	-
Churn-Turbulent Fluid	This Work	**1.53**	0.0092	0.1652	0.2283	0.4701	27.08	16.96	-

The short-cut results are more conservative than the more rigorous results. These results are consistent with the recommendations given by Leung [4, 5] and D'Alessandro [6, 7]. Figure 2 graphically compares the vent rate estimated using D'Alessandro's method and more rigorous method presented in this paper for Co=1.53 and 10% overpressure.

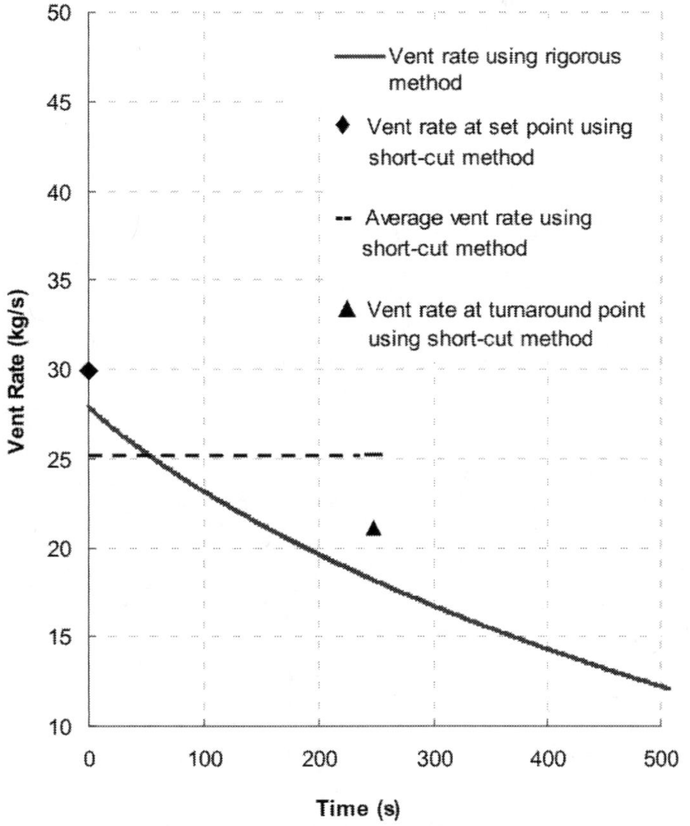

Figure 2: Vent rate using short-cut method and more rigorous method at fixed set pressure of 450kPa and 10 % overpressure with Co=1.5.

3.1 Comparison with SuperChems Solutions

SuperChems^{TM} for DIERS [9] provides a standard tool for designing emergency pressure relief systems which incorporates the latest DIERS pressure relief design techniques and research findings. The software tool has been extensively validated against experimental data to ensure its accuracy.

The transient behaviors for the example of pure propane subjected to external fire are compared to the results generated by SuperChems. The comparisons are plotted in Figure 3. It is clear from the figure that the results are consistent with each other.

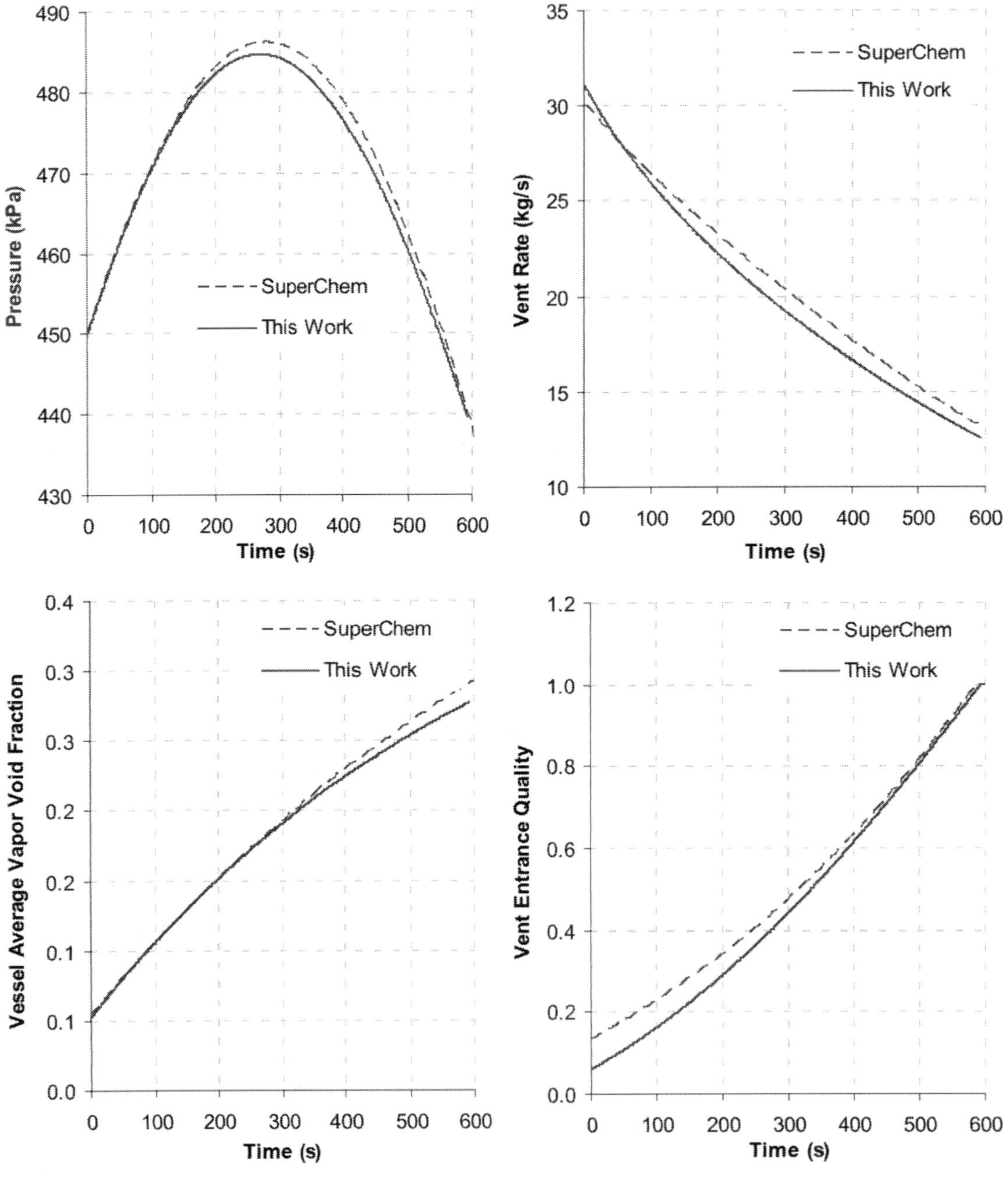

Figure 3: Comparison of more rigorous solutions predicted by method proposed in this paper and solution generated by SuperChems for the example of pure propane with C_o=1.0.

4. Summary and Conclusion

A spreadsheet application for analysis of emergency relief scenarios utilizing an interface to a thermodynamic property package is possible. The spreadsheets were tested for tempered pure and binary systems with relief scenario of external uncontrolled heat input to the fluids exhibiting partial disengagement of bubbly and churn-turbulent flow regime.

The use of a spreadsheet significantly simplifies the tedious coupled differential mass and energy balance calculations while the use of Aspen Properties® provided reliable thermodynamic properties used for numerical evaluation of the temperature and pressure transients. In industry, such a practice ensures the consistency and reliability of the data used. Moreover, in industry, the use of a spreadsheet is practical for essentially all practicing engineers.

5. References

[1] Fisher, H. G., Forrest, H. S., Grossel, S. S., Huff, J. E., Muller, A. R., Noronha, J. A., Shaw, D. A., & Tilley, B. J. *Emergency relief System Design Using DIERS Technology, The DIERS Project Manual*. AIChE, NY (1992).

[2] Boccardi, G., Bubbico, R., Piero Celata, G. & Mazzarotta, B. "Two-Phase Flow through Pressure Safety Valves. Experimental Investigation and Model Prediction." *Chemical Engineering Science*, 60, 5284-5293 (2005).

[3] Darby, R., Self, F. E., & Edwards, V. H. "Properly Size Pressure Relief Valves for Two-Phase Flow. *Chemical Engineering*, 68–74 (2002).

[4] Leung, J. C. "Simplified Vent Sizing Equations for Emergency Relief Requirements in Reactors and Storage Tanks." *AIChE Journal*, 32(10), 1622-1634 (1986).

[5] Leung, J. C. "Overpressure During Emergency Relief Venting in Bubbly and Churn-Turbulent Flow." *AIChE Journal*, 36(6), 952-958 (1987).

[6] D'Alessandro, R. N. "Emergency Venting Requirements for Tempered Systems Considering Partial Vapor-Liquid Separation with Disengagement Parameters Greater than Unity, Part I: Model Development". *Process Safety Progress*, 23(1), 1-15 (2004).

[7] D'Alessandro, R. N. "Emergency Venting Requirements for Tempered Systems Considering Partial Vapor-Liquid Separation with Disengagement Parameters Greater than Unity, Part II: Application". *Process Safety Progress*, 23(2), 86-98 (2004).

[8] Shaw, D.A. "SAFIRE Program for Design of Emergency Pressure Relief Systems." *Chemical Engineering Progress*, 86(7), 14-17 (1990).

[9] Melhem, G. A., and Fisher, H. G. "An Overview of SuperChems™ for DIERS: A Program for Emergency Relief System and Effluent Handling Designs." *Process Safety Progress*, 16, 185-197 (1997).

[10] Excel 2003™, Microsoft Corporation, Redmond, Washington, 2003.

[11] ASPEN PLUS™ Release 9, Volume 2: Physical Property Methods and Models, Volume 3: *Physical Property Data* (1995).

[12] Hoffman, J.D. *Numerical Methods for Engineers and Scientists*. 5th ed., CRC Press, Baca Raton (2001).

Overlooked factors in Pressure Relief Systems Design

Mirna Zamora
Contek Solutions LLC
6221 Chapel Hill Blvd., Ste 300. Plano, TX 75093
mzamora@contekllc.com

Steve Streblow
Contek Solutions LLC
6221 Chapel Hill Blvd., Ste 300. Plano, TX 75093
sstreblow@contekllc.com

Prepared for Presentation at
American Institute of Chemical Engineers
2015 Spring Meeting
11th Global Congress on Process Safety
Austin, Texas
April 27-29, 2015

UNPUBLISHED

Overlooked factors in Pressure Relief Systems design

Mirna Zamora
Contek Solutions LLC
6221 Chapel Hill Blvd., Ste 300. Plano, TX 75093
mzamora@contekllc.com

Steve Streblow
Contek Solutions LLC
6221 Chapel Hill Blvd., Ste 300. Plano, TX 75093
sstreblow@contekllc.com

Keywords: pressure relief system, flare system design, flare system evaluation, relief valve, pressure safety valve, PSV, pressure safety valve outlet lines sizing, design temperature, noise, emissions, dispersion, thermal radiation, smoke.

Abstract

Typically a pressure relief system design is related to relief valve sizing and pressure drop calculations. While these represent the core of the design, usually other important factors are left out and should be considered according to industry standards, codes and recommended practices.

The proper design of a pressure relief system is crucial to the plant's overall safety and the environment. Unfortunately, since this system does not produce revenue, its importance and design is often underestimated.

A comprehensive pressure relief system evaluation ~~requires the~~requires the consideration of some factors that are often neglected and involve the following:

- Sizing the PSV outlet lines based on the relief devices relieving independently and not only on the global relieving scenarios, as is the most common practice.
- Piping and equipment design temperature consideration since it can be exceeded by the relieving temperature in some scenarios.
- Noise evaluation.
- Emissions dispersion and thermal radiation.
- Other factors such as volatile emissions and smoke production.

This paper will identify those factors that are recommended by basic industry standards, codes and practices and the implications of overlooking them. Guidelines such as the API 14E, API 520, API 521 and ASME B31.3 are going to be discussed, as well as some federal and environmental codes. Incorporating these factors will result ~~in a safer and more effective pressure relief systems~~in safer and more effective pressure relief systems.

Note: Do not add page numbers. Do not refer to page numbers when referencing different portions of the paper

1. Introduction

As engineering contractors, we have been involved in many projects that require the evaluation or design of flare and vent systems. Generally the scope of work of those projects was limited only to the pressure drop and velocity calculations through the system to size the pipes piping and the back pressure calculation to evaluate the PSVs, leaving behind the evaluation of factors such as noise, extreme relieving temperatures, dispersion and thermal radiation described in engineering codes, guidelines, industry standards and recommended practices.

Evaluating those factors sometimes requires the identification of different cases of overpressure. Usually the only one considered is the case that produces the maximum flowrate travelling through the system. Even though this can be the worst case for pressure drop calculations, it might be not the worst case to evaluate factors that are going to be described in this article and that can affect the design of the system dramatically.

This paper will present how those factors affect the design of the flare system by presenting some simple case studies with components and conditions we find everyday in the oil and gas industry as well as some of them based on personal experience working in flare system projects.

I have seen how an evaluation of the flare system considering thoseSome of these factors has changedcan change the entire design of the relief system, especially if this the review is done "after built", ". and representsThis can represent a huge large economic investment and deferred production that for some companies. are not willing to pay if this wouldn't compromise the safety of the plant and represent future legal actions. In most of the cases companies were surprised by the findings obtained by those calculations. Let's have a closer look of those factors and the consequences of overlooking them.

2. Overlooked Factors in Pressure Relief Systems Design

2.1 Sizing of the PSV Outlet line

It is common to perform a pressure relief study and size the flare system by evaluating only the global scenarios that overpressure the system by relieving multiple devices at a time. While this consideration is good for the header and sub-header sizing, this evaluation may undersize the individual PSV outlet pipes or tail pipes. The evaluation of PSVs relieving independently will give the most conservative size of the PSV outlet pipe especially if we havethere is a maximum Mach number limit. This is referenced in API 521[1] as: "If the user has established a velocity criterion for tail pipes, the, the maximum velocity in a tail pipe should be calculated with the single source (the relief or depressurization device) as the only source discharging into the disposal system."

When sizing the PSV outlet line using only the flowrate for the one individual device, the back pressure at any point of the system is lower than it would be if the system is evaluated with the discharge of several PSV simultaneously combining their required flowrates. This lower pressure (back pressure) produces lower density fluids that will travel at higher velocities and will require a larger pipe size if one ~~try~~ tries to maintain the velocity below a maximum Mach number. Figures 1 and 2 show the results of a tail pipe sized using the total flowrate of several PSV relieving simultaneously vs. using a single PSV relieving flowrate.

The example was performed using a commercial process simulator available in the market (ProMax® version 3.2) and represents a simple flare system with only two PSVs (PSV-1 and PSV-2) that relieve into a 8" flare header. To make this example more representative and to size the PSV-1 tail pipe, a low flowrate for PSV-1 was used versus a large flowrate used for PSV-2 that may represent the combination of several PSVs.

PSV-1 is a conventional PSV with a flowrate of 1,000 lb/hr, set pressure of 300 psig, fire case relieving pressure of 363 psig and temperature of -30.4 °F. Composition is 80% mole Methane, 15% mole Ethane and 5% mole Propane.

PSV-2 is a conventional PSV with flowrate of 40,000 lb/hr, set pressure of 500 psig, fire case relieving pressure of 605 psig and temperature of 50 °F. Composition is 40% mole Methane, 50% mole Ethane and 10% mole Propane.

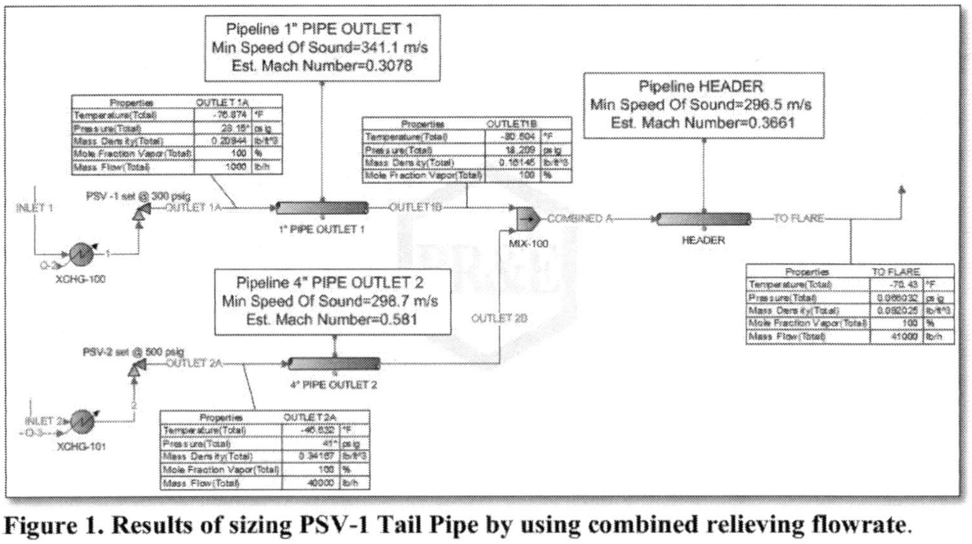

Figure 1. Results of sizing PSV-1 Tail Pipe by using combined relieving flowrate.

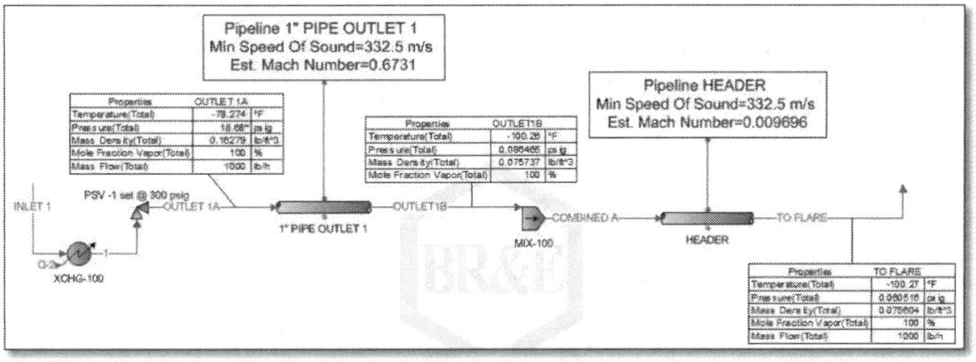

Figure 2. Results of sizing PSV-1 Tail Pipe by using its individual relieving flowrate.

From these figures, it is clear that the density of the fluid in PSV-1 tail pipe decreases to about half from Figure 1 to Figure 2 and the velocity (Mach number) in the PSV-1 outlet pipe (1" PIPE OUTLET 1) is higher, more than double, in Figure 2.

For example, if 0.6 is the maximum Mach number, it will require enlarging the PSV-1 outlet pipe from 1" to a bigger size without affecting the size of the main flare header. The criteria for maximum Mach number for tail pipes varies from company to company and from engineer to engineer, but is commonly 0.5 to a maximum of 1.

For example if the designer is using FlareNet® or UniSimFlare® software to size the tail pipes, sub-headers and headers, the default values are 0.3 for headers and 0.5 for tail pipes, but those values can be customized.

The application of this theory has more impact on the size of the PSV outlet pipe if:

- The PSV whose outlet ~~pipe~~ piping is going to be evaluated is a conventional PSV with a low set pressure and handles a low flowrate relative to the total flowrate of the system, and
- The common flare header size is relatively big compared to the size of the PSV tail pipe to be evaluated.

2.2 Piping and equipment design temperature

Typically a flare system collects the discharge of several relief devices which can handle ~~a~~an extensive range of products with a wide variety of relieving temperatures. Therefore, it can be easy to miss the scenarios that relieve more extreme temperatures into the system. This is especially true if these devices are excluded from the cases that produce larger flowrates into the system, ~~which probably~~which probably are the ~~only~~ cases ~~that are going~~ to be considered if the flare system is ~~going to be~~ evaluated ~~by its~~based on pressure drop and velocity.

Particular attention should be paid to relief devices that handle heavier hydrocarbons whose temperatures in fire case relief event could get extremely high compromising the integrity of the downstream equipment and piping and the safety of the whole unit and people. The additional cost due to the requirement of expensive material in downstream equipment~~s~~ and piping could be very high. Engineers have to be very careful in the evaluation of the piping and equipment materials when working with extreme temperatures. Metal maximum design temperatures for process piping can be found in ASME B31.3[2] Table A-1. In general ~~try to keep it~~temperatures should be below 752°F for ~~Carbon~~ carbon ~~Steel~~ steel and 932°F for ~~Stainless~~ stainless ~~Steel~~steel.~~,~~ ~~although in~~ ASME B31.3[2] can allow~~,~~ ~~it could be found a~~ temperatures as high as 1,100°F for ~~Carbon~~ carbon ~~Steel~~ steel and 1,500°F for ~~Stainless~~ stainless ~~Steel~~ steel but only under certain conditions and for certain grades and specifications.

Table 1 below presents PSV~~'~~s fire case relieving temperatures for different pure components. Calculations were performed in ProMax® and the relieving temperatures were calculated at the bubble point of the fluid at relieving pressure. The higher the molecular weight, the higher the relieving temperature and the less it varies with the relieving pressure. The temperature when handling C15+ is higher than 752°F, which is the generally considered ~~Carbon~~ carbon ~~Steel~~ steel maximum design temperature. For those cases, special evaluation has to be performed on the affected piping since the maximum allowable metal temperature depends on the operating pressure. Tables in ASME B31.3[2] provide this information.

Table 1. Fluids relieving temperatures at different pressures

Fluid	MW	Set Pressure (psig)	Fire Case Relieving Pressure (psig)	Relieving Temperature (°F)
C10	142.28	500	605	652
		100	121	549
C15	212.41	500	605	819
		100	121	753
C20	282.54	500	605	923
		100	121	893
C25	352.68	500	605	1002
		100	121	1000
C30	422.81	500	605	1060
		100	121	1060

When the metal temperature is above the design temperature and to avoid affecting the whole downstream system by the replacement of higher design temperature materials, a recommended solution given by GPSA[3] is the addition of a depressurization system or the inclusion of an additional cooling system~~s~~ as is explained in API 521[1]. In our flare studies we have found temperatures as high as 1,400°F for a fire event relief in a system that handles Heat Medium.

A similar situation is faced with low relieving temperatures, particularly cases where light hydrocarbons are handled. The expansion of those fluids downstream of PSVs ~~could~~ will produce low temperatures that in some cases could be even lower than the material's minimum design temperature. These low temperatures have to be checked to ~~guarantee~~ confirm ~~a~~ the downstream system is designed with special materials that avoid the risk of brittle fracture as well as determine if hydrates or solids CO_2/H_2S/Methane are formed.

Process piping material design specifications ~~are~~ in ASME B31.3[2] ~~where it is~~indicate design limits of -20°F for carbon steel and -92 °F for stainless steel. However, under certain conditions and for certain steel grades and specifications, the minimum metal temperature for carbon steel can be as low as -50 °F and -425°F for stainless steel[2].

As an example, refer to ~~the~~ Figure 2, where PSV-1 handles 80% mole Methane, 15% mole Ethane and 5% mole Propane with fire case relieving conditions at 363 psig and -30~~.4~~ °F reaches temperatures as low as -100~~.3~~ °F at about 0 psig. This temperature is lower than minimum material design temperature for carbon steel and stainless steel. We have found this situation several times, especially close~~d~~ to the knockout drum and flare stack which are the locations with the lowest pressures occur in a flare system. Temperatures as low as -128 °F have been found in our studies.

The isolation of the low relieving temperature system into one specifically designed to handle the low temperatures is recommended in API 521[1] to prevent the low relieving temperatures from affecting the whole flare system.

According to API 520[4] "auto-refrigeration during discharge can cool the outlet of the pressure-relief device and the discharge piping to the point that brittle fracture can occur. Piping design, including material selection, must consider the expected discharge temperature".

~~In any case, the~~ The consequences of missing cases with extreme relieving temperatures could be very expensive, with solutions that involve the modification of the process itself or the replacement of the piping and equipment materials for ones with more extreme design conditions, to avoid its failure with the ~~devasting~~ significant consequences to the ~~whole~~ plant, its people and the environment.

2.3 Noise

High noise level due to an inadequate flare system design can be an environmental and safety concern.

A relief system is undoubtedly associated with high pressure drop and high fluid velocities. Those factors produce turbulence and vibration and ultimately noise in the system. Turbulent flow of compressible fluids, changes in pipe diameters, termination of piping at a manifold or vessel and high flow through devices such as valves, orifices and expansions are some examples of noise sources in a relief flare system.

The noise level in dB, decibels, produced by a single pressure drop device such as a PSV, PRV, orifice ~~or depressurization~~or depressurization valve, when Mach number is 1 or less, is according to API 521[1]:

$$\mathbf{Lw = 10 \log_{10} \{ W^2 [(P1-P2)/P1]^{3.6} (T1/M)^{1.2} \} + 45} \quad \text{[Eq. 1]}^{[1]}$$

Where:
L_W = sound power level, dB
P1 = upstream absolute pressure of pressure letdown source, psia
P2 = downstream absolute pressure of pressure letdown source, psia
T1 = upstream absolute temperature of pressure letdown source, °R
W = gas flowrate, lb/hr
M = molecular weight

Equation 1 is used to calculate the noise level through a device. But, while the fluid travels downstream of the system the noise propagates down the pipe and its level is attenuated by 6 dB for every 100 pipe diameters in length downstream of the sound source[1]. This fact should be considered separately and added to the results of Equation 1.

If more than one device produces noise simultaneously or in series, individual noises calculated using Equation 1 should be added at the pipe junction where those indivuduals devices pipes meets, according to API 521[1]:

$$\sum Lw = 10 \log_{10} [10^{Lw1/10} + 10^{Lw2/10} + ... + 10^{Lwn/10}] \qquad \text{[Eq. 2]}^{[1]}$$

Where:
L_W= individual device (and its downstream piping) noise level, dB.

Specific noise level calculation equations can be found in several PSV vendor's technical information~~,~~ ~~some are~~such as Crosby® [5], GE Consolidated Valve®[6] and Parcol® [7]. Information provided by GE Consolidated Valve was used to produce Table 2, which shows the noise level for some PSVs in a specific flare study project.

Parcol® [7] states that ~~" if~~"if the position of the discharge is high, the noise spread~~ing~~ is spherical, which means a 6 dB decrease for every doubling of distance from the source, so that, if the distance increases from 3.28 ft (1 m) to 98.4 ft (30 m) the noise decreases by approximately 30 dB" therefore, if 136 dB has been allowed at 3.28 ft (1 m), at 98.4 ft (30 m) it should be 106 dB.

Table 2. Noise Level for some GE Consolidated Valve® PSVs[6]

PSV	Rated Flowrate(lb/hr)	Size	Fluid	Set Pressure (psig)	Noise Level (dBA)
PSV- E103	26,183	1.5" H 3"	Propane	250	135 @ 3 ft
PSV- V110	152,110	4" L 6"	HC Vapor	400	145 @ 3 ft
PSV- V201	13,608	1.5" F 2"	Propane	300	135 @ 3 ft
PSV- V202	18,370	1.5" G 3"	Propane	250	136 @ 3 ft
PSV- V204	26,340	2" H 3"	Propane	300	136 @ 3 ft
PSV- V501	115,484	4" G 6"	HC Vapor	150	136 @ 3 ft

From Table 2, it can be noted that most of the PSV's noise levels shown are 136 dBA @ 3 ft for that particular project and vendor. This noise level is ~~pretty~~ high considering that the ceiling limit is 140 dBA[13] [15] for an impulsive noise as is explained in Section 2.3.2 below.

To attenuate the high noise level, the following actions are recommended by API521[1], Parcol®[7] and API 14E[8]:

a) Minimize fluid velocities. This can be done by reducing the flowrate that travels through the system and/or enlarging the line size. API 14E[8] contains recommended velocities that can be followed.
b) Reduce the pressure drop across the system, specifically across the device that produces noise
c) Select devices with low noise trim
d) Use thicker walled piping and fittings to attenuate sound and vibration
e) Avoid fittings and attachments that produce high stress concentrations at the connection, especially those that produces abrupt changes in flow directions.
f) Install the PSV as high as possible away from personnel.
g) Install acoustical insulation to the PSV and discharge piping for at least 32.8 ft (10 m) to 50 ft (15 m) from the discharge
h) Use shielding around pipe and fittings to isolate sound.
i) Use flow stream silencers for extreme cases

Even if these are good solutions for mitigating the noise, the use of acoustic insulation will not avoid the turbulence and vibration in the piping ~~system that are~~system that is the main cause of the noise. The pipe still may have a risk of rupture, see next section *Acoustic Fatigue*. Also, the use of silencers is expensive and adds undesirable extra pressure drop to the system.

Flare system specialized software can report the noise at any point of the flare system helping the engineer to be aware of high noise levels. Some software (i.e. FlareNet®) ~~have~~has 100 ~~dbA~~dBA as a maximum allowable design noise level for tail pipes and headers. It may be a conservative limit, but noise level higher than 90 to 95 dBA may result in hearing loss for sustained exposure[9]. Additionally as can be seen in Table 3 below, workers are allowed to be expose to a noise level of 100 ~~dbA~~dBA for a maximum of 2 hrs according to OSHA[14] and 15 min according to NIOSH[13].

High noise can produce different consequences in the ~~surroundings,~~surroundings; some of them are going to be described in this paper. According to M.D.G. Randall [10] they are:

a) Acoustic fatigue of the components of the system (pipe, fittings, supports)
b) The risk of hearing damage by workers close to the source of noise
c) Disturbing local communities.

Below, there is a brief description of the first and second criteria only.

2.3.1 Acoustic Fatigue

Acoustic fatigue is defined as fractures of the pipe wall due to high noise level in the system and high amplitude vibration[1]. Turbulence and vibration can produce premature fatigue, failure of certain valve parts (relieving devices), inlet and outlet piping, or both, according to API 520[4]. Welded pipe connections such as branches and supports, pipes with big diameters and pipes with thinner pipe shell are more likely to present fractures especially in systems that handle high mass flowrate resulting in severe pipe breaks with large safety and reliability impacts.

Noise level ~~less or equal to~~less than 155 dB should not present a risk for acoustic fatigue for pipes with nominal diameters up to 40 inches according to Carucci and Mueller design curve referenced by Robert D. Bruce, Arno S. Bommer and Thomas E. LePage, CSTI acoustics, Houston, Texas[11] . ~~But~~ However, some other sources have a more restrictive level of 110 dB[12] to avoid the risk of acoustic fatigue, although Jim Cahill refers to a external noise level.

Unfortunately, an easy usable screening criteria for assessing the risk of acoustic fatigue for operating installations has not been readily available.

2.3.2 Risk of workers' hearing damage

Workers that are exposed to loud levels of noise are a risk of suffering hearing damage that in the worst case could be permanent hearing loss that cannot be corrected by a surgery or hearing aids.

As mentioned in previous section, the PSVs relieving could produce high noise level that could be dangerous to the hearing of the personnel close to the source of the noise. This event is considered as an unlikely and non continuous event and therefore regulations, industries protection programs and individuals usually overlook this kind of event.

Organizations like OSHA (Occupational Safety and Health Administration) and NIOSH (National Institute for Occupational Safety and Health) recommend permissible noise exposure limits shown in Table 3. These recommended values are slightly different between these two organizations and the comparison is in Table 3. This table is developed using equation 3 where the allowable time of the exposure (T) is determined based on the exposure level (L). The relationship between these two variables is known as TWA (time-weighted average) and it shall be controlled in the way that workers exposures are less than the values in Table 3. From NIOSH[13] :

$$\mathbf{T\ (min)} = \frac{\mathbf{480}}{\mathbf{2^{(L-Y)/Z}}} \qquad \text{[Eq. 3]}^{[13]}$$

Where:
L= Exposure Level, dBA
Y= Permissible Noise Level (PEL) or Recommended Exposure Limit (REL), dBA. (OSHA's PEL is 90 dBA while NIOSH's REL is 85 dBA) [13][14]
Z= Exchange rate (OSHA's exchange rate is 5 dBA while NIOSH's exchange rate is 3 dBA)[13][14]

Table 3. Comparative table with Permissible Noise Exposures per OSHA and NIOSH

	T, Duration per OSHA[14]			T , Duration per NIOSH[13]		
L (exposure level) (~~dbA~~dBA)	**hrs**	**min**	**sec**	**hrs**	**min**	**sec**
80	32	0	0	25	24	0
85	16	0	0	8	0	0
90	8	0	0	2	31	11
95	4	0	0	0	47	37
100	2	0	0	0	15	0
105	1	0	0	0	4	43
110	0	30	0	0	1	29
115	0	15	0	0	0	28
120	0	7	30	0	0	9
125	0	3	45	0	0	3
130	0	1	53	0	0	0.9
135	0	0	56	0	0	0.3
140	0	0	28	0	0	0.09

Table 3 reflects that OSHA's permissible exposure limit (PEL) is 90 dBA for all workers for an 8 hour day[14]. However, NIOSH recommends a more restrictive 85 dBA[13] for all workers for an 8 hour day, based on significant findings of noise-induced hearing loss at 90 dBA.

The ceiling limit is 140 dBA for a continuous, varying, intermittent or impulsive noise according to NIOSH[13] and according to OSHA, the maximum continuous steady-state noise level is 115 ~~dbA~~dBA and the maximum impulsive noise level is 140 ~~dbA~~dBA[15].

However, both standards recommend the practice of engineering and administrative controls when employees' 8-hr TWA exposure equals or exceeds 85 dBA or any combination of level and duration shown in Table 3, which should be measured without wearing hearing protection.

The controls described in OSHA[14] and NIOSH[13] consist of the followings:
-wearing the appropriate hearing protectors,
-the use of adequately calibrated instruments to measure employee noise exposure, and
-annual audiograms, among others.

If a worker is exposed to different noise levels during different period of time along the day, the daily dose (D) shall be less than 100 and can be calculated according to equation 4[13] :

$$D = (C1/T1 + C2/T2 + \dots\dots\dots + Cn/Tn) \times 100 \qquad \text{[Eq. 4] (Source NIOSH) [13]}$$

Where:
D= Daily dose of exposure to noise of different levels
C= Total time of exposure at a specified noise level
T= Exposure duration for which noise at this level becomes hazardous, per Table 3.

Or, the daily dose converted into an 8-hr TWA as shows in equation 5:

$$\textbf{TWA (dBA)} = \textbf{10} \times \textbf{Log (D/100)} + \textbf{Y} \qquad \text{[Eq. 5] (Source NIOSH)}$$ [13]

2.4 Dispersion

Design, height and location of the flare stack is based on dispersion and thermal radiation evaluation. If these factors are not properly considered, it will affect the environment, personnel and equipments in a negative way.

Flaring is a process where the destruction of organic compounds takes place by a reaction of combustion and the products are sent to the open air. Fuel, steam and air are used to promote mixing and help complete the destruction of the compounds.

Combustion is complete if all the Hydrocarbons are converted to Carbon Dioxide and Water. Incomplete combustion results in some additional non-wanted byproducts such as Carbon Monoxide (CO), Nitrogen Oxides (NO_x), Sulfur Oxides (SOx), also additional noise, smoke, heat radiation and light.

~~Even worse, if~~ If the flame of the flare is lost, the flare is acting like a vent stack sending the compounds directly to the atmosphere without being converted, and the majority of these compounds are hydrocarbons.

It is important to determine the dispersion of those compounds in the air because when hydrocarbon vapors are discharged into the atmosphere the following might occur:

a) Mixtures between the flammable limits,
b) Relieving of toxic materials to workers and local communities,
c) High thermal radiation intensity when a flame is produced, and ultimately
d) Environmental pollution.

Concentrations of certain materials could be harmful even if they are below to the lowest flammable limits of any hydrocarbon. API 521[1] contains a detailed description of the type of emissions and equations used to calculate the concentration of the material at any distance from the discharge. In addition, there is vast literature information about this.

According to API 521[1] : "The design of relief devices vented to atmosphere shall not expose plant personnel and the public to intolerable risks from toxic vapors discharges. The decision to vent to the atmosphere shall comply with local regulations, this standard, and the owner's risk tolerance criteria, whichever is more restrictive". This statement applies to any atmospheric discharge, including through the flare stack with or without combustion.

The main air pollutants produced from burning hydrocarbon gases in flares are: hydrocarbons (HC), Carbon Monoxide (CO), Carbon Dioxide (CO_2), Nitrogen Oxides (NO_x) and Suspended Particulate Matter (SPM). If hydrogen sulfide is presente~~d~~ in ~~enough amount~~in significant anounts in the ~~natural~~ process gas, hydrogen sulfide and SO_2 emissions will result from flaring. Other pollutants that may be released from ~~Natural Gas Flares~~ flares are: acetaldehyde, acrolein, benzene, ethyl benzene, formaldehyde, hexane, polycyclic aromatic hydrocarbons (PAHs, including naphthalene), propylene, toluene and xylenes[16].

~~According to Somayeh Pourhassam and Alireza Taravat[17] one of the disadvantages of flaring is increase of the environmental pollutions, especially greenhouses gases such as CO2, CO, NO and NO2. Although the amount of produced greenhouse gases from flaring process is less than its amount in venting process, usually flaring does not provide complete combustion and the existence of smoke from the flare supports this fact[17]. In addition, the noise and odor from the flaring process are some of dangers which can threaten the health of personnel and local residents[17]. The acid rain, global warming, air pollution, water pollution and destruction of ozone layer are some environmental results of venting and flaring processes[17].~~

Even if the relieving occurs in a remote location, dispersion analysis ~~shall~~ should be used to evaluate the concentration of harmful materials and should be done considering specific operations characteristics such as type of burned gases, the flare efficiency, flame temperature, height of the release above ground level and momentum of the initial material released. Also site characteristics such as ambient temperature, wind speed, wind direction, atmospheric stability, local topography and presence of buildings or nearby structures should be considered. All these factors affect the concentration of pollutants in the air[16][18].

Specialized computer programs are available in the market to perform dispersion model calculations. One of them, ALOHA®, ~~some of them are~~is provided through the US Environmental Protection Agency (EPA) and ~~are~~ is free of cost and available through the internet, like ALOHA®[20]. This dispersion modeling software is able to analyze hazards such as: Toxic Area of Vapor Cloud, Flammable Area of Vapor Cloud and Blast Area of Vapor Cloud explosion.

Criteria for the allowable limit of toxic concentration could vary depending on the way the engineer will approach the situation. If the consequence-based approach is chosen~~, ~~, the concentration of toxic vapors ~~at the company property line~~in the plant area ~~shall~~ should not exceed the ~~ERPGs~~OSHA-PEL (Permissible Exposure Level) levels or NIOSH REL ~~shown in Table 4 that cause life threatening health effects.~~ For personnel inside the plant, the ability to escape is a critical factor in determining the exposure limit[1].

~~According the American Industry Hygiene Association (AIHA)[19]: "The primary focus of ERPGs® is to provide guideline levels for once in a lifetime, short term (typically 1 hr) exposures to airborne concentrations of acutely toxic, high-priority chemicals. Therefore, ERPGs® values should not be used in situations where exposure could be frequent. These levels shouldn't be used to determine safe limits for routine operations or as a basis for quantitative risk assessments"[19].~~

Regarding the flammability thresholds limit and according to OSHA[21]:"Atmospheres with a concentration of flammable vapors at or above 10 percent of the lower explosive limit (LEL) are considered hazardous when located in confined spaces".

On the other hand and according to EPA (ALOHA®)[20]: "Local areas of flame can occur even though the average concentration is below the LEL. ALOHA finds the flammable area buy using 60% of the LEL" that stands for Lower Explosive Limit. At 60% of the LEL, there is a risk of flame pockets formation[20] .

~~In~~ Table 4 below~~, there are shown toxic thresholds limit values (ERPGs or Emergency Response Planning Guidelines) for some components and its~~shows the flammability thresholds limit (LEL, 60% LEL and 10% LEL) for some components. In ~~general~~some cases the~~,~~ toxic limits are ~~more~~ lower than the flammability thresholds and can become the governing factor for design[1].

Table 4. Components toxic and flammability thresholds limit values[20]

Chemical	LEL (ppm)	60% LEL	10% LEL
Acetaldehyde	40,000 ppm	24,000 ppm	4,000 ppm
Acrolein	28,000 ppm	16,800 ppm	2,800 ppm
Ammonia	150,000 ppm	90,000 ppm	15,000 ppm
Benzene	12,000 ppm	7,200 ppm	1,200 ppm
Carbon Dioxide	Non Flammable	NA	NA
Carbon Disulfide	13,000 ppm	7,800 ppm	1,300 ppm
Carbon Monoxide	125,000 ppm	75,000 ppm	12,500 ppm
Chlorine	Non Flammable	NA	NA
Chlorine Dioxide	Unknown[22]	NA	NA
Hydrogen Sulfide	40,000 ppm	24,000 ppm	4,000 ppm
Methane	50,000 ppm	30,000 ppm	5,000 ppm
Nitrogen Dioxide	Non Flammable	NA	NA
Propylene Oxide	22,000 ppm	13,200 ppm	2,200 ppm
Toluene	11,000 ppm	6,600 ppm	1,100 ppm

~~Notes: (1) EPRG's are not available in the ALOHA chemical library for that substance. In this case, the PAC values will be the TEEL values[23].TEEL or Temporary Emergency Exposure Limit, estimates "the concentrations at which most people will begin to experience health effects if they are exposed to a hazardous airborne chemical for a given duration. TEELs are used in similar situations as the 60-minute ERPGs"[24].~~

Formatted: Keep with next
Formatted: Font: 11 pt
Formatted Table
Formatted: Font: 11 pt
Formatted: Keep with next
Formatted: Font: 11 pt
Formatted: Keep with next
Formatted: Font: 11 pt
Formatted: Keep with next
Formatted: Font: 11 pt
Formatted: Keep with next
Formatted: Font: 11 pt
Formatted: Keep with next
Formatted: Font: 11 pt
Formatted: Keep with next
Formatted: Font: 11 pt
Formatted: Keep with next
Formatted: Font: 11 pt
Formatted: Keep with next
Formatted: Font: 11 pt
Formatted: Keep with next
Formatted: Font: 11 pt
Formatted: Keep with next
Formatted: Font: 11 pt
Formatted: Font: 11 pt
Formatted: Font: 11 pt
Formatted: Keep with next
Formatted: Font: 11 pt
Formatted: Keep with next
Formatted: Font: 11 pt
Formatted: Keep with next
Formatted: Font: 11 pt
Formatted: Keep with next
Formatted: Font: 11 pt
Formatted: Keep with next

To understand the terminology for different levels of ERPGs, from AIHA[19]:
"ERPG-1: The maximum airborne concentration below which it is believed nearly all individuals could be exposed for up to 1 hr without experiencing more than mild, transient adverse health effects or without perceiving a clearly defined objectionable odor".[19]
"ERPG-2: The maximum airborne concentration below which it is believed nearly all individuals could be exposed for up to 1 hr without experiencing or developing irreversible or other serious health effects or symptoms that could impair and individual's ability to take protective action."[19]
"ERPG-3: The maximum airborne concentration below which it is believed nearly all individuals could be exposed for up to 1 hr without experiencing or developing life-threatening health effects."[19]

As a case study, we are going to evaluate the dispersion of some typical flare release components using the EPA dispersion model software (ALOHA®) with the inputs described below. Results for toxicity and flammability are shown in Table 5.

Input Conditions for the calculation were assumed as:

a) Site Data:
   Location: BRENHAM, TEXAS
   Building: unsheltered single storied
b) Atmospheric Options:
   Wind Speed: 15 mph measure at 3 m above ground
   Wind is from North
   Ground Roughness: Open Country
   Cloud cover: partly cloudy
   Air temperature: 60F
   Stability Class: D (default value)
   Inversion Height Options: No inversion
   Humidity: Medium
c) Source:
   Type: Direct continuous source
   Amount of pollutant entering the Atmosphere: 50,000 lb/hr
   Release Duration: 60 minutes
   Source Height: ground level
d) Model Run:Gaussian

Table 5. Different Components Toxic and Flammable Threat Zone from ALOHA®

Component	**Flammable Threat Zone**
Methane	10% LEL: 224 yards 60% LEL: 88 yards
Carbon Dioxide (CO_2)	none
Carbon Monoxide (CO)	10% LEL: 103 yards 60% LEL: 41 yards
Nitrogen Dioxide (NO_2)	none

Formatted: Font: 11 pt
Formatted Table
Formatted: Font: 11 pt
Formatted: Font: 11 pt
Formatted: Font: 11 pt
Formatted: Font: 11 pt

Results shown in Table 5 above,above are based on the Toxic and Flammable thresholds limit values shown in Table 4.

As it can be observed, the more toxic is the component the larger is the threat zone. Being required to keep more distance from the source to be safe.

It is recommended to do this evaluation using the plot plan of the plant in study as a guide, to locate the physical limits of the threat zone and being able to identify equipments, buildings, and personnel in risk.

Just to show how threaten could result the Nitrogen Dioxide produced by a inefficient flare, the same example shown in Table 5 before was used but only with Nitrogen Dioxide varying its flowrate to 1,000 lb/hr and 100 lb/hr and results are in Table 6 below.

Table 6. Toxic Threat Zone for Nitrogen Dioxide with different flowrate from ALOHA®

Component	Flowrate (lb/hr)	Toxic Threat Zone
Nitrogen Dioxide	1,000	ERPG-1: 1568 yards ERPG-2:378 yards ERPG-3:263 yards
Nitrogen Dioxide	100	ERPG-1: 304 yards ERPG-2:73 yards ERPG-3:52 yards

According to API 521[1], if dispersion analyses indicate that some harmful levels can be exceeded, it is necessary to improve the design of the flare system by the elimination of the relevant relief cases implementing HIPS, redesign of equipments etc.

2.5 Thermal Radiation

In the flare system design and evaluation, the effect of heat radiating from the combustion of flammables compounds on personnel and equipments must be evaluated. This evaluation will determine the location and the height of the flare stack in the plant and help in the developing of safety procedures inside the installations.

The potential radiation levels should not exceed the permissible radiation levels reported in GPSA[25] and API 521[1] and that are shown in Table 7 below:

Table 7. Permissible Design Level of Radiation[1][3][25]

Permissible Design Level (BTU/ hr ft^2)	Conditions
5,000	For Equipment Protection
3,000	Personnel, few seconds escape only. If reached, urgent emergency actions by personnel is required
2,000	Personnel, less than one minute exposure. If reached, emergency actions lasting up to 30 seconds can be required by personnel without shielding but with

Formatted: Centered

Formatted: Centered

Formatted: Centered

	appropriate clothing.
1,500	Personnel ,- several minutes short exposure. If reached, emergency actions lasting 2lasting 2 to 3 minutes can be required by personnel without shielding but with appropriate clothing.
500	Personnel, continuous exposure with appropriated clothing

Formatted: Centered

Formatted: Centered

To calculate the intensity of the radiation heat from the flame of a flare stack, an equation from GPSA[3] could be used as shown below :

$$I = \frac{(Wf)(NHV)(\varepsilon)}{4\pi(R^2)}$$ [Eq. 6] (Source GPSA)[3]

Where:
I = Radiation Intensity at point X, BTU/hr ft^2
Wf = Flare gas flow rate, lb/hr
NHV = Net Heating Value of Flare Gas, BTU/lb
ε = Fraction of heat radiated, emissivity
R = Distance from Flame center to the point X where it is desired to calculate I, ft

API 521[1] presents a different equation which is:

$$D = \sqrt{\frac{\tau F Q}{4\pi K}}$$ [Eq. 7] (Source API 521)[1]

Where:
D=minimum distance from epicenter of flame to object to be considered, ft
τ= fraction of radiated heat transmitted through the atmosphere, from API 521[1]
F= fraction of heat radiated
Q = heat released based on the lower heating value, BTU/hr
K = radiant heat intensity, BTU/hr ft^2

Results obtained from equation 7 are more conservative than the results obtained by using equation 6 and it has been found that this lastequation 7 one is more accurate for distances as close to the flame as one flame length[3].

To calculate the height of the flare and the length of the flame, additional equations to the intensity calculation equations shown previously need to be considered and can be found in GPSA[3] and API 521[1]. The rigorous calculation considering factors such as wind speed and the angle of inclination of the flame could be laborious if done manually. The use of specialized software available in the market will perform this calculation accurately. Some of the software are particularly focus on flare stack design and will providprovideing all kind ofmore details, but some less sophisticated software used to design and evaluate flare systems might be able to perform this calculation in a simplified way.

If radiation calculation shows that radiation intensity level is higher than the maximum recommended according to Table 7, it is required to modify the flare system by adding protective shields to the affected equipments, control the access of personnel to the area, eliminate the relevant relief cases by implementing HIPS, send the relevant relief cases to another r destination and ultimately, move the flare stack to a safer location or replace it by one that guarantee provide safer radiation levels.

In the several flare studies performed by the company and based on the application of the method described in GPSA[3] and simplified by the assumption of zero wind speed (vertical flame), it was found that:

-Only in one case with a 199 ft height flare stack handling a maximum flowrate of 530,000 lb/hr of majority Ethane, a clearance distance of 429 ft from the flare stack needed to be allowed instead of the 350 ft that was currently allowed for continuous personnel exposure. At 350 ft it was found a radiation intensity of 690 BTU/hr-ft2 higher than the 500 BTU/hr-ft2 required according to Table 7. Distances required for personnel exposure only for shortfor short period of time and equipments protection was adequate. This maximum capacity was below the flare stack design flowrate.

This kind of conclusion proved the importance of calculating the Thermal Radiation from the flare in that project, and even though that this calculation probably was performed in the design phase of the project, it is important to make the evaluation every time that the flare system is modified by the addition of new reliefs, etc.

2.6 Other Factors

Other Factors related to flame properties and that may be worth to evaluate in a flare system are:

2.6.1 Volatile Emissions

Volatile Emissions (VOCs from Volatile Organic Compounds) defined in EPA 40 CFR 51.100[26] as:"any compound of carbon, excluding carbon monoxide, carbon dioxide, carbonic acid, metallic carbides or carbonates, and ammonium carbonate, which participates in atmospheric photochemical reactions." [26]. There is a list of exempted compounds that according to EPA 40 CFR 51.100 shouldn't be considered in the VOC for having negligible photochemical reactivity. Methane and Ethane are exempted along with a list of compounds in 40 CFR 51.1000, therefore VOC only considered C3+ compounds.

If the conditions of the flame are such that produces an incomplete combustion or when the flame is lost, VOCs which include a wide variety of hydrocarbon compounds heavier than ethane are going to be released directly to the atmosphere. They can combine with oxides of nitrogen in the presence of sunlight to create ground-level ozone and smog. In 2000, flaring and incineration accounted for 0.5 per cent of VOC emissions from the

Canadian upstream petroleum industry. One of the VOCs is benzene, which is classified as toxic and is a known cancer-causing compound[27].

2.6.2 Smoke

Smoke production ~~might be~~is related to incomplete combustion as well and an inefficient burning flare operation, although the exact reasons and mechanisms by which smoke is formed are still not fully understood[1].

One of the key design considerations was the need for smokeless operation within the limitations imposed by stack height and radiation. A flare design had to be selected ~~to provide~~to provide optimal mixing of gas and air to produce virtually smokeless operation over the entire flowrate range.

3. Conclusions

1. The design and evaluation of a flare system, is usually performed by considering only the pressure drop across the system produced by the worst case scenario that will relief the maximum flowrate travelling through the system.

Formatted: Line spacing: Multiple 1.15 li

2. Often factors such as noise, emissions dispersion, thermal radiation and extreme relief temperatures are often overlooked in projects that involved a flare system design or evaluation. ~~By personal experience in this kind of projects,~~ When evaluating these factors, it has been found that significant changes ~~of high magnitude~~ in the system ~~are~~ may be required. ~~when evaluating these factors, such~~ These include ~~as~~ changes in the type of PSVs used, changes in pipe diameters, changes of material of construction of ~~pipes~~ piping and equipments, changes in the design of the flare stack, changes in the overall design of the flare system and even changes in the location of the flare stack have been recommended by all the findings.

3. Sizing the PSV outlet line or tail pipe using only the global relief scenarios that produce higher flowrate through the ~~system may~~system may produce undersized results in individual relief scenarios. It is recommended to do this calculation by ~~the evaluation~~ evaluating ~~of~~ the PSVs relieving independently ~~that~~ which will give the most conservative size of the tail pipes.

4. Relief Cases that produce extreme temperature in the system are easy to miss due to the extensive range of products with a wide variety of relieving temperatures that are handled. ~~By considering cases~~ Cases ~~that~~ with relief temperatures outside

the piping material~~s~~ maximum or minimum design temperatures could ~~cause~~ result in fractures in the materials with ~~desvasting~~ significant consequences. Material~~s~~ design specifications have to be carefully examined when evaluating a flare system. If cases with extreme temperatures outside of design conditions are found in the evaluation, changes in the design of flare system are required, such as changes in the process of flaring and changes to the materials of construction. Special materials to avoid the risk of failure might be an expensive investment.

5. High pressure drop and high velocities in the flare system produce turbulent flow and vibration. Those factors produce noise. High noise level in a relief system might produce acoustic fatigue in the materials of construction of the components of the system, risk of hearing damage by workers closed to the source of noise and disturbing of local communities.

6. Relieving of PSVs can be very noisy, with noise levels closed to the ceiling limit according to US ~~Safety~~ safety and ~~Health~~ health organizations. The ~~Peak~~ peak of level noise is produced at the source of the noise, but ~~while~~ the fluid travels downstream of the system the noise is attenuated by 6 dB for every 100 pipe diameters in length downstream of the sound source, according to API 521[1].

7. Although time-weighted average or TWA for permissible noise level vary slightly between the two US ~~Safety~~ safety and ~~Health~~ health ~~Organizations~~ organizations referenced in this article, both recommend the practice of engineering and administrative controls when employees' 8-hr TWA exposure equals or exceeds 85 dBA.

8. Changes in the design of the flare system and the application of occupational procedures followed by workers are required to mitigate high noise level and avoid its consequences.

9. The process of ~~flaring might~~flaring might release dangerous products to the atmosphere, especially if the combustion is ~~incomplete,~~incomplete; the flare burner efficiency is low or ~~occur~~ an~~y~~ upset in the ~~installation~~ process that would cause the loss~~t~~ of the flame in the flare. Therefore, it is important to determine the dispersion of those compounds in the ~~air~~ atmosphere because ~~it might occur:~~ mixtures between the flammable range~~,~~ , and ~~relieving of~~ toxic materials could be released~~, to workers and local communities, heat thermal~~ Thermal radiation with high radiation intensities when a flame is produced can also occur~~, and ultimately environment pollution~~.

10. High intensity of thermal radiation can affect the ~~health~~ safety of workers and the integrity of equipment~~s~~. Recommended limits have to be carefully evaluated and modifications to the flare system, specifically the flare stack and the application of more rigorous regulations have to take place.

3. References

[1] API RP 521. Pressure-relieving and Depressuring Systems. Sixth Edition, January 2014.

[2] ASME B31.3. Process Piping. 2012.

[3] GPSA. Engineering Data Book. Section 5.Volume I. Thirteen Edition – 2012.

[4] API RP 520. *Sizing, Selection, and Installation of Pressure-Relieving Devices in Refineries.* Part II- Installation. Fifth Edition, August 2003.

[5] Crosby®Valve Inc. *Pressure Relief Valve Engineering Handbook.* Technical Document No. TP-V300. Effective May 1997.

[6] GE Consolidated Valves® PSV Technical Information for MarkWest Cadiz Carthage Plant Vent Study– February 2015.

[7] Parcol®. *Pressure Safety Relief Valves* – Technical Bulletin 3-1.

[8] API RP 14E. *Recommended Practice for Design and Installation of Offshore Production Platform Piping Systems.* Fifth Edition, October 1, 1991.

[9] Galen Carol Audio. *How To's. Decibel (Loudness) Comparison Chart.* http://www.gcaudio.com/resources/howtos/loudness.html

[10] MDG Randall, Foster Wheeler Energy Limited, "*PSV noise – criteria, limits and prediction*," Valve World. June 2001

[11] Bruce, Robert D., Bommer, Arno S. and LePage, Thomas E. "*Solving Acoustic-Induced Vibration Problems in the Design Stage*". CSTI acoustics, Houston, Texas.

[12] Jim Cahill, Emerson Process Experts, "*Abating Acoustical Noise in Piping*". *http://www.emersonprocessxperts.com/2010/02/abating_acousti/*

[13] U.S. Department of Health and Human Services. The National Institute for Occupational Safety and Health (NIOSH). *Occupational Noise Exposure.* Revised Criteria 1998.

[14] OSHA 1901.95. *Occupational Noise Exposure.* Dec. 12, 2008.

[15] OSHA. *Noise and Hearing Conservation.* Appendix II:A. General Industry Standard. https://www.osha.gov/dts/osta/otm/noise/standards_more.html.

[16] EARTHWORKS. *Flaring and Venting.* http://www.earthworksaction.org/issues/detail/flaring_and_venting#)

[17] Pourhassam, Somaych and Taravat, Alireza. *Effect of gas flaring on environmental variables in developing countries*. International Journal of Business and Management Vol II (I), 2014.

[18] Crowl, Daniel A. and Louvar, Joseph F. Chemical Process Safety. Prentice Hall. Third Edition. 2011. Pp 185-244

[19] WEEL Handbook. 2013 ERPG. AIHA Guideline Foundation. https://www.aiha.org/get-involved/AIHAGuidelineFoundation/EmergencyResponsePlanningGuidelines/Documents/ERPGIntroText.pdf

[20] US Environmental Protection Agency (EPA). ALOHA® software.

[21] Occupational Safety & Health Administration (OSHA). *Compliance Assistance Guidelines for Confined and Enclosed Spaces and Other Dangerous Atmospheres*. Part No 1915, Subpart B App A.

[22] The National Institute for Occupational Safety and Health (NIOSH). *Chlorine dioxide*. http://www.cdc.gov/niosh/idlh/10049044.html

[23] US Department of Commerce. National Oceanic and Atmospheric Administration (NOAA). *Public Exposure Guidelines*. http://response.restoration.noaa.gov/oil-and-chemical-spills/chemical-spills/resources/public-exposure-guidelines.html

[24] US Department of Commerce. National Oceanic and Atmospheric Administration (NOAA). *Temporary Emergency Exposure Limits (TEELs)*. http://response.restoration.noaa.gov/oil-and-chemical-spills/chemical-spills/resources/temporary-emergency-exposure-limits-teels.html

[25] GPSA. Engineering Data Book. Section 5.Volume I. Eleventh Edition – 1998.

[26] US EPA. *Definition of Volatile Organic Compounds (VOC)*. 31 March 2009. http://www.epa.gov/ttn/naaqs/ozone/ozonetech/def_voc.htm

[27] http://siteresources.worldbank.org/EXTGGFR/Resources/578068-1258067586081/FlaringQA.pdf

Will It Really Make that Much of a Difference?
Broad Effects of Operational Changes on Relief System Design

Ms. Marie Baker
Cognascents Consulting Group, Inc.
908 Town & Country Blvd., Suite 500
Houston, TX 77024
marie.baker@cognascents.com

Mr. Teddy Bucher
Cognascents Consulting Group, Inc.
908 Town & Country Blvd., Suite 500
Houston, TX 77024
teddy.bucher@cognascents.com

Prepared for Presentation at
American Institute of Chemical Engineers
2015 Spring Meeting
11th Global Congress on Process Safety
Austin, Texas
April 27-29, 2015

UNPUBLISHED

Will It Really Make that Much of a Difference?
Broad Effects of Operational Changes on Relief System Design

Ms. Marie Baker
Cognascents Consulting Group, Inc.
908 Town & Country Blvd., Suite 500
Houston, TX 77024
marie.baker@cognascents.com

Mr. Teddy Bucher
Cognascents Consulting Group, Inc.

Keywords: Evergreen, flare system analysis, Hazard and Operability Study (HAZOP), Independent Protection Layer (IPL), Management of Change (MOC), overpressure analysis, pressure relief analysis and design, Process Hazard Analysis (PHA), process modes of operation, Process Safety Information (PSI), relief device design basis, relief rate calculation and sizing, safeguard.

Abstract

Throughout the life of an operating facility, changes to the process are inevitable and potentially affect the systems that keep personnel and equipment safe. The Management of Change (MOC) process is intended to evaluate proposed changes prior to implementation to assess and address any risks that might be introduced as a result of this change. The MOC process typically includes a process hazards analysis (e.g. Hazard and Operability Study, Layer of Protection Analysis), which evaluates safeguards and independent protection layers (IPLs) for the proposed changes.

Of the many IPLs, pressure relief devices are often overlooked and a re-evaluation of the relief system design basis is sometimes not performed. The authors have observed that personnel do not always recognize the operational change may affect the relief system.

In this paper, the authors explore what types of changes should trigger a relief system design review, exploring why minor modifications may have major ramifications. They also provide specific examples of the most common changes that demonstrate how the relief system design can be affected.

The target audience for this paper is anyone whose responsibilities include (1) pressure relief analysis, (2) process safety management, (3) management of an operating facility, (4) process engineering, and (5) process safety information management.

1. Background and Purpose

All operational changes potentially affect relief system design. Some changes have a minor or negligible affect while others can change the entire relief system design basis.

In theory, a facility's relief system design basis should be reviewed before any operational change occurs. However, this is not a practical approach as change is an ever-present aspect of an operating facility. Not all operational changes require a full relief system design review. Within any facility, operational variation is present and the relief system should be designed to account for small variations in operating pressure, operating temperature, and normal vessel liquid levels.

The Management of Change (MOC) process should include a review of affected relief system to assess the adequacy of the design basis or bases. Will the proposed change result in a new credible overpressure scenario or an undersized relief device? What effect might the change have on downstream relief devices? These are questions that should be asked during the MOC process before a change is implemented. These are also questions that will be asked in a Process Hazards Analysis (PHA). Without a good understanding of how changes can affect the relief system, the resulting risk level may be higher if the relief device can no longer be considered an Independent Protection Layer (IPL).

Sometimes the relief system design basis may be reviewed, but the documentation and any required calculations are not updated; thereby, giving the appearance that no review was performed. Two outcomes are possible when documentation is not updated. One outcome could be that the relief device is adequately sized for the operational change, but not documented and not included as an IPL; therefore, the perceived risk is greater than the actual risk. The second outcome could be that the change results in an additional overpressure scenario, for which the relief device is not sized or is undersized. With no documentation, though, the relief device is counted as an IPL and the perceived risk is less than the actual risk.

2. Exploring Operational Changes

There are many operational changes that can occur in a facility. Due to the inherent interconnectivity of all the components and equipment at a facility, sometimes an operational change to one part of the process will result in another operational change to a different part. A throughput change may require modifications to equipment. An equipment re-rate may require an operating envelope change. Accordingly, the total effects of a relatively minor change to a system may be larger than originally anticipated.

These operational changes can affect relief valve sizing in different ways, and some calculations are more susceptible to operational change than others. They can result in the addition or elimination of an overpressure scenario, or a change to the current sizing of the relief device.

In the sections that follow, the authors will provide details for the most common operational changes: throughput change, equipment change, equipment re-rate, feedstock (compositional) change, and operating envelope change. Each section will (1) outline the impact each type of change may have on an affected pressure relief system, focusing on the addition of credible scenarios and increases in the sizing of the relief device, and (2) examine correlations between the magnitude of operational change and resultant change to relief device sizing.

The information provided is not all-inclusive, nor is it absolute. Exceptions always exist, but in the authors' experience the descriptions and guidance are what normally occurs.

Even if the controlling overpressure scenario and relief device size does not change, the relief system design documentation should be updated to reflect the current system configuration.

3. Throughput Change

3.1 Throughput Change

A throughput change describes a change in the rate of material entering the facility or unit. Changes could be for increased or decreased production, or as a result of a change in another part of the process. To simplify things, we will only consider throughput increases as throughput decreases tend to be short-term in nature and rarely result in a higher required relief rate.

Overpressure scenarios can be classified as rate dependent or non-rate dependent. For the purposes of this paper, a rate dependent overpressure scenario will be defined as any calculation which changes based on the amount of fluid passing through the system. Rate dependent cases typically involve specifying the relief rate instead of calculating a relief rate. Calculations which are based on a pressure differential (e.g. flow across a control valve, orifice flow), or which depend on a vaporization rate (e.g. external fire), are considered to be non-rate dependent. Throughput changes directly affect rate dependent overpressure scenarios, the most common of which are blocked outlet and loss of cooling/reflux.

On the surface, a throughput change can seem straightforward, and one might expect it to be easy to check relief valve sizing. As the throughput rate increases, the new throughput should be compared to the capacity of the relief device to determine if the capacity is exceeded. This simplified analysis is indeed effective in systems where boiling or chemical reaction does not occur (for which a more complex approach should be used).

To illustrate a simple analysis, let us look at a basic liquid/vapor separator and the blocked outlet overpressure scenario on the vapor overhead line. Two-phase feed enters the separator at 10,000 lb/hr and produces 2,415 lb/hr of vapor which corresponds to a required relief area of 0.081 in^2. A debottlenecking project is implemented which increases the feed to 15,000 lb/hr having the same composition and operating conditions. The new feedrate produces 3,623 lb/hr of vapor which corresponds to a required relief area of 0.122 in^2. The vapor flowrate and required area both increased by the same percentage the feed was increased, showing a direct correlation between the throughput increase and the relief rate/area increase.

When looking at how the more complex systems change based on throughput, the best way to perform an analysis is using a simulation to mimic relief conditions. An example of a more complex system is a tower with a loss of cooling to the overhead condenser. Consider a tower designed to separate C4s and lighter from a feed stream. Losing overhead cooling produces a certain amount of relief fluid. When the throughput to the original tower is increased by 15% and all other conditions remain the same, the required orifice area also increases by approximately 15%.

Both of these examples show that when other conditions remain the same, there is a direct correlation between throughput increase and relief area increase. If other factors (e.g. pressure, desired separation) change, then the desired correlation may not hold true.

4. Equipment Change

Another potential pitfall is changes to equipment. This can be something as simple as changing a pump impeller or something as complex as replacing a fractionating column. As a general guide, replacements-in-kind typically will not affect the relief device sizing. Examples of equipment replacements-in-kind include replacing a heat exchanger with a similar size and duty exchanger, replacing a control valve with one having the same size and C_V value, or replacing a tower with one having the same dimensions and number of trays so as to keep the separation the same. If the replacement heat exchanger has a longer shell or thicker tubesheet, it would not be considered a replacement-in-kind, relief device sizing could be affected, and the change may require the relief valve to be resized.

4.1 Pump Impeller Change

Consider a process which requires additional flow through a centrifugal pump. The facility currently has one online pump and one spare and would like to remain with only one pump in operation at a time. The process dictates that pump discharge pressure must remain the same, so the decision is made to install a larger diameter pump impeller to achieve the increased flow. With the larger impeller, the pump curve has changed and the deadhead pressure has increased. This example actually has two items that need to be checked.

First, it must be determined whether the previous deadhead pressure of the pump would overpressure the vessel if the liquid outlet was blocked and the vessel overfilled. A review of the relief device protecting the downstream vessel may show the MAWP of the vessel is greater than the deadhead pressure of the pump. In this example, liquid blocked outlet is not a credible overpressure scenario and no sizing was performed. However, with the *increase* in deadhead pressure, a review of the same relief device may show the deadhead pressure has increased sufficiently to result in overpressure of the vessel in the event of a liquid blocked outlet. Sizing must be performed to determine the adequacy of the relief valve.

To put it in terms of actual values, originally, the pump deadhead pressure was 195 psig. The downstream vessel has a MAWP of 200 psig and the relief device is set at the same pressure. Since the deadhead pressure did not exceed the MAWP of the vessel, this would not have been considered a credible overpressure scenario and no relief device sizing would have been performed. With the new impeller the deadhead pressure of the pump is 225 psig which now exceeds the MAWP plus allowable accumulation of the vessel which equates to 220 psig. Now this scenario has become a credible overpressure scenario and requires relief device sizing.

Second, if the relief device was initially sized for the liquid blocked outlet overpressure scenario, the device must be checked to determine if the relief device has enough capacity to pass the new flowrate. This would be similar to the throughput increase and the relief rate change would be in direct proportion to the increase in pump flowrate.

4.2 Control Valve Change

A different type of equipment change involves a control valve used to let down pressure. A given process requires an increase in fuel gas to a vessel with an MAWP of 150 psig and relief device set at the same pressure. A control valve is installed on the fuel gas line to let the pressure down from 200 psig to 100 psig before entering the vessel. The existing control valve is not large enough to allow the increased flow and remain at the downstream operating pressure of 100 psig. The facility decides to install a larger control valve to obtain the required flow at the normal operating pressure. The downstream relief device is properly sized for the flow through the control valve currently installed. Though various factors contribute to the amount of flow that passes through a control valve, the C_V value has the greatest impact. The greater the change in wide open C_V value, the greater the increase in required area.

5. Equipment Re-rate

Equipment re-rating can be done to decrease the MAWP of equipment and is often done to extend its life. When equipment is re-rated, it can have an effect not just on the relief device protecting it, but it could also affect additional relief valves that are not as obvious. It could potentially eliminate credible overpressure scenarios in other downstream relief devices or it may add scenarios to other relief devices. For instance, re-rating the low pressure side of a shell-and-tube heat exchanger may result in tube rupture becoming a credible overpressure scenario where it may not have previously been. Re-rating the high pressure side of a shell-and-tube heat exchanger may eliminate a tube rupture scenario for that exchanger.

5.1 Heat Exchanger Re-rate

In the heat exchanger examples mentioned above, let us consider what happens when each side is re-rated. A heat exchanger has a shell-side (high pressure) MAWP of 300 psig. The tube-side (low pressure) has an MAWP of 200 psig and a hydrotest of 1.5 times (300 psig). For this exchanger, the shell-side MAWP does not exceed the tube-side hydrotest pressure and tube rupture is not considered a credible overpressure scenario. If the tube-side is re-rated to 175 psig MAWP and 1.5 times hydrotest (263 psig), the tube-side hydrotest pressure is now less than the shell-side MAWP. This has created a credible overpressure scenario and the tube-side relief valve should be resized to pass tube rupture flow.

A different heat exchanger has a tube-side (high pressure) MAWP of 400 psig. The shell-side (low pressure) MAWP is 250 psig and has a hydrotest of 1.5 times (375 psig). In this exchanger, the shell-side relief valve is sized for a tube rupture overpressure scenario. Downstream of the shell-side of this exchanger is a vessel also with an MAWP of 250 psig but a hydrotest of only 1.3 times (325 psig). The tube-side of the exchanger is re-rated down to 370 psig, which eliminates the tube rupture case in the exchanger. However, the downstream vessel hydrotest pressure is still less than the tube-side MAWP. Now the downstream vessel has a credible tube rupture scenario and the relief valve should be resized for it.

5.2 Vessel Re-rate

Often, an upstream relief valve is used to limit the pressure to downstream equipment (i.e. both vessels have an MAWP of 300 psig and the upstream relief valve is sized for a blocked outlet case so the downstream relief valve is not needed for a blocked outlet). If the downstream vessel is re-rated to 250 psig, the upstream relief valve cannot be used to limit the pressure to the downstream vessel and the lower set pressure relief valve will need to be re-sized for the blocked outlet scenario.

6. Feedstock (Compositional) Change

Composition changes can occur in any part of a facility. They can be changes to feedstock, addition of a new feedstream, changes to chemical additives, or changes in composition within the facility or process unit. Compositional changes may not have the same effect at the end of a facility or process unit as they do at the inlet. The potential for the most change to the relief valve is where the compositional change is introduced. The fluid phase also contributes to how much the composition will affect the relief valve sizing.

Vapor streams are affected by changes in molecular weight (MW) which is essentially a specific gravity change. The lower the molecular weight and subsequently specific gravity, the greater the required relief area -- assuming the same temperature, pressure, and flowrate for a given stream. However, the change here is not proportional as it is with a throughput change. Consider a vapor hydrocarbon stream entering a gas plant having a molecular weight of 22.5 lb/lbmol. Gas from a different well is routed to this same gas plant, changing the molecular weight to 21.3 lb/lbmol. This change in composition will have some effect on relief valve sizing but will likely be insignificant. If the gas molecular weight changed from the initial down to 18.8 lb/lbmol, the effect on relief valve sizing will likely be greater.

Liquid streams are also affected by changes in molecular weight and liquid specific gravity. However, changes in molecular weight without a subsequent change in specific gravity have little effect on relief valve sizing. For example, MDEA used for removing H_2S from vapor streams is typically 40% MDEA and 60% water and has a molecular weight of 58.47 lb/lbmol. A stream of pure MDEA has a much higher molecular weight (119.16 lb/lbmol); however, the specific gravities of these two streams are approximately the same at the same conditions. The similar specific gravity indicates relief valve sizing will be similar.

A final feedstock change worth considering pertains to production wells. Over the lifetime of a given subsea gas well, for instance, the production fluid from the well will typically develop higher and higher water to gas ratios. If a relief valve is installed on the flowline upstream of any separation to safeguard against a blocked-in scenario, it would need to relieve the full production fluid -- in this case, a two-phase mixture. The relief installation should be adequately sized to account for all expected flow profiles; however, it is significantly more difficult to generalize the effect different flow qualities will have on relief valve sizing, as sizing of two-phase flow requires a much more rigorous analysis than single phase flow. As such, the recommended approach is to ensure that the relief design is adequate for all extremes of the flow profile for the projected lifetime of the well.

7. Operation Envelope Change

Of all potential operational changes, the operating envelope changes have the potential to be the most severe and seem to be the easiest to overlook in relief device sizing. Operation envelope refers to pressure, temperature, liquid level, and alarm setpoints/interlocks. In the authors' experience, a good practice is to size relief devices such that minor fluctuations in operating conditions will not require resizing of the relief valve. For example, if the controlling case for a valve requires an area of 1.270 in^2, good engineering practice when designing the relief system is to install a K orifice (1.853 in^2) instead of a J orifice (1.287 in^2) to allow for fluctuations in the process.

Required relief rates are typically calculated based on the normal operating conditions at the time of the analysis. If those operating conditions change significantly, the required relief rate and subsequently the relief device sizing may change as well. This is especially important if the change in operating conditions results in additional credible overpressure scenarios.

A pressure or temperature change in an upstream vessel may also result in more than a pressure/temperature change in downstream vessels. For example, look at the cold side of a shell-and-tube heat exchanger. If the operational temperature of the hot side is increased, the result of isolating the cold side could be vaporization of the liquid instead of thermal expansion of the liquid. Additionally, increases to the pressure of the high pressure side could render the installed relief device ineffective due to the response time.

7.1 Pressure Change

Many required relief rate calculations are based on pressure differential. As the pressure increases on the high pressure side, the required relief rate increases and subsequently the relief device required area increases.

One of the most frequently occurring examples of pressure change involves gas blowby from a higher pressure vessel down to a lower pressure vessel by way of a wide open control valve. Consider a separator operating at 400 psig. As a result of a change to the process upstream, the separator must run at a higher operating pressure than normal. The operating pressure needs to increase from 400 psig to 475 psig. The downstream vessel has an MAWP of 250 psig with the relief device set at the same pressure. This creates an increased pressure differential from 125 psi to 200 psi.

Operating pressure increases can also affect the stability of the relief device. In order for relief devices to operate properly, a pressure differential must exist between the set pressure of the relief device and the operating pressure of the vessel due to blowdown and tightness requirements. For direct, spring-operated conventional and balanced bellows relief devices, there are general guidelines for the minimum difference.[1]

- 5 psig for set pressures ≤ 70 psig
- 10% for set pressures between 71 psig and 1000 psig
- 7% for set pressures > 1000 psig

Pilot-operated relief valves are a special case as they have different blowdown and tightness requirements and can operate at a much lower differential than spring-loaded valves. The typical differential pressure for pilot-operated valves is 5%. Operating at differentials less than the recommended minimum values may result in chattering or cycling and subsequent damage to the relief valve or failure of the relief valve altogether.

7.2 Temperature Change

Temperature changes do not necessarily have as direct an effect on relief device sizing as pressure changes do. Incompressible fluids are more affected by temperature changes than compressible fluids. Unlike pressure, temperature changes will not create new overpressure scenarios except in the case of thermal expansion due to ambient heating (solar radiation). Temperature has the potential to change the phase of a fluid from all liquid to two-phase or vapor when flashing from higher pressure down to lower pressure. Required relief area is much greater for a two-phase fluid than for a liquid.

For instance, a vessel with a liquid hydrocarbon mixture operates at 250 psig and 180 °F. The liquid flows to a downstream vessel through a control valve to a vessel which has a MAWP of 100 psig. At these upstream conditions, the fluid will remain all liquid as it flows through the control valve and the relief fluid will be all liquid. If the temperature of the upstream fluid is increased to 210 °F, the fluid flashes as it flows through the control valve and the relief fluid becomes two-phase which has a much greater required relief area than liquid.

The other main effect temperature changes have is during a liquid thermal expansion overpressure scenario, specifically thermal expansion in the cold side of a shell-and-tube heat exchanger. This occurs when the cold fluid is isolated while hot fluid continues. The temperature of the hot side fluid determines if the cold fluid remains a liquid or vaporizes at relief pressure. During the initial analysis of a relief device, the temperature of the hot fluid may not have been high enough to vaporize the cold fluid. If the hot fluid temperature is increased, the current analysis may require the relief device to be sized for vaporization of the cold fluid instead of expansion. This could significantly change the sizing of the relief valve.

Another example involves a shell-and-tube heat exchanger having the hot side fluid on the shell-side and cold side fluid on the tube-side. The hot side fluid normally operates at about 250 °F. At this temperature, the colder fluid will remain liquid at relief pressure if the tube side of the exchanger is blocked in while the hot side remains flowing. However, if the temperature of the hot side is raised to 305 °F, the fluid will vaporize when blocked in.

7.3 Liquid Level Change

Vessel liquid levels can be one of the most difficult operational changes to assess against a given relief system design. Liquid levels factor into more than just the required relief rate calculations. They may factor into what types of instrumentation are installed on a vessel and the determination of response time during an upset. A cursory glance through the relief system design may not be enough to understand how a change in liquid level will affect the system.

The most common impact of a liquid level change is on the external fire overpressure scenario. Liquid level affects the wetted surface area calculation, which changes the external fire required relief rate calculation. Since the heat input calculation is based on the wetted area raised to the 0.82 power, it is difficult to determine a correlation between the increase in liquid level and relief valve sizing. Small increases in liquid level are expected to have less effect on relief valve sizing than larger changes.

Another area where adjustments to liquid level can be an issue is in the case of high pressure to low pressure interfaces (e.g. high pressure separator flowing into a low pressure separator) where liquid overfill can occur prior to gas blowby when a bottom level control valve fails open. The liquid levels in both vessels can initially be set such that loss of liquid level in the upstream vessel does not overfill the downstream vessel prior to gas blowby occurring from the upstream vessel. An increase to either the upstream vessel or downstream vessel liquid level may be sufficient that the low pressure vessel will overfill with liquid prior to gas blowby. This liquid full condition can result in the gas displacing the liquid in the vessel and that liquid having to be relieved before the high pressure gas can be relieved.

8. Additional Considerations

The operational changes described above are the most common and have the greatest impact on relief system design. As mentioned above, any operational change has the potential to result in a relief design basis change. Some additional operational changes are listed below:

- Removal of restriction orifices - if the orifice is used in sizing the relief device, removal of the orifice will require a change in the relief device sizing and probably an increase in the size of the relief device;
- Changes to relief device inlet and outlet piping;
- Changes in alarms or interlock setpoints - a note of caution here, changes in alarm setpoints may not require an MOC but the setpoints may be used in the relief device sizing as an upper or lower limit; and
- Installing or removing check vales - API 521 Section 4.4.9.3[2] provides guidance on taking credit for installed check valves to reduce required relief rates.

9. Guidance

Now that the most common operational changes have been described and specific examples have been detailed, how does that translate into knowing if your change will have a large or small effect on the relief system design? The authors have put together some guidance to follow when evaluating operational changes. As a note, in the authors' experience, if the current required area for the overpressure scenario affected is within 10% of the available area of the relief device, the device should be resized for any change.

While each operational change is unique, there are some commonalities that may provide the engineer a way to evaluate the effect on relief system design.

- Throughput changes generally result in a percent change in relief area proportional to the percent change in flow when all other conditions remain the same.

- Equipment changes are varied, and therefore, more difficult to find commonalities. Changes to rotating equipment tend to follow the same guidelines as throughput changes. For globe style control valves, the percent change in wide open C_V is roughly equal to the percent change in relief device required relief area. Changes to vessels and heat exchangers should be evaluated on a case-by-case basis.

- When re-rating heat exchangers, the tube rupture overpressure scenario is the most critical. If the relief device is already sized for the tube rupture scenario, a pressure differential change of approximately 5% corresponds to a less than 10% change in required relief area; however, the percent change in relief area is not proportional to the change in pressure. Re-rating of a vessel should require the entire relief system to be re-evaluated at the new pressure. Any changes that need to be made should follow the other typical operational changes described.

- Feedstock changes generally have the least effect on relief valve design. When evaluating vapor, as the MW decreases, the required relief area increases. There is difficulty in determining a percentage change since the composition affects other properties. With liquid compositions in the same family (e.g. hydrocarbons), as the molecular weight increases, the density increases as well which in turn increases the required relief area. Generally, the required relief area changes at a rate of 50% of the density change.

- Changes in pressure when evaluating vapor show that at pressures less than 200 psig, the percent change in pressure is greater than the percent change in required relief area. At pressures greater than 200 psig, the percent change in pressure is less than the percent change in required relief area. For liquids, the percent change in pressure corresponds to the percent change in required relief area. Changes in temperature have little to no effect on the required relief area of vapors. For liquids, unless a significant change in specific gravity is associated with the temperature change, the required relief area is minimally changed. Changing a vessel's liquid level by 20% creates a 10% - 15% change in required relief area.

10. Conclusion

Change is an integral part of the long term operation of any production facility. Balancing the need for safety of personnel and equipment with the profitability of a facility can be difficult at times. Knowing where to focus efforts will help streamline that process. Relief system design can be complex and confusing, but understanding how operational changes affect the relief system is beneficial (especially in the broader context of PHAs).

The most common operational changes (throughput change, equipment change, equipment re-rate, feedstock/compositional change, and operating envelope change) were detailed and guidelines were provided based on how the change may affect the required relief area. These changes may show different effects based on fluid phase. While the material presented is not all encompassing, it provides a basis for evaluation. Some changes have little to no effect on the relief device. Others can completely change the relief device design basis. Beginning from a starting point is better than not beginning at all.

11. References

[1] ASME Boiler and Pressure Vessel Code. *Section VII, Division I, Rules for the Construction of Pressure Vessels.* American Society of Mechanical Engineers, New York, NY, 2013.

[2] API Standard 521. *Sixth Edition, Pressure-relieving and Depressuring Systems.* American Petroleum Institute, Washington, DC, 2014.

11TH GLOBAL CONGRESS ON PROCESS SAFETY

Auditing Relief Systems Design Basis – Best Practices

Neil Prophet
ioMosaic Corporation
93 Stiles Road | Salem, NH 03079
prophet.n.tx@iomosaic.com

Prepared for Presentation at
American Institute of Chemical Engineers
2015 Spring Meeting
11th Global Congress on Process Safety
Austin, Texas
April 27-29, 2015

UNPUBLISHED

Auditing Relief Systems Design Basis – Best Practices

Neil Prophet / Principal Author
ioMosaic Corporation
93 Stiles Road | Salem, NH 03079
prophet.n.tx@iomosaic.com

Keywords: Auditing, Relief Systems, Pressure Relief

Abstract

Ever since OSHA implemented their National Emphasis Program in 2007, facility's pressure relief systems design basis have come under increasing scrutiny. Recognizing that they may not be fully compliant, many companies are conducting audits of their relief systems design basis to determine their current state, identify gaps, and establish a path forward for compliance. ioMosaic Corporation is often called upon to conduct these audits, and in doing so, has developed a successful methodology to do this efficiently and effectively. This paper outlines how companies can conduct audits of their relief systems in a successful way.

1. Background

The explosion at BP's Texas City Refinery on March 23, 2005 triggered an industry-wide increase in focus on pressure relief systems design basis. Relevant codes and standards, such as API Standard 520 **[1]**, **[2]** and 521 **[3]** were updated and expanded. At the same time, OSHA implemented CPL-03-00-004 (Petroleum Refinery Process Safety Management National Emphasis Program) in 2007, followed by CPL-03-00-014 (PSM Covered Chemical Facilities National Emphasis Program) in 2011. Both of these programs involved onsite auditing activities, with pressure relief systems being one area of particular focus. Since then, there is still a continued awareness of the benefit and need to audit pressure relief systems.

Additionally, Section O of the OSHA PSM standard requires that compliance audits be conducted every three years. Since a facility's relief systems design and design basis is part of its Process Safety Information (PSI), this information should be audited every three years, and revalidated every five years.

The main purpose of auditing relief systems design basis is threefold:

- Ensuring a design basis exists for every relief system
- Ensuring accuracy of existing relief system design basis
- Ensuring adequacy of existing relief systems design basis documentation

Only by conducting all three of these activities can a company develop a complete understanding of their existing relief systems design basis. Additional benefits and reasons for auditing relief systems design basis include:

- Ensuring regulatory compliance
- Maintaining up-to-date process safety information
- Identifying non-conformities, or areas of improvement
- Promotion of industry best practices and Recognized and Generally Accepted Good Engineering Practices (RAGAGEP)

2. Audit Preparation

Preparation is a key part to conducting any audit, including one that is focusing on relief systems design basis. An audit plan should be developed which establishes the scope and schedule of the audit. This plan will take into account that audits are usually time and resource limited, and should aim to focus resources on the areas with highest priority. Scope definition determines whether the audit may be facility-wide, unit-wide, or limited to a specific area of focus. An audit protocol is also a key part of the preparation. The term "protocol" means the checklist used by the relief systems auditor as the guide for conducting the audit activities. The audit protocol considers the following for each item in the scope of the audit:

- Scope Item - what is being evaluated and how is it being measured
- Performance Criteria - what is the item being measured against
- Review Depth - how much need to be measured
- Audit Technique - how is the item being measured

The audit protocol will be based on the appropriate codes and standards that are being considered, and may also be used to define which audit techniques that should be utilized. For example, use of questionnaires, interviews, documentation review, calculation checking, or field verification. A combination of various techniques may be applied. A successful audit protocol should ensure that a consistent approach is followed by each auditor involved.

The preparation stage prior to conducting an audit should also include establishing data requirements for the audit. The analysis of pressure relief systems requires extensive process and equipment information. OSHA 1910.119 Section D outlines data that should be available to support a process hazard analysis study, and the requirements are similar for a relief systems design study. In order to fully conduct a relief systems design audit, a full range of supporting process safety documentation should be available, such as that shown in Table 1. Additionally, the auditor should consider field verification of piping isometrics, relief devices and vessel design parameters, and piping and instrument drawings (P&IDs) to ensure an accurate evaluation. In addition, there are many gray areas in the current state of the art for relief systems evaluation that complicate relief systems design basis. These may need to be considered during development of the audit protocol. Technical issues that are currently the subject of debate and some controversy amongst relief systems experts include, but are not limited to:

- Use of actual (best estimate) flow vs. required flow for (a) inlet pressure loss, (b) backpressure, (c) sub-header/ flare header hydraulics, and (d) effluent handling equipment (knockout drums, flare tips) design
- Use of 3 % inlet pressure loss requirement vs. a more relaxed requirement for existing installations such as blowdown minus 2 %
- Fire exposure and cold temperature development for depressuring systems, especially for gas filled vessels
- Correct usage of two-phase discharge coefficient
- Estimation of two-phase density with slip
- Use of fire flux for dynamic simulations. Decreasing wetted surface area for all gas flow as well as use of total vessel wetted surface area for two-phase flow
- Level of documentation that is sufficient to meet the OSHA PSI requirements

Table 1. Typical Relief Systems Design Study Data Requirement

	General Data Requirements	
Process Design and Description	Piping and Instrumentation Diagrams (P&ID) Heat and Material Balances (H&MB) Process Flow Diagrams (PFD) Process Safety Flow Diagrams (PSFD) Process descriptions / operating procedures Plot plans / elevation plans	
Utility and Piping Design	Utility operating conditions (electrical, instrument air, cooling water, steam, etc.) Electrical one-line diagrams Piping designations and ratings Insulation designations and ratings	
	Data Requirements	
	Required Information	**Data Source Hierarchy**
Fluid and Mixture Properties	Thermophysical properties	1. DIPPr database using modified PR EOS 2. Company generated data 3. Estimates based on structure
	Reaction kinetic models	1. Company provided adiabatic calorimetry data 2. Open literature data 3. Externally generated adiabatic calorimetry data
Pressure Relief Devices	Manufacturer / model number Inlet / outlet / discharge area sizes Opening pressure and temperatures	Relief Device Information: 1. Maintenance records 2. Relief device specification sheets 3. Original design basis 4. P&ID 5. Valve Tag Inlet / Outlet Piping Details: 1. Existing isometric drawings 2. Field sketches
Fixed Process Equipment (General)	MAWP, MAWT, and vacuum rating Design conditions Equipment Dimensions	1. U-1A forms 2. Mechanical drawings 3. Equipment specification sheets 4. Operating Manuals 5. P&IDs 6. Nameplate

General Data Requirements		
Process Design and Description	Piping and Instrumentation Diagrams (P&ID) Heat and Material Balances (H&MB) Process Flow Diagrams (PFD) Process Safety Flow Diagrams (PSFD) Process descriptions / operating procedures Plot plans / elevation plans	
Utility and Piping Design	Utility operating conditions (electrical, instrument air, cooling water, steam, etc.) Electrical one-line diagrams Piping designations and ratings Insulation designations and ratings	
Data Requirements		
	Required Information	**Data Source Hierarchy**
Vessels	Liquid levels	1. Operating procedures 2. P&IDs 3. Equipment design drawings 4. Level alarm set-points 5. Level-gauge tapping locations (from equipment design drawings)
	Elevation	1. P&ID 2. Equipment design drawing
	Insulation type, thickness, firegrade status	1. Maintenance records 2. Equipment design specification 3. P&IDs
Heat Exchangers	Design type Rated and normal duty Tube ID / length	1. U-1A forms 2. Heat exchanger specification sheets 3. P&ID 4. Nameplate
Heaters / Steam boilers	Tube Design Pressures Furnace design duty Boiler dimensions and design duty	1. Heater / Boiler specification sheets 2. U-1 Forms 3. P&ID 4. Nameplate
Rotating Process Equipment (General)	MAWP, MAWT Design conditions	1. Equipment specification sheets 2. P&ID 3. Equipment nameplate
Centrifugal Pumps	Pump capacity curve, rated capacity, and installed impeller size Suction Conditions	1. Performance curves 2. Pump specification sheets 3. Maintenance records (installed impeller and corresponding curve) 4. P&ID 5. Nameplate
Centrifugal Compressors	Compressor capacity curve and rated capacity Suction conditions Isentropic or polytropic efficiencies	1. Performance curves 2. Compressor specification sheet 3. Original design data 4. P&ID 5. Nameplate
Positive Displacement Pumps	Pump casing MAWP / MAWT, design conditions Rated capacity	1. Pump specification sheets 2. P&ID 3. Nameplate
Reciprocating Compressors	Compressor manufacturer/model Cylinder type (double acting, etc.), diameter Stroke length, Rod diameter, Piston displacement, Engine speed,	1. Compressor specification sheets 2. Original design specification 3. P&ID 4. Nameplate

	General Data Requirements	
Process Design and Description	Piping and Instrumentation Diagrams (P&ID) Heat and Material Balances (H&MB) Process Flow Diagrams (PFD) Process Safety Flow Diagrams (PSFD) Process descriptions / operating procedures Plot plans / elevation plans	
Utility and Piping Design	Utility operating conditions (electrical, instrument air, cooling water, steam, etc.) Electrical one-line diagrams Piping designations and ratings Insulation designations and ratings	
	Data Requirements	
	Required Information	**Data Source Hierarchy**
	Volumetric efficiency	
Turbines	Exhaust casing MAWP / MAWT, design conditions, Steam throughput	1. Turbine specification sheets 2. P&ID 3. Nameplate
Control Valves	Sizes (inlet / outlet / port) Manufacturer and model number Fail safe position	1. Control valve data sheets 2. Vendor data 3. Nameplate

3. Codes and Standards

It is important to be aware of the relevant codes and standards that are either directly or indirectly related to a relief systems design basis audit. The audit team should have these documents at their disposal, and should be aware of the content of each. The main codes and standards that should be considered include:

- API Standard 520 Part 1 (9th Edition). "Sizing, Selection, and Installation of Pressure-Relieving Devices in Refineries", 2014.
- API Recommended Practice 520 Part 2 (6th Editions). "Sizing, Selection, and Installation of Pressure-Relieving Devices in Refineries", 2015.
- API Standard 521 (6th Edition). "Guide for Pressure-relieving and Depressuring Systems", 2014.
- API Standard 526 (6th Edition). "Flanged Steel Pressure Relief Valves", 2009.
- API Standard 2000 (7th Edition). "Venting Atmospheric and Low-Pressure Storage Tanks: Non-refrigerated and Refrigerated", 2014.
- API Recommended Practice 576 (3rd Edition). "Inspection of Pressure-Relieving Devices", 2009.
- ASME BPVC Section VIII. "Boiler & Pressure Vessel Code (BPVC), Section VIII, Division 1: Rules for Construction of Pressure Vessels", 2013.
- ASME B31.3. "Process Piping (B31.1-2014) and Process Piping Design", 2014.
- OSHA 29 CFR 1910.119: "Process Safety Management of Highly Hazardous Chemicals".
- OSHA Emphasis Programs

In addition, any company-specific guideline documents should be considered within the audit. Other codes and standards which may be relevant depending on the process being evaluated include:

- ASHRAE Standard 15. "Safety Standard for Refrigeration Systems and Designation and Classification of Refrigerants", 2013.
- OSHA 29 CFR 1910.110. "Storage and Handling of Liquefied Petroleum Gases".
- OSHA 29 CFR 1910.111. "Storage and Handling of Anhydrous Ammonia".
- NFPA 30. "Flammable and Combustible Liquids Code", 2015.

4. Conducting the Audit

The scope of the audit will have been defined during the audit preparation. Having developed the audit protocol, determined what data is required, what standards are being measured against, the auditor can proceed with conducting the audit.

4.1 Sample Selection

For the purposes of the audit, it is necessary to select a subset of the scope for further, detailed analysis. This representative sample should be carried forward to the audit steps that include:

- Ensuring accuracy of existing relief system design basis
- Ensuring adequacy of existing relief systems design basis documentation

It is therefore important to select a sample which is a good cross-representation of all relief devices within the scope, and to ensure that various design criteria are addressed. Various sampling techniques exist for conducting audits, which may be considered for a relief systems design basis audit:

- Random - selection purely by chance
- Block - selection based on clusters of information
- Stratification - selection based on subjective decisions of higher risk categories
- Interval - selection based on every n^{th} item in a list

Regardless of sampling technique used, the sample selection should aim to ensure that the criteria shown in Table 2 are covered:

Table 2. Relief System Criteria

Criteria	Scope
Valve Type	Conventional, Balanced Bellows, Pilot, Rupture Disk
Equipment Type	Column, Vessel, Heat Exchanger, Pump, Compressor, Reactor
Discharge Location	Flare Header, Atmosphere, Process
Flow Regime	Vapor, Liquid, Two-Phase

It is typically possible to meet all the desired criteria, with some systems able to meet several criteria at the same time, e.g. a bellows device protecting a column, relieving to flare, which could experience vapor, liquid or two-phase flow. The CCPS book "Guidelines for Auditing Process Safety Management Systems" **[4]**, typically recommends a sample size of between 10 and 20% for detailed auditing. However, depending on schedule and resource availability, this size may be increased or decreased.

4.2 Ensuring a Design Basis Exists for Every Relief System

While it sounds fairly obvious to ensure a design basis exists for every relief system, it is not uncommon for unprotected equipment to be identified during a relief systems design basis audit. This activity is typically conducted through detailed review of Piping and Instrumentation Diagrams (P&IDs). As the P&IDs are being reviewed, the auditor should be compiling a list of relief devices and protected equipment. While a relief device list may already exist for the facility, it is important for the auditor to independently compile this list. This list should cross-reference which relief devices protect each piece of equipment within the audit scope.

Therefore, the following issues may be uncovered in conducting this P&ID review:

- Potentially unprotected equipment
- Potential obstructions in relief path: Block valves which should be locked open, check valves, control valves
- Incorrect set pressures
- Relief piping issues: The nominal size of the inlet piping must be the same as or larger than the nominal size of the pressure relief valve inlet flange connection
- Atmospheric relief: potential for relieving toxic or flammable fluid to the atmosphere

Another benefit of developing a list of relief devices and protected equipment comes when verifying if a design basis exists for every system. The list compiled by the auditor can be compared with the number of records currently being maintained by the facility. The comparison should yield identical results - if this is not the case, there could be relief devices which do not have any documentation, or which may not be on the facility's inspection and testing schedule.

4.3 Field Verification

Having reviewed the facility P&IDs, identified potential areas of concern, and selected an audit sample; it is very useful to conduct field verification. As well as getting a general feel for the facility and the housekeeping standards that are in place, the field verification includes some specific activities of benefit to the relief systems audit:

- Confirming relief device tags are present and correct
- Comparison of relief device installation with that shown on P&ID:
  - Set pressure, relief device size, piping size, fittings, etc.
- Check block valves in relief piping to confirm if car-seal program is in effect
- Identify any areas of pocketing in relief piping
- Evaluate potential for atmospheric relief devices for toxic or flammable discharges:
  - Ensure no liquid scenarios can relieve to atmosphere
  - Ensure flammable releases have at least 50 feet horizontal separation
  - Check if consequence modeling has been performed for toxic or flammable releases
- Conduct field sketching for relief devices that are selected for the audit sample

4.4 Ensuring Accuracy of Existing Relief System Design Basis

The relief systems audit should also aim to ensure the accuracy of the existing relief systems design basis. This can be achieved by conducting a revalidation of the audit sample - in other words, a complete redo of the design calculations for all the relief devices selected in the audit sample. The main steps involved in a relief systems design basis study include:

- Define scope, basis, and project guidelines (this step will already have been conducted during the audit preparation stage)
- Gather project and process data (this step may also already have been conducted during the audit preparation stage)
- Develop overpressure scenarios for each protected system
- Determine relief requirement for each applicable overpressure scenario
- Calculate relief device capacity for each applicable overpressure scenario
- Check reaction forces, acoustic induced vibration, discharge temperatures, and relief device stability
- Identify relief system deficiencies and formulate options
- Generate report

Overpressure scenario identification is a critical aspect of the relief systems evaluation. For each selected system, a list of all credible potential causes of overpressure should be developed; stating whether an overpressure scenario is applicable or not applicable, and why. The overpressure scenarios considered should include, but are not necessarily limited to, those provided in API Standard 521 **[3]**.

For each potential overpressure scenario, the required relief rate should be calculated. Additionally, the flow capacity should be calculated to determine whether each relief system provides adequate capacity to prevent overpressure. The calculation methods used to determine each required relief rate and flow capacity should be documented in the individual relief systems calculation documentation. Additionally, the required orifice area and actual orifice area should be calculated.

Additional key pieces of data such as inlet pressure drops (between protected equipment and the relief device) and outlet pressure drops (between the relief device and main header) should also be calculated and documented based on actual expected relief capacities. A report for each relief system being analyzed as part of the audit sample should be developed. The report should be consistent with the documentation requirements specified in API Standard 521 **[3]**, 6th Edition, Section 4.7.

The purpose of revalidation is to identify any design issues within the audit sample, which may not have previously been addressed. It is a worthwhile exercise to determine why the original relief systems design basis may differ from the audit sample calculations. This could be due to a number of reasons such as:

- Changes in operating conditions (flowrates, temperatures, pressures)
- Changes in process equipment (equipment sizes, pump impellers, control valve trims)
- Missed scenarios
- Changes in calculation methodologies (for example, two phase modeling has progressed significantly over the past twenty years)

The types of issues that may be uncovered during this stage of the audit include:

- Undersized relief valves for a specific overpressure scenario
- Improper set pressure on relief valves
- Inlet pressure loss which exceeds three percent of set pressure
- Outlet backpressure which exceeds the limit for the type of relief device
- Excessive reaction forces
- Missed overpressure scenario
- Discharge temperatures below Minimum Design Metal Temperature (MDMT)
- Data discrepancies or missing process safety information

Should any design deficiencies such as those listed above be identified, recommendations should be formulated to resolve these issues. The relief systems design basis auditor should also bear in mind that analysis of individual audit sample systems, which may be scattered all across the facility, tends to be more time-consuming than when conducting a unit-wide analysis, which allows the designer to gain efficiency as progress is made throughout the unit.

4.5 Ensuring Adequacy of Existing Relief Systems Design Basis Documentation

Another key part in conducting the audit is to ensure the adequacy of existing relief systems design basis documentation. This step involves reviewing a facility's existing documentation, and determining if the level of documentation meets required criteria. The OSHA PSM Standard, Section d (3)(i)(D) lists relief systems design and design basis, as part of the required process safety information (PSI) for a covered facility. However the OSHA PSM standard is performance based and does not specify the required contents of a relief system design and design basis.

API Standard 521, 6th Edition, Section 4.7 provides guidance on documentation requirements for individual relief systems design basis. While the documentation requirements provided by API Standard 521 are extensive, this tends to be the most commonly accepted set of design criteria used within industry.

Most vintage relief systems are typically not compliant with the documentation requirements put forth by API Standard 521, and the auditor should note these exceptions and determine their significance on the overall quality of the relief systems design basis.

Additionally, the auditor should ensure that there are no open action items or deficiencies that have arisen from any previous relief systems design basis studies. PHA studies should also be reviewed, to ensure that any time a PHA study takes credit for a relief system, it is important to verify that the relief system is designed to handle the scenario in question. A good relief systems design basis will generate fewer PHA action items.

4.6 Further Work

Upon completion of the audit activities, it is important to draw conclusions on the overall adequacy of the relief systems design basis that was audited. Percentage compliant and percentage non-compliant should be shown, with a breakdown of non-compliant statistics provided. The purpose of the audit sample is to act as a representative sample, which could be extrapolated to indicate the broader status of the unit or facility in general. If deficiencies are identified, these need to be assessed and prioritized for mitigation as part of a corrective action plan.

Additionally, in the way that this paper describes auditing individual relief systems design basis, consideration should be given to conducting similar audits of flare systems to verify their adequacy.

5. Conclusions

Auditing is an important and independent element of process safety management systems. It is useful for assuring regulatory compliance, conformance to Recognized and Generally Accepted Good Engineering Practice (RAGAGEP); and can help improve the quality of a company's management systems. Given the increased focus in pressure relief systems, it is important that companies are aware of the completeness and quality of their existing relief systems design basis, before implementing corrective action plans.

6. References

[1] "Sizing, Selection, and Installation of Pressure Relief Devices in Refineries" 9th Edition, API Standard 520 Part I (2014).

[2] "Sizing, Selection, and Installation of Pressure Relief Devices in Refineries" 6th Edition, API Recommended Practice 520 Part II (2015).

[3] "Pressure-relieving and Depressuring Systems" 6th Edition, API Standard 521 (2014).

[4] "Guidelines for Auditing Process Safety Management Systems", 2nd Edition, AIChE/CCPS (2011).

398048 Dynamics of Direct Spring Operated Prv's with Inlet Piping in Gas Service

Abstract:

This paper will discuss the development of a numerical solutions which can be used as a predictor of direct spring operated pressure relief valve stability when connected to inlet piping on gas service and experimental data that correlates with the numerical solution. Topics to be discussed include:

1. Development of a full model.
2. Development of a simplified model (Quarter Wave model) based on assumptions learned from the full model.
3. Validation with test data.
4. Discussion of the 3% rule.
5. Parametric studies and sensitivity of the stability boundary to various parameters.
6. Development of a N Wave model and it's capability to provide a yes/no answer to the question: Is this system stable?

11TH GLOBAL CONGRESS ON PROCESS SAFETY

Effect of Body Bowl Choking on Pressure Relief Valve Stability

Hisao Izuchi
Advanced Process Engineering Unit, Chiyoda Corporation
4-6-2, Minatomirai, Nishi-ku, Yokohama 220-8765, Japan
izuchi.hisao@chiyodacorp.com

Prepared for Presentation at
American Institute of Chemical Engineers
2015 Spring Meeting
11th Global Congress on Process Safety
Austin, Texas
April 27-29, 2015

UNPUBLISHED

Effect of Body Bowl Choking on Pressure Relief Valve Stability

Hisao Izuchi
Advanced Process Engineering Unit, Chiyoda Corporation
4-6-2, Minatomirai, Nishi-ku, Yokohama 220-8765, Japan
izuchi.hisao@chiyodacorp.com

Keywords: Pressure Relief Valve, Stability, Body Bowl Choking, Limited Lift

Abstract

It is known that the velocity at the outlet flange of a PRV (Pressure Relief Valve) reaches sonic velocity in case of larger size of PRV with relatively high set pressure. This phenomenon is called body bowl choking and the pressure in the PRV body rises irrespective of the back pressure at the PRV outlet piping. Though this pressure rise in the PRV body could result in insufficient opening and/or unstable motion for the conventional spring loaded PRV, there is no design guide to prevent the adverse effect of body bowl choking. This paper presents the investigation results of the body bowl choking effect on the conventional spring loaded PRV stability using a simplified pressure calculation formula of one dimensional gas flow with comparison to the experiment data. It is concluded that the calculated pressure rise in the PRV body should be less than 15% of the set pressure to prevent unstable motion of the conventional PRV and recommended changes in PRV outlet flange sizes are introduced with this criterion. Instead of the changes in the PRV outlet flange sizes, restricted lift of the conventional PRV is also effective in preventing PRV instability. From this point of view the critical ratio of PRV outlet area to the orifice area is also obtained as a function of the set pressure of the PRV to prevent the instability of the conventional spring loaded PRV.

1. Introduction

It is known that the velocity at the outlet flange of a PRV (Pressure Relief Valve) reaches sonic velocity in case of larger size of PRV with relatively high set pressure. This phenomenon is called body bowl choking and the pressure in the PRV body rises irrespective of the back pressure at the PRV outlet piping [1]-[9]. This pressure rise in the PRV body could result in insufficient opening and/or unstable motion for the conventional spring loaded PRV in which the pressure rise in the PRV body directly affects the valve disc motion. Dossena [5] studied the effect of choking at the PRV outlet in conjunction with the backpressure effect with experiments. Dossena [6] also executed CFD (Computational Fluid Dynamics) studies for 2J3 and 8T10 PRVs and showed that insufficient flow would occur for the 8T10 PRV due to the lack of lift force caused by the body bowl choking relating to the relatively lower outlet area to the orifice area ratio compared with the 2J3 PRV. Sallet and Sommers [7] executed flow capacity tests for conventional PRVs with saturated and subcooled water and showed that the flow capacities of the conventional PRVs decrease when the ratio of outlet area to orifice area is lower than 6.0. The experiment results of Sallet and Sommers also suggested that the unstable motion of the conventional PRVs would occur in case of lower outlet area to orifice area ratio. Izuchi [8], [9]

Note: Do not add page numbers. Do not refer to page numbers when referencing different portions of the paper

executed the experiment and simulation studies for the conventional PRVs with air service and showed that the unstable motion of the conventional PRVs occur if the ratio of outlet area to orifice area is lower than 6.0 with set pressure of 2.1 MPa. The occurrence of body bowl choking is related to the PRV outlet flange size, which is defined by API standard 526 [10], and several studies have pointed out that there is a size problem defined by this standard [5], [6], [8], [9]. However there might be no design guidance to prevent this adverse effect of body bowl choking for the conventional PRV. API standard 520 Part I [11] mentioned that the built-up backpressure should not exceed 10 % of the set pressure at 10 % allowable overpressure in the conventional PRV application to prevent variations in opening pressure, reduction in flow capacity, instability, or a combination of all three. However, there is no description of the effect of the body bowl choking in the API standards.

This paper presents the investigation results of the body bowl choking effect on the stability of the conventional spring loaded PRV. The pressure rise in the PRV body is calculated by a simplified pressure calculation formula of one dimensional gas flow; the pressure rise ratio of the PRV body to the set pressure in relation to PRV stability is investigated from comparison with the experiment data. Using this approach, the adequacy of the PRV outlet flange sizes defined by API standard 526 [10] is investigated and changes are recommended in the outlet flange sizes for the conventional spring loaded PRV. Instead of the changes in the PRV outlet flange sizes, the restricted lift of the conventional PRV is also effective in preventing the instability of the conventional spring loaded PRV, taking the actual application into account [12]. From this point of view the critical ratio of outlet area to orifice area is also obtained as a function of the set pressure of the PRV to prevent the instability of the conventional spring loaded PRV.

2. Study Method

2.1 Approach and Assumptions to Investigate Effect of Body Bowl Choking

(1) The purpose of this study is to establish an appropriate general design guidance to prevent unstable motion of the conventional spring loaded PRV related to body bowl choking. From this point of view a simplified pressure calculation with one dimensional compressible flow is used to obtain the pressure in the PRV body. Three dimensional CFD can calculate the actual pressure accurately. However, CFD is not suitable as a design tool since it requires a lot of time and cost.

(2) Only the cases with the occurrence of body bowl choking are studied. If the back pressure of the PRV becomes higher than the critical one, the body bowl choking merely does not occur. In such cases similar one dimensional pressure calculations can be used to obtain the pressure in the PRV body to investigate the instability of the conventional spring loaded PRV.

(3) The outlet pressure of the PRV is assumed to be atmospheric since choked flow is readily formed with this condition.

(4) There is a contraction loss from the PRV body to the outlet piping through the PRV outlet flange. This contraction loss depends on the configuration of the PRV internals and differs depending on the manufacturer design. However the contraction loss is neglected in this study. This means the pressure calculation in this study would not be accurate but would be effective to obtain the order of the magnitude of the pressure rise in the PRV body. This assumption is useful for general guidance since the calculation method does not require the manufacturer

detailed information, which would be difficult to obtain. The fact that the contraction loss is usually lower than the expansion loss would also support the adequacy of this assumption.

(5) The pressure in the valve body is assumed to be equal to the stagnation pressure corresponding to the choking flow condition at the PRV outlet flange. After the flow through the PRV orifice, the flow distributes in various directions so that the average velocity toward the outlet direction at this portion would be approximately zero. Then the fluid is accelerated toward the PRV outlet from zero velocity to the sonic flow. Therefore this assumption with the stagnation pressure would be appropriate.

(6) The calculated pressure difference between those in the PRV body and the PRV outlet is divided by the set pressure of the PRV to obtain the pressure ratio called "PRV body pressure to set pressure ratio" in this paper hereinafter. This PRV body pressure to set pressure ratio is a key factor for the effect of body bowl choking.

(7) The calculated stagnation pressures in the PRV body are compared to the experiment data [8], [9] in order to confirm the adequacy of this study method. The measured pressures at the full lift positions fixed by a stopper are available in previous experiments for the cases with smaller PRV outlet area to orifice ratio.

(8) The critical calculated ratio of PRV body pressure to set pressure is determined from the experiment results when PRV instability occurs.

(9) Using this criterion for the critical PRV body pressure to set pressure ratio, the adequate PRV outlet sizes are studied for all the orifice sizes defined by API 526 [10]. Also the ratio of critical outlet area to orifice area for preventing PRV instability is obtained as a function of the set pressure of PRV. This critical outlet area to orifice area ratio can be used to determine the adequate limited lift of a PRV to keep stable PRV operation.

2.2 Expected Pressure Profiles

Figures 1 and 2 show the expected pressure profiles for the cases of non-choking and choking at the PRV outlets, respectively. In this study the pressure profile of the choking occurrence at the PRV outlet depicted in Figure 2 is assumed and the contraction loss from the PRV body to the PRV outlet is neglected as explained in section 2.1. The reducer size is assumed to be selected so that the velocity at the outlet pipe becomes subsonic.

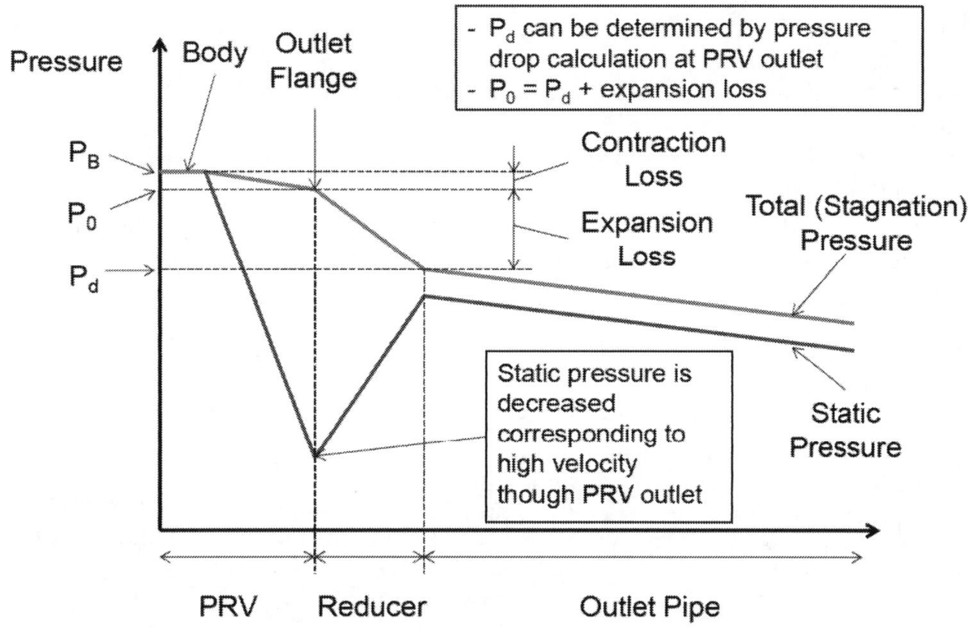

Figure 1 Expected Pressure Profile (No Choking at PRV Outlet Flange)

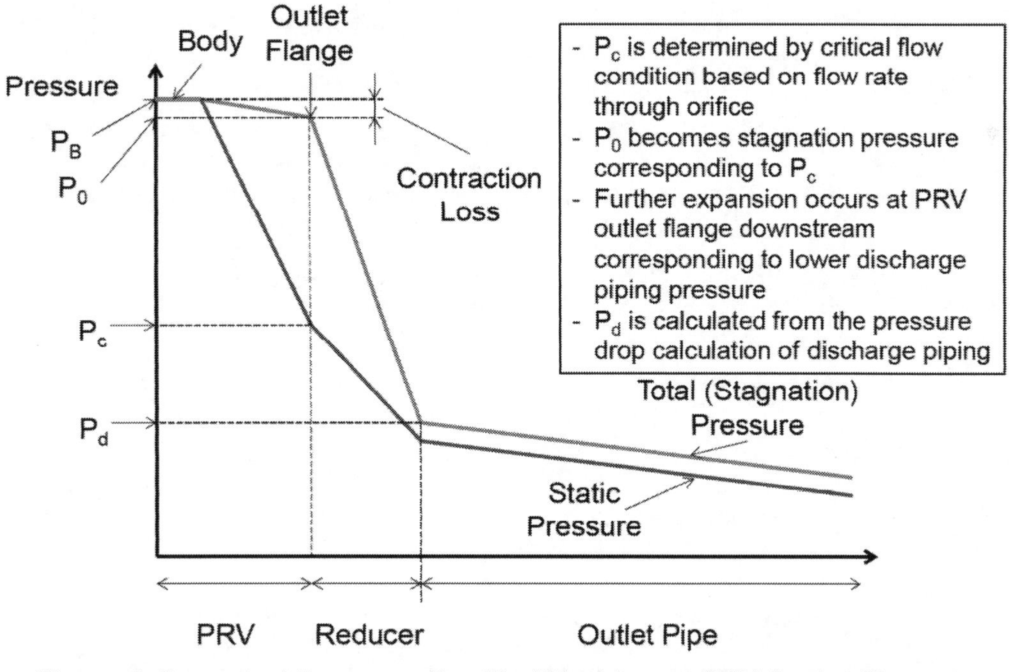

Figure 2 Expected Pressure Profile (Choking at PRV Outlet Flange)

2.3 Study Cases

Figure 3 shows the calculation model in this study. As shown in Figure 3 the following two cases are studied:

(1) Original Case: A block valve is installed at the reducer downstream. Choked flow occurs at the PRV outlet flange.

(2) Alternate Case: A block valve is installed at the outlet of the PRV just before the reducer. Choked flow occurs at the block valve outlet and the velocity at the PRV outlet flange becomes lower than sonic, corresponding to the pressure rise related to the pressure drop of the block valve.

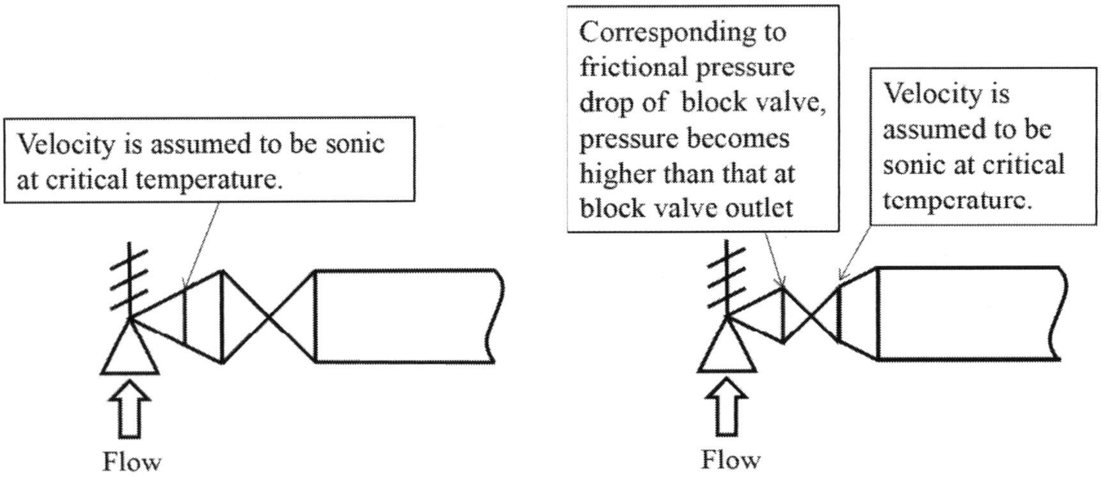

<u>Original Case</u>
(Block valve is installed at reducer outlet)

<u>Alternate Case</u>
(Block valve is installed at just PRV outlet before reducer)

Figure 3 Calculation Model (Study Cases)

2.4 Calculation Method

<u>2.4.1 Original Case</u>

The ratio of PRV body pressure to set pressure can be obtained by the flowing step by step approach:

(1) Step 1; Calculate Rated Flow

$$W = A_O P_1 C K_d K_b \sqrt{\frac{M_W}{TZ}} \times \frac{10^3}{3600}, \quad C = 0.03948\sqrt{k\left(\frac{2}{k+1}\right)^{\frac{k+1}{k-1}}} \qquad \text{[Eq. 1]}$$

Here, W : Rated Flow through PRV (kg/s)
A_O : Orifice Area of PRV (m^2)
P_1 : Relieving Pressure (usually 1.1 times of set pressure + atmospheric pressure, Pa)
K_d : Discharge Coefficient (assumed to be 0.975 in this study)
K_b : Capacity Correction Factor due to Backpressure (=1.0 in this study)
M_w : Molecular Weight of Gas (kg/kmol)
T : Relieving Temperature (K)
Z : Compressibility Factor (-)
k : Ideal Gas Specific Heat Ratio at Relieving Temperature (-)

(2) Step 2; Calculate Temperature and Sonic Velocity at Critical Condition

$$T_c = \frac{2}{k+1}T, \quad c_c = \sqrt{\frac{kZRT_c}{M_W}} \quad \text{[Eq. 2]}$$

Here, T_c : Critical Temperature (K)
c_c : Sonic Velocity at Critical Temperature (m/s)
R : Universal Gas Constant (=8314 J/kmol/K)

(3) Step 3; Calculate Density and Pressure at Critical Condition

$$\rho_c = \frac{W}{Ac_c}, \quad P_c = \frac{\rho_c ZRT_c}{M_W} \quad \text{[Eq. 3]}$$

Here, ρ_c : Density at Critical Condition (kg/m^3)
A : Flow Area at PRV Outlet Flange (m^2)
P_c : Critical Pressure (Pa)

Pressure at the PRV outlet becomes critical pressure, P_c corresponding to the choking flow condition with relieving flow capacity, W. If P_c is lower than atmospheric pressure, there is no choking at the PRV outlet and the pressure at the block valve outlet should be determined by the back pressure calculation for the PRV outlet piping.

(4) Step 4; Calculate Stagnation Pressure at PRV Outlet

$$P_0 = P_c \Big/ \left[\frac{2}{k+1}\right]^{\frac{k}{k-1}} \quad \text{[Eq. 4]}$$

Here, P_0 : Stagnation Pressure at PRV Outlet (Pa)

(5) Step 5; Calculate PRV Body Pressure to Set Pressure Ratio

$$\Delta P = P_0 - P_{atm}, \quad \Delta P_r = \Delta P / \left(P_{set} - P_{atm}\right) \quad \text{[Eq. 5]}$$

Here, ΔP : Difference between Pressure in PRV Body and PRV Outlet (Pa)

ΔP_r : PRV Body Pressure to Set Pressure Ratio (-)
P_{atm} : Atmospheric Pressure (Pa)
P_{set} : Set Pressure of PRV (Pa)

2.4.2 Alternate Case

Steps 1, 2, 3 and 5 are same as those for the original case. Step 4 for the original case is replaced by the following Steps 4A-1 and 4A-2;

(1) Step 4A-1; Calculate Fanno Flow Equation for Block Valve

The flow through the block valve could be adiabatic compressible flow and the friction loss coefficient of the block valve, K, can be expressed by the following Fanno flow equation with the condition that the Mach number at the block valve outlet becomes 1.0:

$$K = \frac{\frac{1}{Ma_2^2} - 1 - \left(\frac{k+1}{2}\right) \ln\left(\frac{1}{Ma_2^2}\frac{Y_2}{Y_c}\right)}{k}, \; Y_2 = 1 + \frac{k-1}{2}Ma_2^2, \; Y_c = 1 + \frac{k-1}{2} \quad \text{[Eq. 6]}$$

Here, Ma_2: Mach Number at Block Valve Inlet (-)

The Mach number at the block valve inlet Ma_2 can be obtained by solving Eq.6 numerically. K is equal to $8f_T$ for typical gate valves based on Crane [13], here, f_T is the friction factor for the clean commercial steel pipe with flow zone in complete turbulence.

(2) Step 4A-2; Calculate Static Pressure & Stagnation Pressure at Block Valve Inlet, i.e., PRV Outlet

$$P_2 = \frac{P_c}{Ma_2}\sqrt{\frac{1+\frac{k-1}{2}}{1+\frac{k-1}{2}Ma_2^2}}, \quad P_0 = \frac{P_c}{Ma_2\left(\frac{2}{k+1}\right)^{\frac{k}{k-1}}}\sqrt{\left[\frac{2\left(1+\frac{k-1}{2}Ma_2^2\right)}{k+1}\right]^{\frac{k+1}{k-1}}} \quad \text{[Eq. 7]}$$

Here, P_2 : Pressure at Block Valve Inlet (Pa)

3. Calculation Results

3.1 Comparison with Experiment Data

3.1.1 Pressure in PRV Body

In the experiment to investigate PRV stability [8], [9], the pressure in the PRV body is measured using a pressure tap. The lift of the PRV was fixed at the full lift position by an attachment and the ratio of PRV outlet area to orifice area is adjusted by a reducer installed at the PRV outlet. There is no block valve at the PRV outlet and this experimental facility is equivalent to the system for Case 1 (original case) in this paper. Table 1 shows the comparison of the measured pressures and the calculated pressures in the PRV body. The calculated pressures are obtained by the step by step method described in section 2.4. As shown in Table 1 the measured pressures in the PRV body becomes higher than the calculated pressure to a certain extent. These differences between the measurement and calculation could be caused by neglecting the contraction loss from the PRV body to the PRV outlet. However, the simplified calculation with one dimensional compressible flow described in section 2 is useful for estimating the magnitude of the pressure rise in the PRV body.

Table 1 Comparison between Calculation and Experiment Results for Pressure in PRV Body

PRV Size	Outlet Size (w/ Reducer)	Outlet Area to Orifice Area Ratio	Pressure in PRV Body		
			Calculation (Stagnation Pressure) (1)	Measurement (2)	Ratio (2)/(1)
1.5F2	1"	2.5	0.85 MPa	1.30 MPa	1.53
1.5F2	1-1/4"	4.1	0.52 MPa	0.66 MPa	1.28
1.5F2	1-1/2"	5.6	0.38 MPa	0.46 MPa	1.21
1E2	1"	3.3	0.64 MPa	0.93 MPa	1.45
1E2	1-1/4"	5.5	0.38 MPa	0.45 MPa	1.16

3.1.2 Condition for Stable Operation of Conventional PRV

In the experiment [8], [9], the PRV stability is investigated by adjusting the ratio of PRV outlet area to orifice area with a reducer installed at the PRV outlet. The experiment was executed with air at a set pressure of 2.1 MPa. Due to a lack of air compression capacity in the experimental facility, the pressure at the PRV inlet was almost kept at the set pressure and there was no pressure rise after the PRV working. Figure 4 shows a typical measured time history in case of PRV instability occurrence caused by the body bowl choking. As shown in Figure 4 the differential pressure across the valve disc decreases as the valve lift increases. Since the PRV inlet pressure is almost constant, the differential pressure decrease across the valve disc is mainly caused by the pressure accumulation in the PRV body and as a result, the occurrence of PRV instability is attributed to the body bowl choking. Table 2 shows the summarized results of the experiment to investigate the conventional PRV instability relating to the ratio of PRV outlet area to orifice area. "Yes" of instability occurrence in Table 2 means that cyclic lift movements were observed several times a seconds in the experiment. As shown in Table 2, the conventional PRV becomes unstable under the condition that the PRV outlet area to orifice area ratio is smaller than 6.0 at the set pressure of 2.1 MPa.

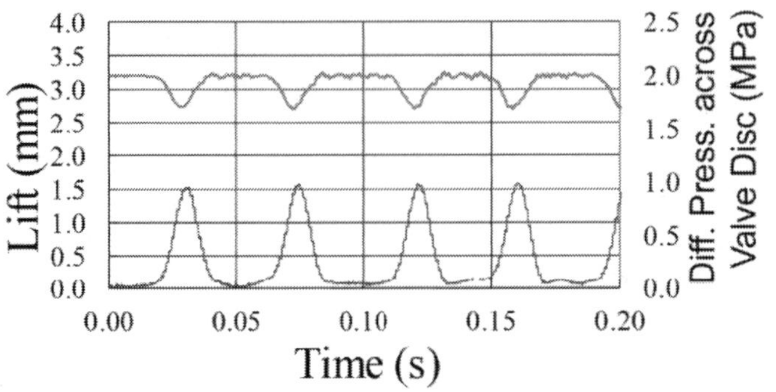

Figure 4 Typical Time History in Case of PRV Instability Occurrence Caused by Body Bowl Choking

Table 2 Summarized Results of Experiment

PRV Size	(1) Orifice Area	Outlet Size	(2) Outlet Area	Ratio (2) / (1)	Instability Occurrence
1E2	1.82 cm2	2"	20.3 cm2	11.2	No
		1-1/2"	13.6 cm2	7.5	No
		1-1/4"	10.0 cm2	5.5	No
		1"	6.0 cm2	3.3	Yes
1.5F2	2.43 cm2	2"	20.3 cm2	8.4	No
		1-1/2"	13.6 cm2	5.6	Yes
		1-1/4"	10.0 cm2	4.1	Yes
		1"	6.0 cm2	2.5	Yes

Table 3 shows the calculated variation of the PRV body pressure to set pressure ratio with the PRV outlet area to orifice area ratio at the experimental set pressure condition of 2.1 MPa. As shown in Table 3 the ratios of PRV body pressure to set pressure become 13.7% and 15.4% for the PRV outlet area to the orifice area ratios of 6.0 (stable) and 5.5 (unstable), respectively. Comparing these calculations to the experiment results, 15% of the PRV body pressure to set pressure ratio would be adequate to maintain stable operation of a conventional spring loaded PRV. This criterion is based on the simplified pressure calculation method explained above relating to the body bowl choking.

Table 3 Calculated PRV Body Pressure to Set Pressure Ratio
(Class 150, Set Press=2.1 MPa, Air)

Orifice Size	Orifice Area	Outlet Area to Orifice Area Ratio	ΔPr
	(cm2)	(-)	(%)
E	1.265	10.0	6.2
		9.0	7.4
		8.0	9.0
		7.0	11.0
		6.0	13.7
		5.5	15.4
		5.0	17.4
		4.0	23.0
		3.0	32.4

3.2 Adequacy of PRV Outlet Size Defined by API526

Table 4 shows the calculation results for the ratio of PRV body pressure to set pressure for the ANSI class 150 PRVs in which the outlet and orifice sizes are specified by API standard 526 [10]. In this calculation the set pressures are assumed to be equal to the maximum pressure limit specified in API526. Both the original and alternate cases are studied. As shown in Table 4, the PRV body pressure to set pressure ratios become larger than 15% for the PRVs with orifice sizes of J, L, P, Q, R and T, corresponding to lower PRV outlet area to orifice area ratios. This study result suggests that the enlargements of the outlet flange sizes are required to avoid instability for orifice sizes of J, L, P, Q, R and T in the case of ANSI class 150 PRVs specified in API standard 526. Table 5 shows the calculation results for the PRV body pressure to set pressure ratios for the ANSI class 1500 PRVs. As shown in this table, adequate PRV outlet sizes for avoiding instability are specified in API standard 526 for the class 1500 pressure rating. Table 6 shows the recommended PRV outlet sizes obtained by this study to keep stable operation for the conventional spring loaded PRVs. Table 6 indicates that the outlet flange size should be enlarged for relatively-large PRVs at lower pressure rating to prevent unstable motion of the conventional spring loaded PRVs.

Table 4 Calculation Results for PRV Body Pressure to Set Pressure Ratio (Class 150, Air)

Orifice Size	Orifice Area	Outlet Nominal Size	Outlet Area to Orifice Area Ratio	Set Pressure	ΔPr	
					Original Case	Alternate Case
	(cm2)	(inches)	(-)	(MPa)	(%)	(%)
D	0.710	2	27.7	2.07	No Choke at Outlet	
E	1.26	2	15.5	2.07	2.1	2.6
F	1.98	2	9.9	2.07	6.2	7.0
G	3.25	3	13.6	2.07	3.1	3.7
H	5.06	3	8.7	2.07	7.7	8.6
J	8.30	3	5.3	2.07	***16.0 ****	***17.4 ****
K	11.9	4	6.6	2.07	11.8	12.9
L	18.4	4	4.3	2.07	***21.2 ****	***22.9 ****
M	23.2	6	7.6	2.07	9.6	10.5
N	28.0	6	6.3	2.07	12.7	13.7
P	41.2	6	4.3	2.07	***21.0 ****	***22.6 ****
Q	71.3	8	4.4	1.24	***17.4 ****	***18.9 ****
R	103	8	3.0	0.79	***25.3 ****	***27.5 ****
T	168	10	2.9	0.55	***21.6 ****	***24.1 ****

* large than 15%

Table 5 Calculation Results for PRV Body Pressure to Set Pressure Ratio (Class 1500, Air)

Orifice Size	Orifice Area	Outlet Nominal Size	Outlet Area to Orifice Area Ratio	Set Pressure	ΔPr	
					Original Case	Alternate Case
	(cm2)	(inches)	(-)	(MPa)	(%)	(%)
D	0.710	2	27.7	25.65	3.5	3.8
E	1.26	2	15.5	25.65	6.5	7.0
F	1.98	3	22.3	25.65	4.4	4.8
G	3.25	3	13.6	25.65	7.5	8.1
H	5.06	3	8.7	20.79	11.9	12.7
J	8.30	4	9.5	20.79	10.9	11.7
K	11.9	6	14.9	20.79	6.7	7.2
L	18.4	6	9.6	20.79	10.7	11.4

Table 6 Recommended PRV Outlet Sizes to Keep Stable Operation for Conventional Spring Loaded PRVs

Orifice	Pressure Rating; Flang Class					
	150	300	600	900	1500	2500
D	2"	2"	2"	2"	2"	3"
E	2"	2"	2"	2"	2"	3"
F	2"	2"	2"	3"	3"	3"
G	3"	3"	3"	3"	3"	3"
H	3"	3"	3"	3"	3"	-
J	**3" => 4"**	**3" => 4"**	4"	4"	4"	-
K	4"	**4" => 6"**	**4" => 6"**	6"	6"	-
L	**4" => 6"**	**4" => 6"**	6"	6"	6"	-
M	6"	6"	6"	6"	-	-
N	6"	**6" => 8"**	**6" => 8"**	**6" => 8"**	-	-
P	**6" => 8"**	**6" => 8"**	**6" => 8"**	**6" => 8"**	-	-
Q	**8" => 10"**	**8" => 10"**	**8" => 10"**	-	-	-
R	**8" => 10"**	**10" => 12"**	**10" => 12"**	-	-	-
T	**10" => 12"**	**10" => 16"**	-	-	-	-

3.3 Minimum Required Outlet Area to Orifice Area Ratio

Figure 5 shows the relation between the minimum required outlet area to orifice area ratios and the set pressure obtained by this study for both the original and alternate cases. As shown in this figure, the minimum required PRV outlet area to orifice area ratio tends to become higher as the set pressure increases. And the minimum required outlet area to orifice area ratios become 7.5 and 7.0 for the original and alternate cases, respectively, at a set pressure of 50 MPa. Instead of the changes in the PRV outlet flange sizes the restricted lift of the conventional spring loaded PRV is quite effective to prevent the PRV instability taking the actual application into account [12]. The relation shown in Figure 5 would be useful for determining the position of the restricted lift to prevent unstable motion of the conventional spring loaded PRV. The capacity calculation method for restricted lift PRV is defined in ASME BPVC sec.VIII [14] and the equivalent orifice area corresponding to the restricted lift can be determined in accordance with this ASME standard. Using this equivalent orifice area and flange area, the possibility of the PRV instability caused by body bowl choking can be evaluated by Figure 5.

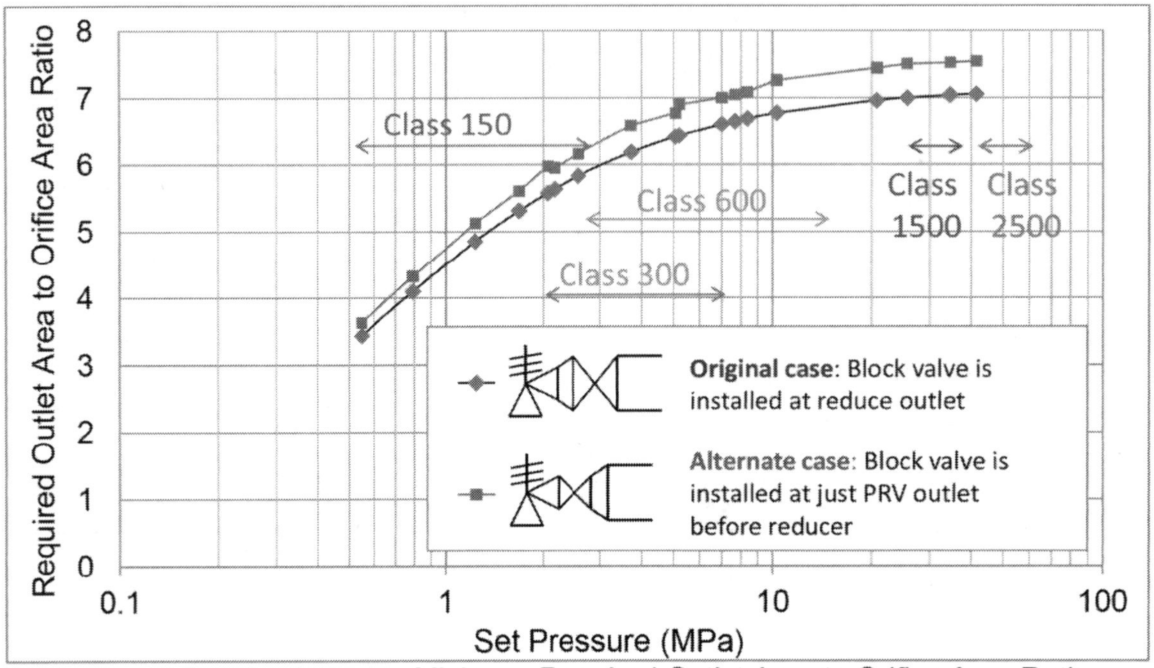

Figure 5 Relation between Minimum Required Outlet Area to Orifice Area Ratio

4. Conclusions

The effect of body bowl choking relating to the unstable motion of the conventional spring loaded PRV has been studied. As a result of the study, the following can be concluded:

(1) A simplified pressure calculation with one dimensional compressible flow is useful for obtaining the pressure in the PRV body. Comparison with the experiment data supports this approach though the actual pressure rise in the PRV body is found to be higher than the calculated one to a certain extent.

(2) The conventional spring loaded PRV becomes unstable if the calculated ratio of PRV body pressure to set pressure becomes equal to or larger than 15%. This criterion is derived from the experiment data and comparison with the pressure calculation with one dimensional compressible flow.

(3) The adequacy of the PRV outlet flange size specified in API standard 526 is investigated using this criterion for the calculated ratio of PRV body pressure to set pressure. The outlet flange size should be enlarged for relatively-large PRVs at lower pressure rating to prevent unstable motion of the conventional spring loaded PRVs.

(4) The minimum required ratio of outlet area to orifice area to prevent unstable motion of the conventional spring loaded PRVs is obtained as a function of the set pressure. This minimum required ratio of outlet area to orifice area could be useful for determining the position of the restricted lift to prevent unstable motion of the conventional spring loaded PRVs.

The above conclusions are derived from the experiment data for the conventional spring loaded PRV prepared by one manufacturer. The pressure rise in the PRV body might be affected by the configuration of the PRV internals which would differ depending on the manufacturer design. From this point of view further investigation would be desired to confirm the adequacy of this proposed calculation approach and the criteria to assess the stability of the conventional spring loaded PRV relating to the body bowl choking.

5. References

[1] G.A. Melhem, Pressure Relief Valve Stability with SuperChems, API Fall Meeting, Nov. 2011. Available at http://mycommittees.api.org/standards/cre/scprs/default.aspx (accessed Jan. 2015).

[2] G.A. Melhem, Effective Modeling of Body Bowl Choking, Fall DIERS Users Group Meeting, Oct. 2014.

[3] R. D'Alessandro, Body Bowl Choking in Pressure Safety Valves, DUG/EDUG Joint Meeting, June 2011.

[4] R. D'Alessandro, Pressure Safety Valve Thrust Forces for Compressible Gas or Vapor Flow, International Symposium on Runaway Reactions and Pressure Relief Design, Oct. 2005.

[5] V. Dossena, P. Gaetani, F. Marinoni and C. Osnaghi, On the Influence of Back Pressure and Size on the Performance of Safety Valves, ASME PVP, 2002.

[6] V. Dossena, F. Marinoni, B. Paradiso, Valve Size Influence on the Discharge Capacity of Spring Loaded Safety Valves, Paper 722, Valve World Conference, 2007.

[7] D.W. Sallet and G.W. Somers, Flow Capacity and Response of Safety Relief Valves to Saturated Water Flow, Plant/Operations Progress 4-4 (1985), 207-216.

[8] H. Izuchi, Stability Analysis of Safety Valve, 10th Topical Conference on Gas Utilization, AIChE Spring Meeting, April 2010.

[9] H. Izuchi, Chatter of Safety Valve, API Spring Meeting, April 2008. Available at http://mycommittees.api.org/standards/cre/scprs/default.aspx (accessed Jan. 2015).

[10] API Standard 526, Flanged Steel Pressure Relief Valves, 5th Edition, 2002.

[11] API Standard 520, Sizing, Selection, and Installation of Pressure-relieving Devices, Part I - Sizing and Selection, 9th Edition, 2014.

[12] D. Smith and S. Yoram, Address Inlet Pressure Loss Concerns with Restricted Lift Relief Devices, Hydrocarbon Processing, March 2014, 71-73.

[13] Flow of Fluids through Valves, Fittings, and Pipe, Crane Co., 1986.

[14] ASME Boiler and Pressure Vessel Code, sec.VIII, Rules for Construction of Pressure Vessels, Division 1, 2013

11TH GLOBAL CONGRESS ON PROCESS SAFETY

Modeling and Computation of Reaction Forces on Relief Piping During Depressurization

Jens Conzen
Fauske & Associates, LLC
16W070 83rd Street, Burr Ridge, IL 60527
conzen@fauske.com

Gabe Wood
Fauske & Associates, LLC
16W070 83rd Street, Burr Ridge, IL 60527
wood@fauske.com

Prepared for Presentation at
American Institute of Chemical Engineers
2015 Spring Meeting
11th Global Congress on Process Safety
Austin, Texas
April 27-29, 2015

Modeling and Computation of Reaction Forces on Relief Piping During Depressurization

Jens Conzen
Fauske & Associates, LLC
16W070 83rd Street, Burr Ridge, IL 60527
conzen@fauske.com

Gabe Wood
Fauske & Associates, LLC
16W070 83rd Street, Burr Ridge, IL 60527
wood@fauske.com

Keywords: Reaction forces, structural verification, relief piping analysis, relief system design.

Abstract

The sudden opening of a relief device during an overpressure event will cause the relief piping to experience dynamic loads. A pressure wave may be generated that travels at the speed of sound through the fluid. This has led to force imbalances, particularly on long pipe segments, that can potentially damage the relief system or prevent the relief system from operating properly. In order to design and properly protect the relief piping with the necessary supports, it is critical that a detailed analysis of the dynamic loads (e.g. thrust, momentum, and wave) following the opening of a pressure relief device is performed. In many cases, simple analytical techniques may not be appropriate and a time history analysis of the loads acting on the relief system may be required. To address this phenomenon, an evaluation of the dynamic loads using RELAP5 (Reactor Excursion and Leak Analysis Program) software will be discussed. The RELAP5 software can be used to analyze fluid transients and compute resulting loads. It is possible to evaluate systems undergoing single phase or two phase relief using water as the fluid.

1. Introduction

This paper summarizes a methodology that can be used to size and verify the design of emergency relief systems for chemical reactors, with a focus on reaction force calculations. The methodology consists of multiple steps that include calorimetric trials, fluid dynamics, and structural calculations. The example plant is a three reactor site that produces phenolic resins. During a hypothetical process hazard analysis (PHA), it was determined that a loss of cooling prior to the catalyst addition could lead to overpressurization of the vessels. In response, calorimetric trials were performed and the data obtained were used to evaluate the required size of the relief device.

The characterization of the runaway reaction was then used in a fluid dynamic assessment which provided the reaction forces. The reaction forces were used as input to the structural analysis. The fluid forces were computed with the RELAP5 software and the structural assessment was performed in PIPESTRESS. The adequacy of the structural design was finally evaluated in accordance with the rules of an appropriate piping code.

2. Testing

The data to be used for relief sizing was collected using the Vent Sizing Package 2™ (VSP2) which is designed to acquire thermal and vent sizing data [2,3,4,5]. The VSP2 utilizes a low phi-factor test cell, thus providing near adiabatic runaway conditions. As a result, the temperature and pressure data measured in the VSP2 are directly scalable to any process vessel.

The system that was evaluated in this example is a base-catalyzed phenol-formaldehyde reaction. The recipe uses a 2.2:1 molar ratio of formaldehyde to phenol. The upset scenario that was tested was a loss of cooling prior to the addition of the catalyst (50% caustic). The phenol and 34% formaldehyde solution were added to the test cell. The contents were then heated to the process temperature of 50°C. The catalyst was injected into the test cell and the sample was kept adiabatic. The temperature rise rate data are shown in Figure 1. The initial exotherm was caused by the heat of mixing of the catalyst itself, followed by the runaway cross-linking polymerization reaction which produces the phenol-formaldehyde resin. The pressure vs. temperature data are compared with vapor pressure data for water, taken from DIPPR® [1], and shown in Figure 2. The pressure profile indicates that the system very closely follows the vapor pressure of water as the temperature rises. The test data indicate that a possible overpressure hazard exists for the process vessel. Consequently, the process vessel should be evaluated for proper overpressure protection provided by the pressure relief system.

3. Relief Size Evaluation

The required relief size was evaluated for vessel R-3 which is used for the production of phenolic resin. The vessel has a volume of 5 m^3 and contains 4 m^3 of reactant. The vessel maximum allowable working pressure (MAWP) is 30 psig and the rupture disk set pressure is 10 psig. The relief system consists of a rupture disk located just downstream of the reactor nozzle, which leads to a common header and then to a catch tank. The relief system is assumed to have an equivalent length of 300 feet. Based on the upset scenario tested, a vent sizing calculation was conducted using Leung's ω-method within Fauske & Associates LLC's PrEVent software assuming homogeneous vessel venting [6]. The data indicate that the system behaves as a vapor system in the venting region of interest. The PrEVent output confirms that the 16 inch diameter rupture disk and pipe would be adequate to prevent overpressurization of the vessel. After the size of the relief system pipe has been determined, the evaluation can proceed to the mechanical analysis, which begins with calculation of the fluid forces.

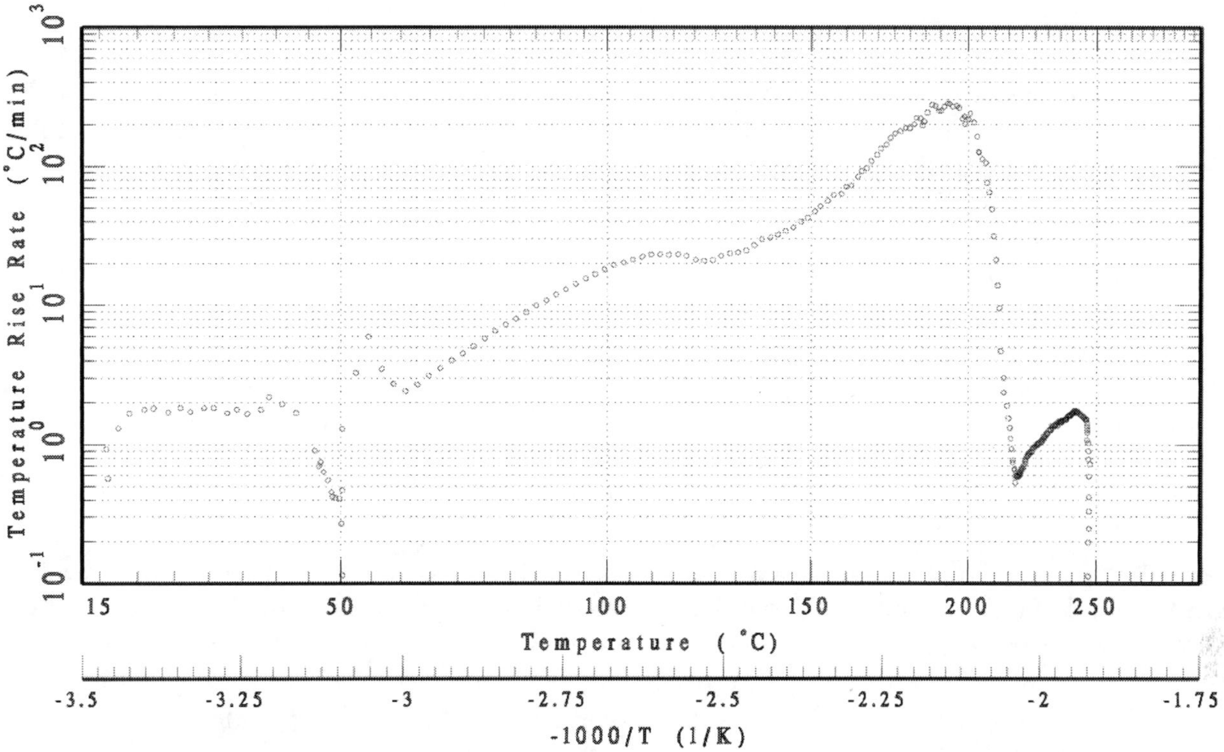

Figure 1. Temperature Rise Rate vs. Temperature Data

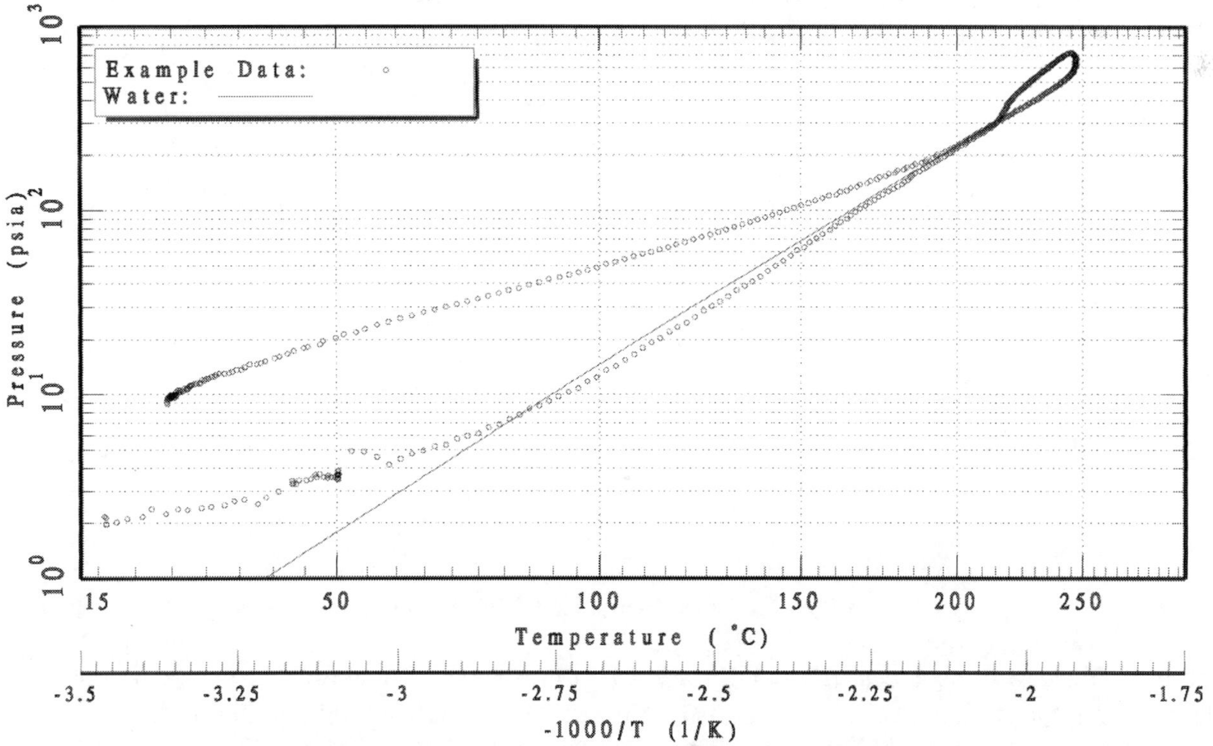

Figure 2. Pressure vs Temperature Data

4. Evaluation of Fluid Forces

Once the upset scenario has been quantified, a fluid model is used to evaluate the loads resulting from the postulated rupture disk bursting. The fluid forces are calculated using the RELAP5 computer code. RELAP (Reactor Excursion and Leakage Analysis Program) is a light water reactor transient analysis code developed for the United States Nuclear Regulatory Commission (US NRC) for simulation of a wide variety of hydraulic and thermal transients in both nuclear and nonnuclear systems involving mixtures of steam, water, non-condensable gases and solutes under single phase and two phase conditions. For this example, it was found that it is reasonable to assume steam/water properties due to the absence of specific knowledge of the fluid properties during the transient. The chemical reaction in the vessel may cause the effluent properties to rapidly change during the transient. The accuracy of the reaction force prediction depends on the density and sonic velocity (in the liquid and gas phase) of the fluid. Therefore, the use of steam/water properties can provide a best estimate solution for the reaction forces.

The capability of RELAP5 has been validated for the analysis of hydrodynamic loads in steam/water induced by large and rapid pressure waves propagating with the speed of sound along piping systems [7]. The code was also validated by the US NRC for use in analyzing the propagation of pressure waves in piping systems [9]. Comparison of the results demonstrated suitability of RELAP5 to be used for computational problems involving propagation of pressure waves in piping systems such as reactor relief system design.

At the example plant, the emergency relief lines from each reactor are tied into a header pipe that feeds the catch tank. The catch tank is vented to the atmosphere. The pipe diameters are 8 inches for reactor 1, 12 inches for reactor 2, and 16 inches for reactor 3 and the header pipe. Figure 3 provides a schematic of the plant layout.

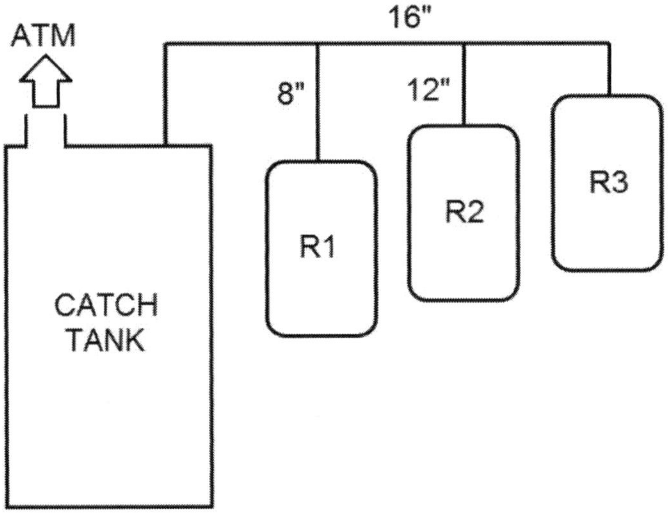

Figure 3. Example Plant Schematic

The burst disks are located at the reactor's extraction nozzles. A numerical model is then built that presents the outlined relief system. It is important to note that the calculation time step and node size must be properly selected. Therefore, the model was finely nodalized to allow the detailed computation of hydrodynamic loads on pipe segments throughout the system. For example, the length-to-diameter ratios (L/D) were set close to unity for each node. A maximum time step was selected that is below the acoustic courant limit to ensure the fidelity of the transient hydrodynamic loads simulation. The maximum time step must be less than the time for the acoustic wave to travel with the speed of sound through a nodal volume.

The runaway reaction is assumed to occur in a single reactor. However, simulation of global venting scenarios is possible. The reactor was initialized with a two phase mixture. The piping and catch tank downstream of the rupture disk were assumed to contain air at ambient conditions. Before the transient calculation begins, a steady state calculation with no flow is performed so that the model can initialize. The rupture disk burst is postulated by opening an arbitrarily modeled valve. Thereby, it is possible to change the rupture disk opening time. In this case, it was assumed that the rupture disk fully opens within 10 ms. Given the short duration of the depressurization transient, no heat transfer was assumed across the piping wall. However, heat transfer can be considered, if needed. The resulting fluid densities, void fractions, velocities, and pressure time histories are used to compute the reaction forces. A schematic of the fluid model is shown in Figure 4.

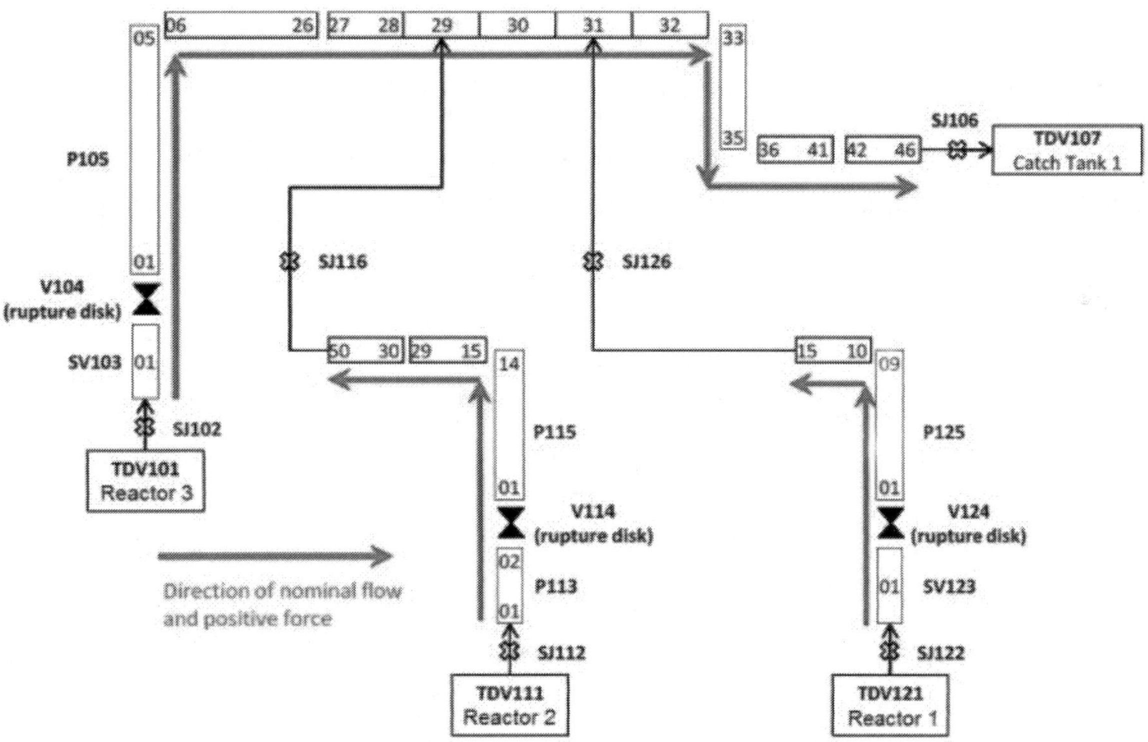

Figure 4. Schematic of Fluid Model

The development of the transient force-time history information for application to structural analysis models is based on the general force equations for a container [10]. The generalized force equation in one-dimensional form can be resolved by a pipe segment bounded by two elbows as

$$F_w = -\frac{1}{g_c}\frac{d}{dt}\int \rho A V dz \qquad \text{[Eq. 1]}$$

This is the unsteady reaction force caused by the rate of fluid momentum change within the control volume represented by the pipe segment (so-called wave load) and the wave load (F_w) approaches zero when the flow approaches a steady-state condition. RELAP5 employs a two fluid treatment, and with consideration of the vapor components of the flow, this equation becomes

$$F_w = -\frac{1}{g_c}\frac{d}{dt}\int \left(\rho_f A V_f \alpha_f + \rho_g A V_g \alpha_g\right) dz \qquad \text{[Eq. 2]}$$

where, ρ = density, A = flow area, V = velocity, α = void fraction, z = distance along piping axis, g_c = gravitational constant, and subscripts f and g refer to the liquid and gas phases.

In addition to the wave load, the system will have thrust loads at the discharge. The thrust loads will include the momentum flux of the discharging fluid (F_m) as well as the differential pressure at the pipe exit (F_p).

$$F_m = -\frac{\dot{m}_e V_e}{g_c} \text{ and } F_p = -(P_e - P_a) A_e \qquad \text{[Eq. 3]}$$

where $\dot{m}_e$ = mass flow rate at the pipe exit, V_e = exit velocity, g_c = gravitational constant, P = pressure, A = flow area, and subscripts e and a refer to the exit and ambient. Therefore, the pipe reaction force during unsteady discharge (F_t) becomes

$$F_t = \underbrace{-\left\{(P_e - P_a) A_e + \frac{\dot{m}_e V_e}{g_c}\right\}}_{\text{thrust load}} + \underbrace{\left(-\frac{1}{g_c}\frac{d}{dt}\int \rho A V dz\right)}_{\text{wave load}} \qquad \text{[Eq. 4]}$$

The generated time history data is now used to compute the force time histories on each pipe segment bound by elbows. Figure 5 shows example force output for the vent pipe at the burst disk at reactor 3. All force time histories are then applied in the PIPESTRESS model as force vectors for final design verification.

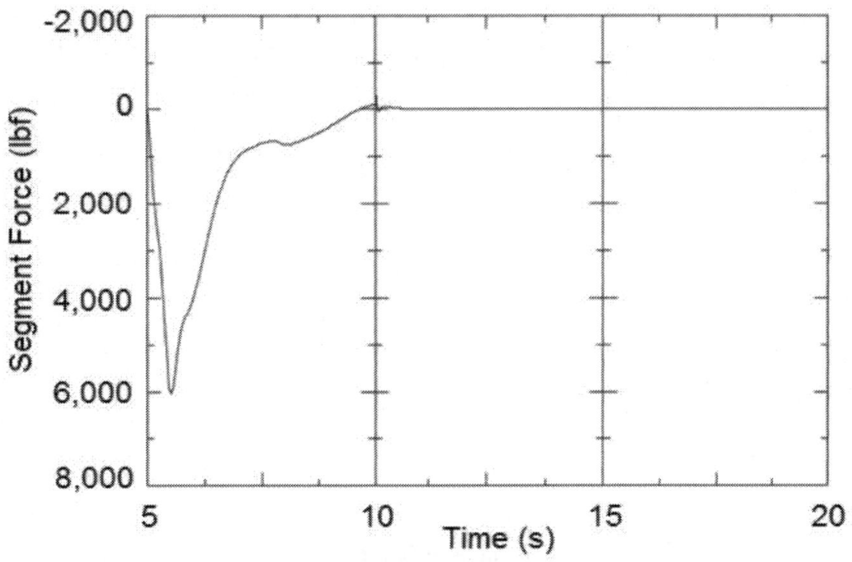

Figure 5. Example Force Output

The catch tank pressure (Figure 6) is also monitored to verify that the MAWP is not exceeded. The backpressure can be monitored in any location of the relief system including attached vessels.

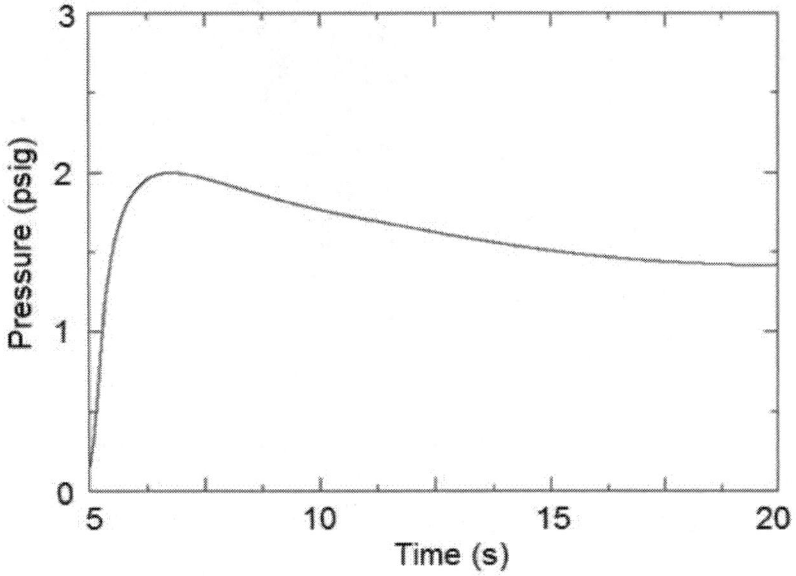

Figure 6. Catch Tank Pressure

5. Pipe Stress Analysis

A structural computer model of the example plant is generated in the PIPESTRESS software. All information required for the model can typically be found on pipe isometric drawings. PIPESTRESS is an analysis program for nuclear and industrial piping. It can be used to evaluate the load cases that are needed to meet compliance with piping codes such as ASME B31.3. In addition to the pipe sizes and pipe segment lengths (from the previous step) one must now specify the pipe support systems as well as material properties. The example plant has one sliding support (one degree of freedom restricted) at the high point near reactor 1 and one sliding support above the catch tank. In addition, bellows are used at the reactor exit nozzles and the catch tank entrance nozzle. The bellows help to compensate displacements caused by the depressurization transient. The pipe supports are modeled as springs with constant stiffness. In that case, mounting hardware, such as anchor bolts, should also be evaluated. Force, pressure, and temperature time histories are imported from the previous analysis step. The example plant is then verified for thermal expansion, dead weight, and the hydrodynamic depressurization loads resulting from the rupture disk bursting. The resultant stresses from the different load cases are divided by the allowable stress for the material to obtain the usage factors for the pipe segments and pipe supports. The highest allowable value for a usage factor is 1. Figure 7 shows the final result of the structural analysis.

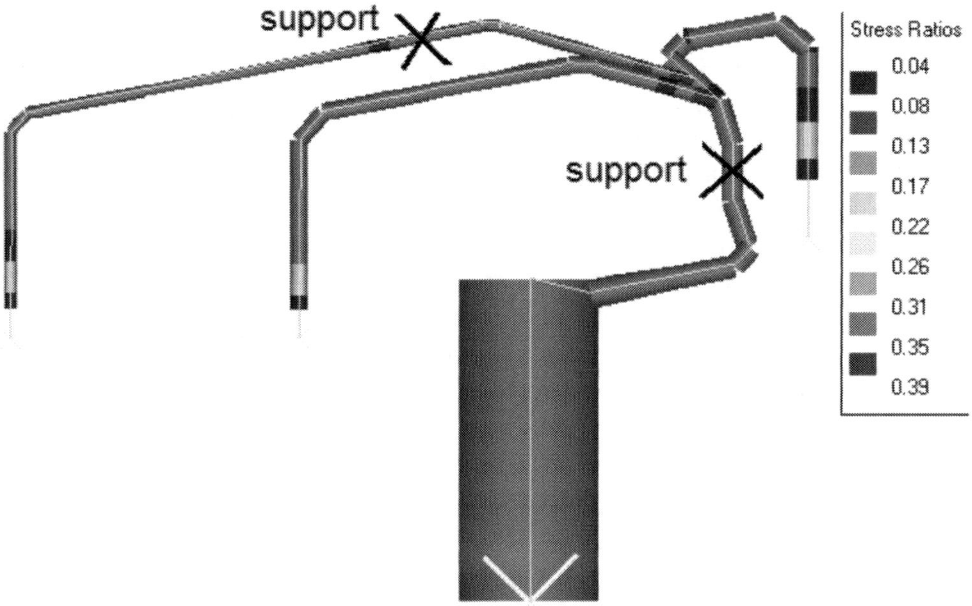

Figure 7. Structural Results

One can see that the maximum usage factor (stress ratios) is below 1, which indicates that the relief system design is adequate. The highest stress ratio (0.39) occurs at the tee connections, which is due to bending stresses and stress concentrations.

6. Conclusion

A solution for relief system design verification was presented, with a focus on demonstrating that RELAP5 can be used to model and quantify the reaction forces on the relief system during depressurization. The methodology utilizes a combination of laboratory testing, fluid dynamics and structural mechanics analyses. The engineering portion of this methodology uses established industry tools that have been extensively validated for these applications. Hence, application of this methodology provides confidence in the relief system design.

7. References

[1] Daubert, T. E., Danner, R. P., Sibul, H. M., and Stebbins, C. C., DIPPR® Data Compilation of Pure Compound Properties, Design Institute for Physical Property Data, AIChE, New York, NY (1995).

[2] Fauske, H. K., "Emergency Relief Systems (ERS) Design," Chemical Engineering Progress, pp. 53-56 (August, 1985).

[3] Fauske, H. K. and Leung, J. C., "New Experimental Technique for Characterizing Runaway Chemical Reactions," Chemical Engineering Progress, Vol. 81, No. 8, pp. 39-46 (August, 1985).

[4] Leung, J. C., Fauske, H. K., and Fisher, H. G., "Thermal Runaway Reactions in a Low Thermal Inertia Apparatus," Thermochimica Acta, 104, pp. 13-29 (1986).

[5] Leung, J. C. and Fauske, H. K., "Runaway System Characterization and Vent Sizing Based on DIERS Methodology," Plant Operations Progress, Vol. 6, No. 2, pp. 77-83 (April, 1987).

[6] PrEVent Version 1.2.14.423, Fauske & Associates, LLC, (2014).

[7] Lukasz Sokolowski and Zbigniew Koszela (2012), "RELAP5 Capability to Predict Pressure Wave Propagation Phenomena in Single- and Two-Phase Flow Conditions", Journal of Power Technologies 92 (3), page 150 – 165.

[8] Joseph, S. Miller and Kevin B. Ramsden (2005), ICONE13-50297 "Validation of a Thermal Hydraulic Computer Code to Perform Two-Phase Multi-Component Force Calculations for Structural Evaluations", Proceedings of 13th International Conference on Nuclear Engineering, page 1-8.

[9] International Agreement Report, "Simulation of the Propagation of Pressure Waves in Piping Systems with RELAP5/MOD 3.2.2: Comparison of Computed and Measured Results", NUREG/IA-0206, (June 2007).

[10] Frederic J. Moody (1990), "Introduction to Unsteady Thermofluid Mechanics", John Wiley and Sons, New York, Section 7.4, pp. 365-366.

Phi Correction for Exothermic Gas Generation Rate for Pressure Relief System Design

Guibing Zhao
Chilworth Technology Inc.
113 Campus Drive, Princeton, NJ 08540
Guibing.Zhao@dekra.com
www.chilworth.com

Prepared for Presentation at
American Institute of Chemical Engineers
2015 Spring Meeting
11th Global Congress on Process Safety
Austin, Texas
April 27-29, 2015

UNPUBLISHED

Phi Correction for Exothermic Gas Generation Rate for Pressure Relief System Design

Guibing Zhao
Chilworth Technology Inc.
113 Campus drive, Princeton, NJ 08540
Guibing.Zhao@dekra.com

Keywords: Pressure relief system, runaway reaction, phi correction, gas generation rate

Abstract

Design of pressure relief systems to accommodate runaway reactions often requires using bench-scale adiabatic calorimeters to obtain the kinetic data for a full-scale reactor. The method to correct dT/dt from bench-scale experimental data to a full-scale reactor, phi correction, has been well established. This study developed a similar method to correct exothermic gas generation rates from bench-scale data with a high phi factor to full-scale reactors with phi factor of unity, which is routinely required for the pressure relief system design. The technique was applied to a study of a real case of pressure relief system design for a gassy reaction system.

1. Introduction

The design of a pressure relief system (for example, a rupture disk with associated piping) for process vessels—in which runaway reactions might occur—requires knowledge of the kinetics of the reactions involved. The use of adiabatic calorimetry for the quantification of the hazards of exothermic runaway reactions has considerably increased over the last 30 years or so. To a great extent, this has been the result of process hazards analyses that have identified the need for pressure relief systems, and the design of such systems involves calculations which require adiabatic data input.

A plant-scale reactor has a smaller surface-to-volume ratio as compared to a laboratory-scale reactor (or to an adiabatic test cell), since this ratio is inversely proportional to radius for a spherical vessel. As a result, the heat loss to the surrounding environment (which is proportional to the heat-transfer surface area) and the heat loss to the body of the reactor (proportional to the mass of the reactor) per unit mass of reactants loaded are smaller than those characteristics in laboratory equipment. Thus, great efforts have been made throughout the industry to develop an adiabatic calorimeter that can simulate exothermic runaway reactions in a large plant reactor, using a small-scale cell that has similar heat losses to both the environment and to the mass of the cell. Such widely accepted and commercially available adiabatic calorimeters include, but are not limited to, the Venting Sizing Package II (VSP II) and the Automatic Pressure Tracking Adiabatic Calorimeter. Both of them have adopted automatic pressure-tracking technology to balance the pressures inside the test cell and outside the test cell, and this allows the use of a thin-wall test cell at high pressures. This is critical to reduce the cell mass and, therefore, reducing the heat loss to cell wall. Adiabaticity is accomplished by the surrounding guard heaters which track the cell temperature and thus maintain a near-zero temperature difference between cell wall and surroundings, if the rates are not too high/fast. Therefore, temperature rise rate and pressure rise rate can be measured accurately under the operating conditions of a full-scale reactor. Reaction kinetic parameters such as the activation energy, the frequency or pre-exponential factor, and the reaction order can also be extracted from the data if a single reaction is involved [1].

The thermal inertia, or Phi factor, is a well-known parameter to describe the heat loss to the test-cell wall. A higher Phi factor means that more energy is absorbed by the test cell, while a lower Phi factor means more energy is retained in the sample being tested. A Phi factor of unity means that all of the reaction energy is retained in the tested material, which closely represents the plant-scale reactor. Therefore, the temperature rise rate and the pressure rise rate that are obtained from tests at a higher phi factor is routinely required to be adjusted to those at phi factor of unity – that is, a phi correction – in order to properly design a pressure relief system for a plant-scale reactor [2].

For a reaction system, the overpressure source can be from vapor(s) and/or gas(es). The reaction system is therefore categorized as a vapor ["tempered"] system, a hybrid system (both vapor and gas), or a gassy system. Phi correction for temperature rise rate has been well established (see the DIERS recommendation as shown in reference [2]). For a vapor reaction system, Phi correction for temperature rise rate is adequate to cover the vapor overpressure effect due to the inherent relationship of pressure and temperature through – for example – the Antoine equation. However, Phi correction for temperature rise rate alone is usually not enough for a hybrid reaction system or a gassy reaction system. Therefore, Phi correction is required for gas generation rate, or pressure rise rate, to properly design a pressure relief system for a plant-scale reactor. There have not been many literature sources available for direct Phi correction of gas generation rate until the recent publication by Kossoy et al. [3].

This study independently developed a phi correction method for exothermic gas generation rates from bench-scale data (with high phi factors) to a full-scale reactor (with a phi factor of unity). The technique was applied to a study of a real case of pressure relief system design for a gassy reaction system.

2. Exothermic Gas Generation Reaction Kinetics

For an exothermic gas generation reaction: $A \rightarrow g\,G + \cdots$ in an adiabatic test (no energy loss to environment, but with energy loss to the testing cell), the following equations can be established from energy balances if the temperature of the testing cell is assumed to be the same as that of the reaction mixture:

$$\emptyset \cdot m_s \cdot Cp_s \cdot (T_m - T_{m0}) = \Delta H_{r,A} \cdot (C_A - C_{A0}) \cdot V_l \qquad \text{[Eq. 1]}$$

$$\emptyset \cdot m_s \cdot Cp_s \cdot (T_{mf} - T_{m0}) = \Delta H_{r,A} \cdot (0 - C_{A0}) \cdot V_l \qquad \text{[Eq. 2]}$$

where $\emptyset$ is phi factor ($\emptyset = 1 + \frac{m_c \cdot Cp_c}{m_s \cdot Cp_s}$), m_c and m_s are the mass of testing cell and testing sample, respectively, Cp_c and Cp_s are the heat capacity of testing cell and testing sample, respectively, $\Delta H_{r,A}$ is the reaction heat (negative for an exothermic reaction), V_l is the liquid volume, C_{A0} and C_A are the initial molar concentration and the decreasing molar concentration of reactant A, respectively, T_{m0} and T_{mf} are the measured onset temperature and the final temperature of the exothermic gas generation reaction, respectively. The relationship of temperature and concentration of A can be established by dividing [Eq. 1] by [Eq. 2]:

$$C_A = C_{A0} \cdot \frac{T_{mf}-T_m}{T_{mf}-T_{m0}} \qquad \text{[Eq. 3]}$$

From stoichiometric relationship of reaction $\rightarrow g\ G + \cdots$:

$$\frac{(C_{A0}-C_A)\cdot V_l}{m_s} = \frac{n_G}{g} \qquad \text{[Eq. 4]}$$

where n_G is the amount of gas that is generated in mol/g sample. The gas generation rate can be established by differentiating [Eq. 4] and [Eq. 3]:

$$\left(\frac{dn_G}{dt}\right)_\emptyset = -g \cdot \frac{V_l}{m_s} \cdot \frac{dC_A}{dt} = g \cdot \frac{V_l}{m_s} \cdot C_{A0} \cdot \frac{1}{T_{mf}-T_{m0}} \cdot \frac{dT_m}{dt} = \frac{K}{T_{mf}-T_{m0}} \cdot \left(\frac{dT_m}{dt}\right)_\emptyset \qquad \text{[Eq. 5]}$$

where K is a constant and is given by $K = g \cdot \frac{V_l}{m_s} \cdot C_{A0}$. Thus, [Eq. 5] indicates that gas generation rate, which can be measured from either a closed-cell test or an open-cell test, is proportional to the measured temperature rise rate of $\left(\frac{dT_m}{dt}\right)_\emptyset$.

At a phi factor of unity (which means that all of the reaction energy is retained in the tested material), [Eq. 5] becomes:

$$\left(\frac{dn_G}{dt}\right)_{\emptyset=1} = \frac{K}{T_{Af}-T_{A0}} \cdot \left(\frac{dT_A}{dt}\right)_{\emptyset=1} \qquad \text{[Eq. 6]}$$

where the subscript of A indicates no energy loss to the testing cell, and which represents most reaction systems of industrial scale. The DIERS recommendation for phi correction [2] is used, with the following equations:

$$\frac{1}{T_{A0}} = \frac{1}{T_{m0}} + \frac{R}{E} ln\emptyset \qquad \text{[Eq. 7]}$$

$$T_A = T_{A0} + \emptyset(T_m - T_{m0}) \qquad \text{[Eq. 8]}$$

$$\left(\frac{dT_A}{dt}\right)_{\emptyset=1} = \emptyset \cdot exp\left[\frac{E}{R}\left(\frac{1}{T_m} - \frac{1}{T_A}\right)\right] \cdot \left(\frac{dT_m}{dt}\right)_{\emptyset>1} \qquad \text{[Eq. 9]}$$

where E is the activation energy of the reaction, and R is the gas constant. Combining [Eq. 5], [Eq. 6], and [Eq. 9] gives the following corrected gas generation rate at a phi factor of unity, as phi-corrected from the measured gas generation rate with phi factor greater than one:

$$\left(\frac{dn_G}{dt}\right)_{\emptyset=1} = exp\left[\frac{E}{R}\left(\frac{1}{T_m} - \frac{1}{T_A}\right)\right] \cdot \left(\frac{dn_G}{dt}\right)_{\emptyset>1} \qquad \text{[Eq. 10]}$$

or

$$\left(\frac{dn_G}{dt}\right)_{\emptyset=1} = \frac{K}{T_{mf}-T_{m0}} \cdot exp\left[\frac{E}{R}\left(\frac{1}{T_m} - \frac{1}{T_A}\right)\right] \cdot \left(\frac{dT_m}{dt}\right)_{\emptyset>1} \qquad \text{[Eq. 11]}$$

Therefore, once the constant, K, is found from experimental data – from [Eq. 5] – the gas generation rate can be easily adjusted from the temperature rise rate of the laboratory-scale test cell to a plant-scale reactor, as shown in [Eq. 11].

3. Case Study: Pressure Relief System for an Exothermic Gas Generation Reaction

A reactor of volume of 3.5 m^3 has a design pressure of 10 barg. A worst-case relief scenario study has identified a gassy decomposition reaction which could occur. The mass of reactants in the reactor would be 2464 kg. A closed-cell test has been performed in a VSPII testing apparatus. The liquid density at the relief condition is about 880 kg/m^3. The generated gas has a specific heat ratio of 1.3. The emergency relief system is to be via a rupture disk with an overall discharge length of 12 meters of constant-diameter piping and two 90° bends and with a velocity head loss of 0.2 per bend. The elevation change from the relief inlet to outlet is about 9.1 meters. The rupture disk has an equivalent length to diameter ratio of 20.

To conduct the test, 22.6 g of sample was charged to a clean dry 110-ml Hastelloy test cell of having a weight of 39.64g. The cell was then placed inside the heater assembly, sealed inside the containment vessel, and connected to all relevant logging and control systems. The test cell was flushed with nitrogen. The test cell heater was then used to heat the sample. The calculated phi factor at the testing condition was about 1.35.

The experimental traces of temperature and the corrected pressure (deducting the initial pad gas) versus time are shown in Figure 1. The onset temperature of the exothermic runaway reaction was 244 °C. Apparently, this was a complicated reaction system, and multiple reactions were involved. However, it was determined that a single gas-generating reaction was involved since the final pressure did not increase significantly even though the temperature kept on increasing. Therefore, some key temperature data can be extracted from the pressure rise rate versus temperature, as shown in Figure 2.

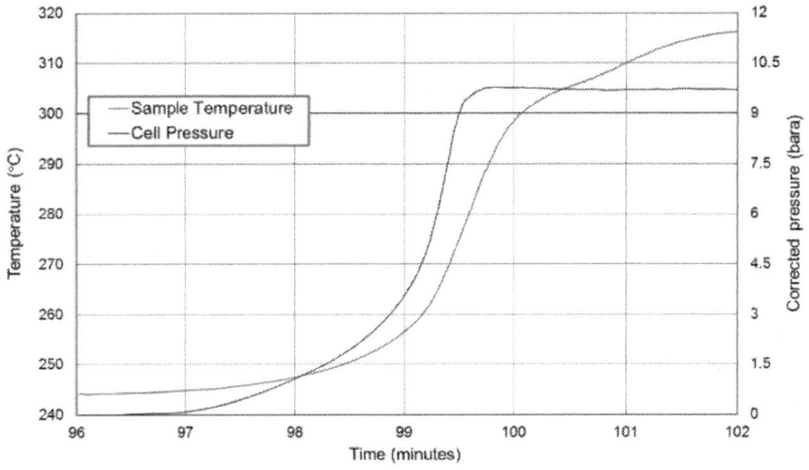

Figure 1: Temperature and corrected pressure versus time

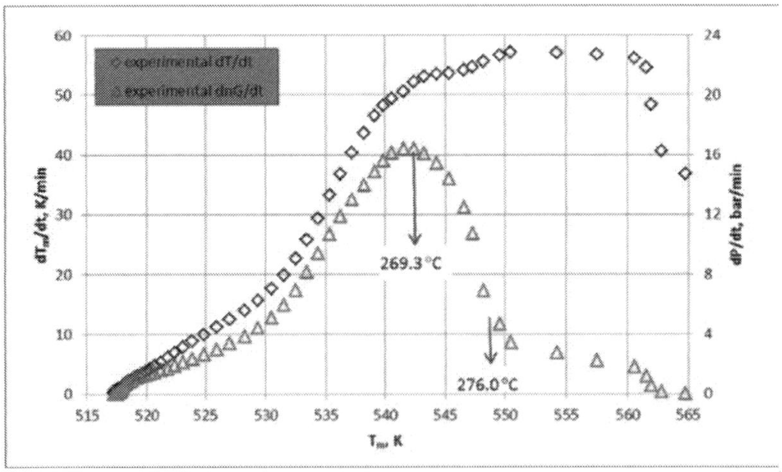

Figure 2: Rate of pressure rise and temperature rise versus time

From the pressure rise rate data, it was determined that the maximum gas generation rate occurred at 269.3 °C, and the exothermic gas generation reaction ended at about 276 °C. With the known final temperature of the gas generation reaction, the reaction kinetic parameters of activation energy and frequency factor can be deduced from a plot of $\ln\left[\frac{dT_m}{dt}/(T_{mf}-T_m)\right]$ versus $-\frac{1000}{T_m}$ with an assumption of a first-order reaction of gas generation, as shown in Figure 3. A verification calculation of the reaction order can be obtained by using the reaction activation energy obtained from Figure 3:

$$n=\frac{E}{R}\cdot\frac{(T_{mf}-T_{mm})}{T_{mm}^{2}} \qquad \text{[Eq. 12]}$$

where T_{mm} is the temperature at the maximum gas generation rate. The calculated reaction order is 1.1, which is close to unity. Therefore, the assumed first-order reaction of gas generation is valid. The activation energy and frequency factor for the gas generation reaction are 406.5 kJ/mol and 1.10 $\times 10^{40}$ 1/s, respectively.

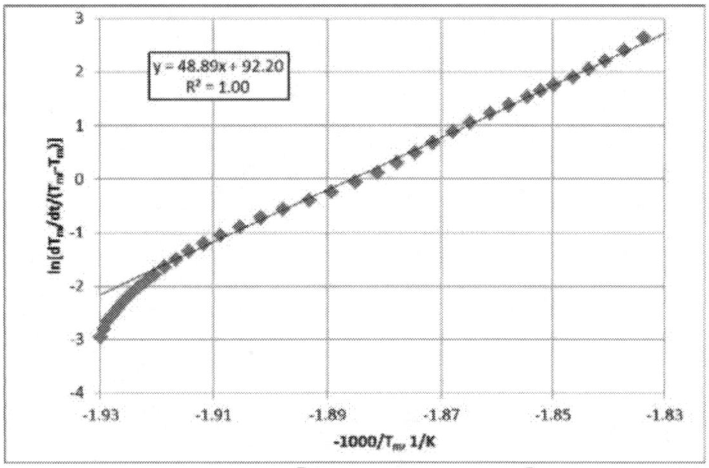

Figure 3: Deduction of kinetic data from $\ln\left[\frac{dT_m}{dt}/(T_{mf}-T_m)\right]$ versus $-\frac{1000}{T_m}$ by assuming a first-order reaction for gas generation

For a closed cell test, the gas generation rate can be calculated from:

$$\frac{dn_G}{dt}=\frac{V_v}{RT_m^{2}}\left(T_m\cdot\frac{dP}{dt}-P\cdot\frac{dT_m}{dt}\right)\cdot\frac{1}{m_s} \qquad \text{[Eq. 13]}$$

where V_v (84.3 ml in the test) is the volume of the head space of the testing cell, and P is the pressure of the generated gas. The $P\cdot\frac{dT_m}{dt}$ term is normally much smaller than $T_m\cdot\frac{dP}{dt}$ and is frequently neglected [4]. Therefore, the gas generation rate is routinely calculated from:

$$\frac{dn_G}{dt}=\frac{V_v}{RT_m}\cdot\frac{dP}{dt}\cdot\frac{1}{m_s} \qquad \text{[Eq. 14]}$$

[Eq. 5] indicates that the measured gas generation rate, which can be calculated from [Eq. 14], should be proportional to the measured temperature rise rate of $\frac{dT_m}{dt}$. Figure 4 shows a good linear relationship between $\frac{dn_G}{dt}$ and $\frac{dT_m}{dt}$. Therefore, the constant K in [Eq. 5] is 8.51×10^{-4} mol/g. For a first-order reaction, the temperature rise rate can be calculated from an Arrhenius relationship [1]:

$$\left(\frac{dT_m}{dt}\right)_{\emptyset>1} = (T_{mf} - T_m) \cdot A \cdot exp\left(-\frac{E}{RT_m}\right) \qquad \text{[Eq. 15]}$$

From the previously-determined activation energy and frequency factor, $\frac{dT_m}{dt}$ and $\frac{dn_G}{dt}$ can be calculated from [Eq. 15] and [Eq. 5], respectively, as shown in Figure 5. The calculated $\frac{dn_G}{dt}$ matches very well with the measured gas generation rate. Also, the calculated temperature rise rate matches very well from the onset temperature to the temperature at the maximum gas generation rate, at about 269.3 °C. At temperature greater than 269.3 °C, another exothermic reaction apparently is initialed, but without gas generation.

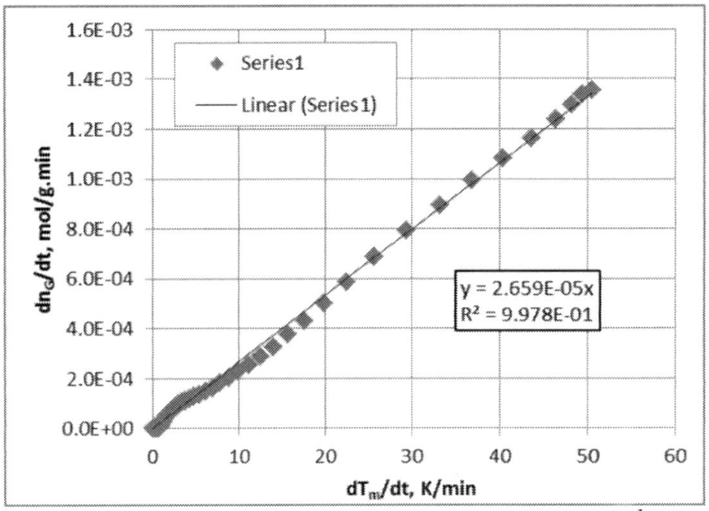

Figure 4: Constant K from linear relationship between $\frac{dn_G}{dt}$ and $\frac{dT_m}{dt}$

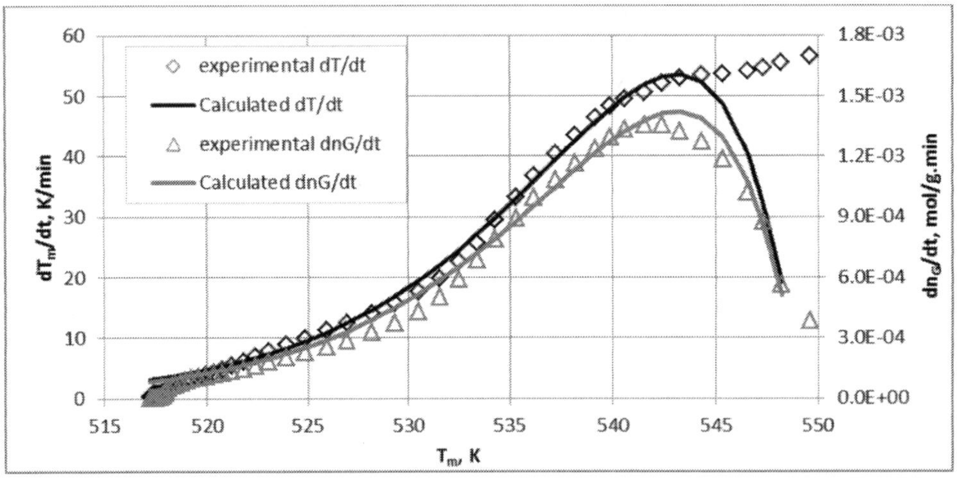

Figure 5: Comparison of the measured $\frac{dT_m}{dt}$ and calculated $\frac{dT_m}{dt}$, and the measured gas generation rate and calculated gas generation rate for the test with a phi factor of 1.35

The kinetic data extracted as shown above are from a VSP test with a phi factor of 1.35. For the purposes of emergency relief system design for an industrial-scale reactor, the measured temperature rise rate $\left(\frac{dT_m}{dt}\right)_{\emptyset=1.35}$ and the gas generation rate $\left(\frac{dn_G}{dt}\right)_{\emptyset=1.35}$ should be adjusted to those for a phi factor of unity. Such an adjustment can be completed through [Eq. 9] and [Eq. 10] with the extracted activation energy from the test data. Figure 6 shows the adjusted temperature rise rate and gas generation rate at a phi factor of unity and the calculated temperature rise rate and gas generation rate at a phi factor of 1.35.

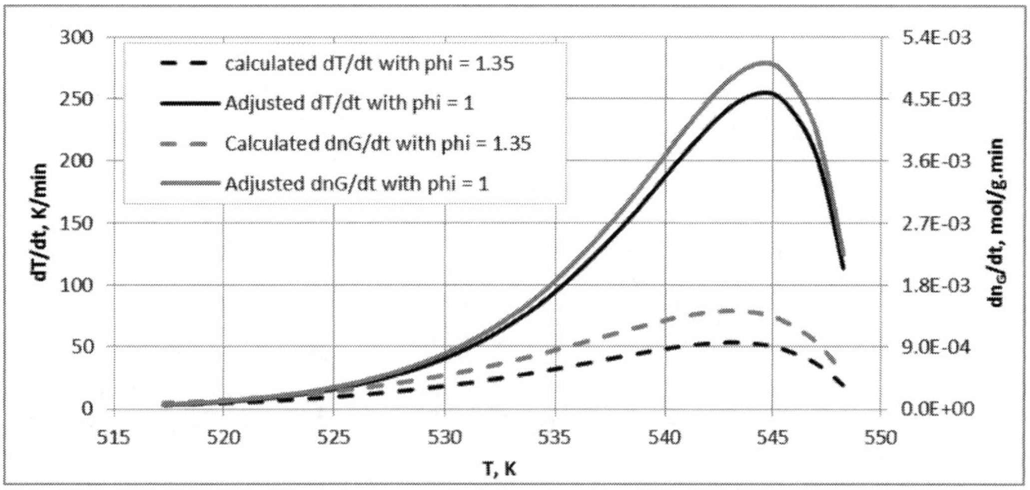

Figure 6: Calculated temperature rise rate and gas generation rate with a phi factor of 1.35 and 1.

As shown in the above graph, the maximum rate of gas generation would be about 0.005 mol/g-min. for a phi factor of 1. The peak rate of gas evolution in a reactor can then be calculated based on the maximum gas generation rate at a phi factor of unity:

$$Q_{G,max} = \frac{RT_{am}}{P_m}\left(\frac{dn_G}{dt}\right)_{max,\emptyset=1} \cdot Mr \cdot 1000$$

$$= \frac{8.314 \cdot 552.2}{12 \cdot 10^5} \cdot \frac{0.00501}{60} \cdot 2464 \cdot 1000 = 0.787\ m^3/s \qquad \text{[Eq. 16]}$$

where Mr is the loaded mass in the reactor (2464 kg), T_{am} is the temperature at maximum gas generation rate at phi factor of unity (279.0 °C), P_m is the maximum allowable accumulated pressure of 12 bara (10 × 1.1 + 1). The require relief rate can now be calculated from:

$$W = Q_{G,max}\frac{m_r}{V} = 0.787 \cdot \frac{2464}{3.5} = 554\ kg/s \qquad \text{[Eq. 17]}$$

By using the OMEGA method proposed by Leung [5], the homogeneous two-phase flow capacity G, in kg/m^2s, can be calculated through a trial and error of the size of a rupture disk (with a flow resistance given by of 4fL/D and a flow inclination number Fi).

The calculated flow capacity at 12 bara would be about 15,500 kg/m^2s. The calculated rupture size would therefore be about 8.4 inches. A 9-inch-diameter rupture disk would therefore be required to adequately provide pressure relief for a potential runaway reaction.

4. Conclusion

This article independently developed a method for adjusting exothermic gas generation rate, or pressure rise rate, from bench-scale data with a high phi factor to a full-scale reactor with phi factor of unity. Gas generation rates for plant-scale reactors can be calculated from the laboratory-test gas-generation rate after the constant K is determined from the linear relationship of dn_G/dt and dT_m/dt. The technique was applied to a study of a real case of pressure relief system design for a gassy reaction system.

5. References

[1] D.I. Townsend and J.C. Tou, Thermal hazard evaluation by an acceleration rate calorimeter, Thermochimica Acta 37 (1980), 1-30.

[2] H.G. Fisher, et al., *Emergency relief system design using DIERS technology*, The Design Institute for Emergency Relief Systems, American Institute of Chemical Engineers, New York, NY, 1992.

[3] A.A. Kossoy, J. Singh, and E.Y. Koludarova, Mathematical methods for application of experimental adiabatic data – an update and extension, J Loss Prev Process Ind 33 (2015), 88-100

[4] J. Singh, Venting sizing for gas generating runaway reactions, J Loss Prev Process Ind 7 (1994), 481-491

[5] J.C. Leung, Easily size relief devices and piping for two-phase flow, Chem Eng Prog (92) 1996, 28-50

Calorimetric Study of the Exothermic Decomposition of Dimethyl Sulfoxide

B. Todd Brandes
SI Group, Inc.
725 Cannon Bridge Road, Orangeburg, SC 29115
todd.brandes@siigroup.com

Dan K. Smith
Albemarle Corporation
Gulf States Utilities Road, Baton Rouge, LA 70805
dan.smith@albemarle.com

Prepared for Presentation at
American Institute of Chemical Engineers
2015 Spring Meeting
11th Global Congress on Process Safety
Austin, Texas
April 27-29, 2015

UNPUBLISHED

Calorimetric Study of the Exothermic Decomposition of Dimethyl Sulfoxide

B. Todd Brandes
SI Group, Inc.
725 Cannon Bridge Road, Orangeburg, SC 29115
todd.brandes@siigroup.com

Dan K. Smith
Albemarle Corporation
Gulf States Utilities Road, Baton Rouge, LA 70805
dan.smith@albemarle.com

Keywords: dimethyl sulfoxide, DMSO, decomposition, autocatalytic, exothermic, calorimetry

Abstract

Dimethyl sulfoxide (DMSO) is a widely-used solvent often employed for a variety of organic syntheses. It is stable at room temperature, can dissolve many types of organic materials, and is miscible in water. However, upon heating in a closed system under conditions typically achievable in a commercial chemical plant, DMSO undergoes significantly exothermic and hazardous decomposition which could realistically lead to a pressure vessel explosion unless adequate protection layers were installed, possibly including a pressure relief system. This study provides calorimetry data from several apparatus and methods to characterize the decomposition. The results are described, and a simplified kinetic model is presented to correlate the low temperature end of the data.

1. Introduction

Dimethyl sulfoxide (DMSO) is a widely-used solvent often employed for a variety of organic syntheses in a range of industry sectors, including pharmaceuticals, petrochemicals, electronic materials, polymer coatings, agricultural chemicals, industrial cleaners, and specialty chemicals. It has the favorable properties of being highly polar, water miscible, clear, nearly odorless, low in toxicity, low in vapor pressure, and a great solvent for many compounds difficult to dissolve. A major manufacturer has provided a considerable amount of physical property data in their product literature [1].

DMSO is known, however, to be subject to decomposition, particularly above its normal boiling point, and accelerated in the presence of a range of other materials, such as acids and bases, halogenated organics and inorganics, and alkalis [1-6]. Previous studies have provided some mechanistic explanations and a variety of calorimetry and chemical analyses to characterize the phenomenon [3, 6]. Unfortunately, the data are inconsistent,

Note: Do not add page numbers. Do not refer to page numbers when referencing different portions of the paper

sometimes explaining lower energy than is fully observed, and sometimes based upon a limited experimental study. A discussion of these issues is provided in section 4.2. In this work, additional calorimetric studies by four methods were conducted to supplement the literature and clarify some of the inconsistencies. Chemical analysis was performed on the residue and vapor decomposition products to identify the species present. The results were then used to develop a limited, simplified model to correlate the heat generation. Pressure generation was shown to closely follow DMSO vapor pressure up to about 250 °C.

2. Experimental

Several calorimetric tests were performed for DMSO for this study, and chemical analysis was performed on the decomposition gas and residue from one of the tests. ACS reagent grade DMSO, > 99.9 wt% purity, was sourced from Fisher Scientific and used without any further purification. Following is a brief description of the various lab tests performed.

2.1 Ramp Differential Scanning Calorimetry

A PerkinElmer® Pyris 6 differential scanning calorimeter was used in this study for ramp tests. Swissi M20 high pressure, 20 µL, gold-plated crucibles were used for the sample and reference cells, capable of containing up to 200 bar. Coarse calibration was achieved by adjusting the heat flow conversion for an indium standard melt point. The furnace constant was set based upon the slope of a series of tests of the indium standard versus its melt point for ramp rates of 20, 10, 5, 2, and 1 °C/min. The baseline for an empty sample cell versus the reference cell was determined from repeating tests to verify consistent results. Lastly, fine calibration was achieved using both indium and zinc standard melt points, all in the equivalent model stainless steel crucibles, from 30 to 450 °C. Once the DSC was calibrated, ramp tests with DMSO were performed with approximately 5 mg of sample, with the sample cells prepared and sealed in a nitrogen-purged drybox. The crucibles were carefully weighed before and after testing to ensure that no mass was lost during the experiment. The raw test results were then integrated using TA Instruments Universal Analysis software.

2.2 Isothermal Differential Scanning Calorimetry

The same PerkinElmer® Pyris 6 differential scanning calorimeter as described above was used in this study for isothermal tests. Calibration was achieved as described, except for isothermal tests it is not necessary to provide a value for the furnace constant since there is no ramping in temperature after the brief initial heating. Consequently, for isothermal tests the furnace constant was set to zero. Samples were loaded in cells as described above then loaded in the DSC where the sample was ramped at 100 °C/min to the testing temperature and held for an extended time.

2.3 Accelerating Rate Calorimetry

A Thermal Hazard Technology, Inc., esARC was used in this study for adiabatic self-heat tests. Prior to the experiments the instrument was calibrated with an empty titanium test cell to a self-heat sensitivity of 0.01°C/min at 25°C intervals over the temperature range of 50 to 450°C. Each experiment was conducted with approximately 3 g of sample. Prior to the experiment the sample was loaded into the test cell in a nitrogen purged dry box to exclude oxygen from the test. Hastelloy C and titanium test cells were used, although the titanium cells experienced leaks in both tests run.

ARC experiments were conducted in both heat-wait-search mode and in isothermal age mode. During heat-wait-search experiments the sample and test cell are heated to an initial starting temperature. The system is allowed to equilibrate ("wait") and then held under adiabatic conditions while temperature and pressure are recorded ("search"). An exothermic reaction is detected if the rate of temperature rise of the test cell increases above a specified sensitivity during the search period. If an exothermic reaction is not detected the sample and test cell are heated to a slightly higher temperature and the heat-wait-search process is repeated. If an exothermic reaction is detected the calorimeter would maintain an adiabatic environment surrounding the test cell while the temperature and pressure are recorded as the reaction progresses. During isothermal age mode the sample and test cell are heated to an initial starting temperature. The system is allowed to equilibrate and then held under adiabatic conditions for an extended period of time while the temperature is monitored to determine if an exothermic reaction is observed.

Heat-wait-search mode experiments were conducted with a starting at temperature of 150°C utilizing 5°C temperature steps. The self-heating sensitivity used in all experiments was 0.02 °C/min.

One isothermal ageing experiment was conducted to further study autocatalyst formation. In that test, the sample was heated at about 1.3 °C/min from ambient to an isothermal ageing temperature of 200 °C. The sample was held adiabatically at this initial temperature. In this experiment a self-heating rate greater than 0.02°C/min was observed after approximately 400 minutes.

The esARC has a high pressure interlock to prevent excessive pressure and possible test cell rupture. This interlock was set at 200 bar.

2.4 Pressure Screening Tool

The pressure screening tool is a reactive hazard screening test utilized to detect the generation of non-condensable gas. The apparatus consists of a closed 9.3 mL spherical test cell placed inside of a cylindrical aluminum tube fitted with an electric heater and thermocouple. The closed test cell is fitted with a 1/32-in diameter thermocouple and an oil filled pressure transducer to measure temperature and pressure of the sample during the experiment. The test cell and heater assembly are placed inside of an autoclave to provide protection from shrapnel in the event of a bomb rupture (and screening tests are not conducted without prior DSC screening). A high pressure interlock stops heating if excessive pressure were to be reached.

Prior to each experiment approximately 3 g of sample was loaded into the test cells in a nitrogen purged dry box. The temperature of the aluminum tube was increased at a ramp rate of 2°C/min while temperature and pressure of the sample were recorded. After the desired maximum temperature was reached the heater was turned off and the sample was allowed to cool back to ambient temperature. The test was used for quick preparation of residue and gas samples for analysis, determination of pressure generation potential, and determination of condensing curves of the intermediate or final reaction products.

2.5 Gas Chromatography–High Resolution Mass Spectrometry

An Agilent 6890A GC-HRMA was used in this study to provide chemical analysis of samples of the decomposition gas product and final liquid residue from the PST tests to identify components involved in the decomposition chemistry.

3. Data Discussion

Four isothermal DSC tests, four ramp DSC tests, four ARC tests, and three PST tests were conducted for pure DMSO. In addition, ramp DSC tests were conducted for DMSO spiked with 98 wt% H_2SO_4 and methansulfonic acid to evaluate whether these materials could be part of the overall decomposition mechanism.

3.1 Ramp DSC

Ramp DSC experiments were conducted at 10, 5, 2, and 1 °C, with results provided in Figures 1-4. Clearly, DMSO liberates a large amount of energy when heated in the DSC, in the range of 500-700 J/g. The shape of the curves clearly show several inflections, indicating the exothermic events are complex with at least three steps. Two additional tests were conducted to identify if either of two acidic species could accelerate the exothermic behavior of DMSO. The results confirm the hypotheses. Figure 5 shows ramp test results for DMSO spiked with 1 wt% of 98 wt% H_2SO_4, and Figure 6 shows ramp test results for DMSO spiked with 1 wt% of methanesulfonic acid. The data from each of these experiments shows a significantly lower onset and peak heat flow temperatures as compared to the 10°C/min ramp experiment with neat DMSO. In addition the evolved heat is somewhat larger than that observed with neat DMSO possibly suggesting an altered decomposition path. A summary of the ramp DSC results is provided in Table 1.

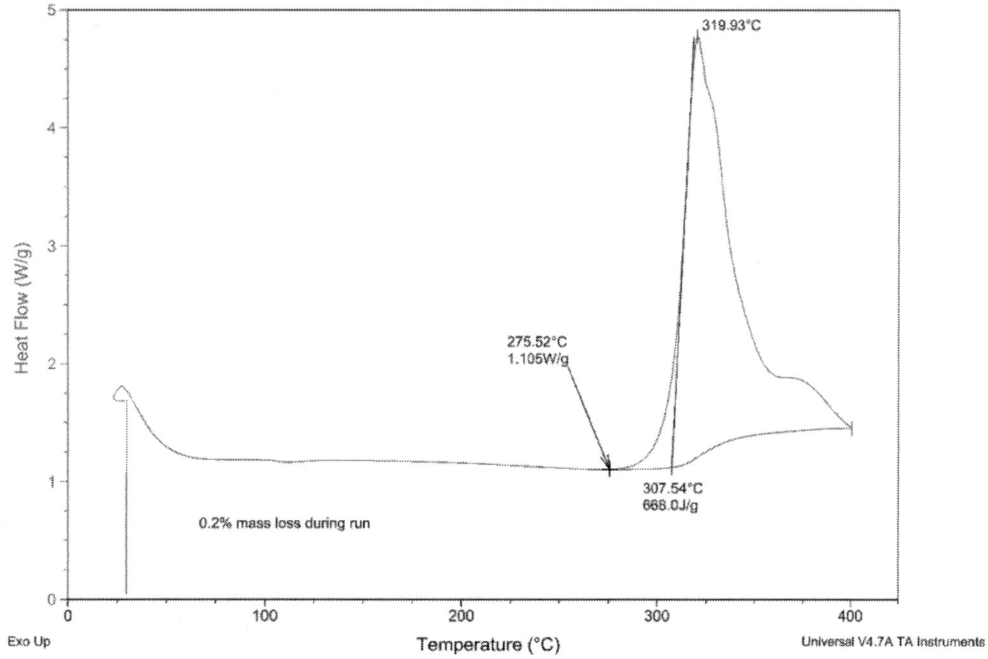

Figure 1. Ramp DSC Test Results for DMSO at 10 °C/min.

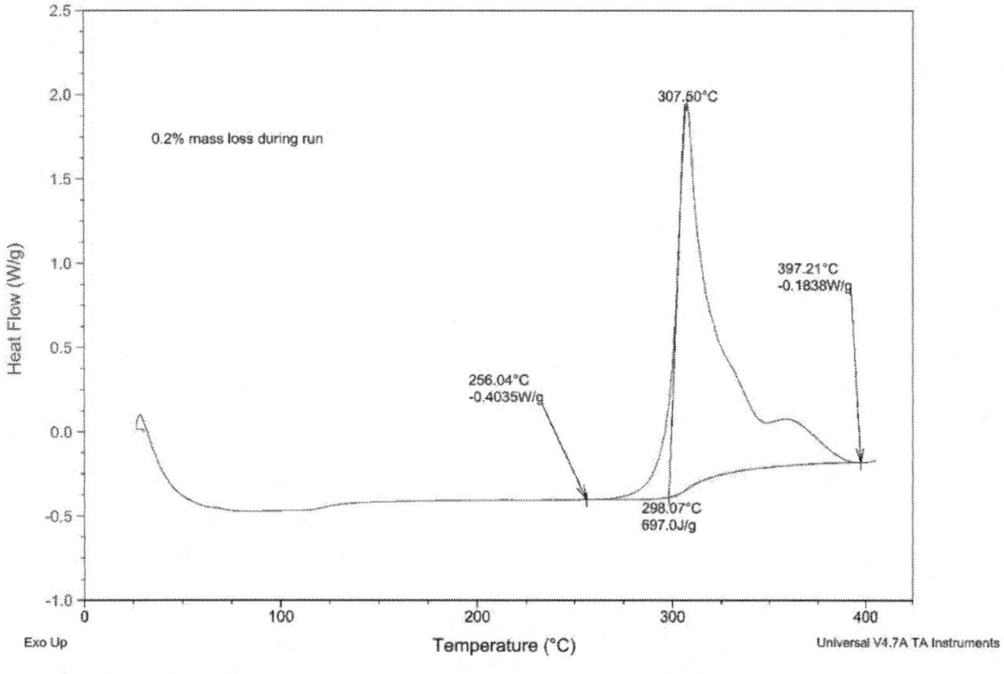

Figure 2. Ramp DSC Test Results for DMSO at 5 °C/min.

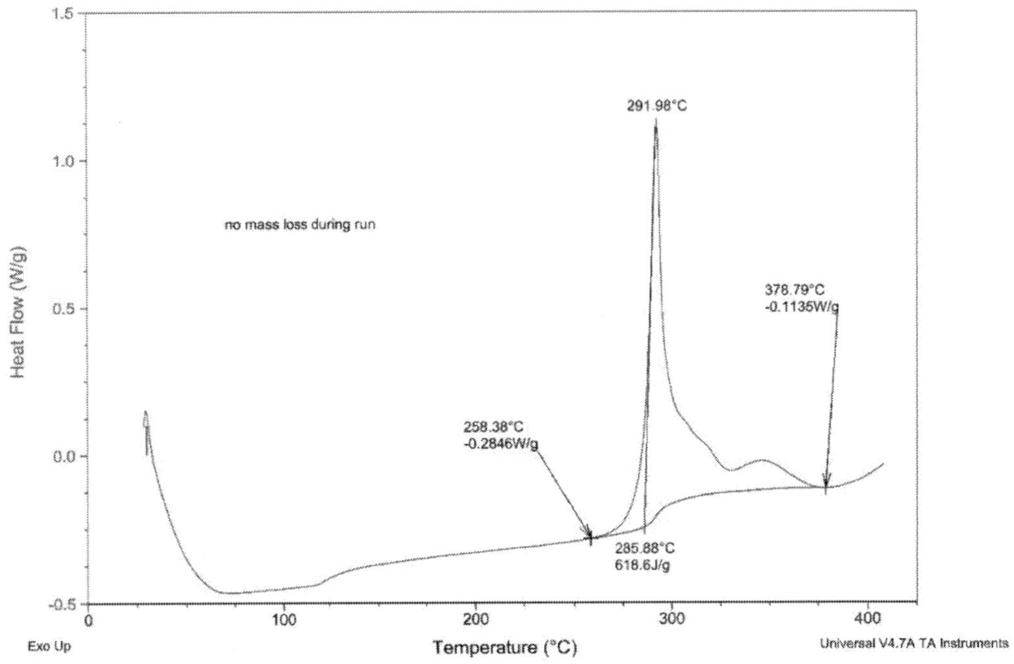

Figure 3. Ramp DSC Test Results for DMSO at 2 °C/min.

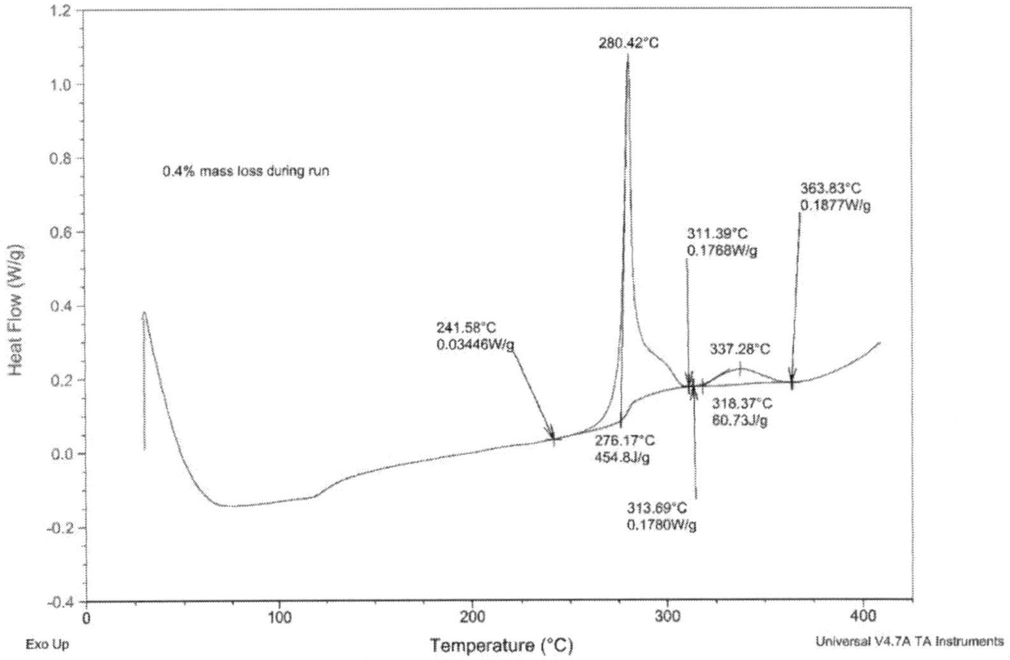

Figure 4. Ramp DSC Test Results for DMSO at 1 °C/min.

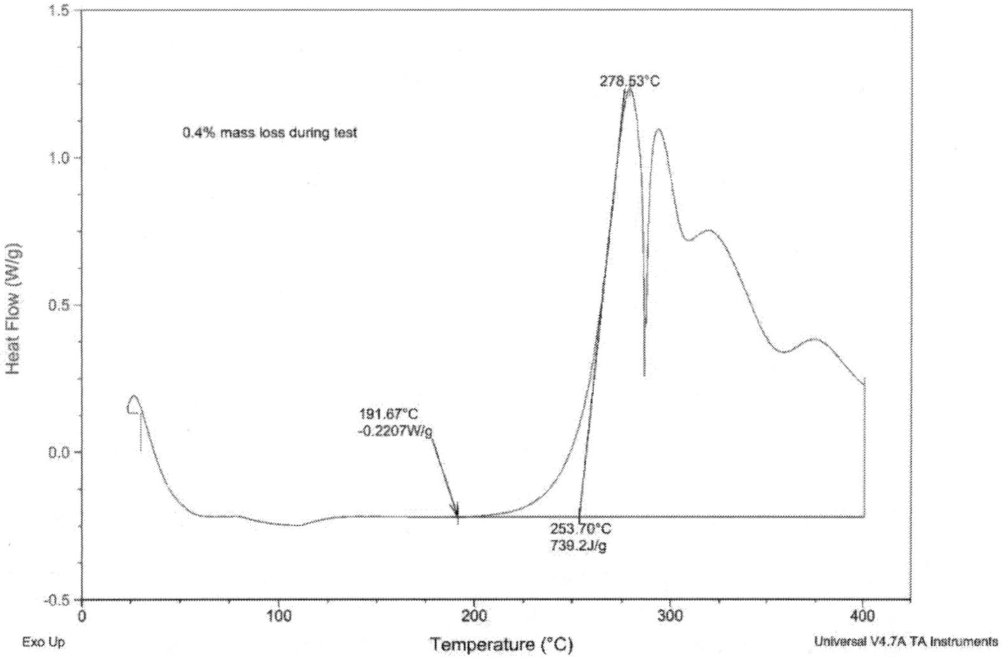

Figure 5. Ramp DSC Test Results for DMSO Spiked with 1 wt% of 98 wt% H_2SO_4 at 10 °C/min.

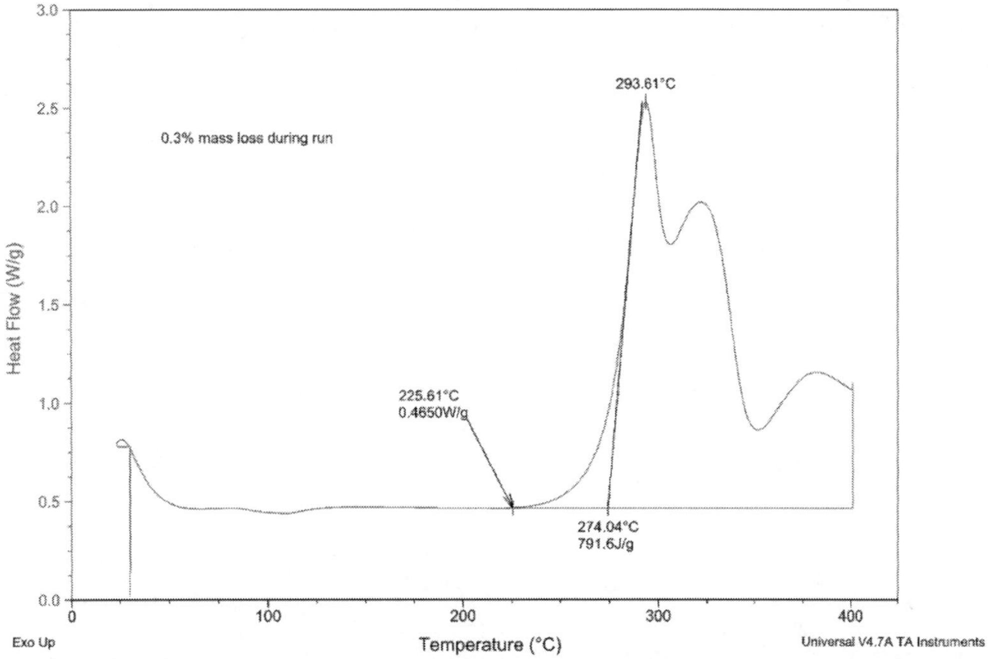

Figure 6. Ramp DSC Test Results for DMSO Spiked with 1 wt% of Methanesulfonic Acid at 10 °C/min.

Table 1. Summary of Ramp DSC Tests.

Description	Ramp Rate dT/dt (°C/min)	Observable Onset T T_o (°C)	T at Peak Heat Flow T_{peak} (°C)	Integrated Energy -ΔU/mw (J/g)
DMSO	10	276	320	668
DMSO	5	256	308	697
DMSO	2	258	292	619[1]
DMSO	1	242	280	516[2]
DMSO+1wt% H_2SO_4	10	192	279	>739
DMSO+1wt% MSA	10	226	294	>792

3.2 Isothermal DSC

Isothermal DSC experiments were conducted at 210, 230, 250, and 270 °C, with results provided in Figures 7-10. The shape of each curve shows increasing rate of evolution of heat initially, reaching a peak, followed by decaying heat flow that eventually becomes undetectable. This shape is characteristic of the general autocatalytic form. At 270 °C, the initial baseline is not well-established so the typical induction portion of the autocatalytic behavior is somewhat obscured. In each case, increasing temperature leads to the detectable onset and peak heat flows occurring at an earlier time during the test. A summary of the isothermal DSC results is provided in Table 2.

[1] Additional heat evolution event observed starting at 379 °C but not fully resolved in test
[2] Additional heat evolution event observed starting at 364 °C but not fully resolved in test

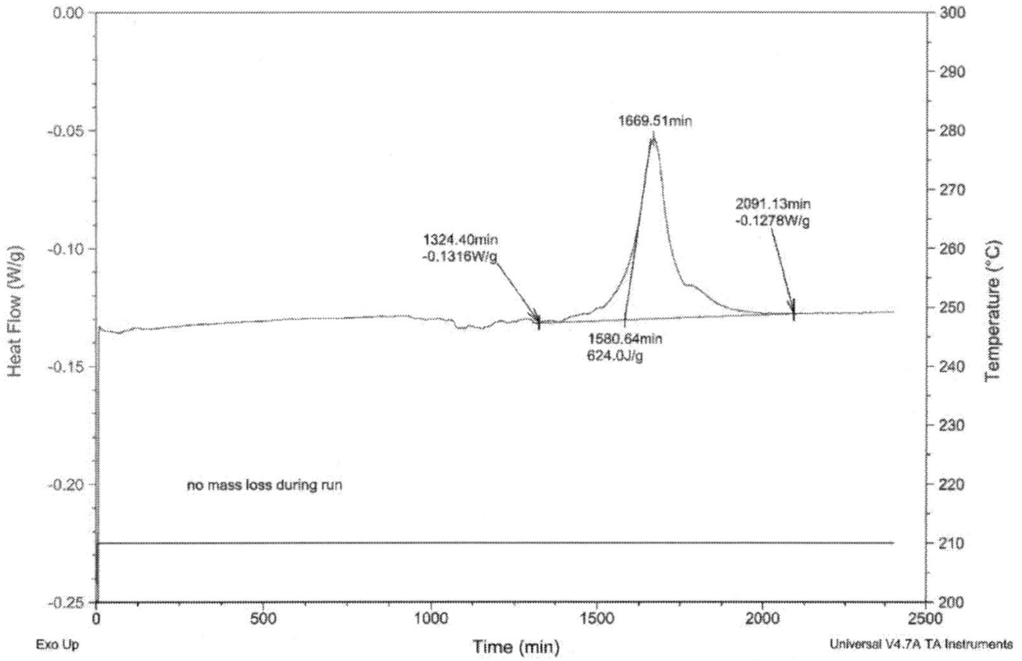

Figure 7. Isothermal DSC Test Results for DMSO at 210 °C.

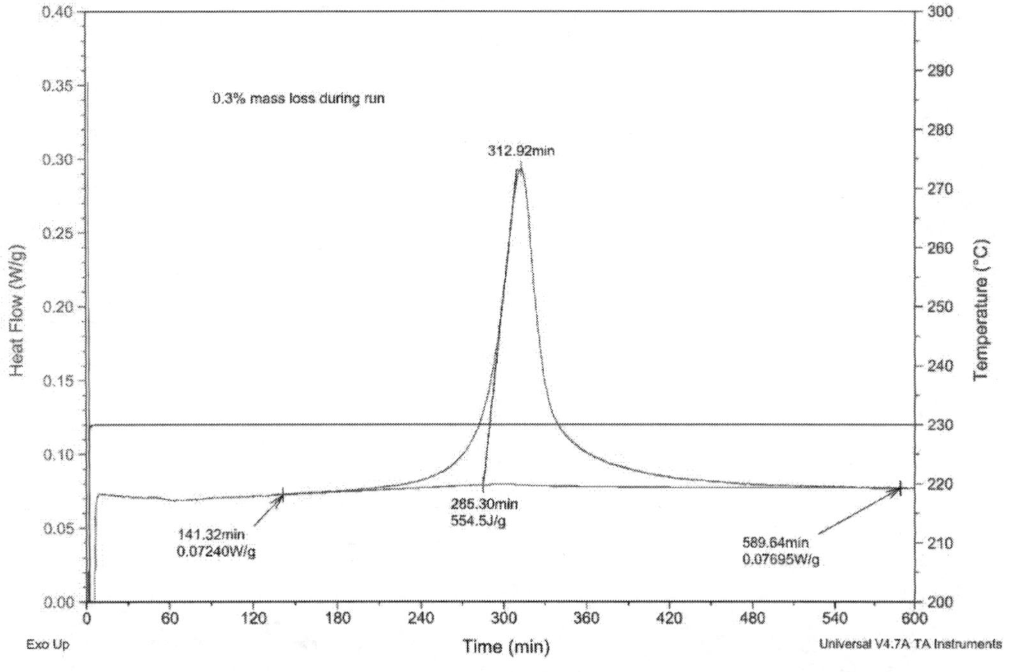

Figure 8. Isothermal DSC Test Results for DMSO at 230 °C.

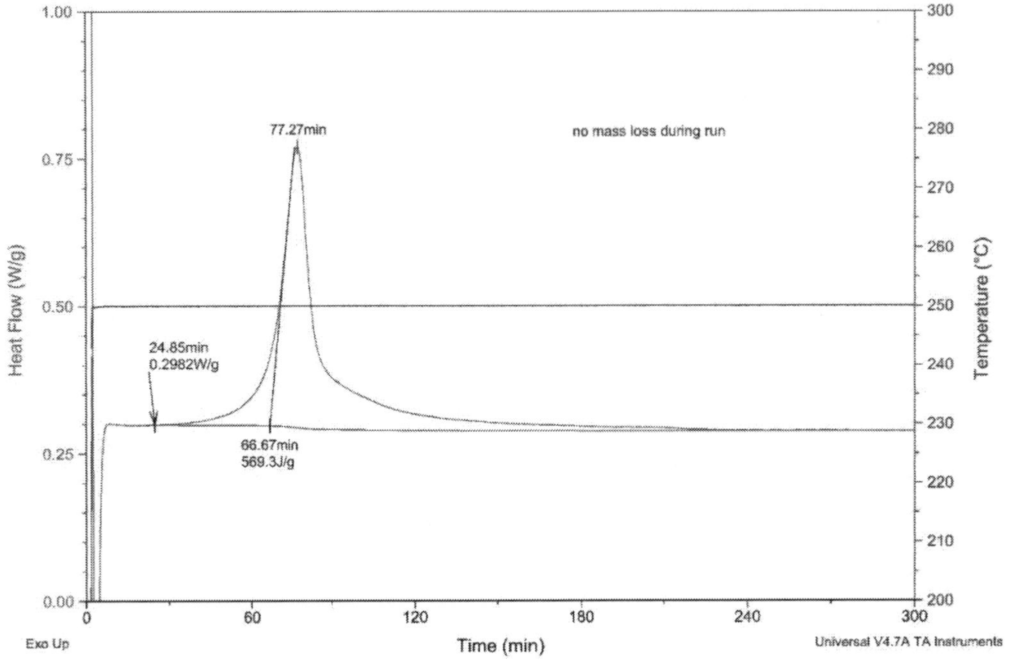

Figure 9. Isothermal DSC Test Results for DMSO at 250 °C.

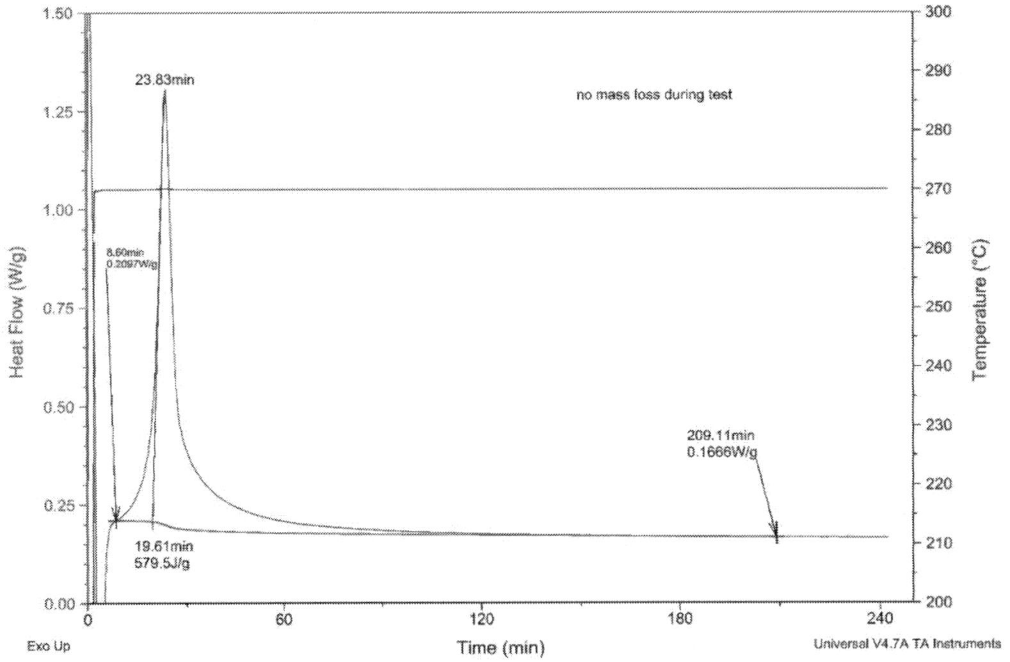

Figure 10. Isothermal DSC Test Results for DMSO at 270 °C.

Table 2. Summary of Isothermal DSC Tests.

Description	Test Temperature (°C)	Ramp Rate to Test T dT/dt (°C/min)	Observable Onset Time t_o (min)	Time at Peak Heat Flow t_{peak} (min)	Integrated Energy $-\Delta U/m_w$ (J/g)
DMSO	210	100	1324	1670	624
DMSO	230	100	141	313	555
DMSO	250	100	24.9	77.3	569
DMSO	270	100	8.6[3]	23.8	580

3.3 Accelerating Rate Calorimetry

Three heat-wait-search tests were conducted for DMSO: two tests in a titanium test cell, and one in a Hastelloy C test cell. Also, a single isothermal ageing test was conducted with an aging temperature of 200°C in a Hastelloy C test cell. The recorded test results are provided in Figures 11-14, and sample sizes and cell details are given in Table 3. During each of the tests in the titanium cells, a pin hole leak developed, the tests terminated prematurely, and pressure-temperature cool down data were lost. The results might be considered to be influenced by the cell material of construction, and the leak may indicate that the decomposition products are corrosive to titanium. In addition, adiabatic tracking in each of the tests with the Hastelloy C test cells terminated when the high pressure interlock of 200 bar was reached.

The results for all four tests were similar in that multiple distinct reaction zones were visible: the first one being the largest, followed by one or two smaller events. This observation is qualitatively consistent with the inflections seen in the ramp DSC test results. A summary of the ARC results is provided in Table 4.

With the third and fourth tests progressing to near completion, the residual pressure measured in the test cell after cooldown was used to estimate the overall amount of light gas formation due to decomposition. For the third test, matching an initial temperature with a final one, at 31.2°C the initial pressure was 15.9 psia, and at 31.1°C, the final pressure was 490 psia. Using the DMSO charge and cell volume for the third ARC test as shown in Table 3, making the simplifying assumption that the liquid volume before and after the test (i.e., at 31.2 and 31.1 °C) is the same, and presuming the liquid mole fraction of the light gas(es) is negligibly small, the molar quantity of gas was estimated. At the final conditions, the compressibility factor was estimated using Pitzer's generalized virial coefficient correlation [7], which ranged from 0.84 for CO_2 to 0.95 for CH_4. Using an approximately average value of 0.9, the results of the calculation showed 0.26 moles of light gas was produced per mole of initial DMSO. Similarly, for the fourth test, matching an initial temperature with a final one, at 32.2°C the initial pressure was 13.0 psia, and at 32.2°C, the final pressure was 469 psia. Using the same method as

[3] Flat baseline not established before observable heat flow was detected

described above, the results of the calculation showed 0.23 moles of light gas was produced per mole of initial DMSO.

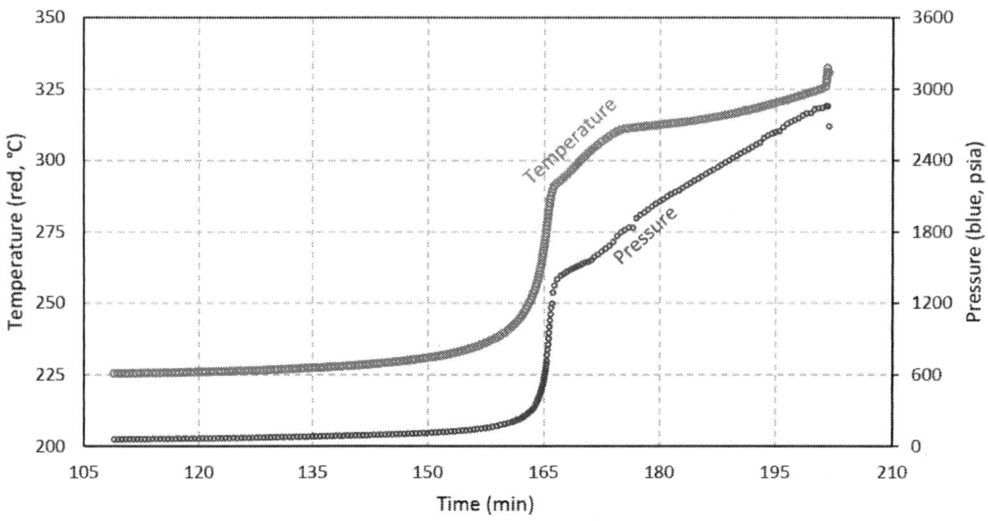

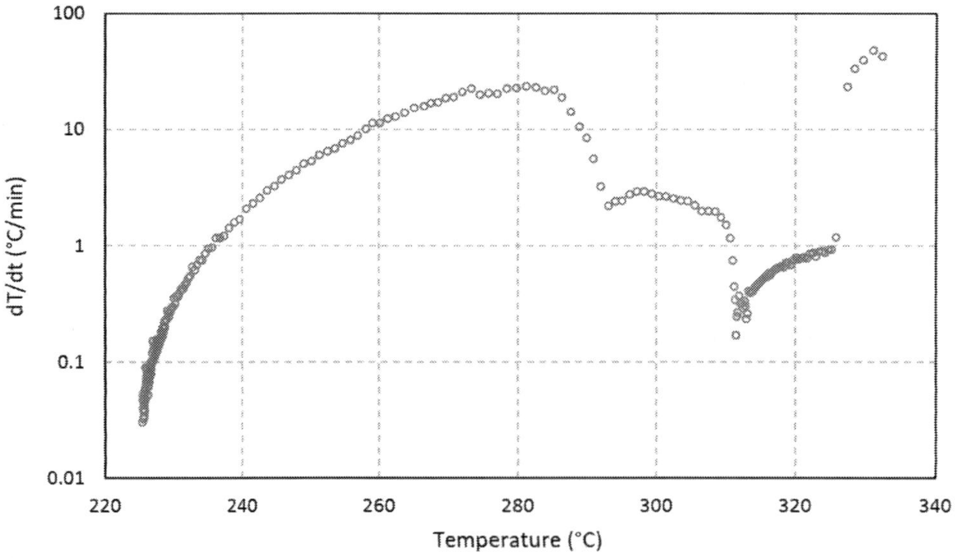

Figure 11. ARC Test 1 Results for DMSO in Titanium Test Cell (leak formed near 325 °C).

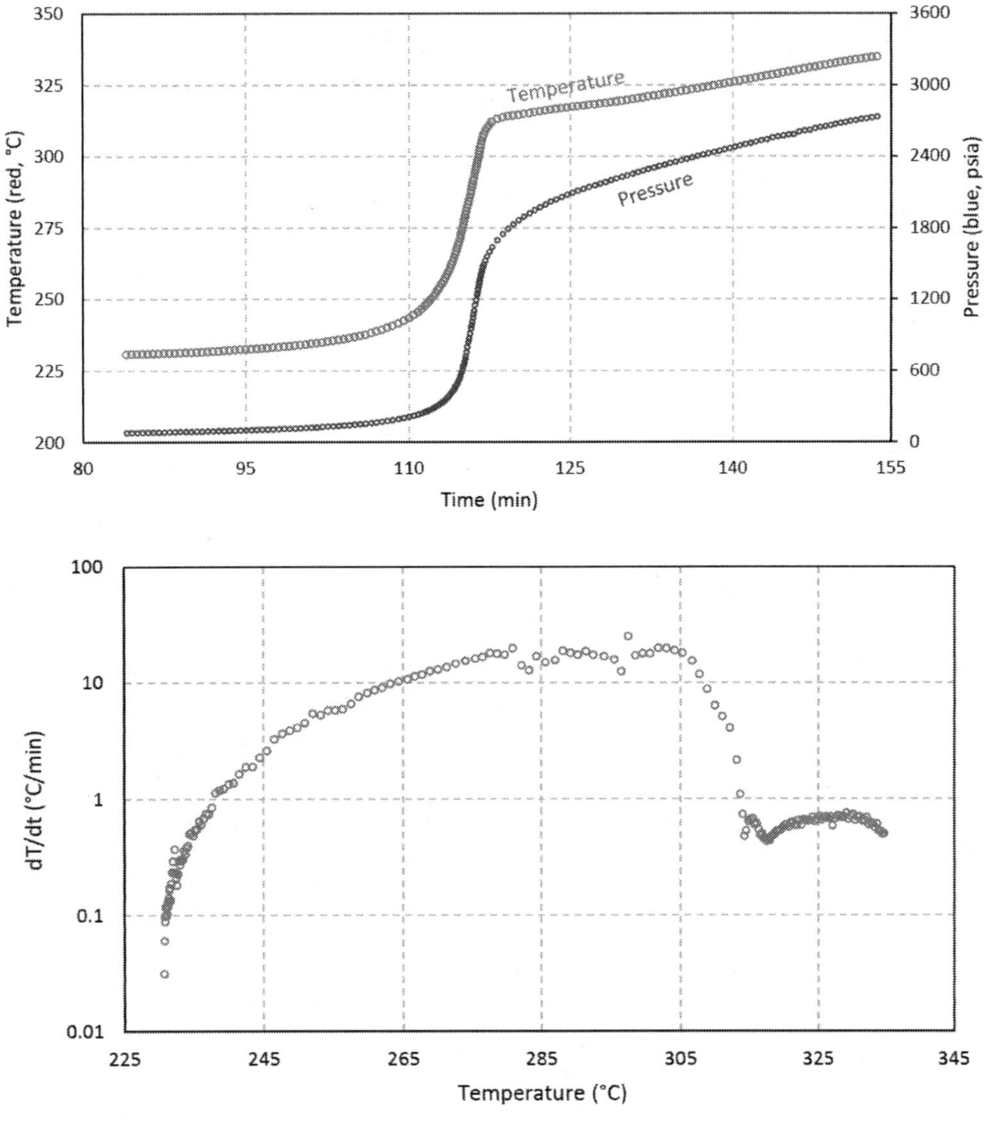

Figure 12. ARC Test 2 Results for DMSO in Titanium Test Cell (leak formed near 328 °C).

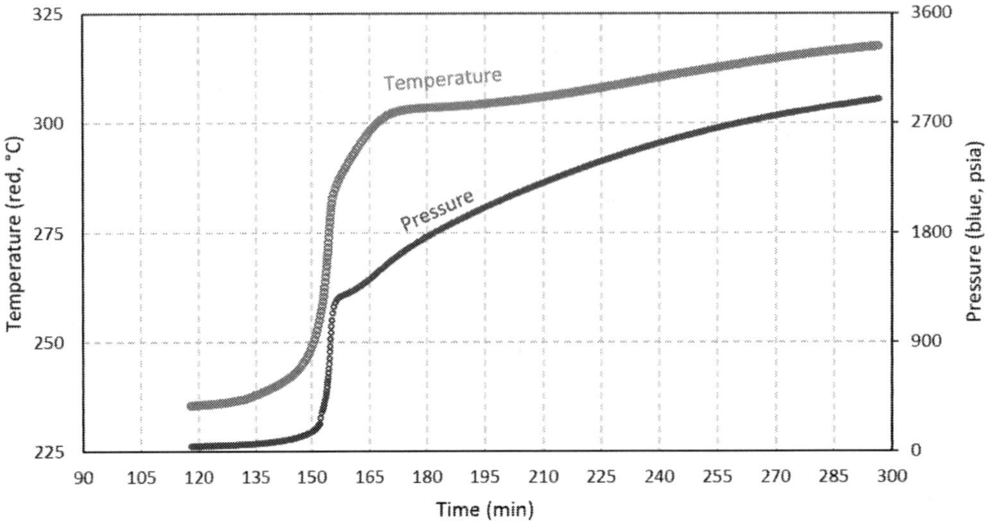

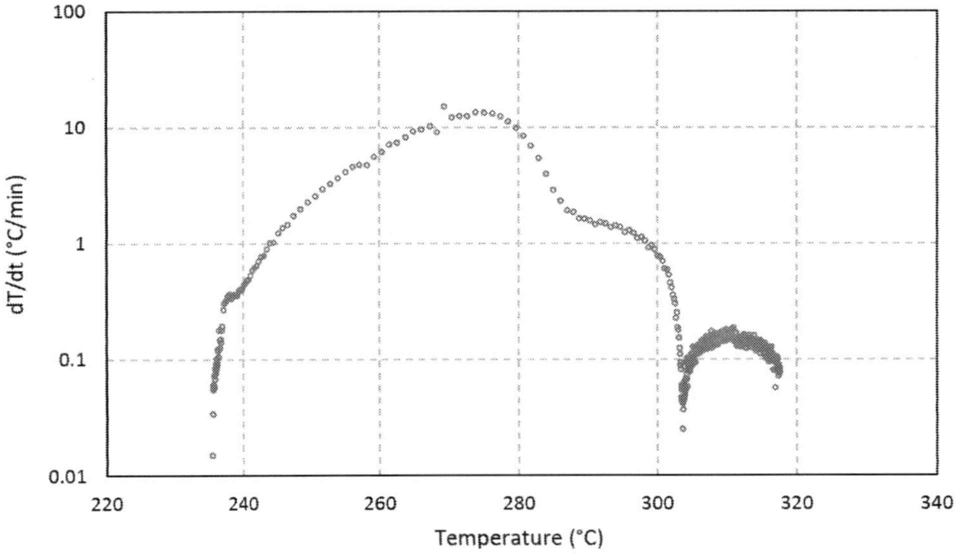

Figure 13. ARC Test 3 Results for DMSO in Hastelloy C Test Cell.

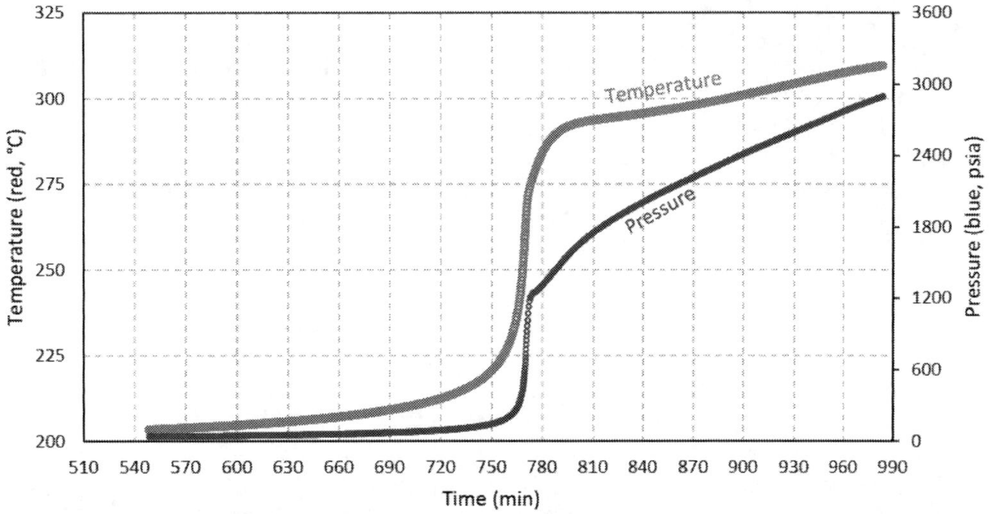

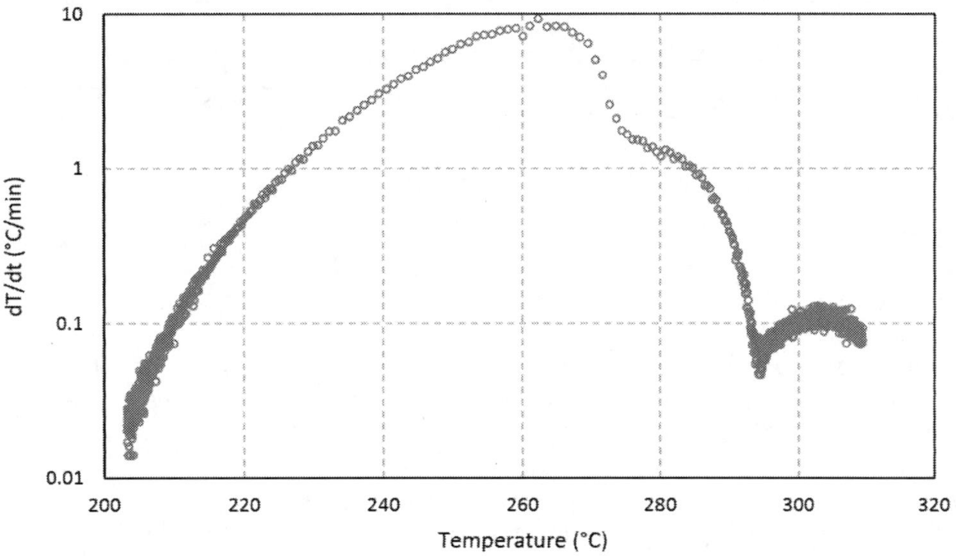

Figure 14. ARC Test 4 Results for DMSO in Hastelloy C Test Cell.

Table 3. Summary of ARC Test Conditions.

ARC Test	Sample Mass (g)	Sample Heat Capacity[4] (J/g-C)	Test Cell Type	Cell Volume (mL)	Test Cell Mass (g)	Test Cell Heat Capacity (J/g-C)	ϕ
1	3.2323	2.6	titanium	9.3	6.7673	0.609	1.490
2	2.6583	2.6	titanium	9.3	6.7203	0.609	1.592
3	2.9996	2.6	Hastelloy C	9.3	21.2759	0.473	2.290
4	3.0015	2.6	Hastelloy C	9.3	21.1064	0.473	2.279

Table 4. Summary of ARC Test Results.

ARC Test	Observable Onset T T_o (°C)	Observed Adiabatic T rise $\Delta T_{ad}/\phi$ (°C)	Peak Observed Heating Rate $(dT/dt)_{peak}$ (°C/min)	Final Pressure after cooldown (psia)	Calc'd Energy $-\Delta U/mw$[5] (J/g)	Comment
1	225.5	100.3	15.22 at 269.3°C	-	389	Test incomplete
2	230.8	104.1	25.11 at 297.4°C	-	430	Test incomplete
3	235.4	82.1	15.22 at 269.3°C	490	489	
4	203.3	106.3	9.28 at 262.3°C	469	630	

3.4 PST

Three pressure screening tests were conducted to obtain intermediate samples of the decomposition and to isolate which temperature regions experience light gas formation as a result of decomposition. Experiments were conducted with a 2°C/min ramp rate with final temperatures of 250, 275, and 400°C. The recorded test profiles are provided in Figures 15-17, and the results are summarized in Table 5.

For the test to 250 °C, the pressure generally increased as explained by the vapor pressure of DMSO with a correction for the increase in partial pressure of the initial vapor space with test temperature. The high temperature end of the test showed pressure below this prediction. Upon cooldown, the pressure followed back down the same profile with only a small amount of residual generated pressure observed upon reaching the initial temperature. For the test to 275 °C, the pressure increase well followed the vapor pressure of DMSO with a correction for the increase in partial pressure of the initial vapor space with test temperature. Upon cooldown, however, the pressure followed a path above the pressure observed during heat-up, with more residual pressure observed upon reaching the initial temperature, corresponding to about 0.01 mole of gas per mole of DMSO. As such, it can be concluded that light gas formation as a result of

[4] All heat capacities are best estimate at 240°C.

[5] Calculated energy = $(\phi)\cdot(\Delta T_{ad}/\phi)\cdot$(sample heat capacity)

decomposition becomes detectable somewhere between 250 and 275 °C. For the test to 400 °C, the pressure increase well followed the vapor pressure of DMSO with a correction for the increase in partial pressure of the initial vapor space with test temperature only up to about 250 °C. Above that temperature, the observed pressure was significantly above the prediction, clearly indicating light gas formation. Upon cooldown, the pressure followed a path above the pressure observed during heat-up, with considerably more residual pressure observed upon reaching the initial temperature, corresponding to about 0.27 mole of gas per mole of DMSO. This result is consistent with the observations from the ARC testing.

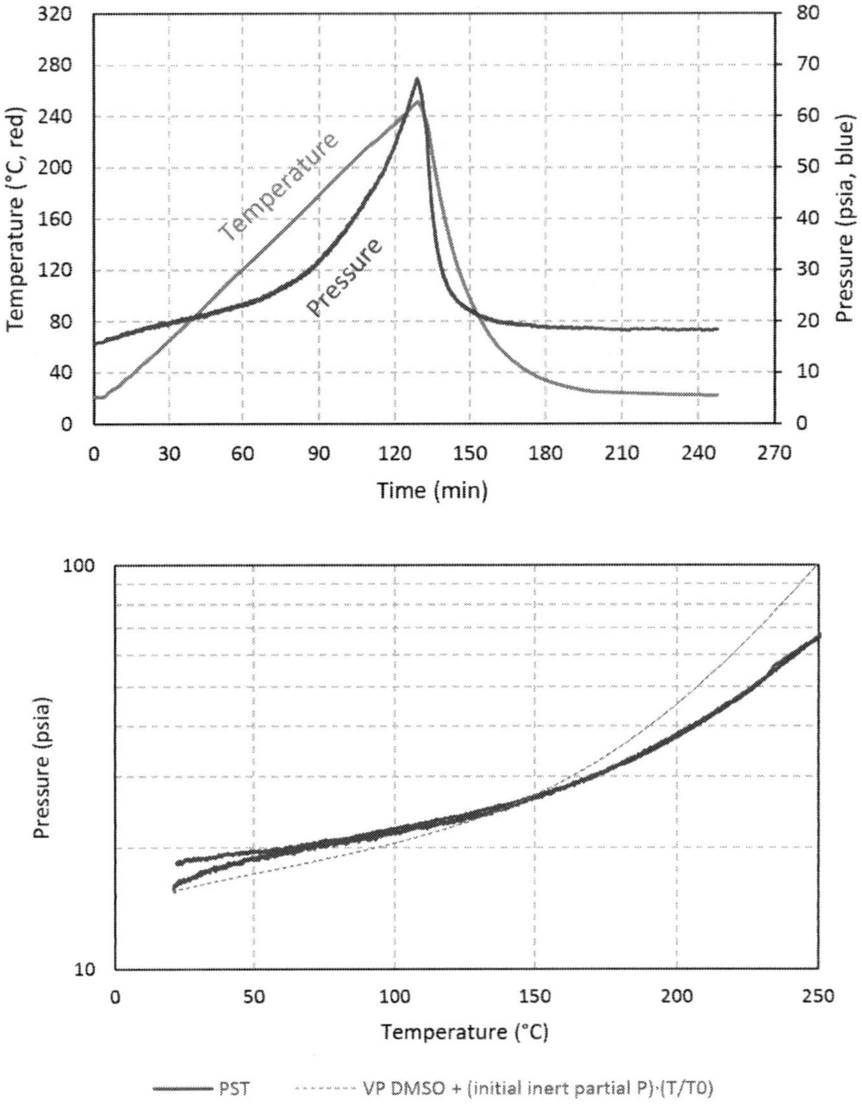

Figure 15. PST Test Results for DMSO (2 °C/min heating rate to 250 °C, followed by cooldown back to initial temperature).

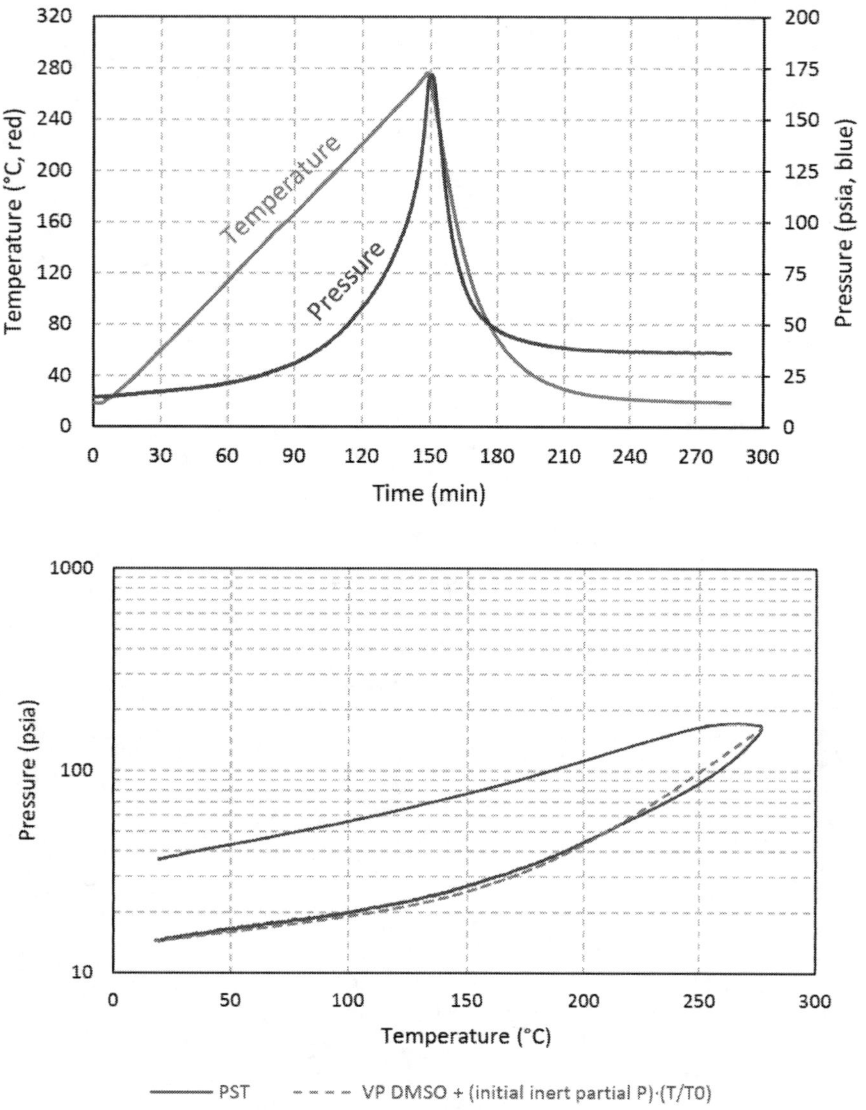

Figure 16. PST Test Results for DMSO (2 °C/min heating rate to 275 °C, followed by cooldown back to initial temperature).

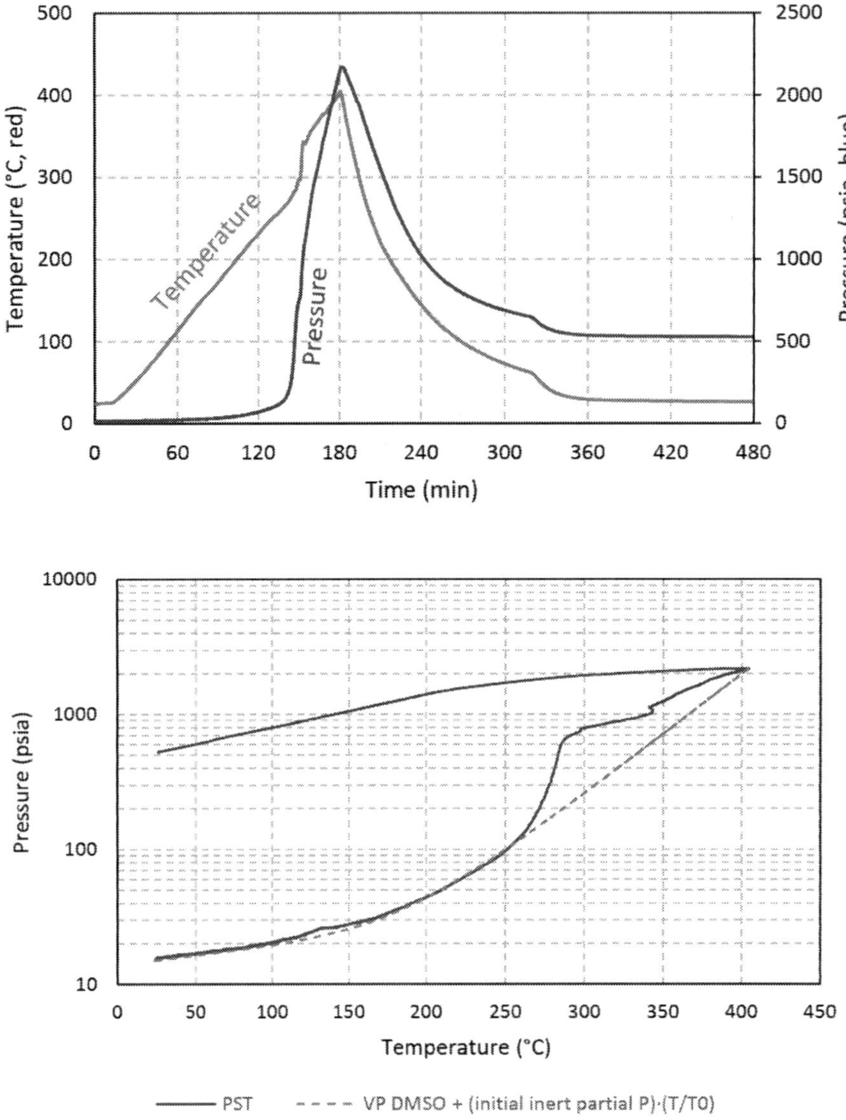

Figure 17. PST Test Results for DMSO (2 °C/min heating rate to 400 °C, followed by cooldown back to initial temperature).

Table 5. Summary of PST Test Results

PST Test	Sample Mass (g)	Recovered Residue Mass (g)	Maximum Temp. (°C)	Peak Pressure (psia)	Final Pressure after cooldown (psia)	Final Temp. (°C)	Residual Pressure at Final T (psia)	Z at final P	Light Gas per mole DMSO[6] (mol/mol)
1	2.4967	0.7	251	67.3	18.4	21.9	2.04	1	0.0013
2	3.0826	1.9782	285	171.9	36.2	19.2	21.7	1	0.010
3	3.0554	3.0534	404	2172	527.2	26.1	511.5	0.88	0.27

3.5 Gas Chromatography–High Resolution Mass Spectrometry

The residue and decomposition gas products from the first PST test were analyzed to identify the chemical species involved. Each analysis provided a list of many compounds present, with some cases being unclear due to cross interference of the signals. The major components identified for the gas decomposition products included:

- Carbon dioxide CO_2
- Methane CH_4
- Ethane C_2H_6
- Ethylene C_2H_4
- Propane C_3H_8
- Propylene C_3H_6
- Butane C_4H_{10}
- Butenes C_4H_8
- Hydrogen sulfide H_2S
- Methyl mercaptan CH_3S
- Dimethyl sulfide C_2H_6S
- Hexane C_6H_{14}
- Dimethyl disulfide $C_2H_6S_2$
- Dimethyl sulfone $C_2H_6O_2S$
- Bis(methylthio)methane (trace) $C_3H_8S_2$

and residue:

- Dimethyl sulfoxide C_2H_6OS
- Bis(methylthio)methane (trace) $C_3H_8S_2$
- Dimethyl sulfone $C_2H_6O_2S$
- Dodecane $C_{12}H_{26}$

Based upon work provided in the literature [3, 8] and by a DMSO manufacturer [1], the following additional stable components are candidates for this system:

[6] Calculated as described in section 3.3.

- Hydrogen H_2
- Carbon disulfide CS_2
- Sulfur dioxide SO_2
- Formaldehyde CH_2O
- Paraformaldehyde $HO\text{-}(CH_2O)_n\text{-}H$
- Water H_2O

A few more components are proposed to be included in the list since they are related to the ones above:

- 1,3,5-trioxane $C_3H_6O_3$
- Sulfuric acid H_2SO_4
- Methane sulfonic acid CH_3SO_3H
- Carbon monoxide CO
- Dimethyl sulfite $(CH_3O)_2SO$
- Oxygen O_2
- Sulfur trioxide SO_3
- Dimethyl sulfate $(CH_3)_2SO_4$

4. Model development

With all the experimental results provided in section 3, several conclusions can be made regarding the decomposition of DMSO. It is clear that the overall system is autocatalytic in shape, accelerated by the presence of acids, mechanistically complex with at least three types of reactions, and experiences a significant amount of light gas formation occurring at higher temperature. Based upon literature data and chemical analysis in this study, a rather large list of components exist in this system, ranging from alkyl sulfates, thioalkyls, alkyl sulfonates, acids, alkanes, alkenes, aldehydes, and CO_2, which suggests the decomposition pathway is very complex with many reactions. In this section, several candidate mechanisms and simplifications are evaluated based upon all the data, with a first-pass model provided. A second paper will be published at a later date with a detailed mechanistic model.

4.1 Thermochemical Analysis

A common practice employed in evaluating and explaining calorimetry data is to screen candidate components and reactions among them, estimate the theoretical heat liberated from those reactions, and compare to the experimental observations. This analysis allows one to establish bounds as to what chemical reactions are thermodynamically consistent with the calorimetry data. The majority of the large amount of tabulated thermochemical data are expressed as enthalpy, *H*, since they were measured or calculated for systems of constant pressure, and most enthalpy of formation data are for standard state conditions (i.e., at 25 °C, 1 atm, and for the naturally-occurring physical state at those conditions). For simplicity, the heat of reaction for simple screening of candidate reactions is often estimated as the enthalpy change at standard conditions. However, the system contained within a calorimeter test cell is of constant volume, so it is isochoric (or isometric) in nature, making internal energy, *U*, the more characteristic thermochemical expression. In addition, most exothermic events as observed in a calorimeter occur at significantly elevated temperature, so appropriate internal energy evaluation should be at an average temperature representative of the test data. In general, the difference between ΔH and ΔU of reaction is not large, and similarly the heat of reaction at an elevated temperature deviates only moderately from the theoretical reaction at 25 °C (while expressing the components at the physical states at the actual temperature and pressure), so satisfactory early evaluation of candidate reactions can be conducted using the standard enthalpy of reaction.

When additional accuracy is needed beyond simply considering standard enthalpy of reaction, with the availability of computerized databases such as DIPPR [9] and modern computer power, it is straightforward to obtain tabulated internal energy data for a large number of components for reaction screening. Starting with the definition of enthalpy on a specific (i.e., per mole) basis for a component in the ideal gas state:

$$\overline{U} = \overline{H} - \left(P\overline{V}\right) = \overline{H} - RT \quad \text{[Eq. 1-2]}$$

Using the standard state of the elements as the reference state, for an ideal vapor component the internal energy can be calculated as:

$$\overline{U}^{vap}(T) = \left[\Delta_f \overline{H}_{IG}^0 + \int_{25°C}^{T} \overline{C}_p^v dT \right] - RT \quad \text{[Eq. 3]}$$

Similarly, for a liquid component:

$$\overline{U}^{liq}(T) = \left[\Delta_f \overline{H}_{IG}^0 + \int_{25°C}^{T} \overline{C}_p^v dT - \Delta\overline{H}_{vap}(T) \right] - P\overline{V}_{liq} \quad \text{[Eq. 4]}$$

With the internal energy calculated in this manner for the candidate components for a system, the stoichiometry and the internal energy analog of Hess' Law can be applied to calculate the internal energy change of proposed reactions.

When the highest degree of accuracy is needed, particularly when the calorimetry test experiences high pressure and with the temperature exceeding the critical temperature of some of the components, the distribution of components among the vapor and liquid states must be taken into account, and the internal energy of each component is evaluated on an appropriate partial molar basis. Typically, a complex equation of state capable of accurately predicting both the liquid and vapor states is used for this purpose, and often a commercial software package is utilized to evaluate the calorimetry data and to apply the results to commercial scale.

4.2 Review of Literature Data

Several references to the thermal decomposition of DMSO are available in the literature which provide calorimetric data, kinetic activation energy, and proposed reaction pathways. The information is useful to describe parts of the overall system, but generally not comprehensive enough due to the complex nature of the decomposition of DMSO. Short reviews and evaluation of these literature data are provided below.

4.2.1 Bretherick's Data

Brief information is provided in Bretherick's Handbook [5] for DMSO, stating it to be energetic with an apparent activation energy of 243 kJ/gmol and with a 24-hour adiabatic induction temperature, T_{AIT24} [10], of 213 °C. This high value of the reported activation energy clearly indicates the mechanism is complex and not simply zero- or first-order, since typical values of the activation energy are in the range of 110 to 170 kJ/gmol [11, 12].

It is important to point out that, for an autocatalytic system, the thermal history affects the T_{AIT24}. Qualitatively looking at the curves of dT/dt vs. T for the ARC tests 3 and 4 (Figures 13 and 14), for the ARC test 4 the heating rate is clearly shifted to lower temperature, indicating the T_{AIT24} calculated from test 4 would have a lower value than for ARC test 3. So the basis of safety for a material that decomposes autocatalytically should not be based upon T_{AIT24} calculated from a single test, but should incorporate realistic aging studies and other factors that could accelerate the decomposition.

4.2.2 Thyrion and Debecker Data

A kinetic study for the thermal decomposition of gas-phase DMSO was available [8] that provided detailed identification of many reaction pathways, several components and intermediates, and a rationalization of which pathways are most likely to occur based upon isothermal testing at several temperatures with mass spectroscopic analysis of the volatile products. The testing was conducted under vacuum and no indication of liquid or solid residue was mentioned. A detailed chemical analysis was provided for the test at 300 °C and is provided in Table 6. The data as given do not achieve an exact atom balance and were slightly adjusted accordingly. The resulting overall gas phase reaction with the adjusted composition would be:

$$\text{DMSO} \longrightarrow 0.744CH_4 + 0.03C_2H_6 + 0.162C_2H_4 + 0.045C_3H_8 + 0.082C_3H_6 + 0.015H_2 + 0.5SO_2 + 0.105CS_2 + 0.052H_2S + 0.149(CH3)_2S + 0.052CH_3SH + 0.019(CH_3)_2S_2 \quad \text{[Eq. 5]}$$

having an equivalent standard state ideal gas phase enthalpy of reaction of -46.5 kJ/gmol (595 J/g exothermic), which is in the range of the internal energy of reaction observed in the tests in this study.

Table 6. Literature Data for Gas-Phase DMSO Thermal Decomposition Products at 300°C [8].

Component	Relative Yield (mol/mol DMSO)	Adjusted Relative Yield to Achieve Atom Balance (mol/mol DMSO)	Standard Ideal Gas Enthalpy of Formation	
			$\Delta_f H^\circ_{IG}$ (kJ/gmol)	Reference
DMSO			-150.5	[13]
Hydrogen	2	2	0	definition
Methane	100	100	-74.6	[14]
Ethylene	28	21.75	52.5	[14]
Ethane	4	4	-84	[14]
Propane	6	6.09	-103.8	[14]
Propylene	11	11	20	[14]
Hydrogen sulfide	7	6.95	-20.6	[14]
Carbon disulfide	20	14.12	116.9	[13]
Sulfur dioxide	54	67.24	-296.8	[14]
Methyl mercaptan	7	7.06	-22.8	[13]
Dimethyl sulfide	20	20	-37.5	[14]
Dimethyl disulfide	2.5	2.5	-24.1	[13]

4.2.3 Traynelis and Hergenrother and Gaylord Data

A simpler series of reactions to describe a mechanistic pathway for the thermal decomposition of DMSO was developed and presented [1, 3]. They showed five parallel reaction schemes, which are duplicated below:

$$(CH_3)_2SO \longrightarrow CH_3SH + CH_2O \quad \text{[Eq. 6]}$$

$$CH_2O \longrightarrow (1/n)(\,CH2O)_n \quad \text{[Eq. 7]}$$

$$2CH_3SH + CH_2O \longrightarrow (CH_3S)_2CH_2 + H_2O \quad \text{[Eq. 8]}$$

$$2CH_3SH + (CH_3)_2SO \longrightarrow CH_3SSCH_3 + CH_3SCH_3 + H_2O \quad \text{[Eq. 9]}$$

$$2(CH_3)_2SO \longrightarrow CH_3SO_2CH_3 + CH_3SCH_3 \quad \text{[Eq. 10]}$$

Internal analytical work provided by Gaylord [15] confirmed the presence of dimethyl sulfide, dimethyl disulfide, and bis(methylthio)methane. To explore whether these reactions could liberate the range of energy observed in the calorimetric tests, the extents of each of these reactions were optimized, maintaining an atom balance, to yield the most exothermic overall change in enthalpy (with the simplification of using standard state enthalpy of formation as given in Table 7). The resulting overall reaction at DMSO atmospheric reflux, 189 °C, would be:

$$\text{DMSO (liq)} \longrightarrow (1/2)H_2O\text{ (liq)} + (1/6)C_3H_6O_3\text{ (liq)} + (1/2)(CH_3S)_2CH_2\text{ (liq)} \quad \text{[Eq. 11]}$$

having an equivalent standard state enthalpy of reaction of -52.8 kJ/gmol (675 J/g exothermic), which is in the range of the internal energy of reaction observed in the tests in this study.

Table 7. Optimization of Literature Reactions for DMSO Thermal Decomposition at 189°C [1].

Component	Formula	Stoichiometric Coefficient in Overall Reaction Equation 11	Standard Enthalpy of Formation		
			State	$\Delta_f H^\circ$ (kJ/gmol)	Reference
DMSO	$(CH_3)_2SO$	1	liq	-203.4	[13]
Dimethyl sulfone	$(CH_3)_2SO_2$	0	liq	-446.4	[13]
Dimethyl sulfide	$(CH_3)_2S$	0	liq	-65.4	[13]
Water	H_2O	0.5	liq	-285.83	[14]
Methyl mercaptan	CH_3SH	0	liq	-46.65	[13]
Dimethyl disulfide	$(CH_3)_2S_2$	0	liq	-62.6	[13]
Formaldehyde	CH_2O	0	gas	-115.9	[13]
1,3,5-trioxane	C_3H_6O3	1/6	liq	-505.76	[13]
Bis(methylthio) methane	$(CH_3S)_2CH_2$	0.5	liq	-57.88	[16]

4.2.4 Lam, Vickery, and Linda Data

A series of ARC studies for the thermal decomposition of DMSO were available [4]. The work mostly was focused on characterizing how the decomposition of DMSO is accelerated by the presence of other materials commonly involved with pharmaceutical reaction masses, but a test of neat DMSO was included and is directly comparable to this study. The results are duplicated in Figure 18. The shape of their results is similar to the results in this study in that multiple events are evident, and the total liberated energy, -$\Delta U/mw$, is 539 J/g, which also compares well to the results of this study. Three main differences are observed, however. First, the data from Lam, *et al.*, do not clearly show the autocatalytic shape as evidenced by this study (the slope of the low-temperature end of their ARC test is not as high as would be expected for an autocatalytic system). The simple first-order mechanism provides a good mathematical fit to the first exothermic event in their data as shown in Figure 19, although the resulting high activation energy (260 kJ/gmol) indicates the mechanism is likely not simple first-order. The basic autocatalytic model fits the data above 205 °C, but does not explain the data below that temperature. Second, the observable onset temperature of their test is much lower than in this study, suggesting some influence such as contamination or prior isoaging of the material tested by Lam, *et al.* Third, the slope of their test results is generally lower than the results in this study, similarly suggesting prior contamination or isoaging of the material they tested.

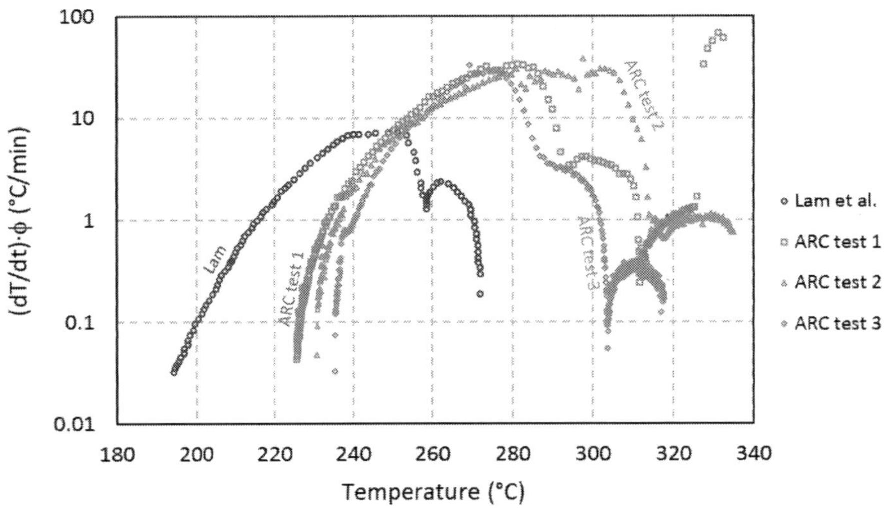

Figure 18. ARC Test Results from [4] (ϕ = 2.67; also shown are results from this study; note vertical axis is heating rate times ϕ, which allows for comparison among tests of differing thermal inertia).

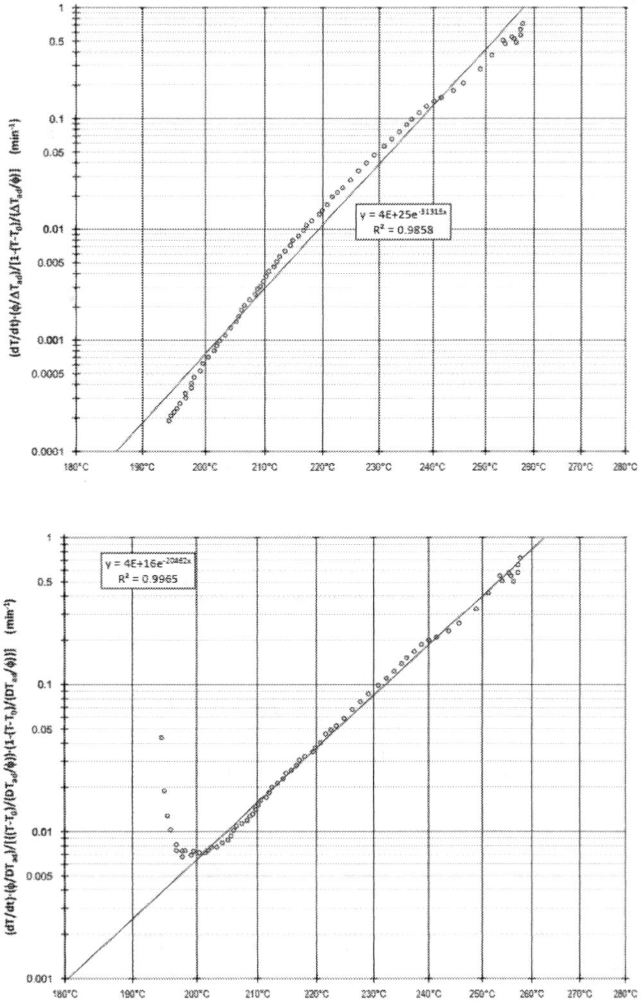

Figure 19. Regression Fit of the First Exothermic Event in the ARC Test Results for [4] – (a) 1st-order model, (b) basic autocatalytic model.

4.2.5 Yang, *et al.*, Data

Another series of ARC studies for the thermal decomposition of DMSO were available [6]. Similar to the section above, the work mostly was focused on characterizing how the decomposition of DMSO is accelerated by the presence of other materials, and a test of neat DMSO was also included. The results are duplicated in Figure 20, which also shows the results from this study. It appears their test of neat DMSO did not complete in that one main thermal event was observed, the measured heating rate at the end of the test was still significant (near 1 °C/min), and the total heat liberated ($-\Delta U/mw$ = 133 J/g) was much lower than in the tests in this study. Titanium test cells were used in their apparatus, so likely they experienced a leaking cell as in this study. The low-temperature end of the test did appear to show the autocatalytic shape, but similar to the test by Lam, *et al.*, the observable onset temperature is much lower than in this study, suggesting some influence such as contamination or prior isoaging of the material tested by Yang, *et al.*

For evaluation purposes, kinetic model regression was applied to the DMSO ARC test by Yang, *et al.* The classical autocatalytic model provided a good fit of the data assuming the test would have continued to higher temperatures, as shown in Figure 21. Allowing simultaneous regression of the Arrhenius parameters for both rate constants and the internal energy of reaction, taking into account the heat-wait-search portion of the test during which some autocatalyst theoretically was forming, reasonable parameters were obtained. But, the results are not completely reliable in that the assumed shape of the part of the curve after the premature end of the test is subjective.

The basic autocatalytic model that was applied to this system is as follows, based upon two generic reactions in the form of A→B and A+B→2B, respectively:

$$\frac{dX_1}{dt} = k_1 \cdot (1 - X_1 - X_2) \quad \text{[Eq. 12]}$$

$$\frac{dX_2}{dt} = \left[k_2 \cdot C_{A_0}\right] \cdot (1 - X_1 - X_2) \cdot \left(\frac{C_{B_0}}{C_{A_0}} + X_1 + X_2 \right) \quad \text{[Eq. 13]}$$

with the rate constants given by the Arrhenius expression:

$$k_1 = A_1 \cdot e^{E_1/(RT)} \quad \text{[Eq. 14]}$$

$$\left[k_2 \cdot C_{A_0}\right] = A_2 \cdot e^{E_2/(RT)} \quad \text{[Eq. 15]}$$

A simple energy balance was applied with the kinetic model, assuming an adiabatic system including the sample and the sample cell:

$$\frac{dT}{dt} = \frac{\Delta T_{ad}}{\phi} \cdot \left(\frac{dX_1}{dt} + \frac{dX_2}{dt} \right) \quad \text{[Eq. 16]}$$

The boundary condition for equation 13 is $X_2 = 0$ at the observable onset temperature T_o. But, since the conversion to the autocatalyst is often a small but significant quantity, the small amount of conversion during the heat-wait-search steps needs to be taken into account by integrating equation 12 all along those steps. To simplify this part of the calculation, equation 12 can be written as a simple zero-order approximation:

$$\frac{dX_1}{dt} = k_1 \quad \text{[Eq. 17]}$$

The incremental conversion during each step can be estimated as simply the product of the rate constant at the temperature of the step times the duration of the step. Finally, the total conversion to the autocatalyst at the observed onset temperature is simply the sum of the incremental conversion of each of the steps:

$$X_1(T_o) = \sum_{i=1}^{n} [k_1(T_i)] \cdot \Delta t_s \quad \text{[Eq. 18]}$$

In the work by Yang, *et al.*, they provided details indicating the heat-wait-search started at 100 °C, with 10 minute holds during the search, and 3 °C increments during each heat, all the way to the observable onset temperature. Optimizing the parameters for equations 12-18 to best fit the data given in Figure 20, presuming the test was ended prematurely, the parameters in Table 8 were obtained, with the fit illustrated in Figure 21. The regressed internal energy of reaction was closer to the range observed in this study, but the reduction in heating rate observed in the test in the range of 225-230°C was ignored, and the extrapolation beyond the test is speculative since no data were available. As such, the parameters in Table 8 are unreliable without further data.

Table 8. Kinetic Parameter Regression of ARC Test Data for DMSO from [6].

		Rxn 1	Rxn 2
ΔT_{ad}	°C	80.3	
$-\Delta U/mw$	J/g	209	
A	min^{-1}	1.37E+15	1.33E+10
E/R	K	20141.7	13103.6
E	kJ/gmol	167	109

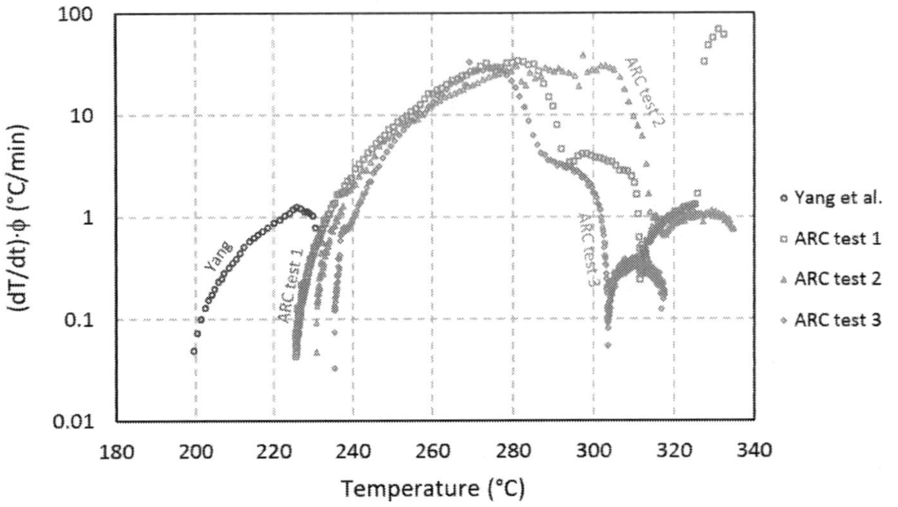

Figure 20. ARC Test Results from [6] ($\phi = 1.65$; also shown are results from this study; note vertical axis is heating rate times ϕ, which allows for comparison among tests of differing thermal inertia).

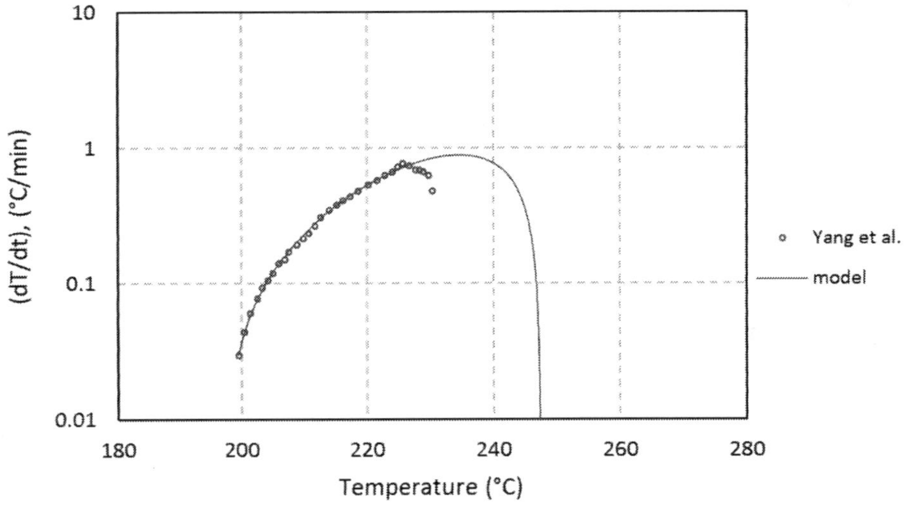

Figure 21. Autocatalytic Model Fit to the ARC Test Results for DMSO from [6] ($\phi = 1.65$).

4.3 General Modeling of Calorimetry Data

The best practice for identifying a reliable model for a complex exothermic system, such as the decomposition of DMSO, is to use several methods in validating the proposed mechanism, hopefully with agreement among all the information available. Since complex reaction pathways are multi-dimensional systems, several calorimetric methods should be used, each operated under several conditions, in order to observe planes or regions of behavior. Recorded data should include temperature, heating flow or rate, and pressure where possible. If feasible, chemical analysis of residues and gas products should be conducted (noting that sometimes samples of these materials might be toxic and should be handled with appropriate precaution) to reliably identify components and some information about their relative concentrations. Sometimes literature data are available that provide explanation of or insight into the reaction pathway(s), and basic heat of reaction calculations or optimizations can be used to help screen candidate reactions.

Several general techniques are developed for utilizing calorimetry data. Isothermal DSC tests can conclusively show if a system has autocatalytic behavior. Since a single test represents one temperature (except for the slightly confounding effect of the initial rapid heat-up), regressed reaction rate constants must be compared among a series of tests at other temperatures to extract Arrhenius parameters. Long runs could be impacted by instrument sensitivity and slow drift, as well as environmental impact. If the reaction products have significantly different heat capacity than the reactant(s), a sigmoid baseline is typically applied to interpret the raw data. Ramp DSC is generally a quick test that provides more information about the system because the temperature changes linearly throughout, giving information about the system's activation energy. If the test passes the critical point of the reactant(s) or products, an often abrupt change in the baseline occurs that is not kinetic in nature and must be appropriately filtered from the test results. The use of a sigmoid baseline is also typically used for this purpose. Repeats of the test are often performed with different ramp rates to validate observed kinetic behavior. Note, for autocatalytic systems, at the observed onset temperature a significant amount of autocatalyst may have formed, so analysis of the results should include information from the region below this point (recommend from start of test). Since simple DSC testing does not provide pressure measurement, the PST (or commercially available pressure DSC apparatus) provides an additional and very useful dimension with pressure data. Light gas formation is clearly identified by comparing the initial and residual pressure after the test (after cooldown back to the initial temperature). Pressure profiles during the test also provide valuable information about mechanisms when there are more than one reaction, although the thermodynamic description of the vapor-liquid equilibrium could be very complex due to multiple components, high pressure, and non-ideal behavior. Finally, adiabatic calorimetry, such as ARC, is extremely valuable in that more closely approximates a large scale runaway reaction since such incidents on the industrial scale typically approach the adiabatic limit. If the system is highly exothermic, very high temperatures can be achieved in ARC testing compared to usual DSC testing ranges since the temperature profile in ARC is controlled by the reactive energy, and the temperature profile in DSC is controlled by the program of the test apparatus. As such, sometimes

DSC testing has to be repeated to higher temperatures to ensure all reactions have occurred and a baseline is re-established. As with the PST, pressure generation due to light gas formation can be quantified and used to identify candidate reactions. The most comprehensive analysis of ARC data involves detailed kinetic and thermodynamic modeling of the results, yielding robust identification of reaction pathways, although the technique can be quite complex mathematically. Where possible, the test should be designed to minimize the thermal inertia within practical limits. Isothermal ageing tests should be conducted if a detailed characterization is necessary for reactions which exhibit autocatalytic behavior.

Without following these general guidelines, it is possible (or likely) that an inadequate model for complex decomposition pathways could be concluded from test data. As shown in Figure 19 above, a simple first-order model was mathematically fit to an ARC test from the literature. Certainly, this model is not able to characterize the autocatalytic behavior of the system. Similarly, the simple first-order model could be fit to a single ramp DSC test for DMSO, which would fail to predict the autocatalytic behavior. For illustration, applying the following form of the nth-order rate model:

$$\frac{dX}{dt} = \left[C_{A_0}^{n-1} \cdot k\right] \cdot (1-X)^n \quad \text{[Eq. 19]}$$

with the rate constant, k, correlated to temperature as in equation 14 above, with a constant heating rate:

$$\frac{dT}{dt} = \beta \quad \text{[Eq. 20]}$$

and with heat flow determined from the rate expression as follows:

$$\frac{\dot{Q}_r}{M} = \frac{dX}{dt} \cdot \left(-\frac{\Delta U_r}{mw}\right) = \frac{dX}{dt} \cdot \left(C_v \cdot \Delta T_{ad}\right) \quad \text{[Eqs. 21-22]}$$

the kinetic parameters were regressed for the first exothermic region in the 10 °C/min ramp DSC data in Figure 1. The parameters are provided in Table 9 and the fit is illustrated in Figure 22. Similarly, the reaction order could be regressed along with the other parameters to fit the entire heat flow profile to arrive at the results in Table 10 and Figure 23. Both of these models appear to well-fit a region or nearly all of the data. But, their activation energies are too high to be realistic, and neither are the autocatalytic model that would be appropriate as shown by the isothermal DSC tests in Figures 7-10.

Table 9. Kinetic Parameter Regression of 10 °C/min Ramp DSC Test Data for DMSO as Shown in Figure 22 (unreliable conclusion).

ΔT_{ad}	°C	127
$-\Delta U/mw$	J/g	330
n		1
A	min^{-1}	9.024E+45
E/R	K	62486
E	kJ/gmol	520

Table 10. Kinetic Parameter Regression of 10 °C/min Ramp DSC Test Data for DMSO as Shown in Figure 23 (unreliable conclusion).

ΔT_{ad}	°C	233
$-\Delta U/mw$	J/g	607
n		4
A	min^{-1}	9.9073E+55
E/R	K	76217
E	kJ/gmol	634

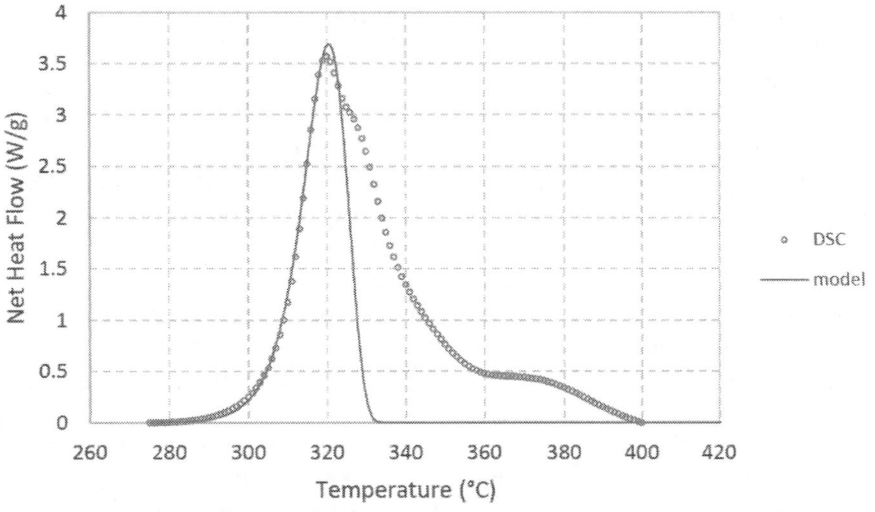

Figure 22. Simple First-Order Model Fit to the Early Exothermic Region of the 10°C/min Ramp DSC Test Results for DMSO (unreliable conclusion).

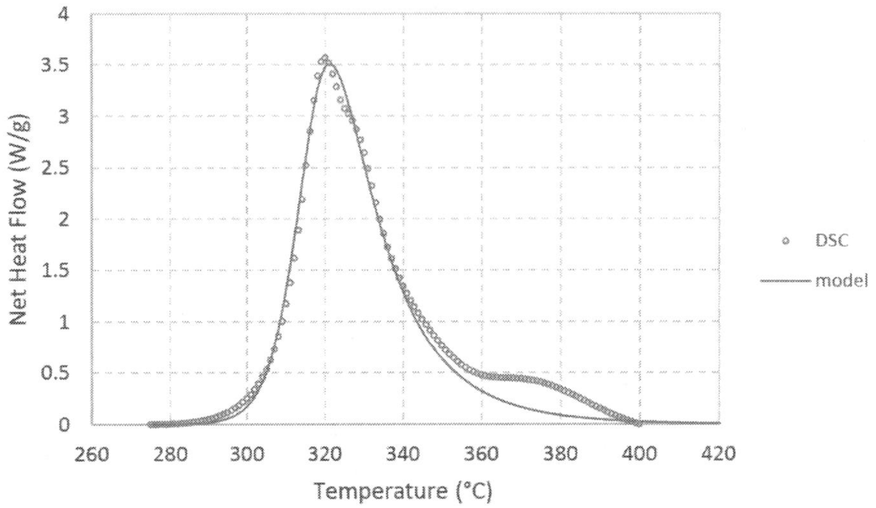

Figure 23. nth-Order Model Fit to the 10°C/min Ramp DSC Test Results for DMSO (unreliable conclusion).

4.4 First-Pass Simplified Kinetic Model

As explained above, all the testing for this study collectively describe the complex behavior of DMSO decomposition. It is a typical practice to attempt to develop a simplified kinetic model that well describes at least a portion of the calorimetric data. For example, such a model may only describe the leading edge of the thermal behavior, not accounting for the complexity of the higher temperature and higher pressure behavior, providing a result that is useful for defining operating limits and assessing risk of upset scenarios [11]. From the results of this study, a simplified model was developed. As clearly shown by the isothermal DSC testing, and also observed in the ARC testing, the system is autocatalytic is shape. Equations 12-15 provide the basic form of the simple autocatalytic model, which does not distinguish multiple species and multiple reaction pathways other than the formation of a hypothetical autocatalyst. The model also assumes constant liquid density and does not include a vapor-liquid equilibrium calculation to account for the distribution of components among the vapor and liquid phases. For the ARC tests, a simple energy balance is applied to relate heat generation from reaction, rate of temperature rise, and the thermal inertia of the test apparatus as expressed in equation 16. For the ramp DSC tests, equations 20-22 are applied to account for temperature ramping and heat flow generation, and the sigmoid baseline is applied to the data to obtain the net heat flow rate. Likewise, for the isothermal DSC data, equations 21 and 22 are applied with the temperature being constant. The internal energy of reaction and the Arrhenius parameters were then obtained by regressing these models to all the test data simultaneously. For all three types of tests, it is very important to begin applying these equations well before the observable onset of heat generation to account for small but significant formation of the autocatalyst. As such, the data during all the heat-up or early ramping were included in the regression.

The final regression of the early exothermic event data from the ramp DSC tests, isothermal DSC tests, and ARC tests in Figures 1-4, 7-8, 13, and 14 gave the parameters provided in Table 11 optimized using downhill simplex by Nelder and Mead [17], with the model fit as shown in Figures 24-31. The model is in good agreement with all the early portion of the ramp DSC and ARC data, with some deviation from the isothermal DSC data. The activation energy for the A→B reaction, 307 kJ/gmol, is clearly too high for a mechanistically correct model, indicating this simplified model is only an approximate characterization of the system, and valid only on the low temperature end. Regarding pressure, based upon the PST tests, it is reasonable to take the vapor pressure of DMSO as the system's partial pressure up to about 250 °C.

To improve the results, a detailed simulation and regression involving identified components, defined reactions, and detailed vapor-liquid equilibrium calculations is required. This work is currently underway and will be featured in a second paper.

Table 11. Kinetic Parameter Regression of Autocatalytic Model for Ramp DSC, Isothermal DSC, and ARC Tests for DMSO.

	ΔT_{ad}	°C	115[7]
	$-\Delta U/mw$	J/g	299
A→B	A	min^{-1}	2.801E+26
	E/R	K	36876
	E	kJ/gmol	307
A+B→2B	A	min^{-1}	8.012E+07
	E/R	K	10113
	E	kJ/gmol	84

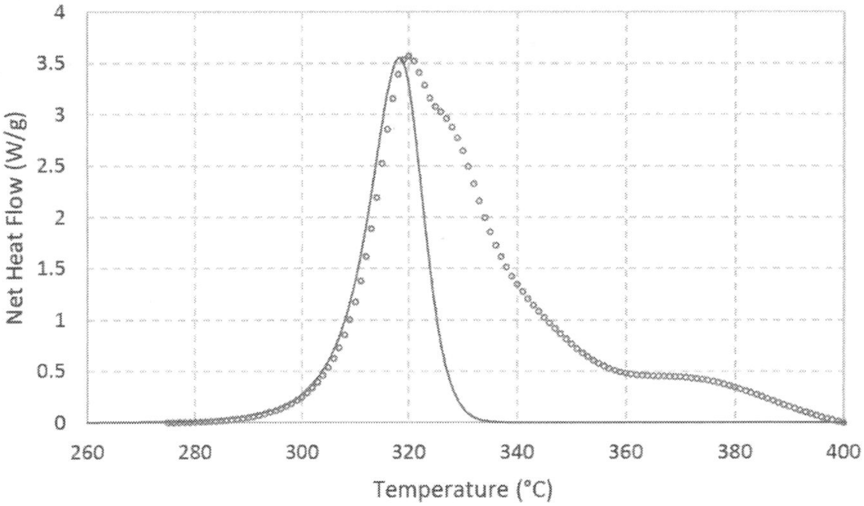

Figure 24. Autocatalytic Model Fit to 10 °C/min Ramp DSC Data for DMSO.

[7] For the ARC test 4, the observed ΔT_{ad} was higher than in the rest of the tests. The regressed value was 160 °C for the early exothermic region.

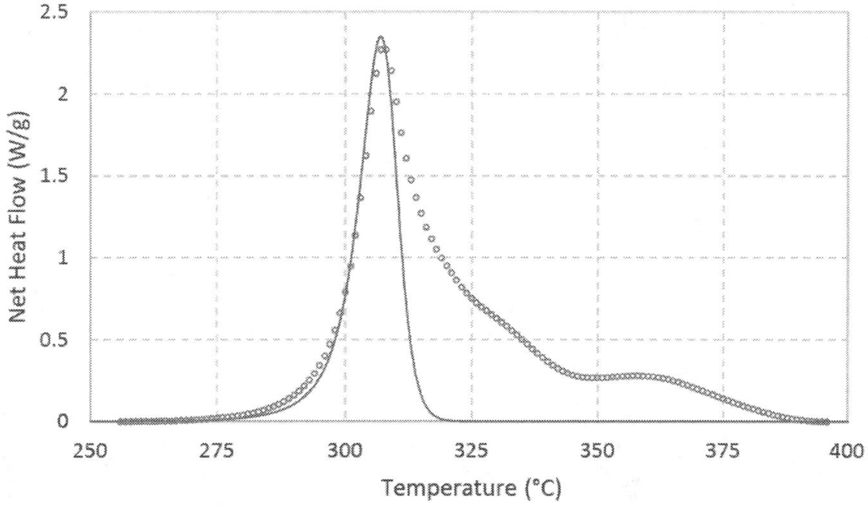

Figure 25. Autocatalytic Model Fit to 5 °C/min Ramp DSC Data for DMSO.

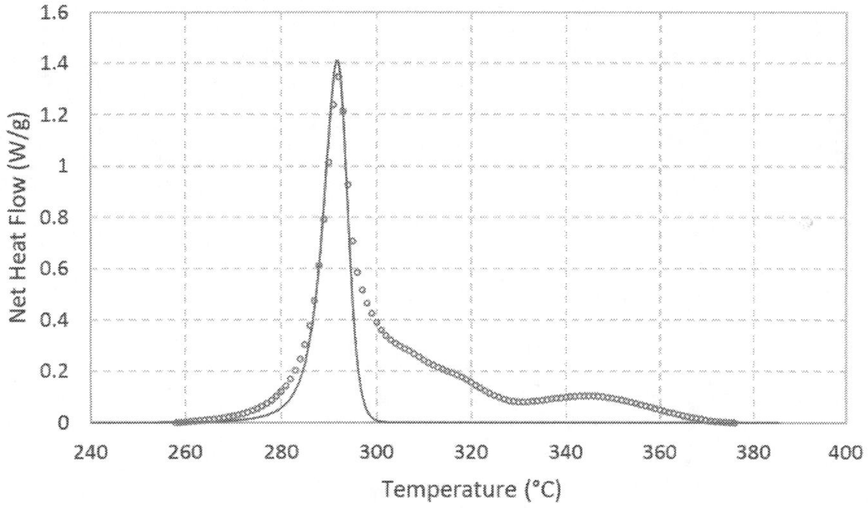

Figure 26. Autocatalytic Model Fit to 2 °C/min Ramp DSC Data for DMSO.

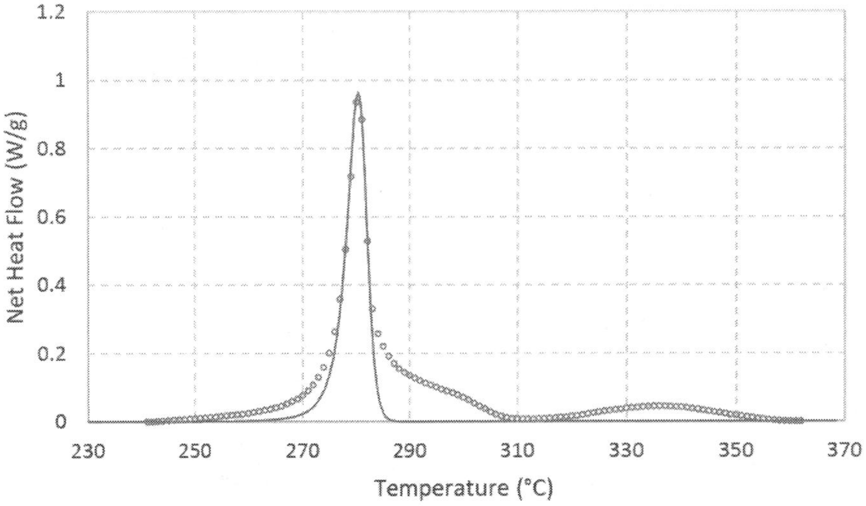

Figure 27. Autocatalytic Model Fit to 1 °C/min Ramp DSC Data for DMSO.

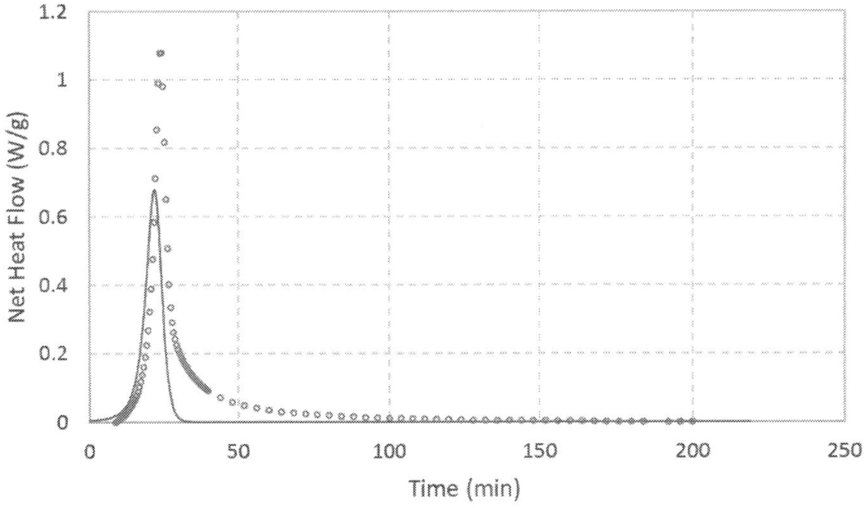

Figure 28. Autocatalytic Model Fit to 270 C Isothermal DSC Data for DMSO.

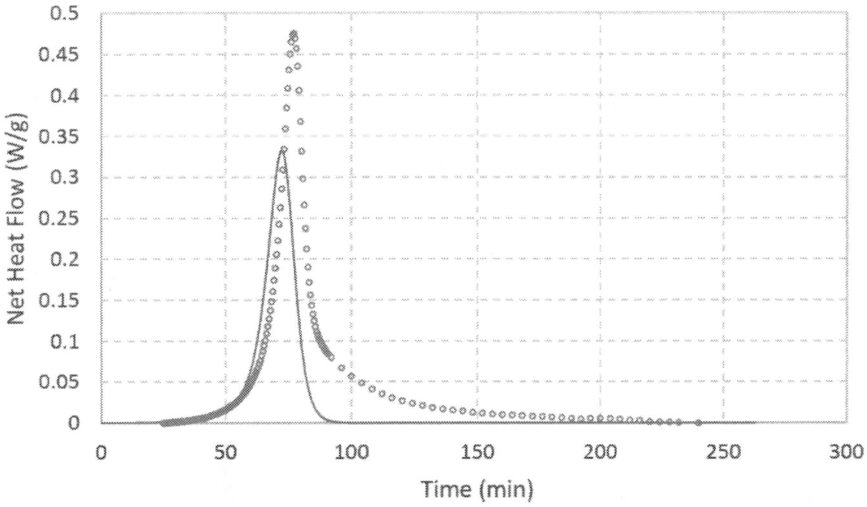

Figure 29. Autocatalytic Model Fit to 250 C Isothermal DSC Data for DMSO.

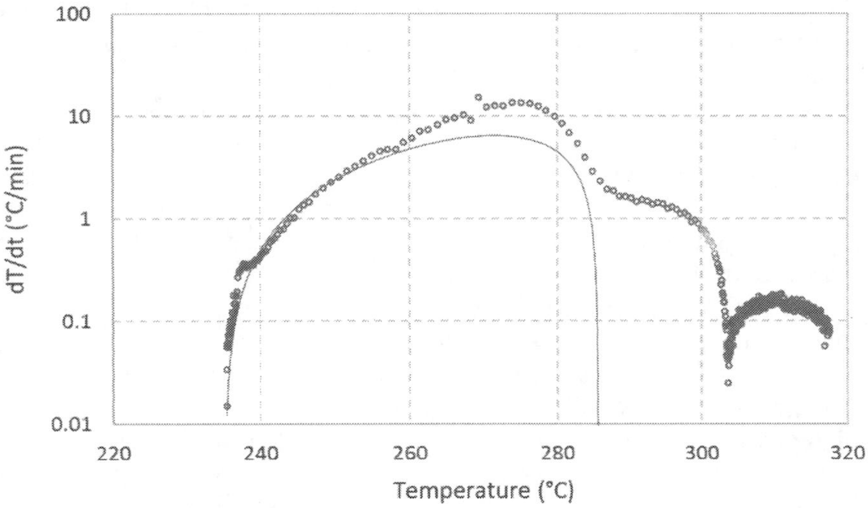

Figure 30. Autocatalytic Model Fit to ARC Test 3 Data for DMSO.

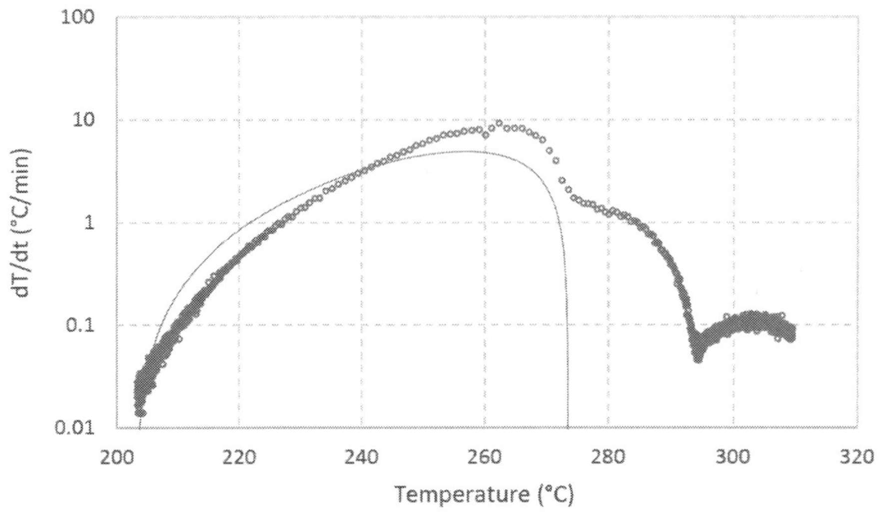

Figure 31. Autocatalytic Model Fit to ARC Test 4 Data for DMSO.

6. Conclusion

DMSO is stable at room temperature, but upon heating in a closed system under conditions typically achievable in a commercial chemical plant, DMSO undergoes significantly exothermic and hazardous decomposition which could realistically lead to a pressure vessel explosion unless adequate protection layers were installed, possibly including a pressure relief system. Literature data that were available explain that the hazard potential exists, but the data are somewhat limited and have some inconsistencies among the sources. With the calorimetry data in this study from several apparatus and methods to characterize the decomposition, a more complete picture of the system was provided. The thermal decomposition is autocatalytic in shape, sensitive to the presence of acids, mechanistically complex with at least three types of reactions, and with a significant amount of light gas formation occurring at higher temperature. A rather large list of components exist in this system, ranging from alkyl sulfates, thioalkyls, alkyl sulfonates, acids, alkanes, alkenes, aldehydes, and CO_2, which suggests the decomposition pathway is very complex with many reactions. Based upon all the results, a simplified and approximate kinetic model was developed to correlate the data for the low temperature end. Additional work is also underway to develop an improved, detailed mechanistic model for the entire range, which will be presented in a future paper.

7. Nomenclature

A	=	rate constant pre-exponential factor
C_{A_0}	=	initial molar concentration of the reactant

$\frac{C_{B_0}}{C_{A_0}}$ = initial concentration of autocatalyst per initial concentration of DMSO

$\overline{C}_p^v$ = isochoric heat capacity

E_i = reaction rate constant activation energy

H = enthalpy

$\overline{H}$ = specific enthalpy

$\Delta_f \overline{H}_{IG}^0$ = enthalpy of formation at 25°C and in the ideal gas state

$\Delta\overline{H}_{vap}$ = enthalpy change of vaporization

k_i = reaction rate constant, given by the Arrhenius equation

mw = molecular weight

n = reaction order

n_v = moles of vapor generated by reaction

P = pressure

$\dot{Q}/M$ = specific heat flow rate

R = ideal gas constant

t = time

Δt_s = duration of a search step

T = temperature

T_o = observed onset temperature

ΔT_{ad} = ϕ-corrected adiabatic temperature rise

U = internal energy

$\overline{U}$ = specific internal energy

$\overline{U}^{liq}$ = specific internal energy of liquid

$\overline{U}^{vap}$ = specific internal energy of vapor

ΔU_r = internal energy change of reaction

$\overline{V}$ = specific volume

$\overline{V}_{liq}$ = specific volume of liquid

X_i = molar conversion

Greek symbols

β = ramp DSC heating rate

ϕ = thermal inertia

8. References

[1] "Dimethyl Sulfoxide (DMSO) Physical Properties," Gaylord Chemical Corporation Bulletin # 101, Nov 2005.

[2] J. Hall, Articles and Case Histories from Process Industries Throughout the World, Loss Prev Bull, 114, (1993), 9-15.

[3] V. J. Traynelis, and W. L. Hergenrother, Decomposition of Dimethyl Sulfoxide Aided by Ethylene Glycol, Acetamide, and Related Compounds, J. Org. Chem., 29, (1964), (1964), 221-222.

[4] T. T. Lam, T. Vickery, and L. Tuma, Thermal Hazards and Safe Scale-Up of Reactions Containing Dimethyl Sulfoxide, J. Therm. Anal. Calorim., 85, (2006), 25-30.

[5] P. G. Urben, ed., *Bretherick's Handbook of Reactive Chemical Hazards*, 7th ed, vol 1. Elsevier, New York, 2007.

[6] X-W Yang, X-Y Zhang, Z-C Guo, W-S Bai, L. Hao, and H-Y Wei, Effects of Incompatible Substances on the Thermal Stability of Dimethyl Sulfoxide, Thermchim. Acta, 559, (2013), 76-81.

[7] J. M. Smith and H. C. Van Ness, *Introduction to Chemical Engineering Thermodynamics*, 4th ed., McGraw-Hill Book Co., New York, 1987.

[8] F. C. Thyrion and G. Debecker, Thermolysis of Dimethyl Sulfoxide, Int. J. Chem. Kinet., 5, (1973), 583-592.

[9] Design Institute for Physical Property Data (DIPPR), American Institute of Chemical Engineers (AIChE), Version March 1995, Technical Database Services, Inc.

[10] F. Stoessel, *Thermal Safety of Chemical Processes – Risk Assessment and Process Design*, Wiley-VCH Verlag GmbH & Co. KGaA, Weinheim, 2008.

[11] T. Grewer, *Thermal Hazards of Chemical Reactions*, Elsevier, Amsterdam, 1994.

[12] C. H. Bamford and C. F. H. Tipper, ed., *Comprehensive Chemical Kinetics*, Vol. 5 – Decomposition and Isomerization of Organic Compounds, Elsevier, Amsterdam, 1972.

[13] National Institute of Standards and Technology, NIST Chemistry WebBook – Group , Available at http://webbook.nist.gov/chemistry/ Accessed on December 30, 2014.

[14] J. A. Dean, *Lange's Handbook of Chemistry*, 15th ed., McGraw-Hill Inc., New York, 1999.

[15] A. McKim, Gaylord Chemical Corporation, personal communication, November 14, 2014.

[16] National Institute of Standards and Technology, NIST Chemistry WebBook, Available at http://webbook.nist.gov/chemistry/grp-add/ Accessed on November 20, 2014.

[17] W. H. Press, B. P. Flannery, S. A. Teukolsky, and W. T. Vetterling, *Numerical Recipes*, Cambridge University Press, New York, 1987.

Relief System Sizing for Runaway Chemical Reactions: A Simple Comprehensive Approach

Ken Kurko
Fauske & Associates, LLC
16W070 83rd Street
Burr Ridge, IL 60527
kurko@fauske.com

Prepared for Presentation at
American Institute of Chemical Engineers
2015 Spring Meeting
11th Global Congress on Process Safety
Austin, Texas
April 27-29, 2015

Relief System Sizing for Runaway Chemical Reactions: A Simple Comprehensive Approach

Ken Kurko
Fauske & Associates, LLC
16W070 83rd Street
Burr Ridge, IL 60527
kurko@fauske.com

Keywords: Vent sizing, runaway reactions, source terms, adiabatic calorimetry, flow regime

Abstract

When conducting relief sizing calculations for reactive systems, experimental data is of paramount importance. Experimental techniques for gathering the necessary kinetic data highlighting VSP2 (Vent Sizing Package 2) and ARSST (Advanced Reactive System Screening Tool) testing will be outlined. In addition, the principles behind the system classifications (vapor, gassy and hybrid) will be outlined with example experimental data sets. Relief sizing calculations and source term considerations for each system will be discussed. Two-phase flow and experimental techniques for determining flow regime (vapor, churn, bubbly or homogeneous flow) will be presented with the goal of providing the audience with a simple comprehensive experimental approach to designing relief systems.

Upset Scenario Selection

Evaluation of reactive chemistries for relief system design requires consideration of plausible upset scenarios. What exactly goes wrong (and when it goes wrong) in a chemical process can have a great affect on the rates at which a reaction runs away. This in turn affects relief requirements and whether or not a particular relief system design is adequate to handle the worst potential scenario. Common upset scenarios include, but are not limited to, loss-of-cooling during normal process conditions, inadvertent heating, wrong reagent addition, and addition at wrong temperature. Upset scenario selection is typically done in a process hazard analysis (PHA) that includes process engineers, chemists, reactive hazards experts, operators, and maintenance workers.

Experimental Techniques

Once plausible upset scenarios have been determined, it is critical that adiabatic calorimetry experiments be performed to characterize the reactive chemistry in question. The VSP2 and ARSST adiabatic calorimeters provide a means to directly simulate a specific upset scenario on the bench scale and do so in a safe manner. Due to the low phi-factor, the measured reaction rates are directly applicable to the process scale.

Source Term Considerations

During a reactive upset scenario, overpressurization of a process vessel can be caused by an increase in vapor pressure, generation of non-condensable gas, or a combination of the two. These overpressure sources are classified as vapor, gassy, or hybrid. Adiabatic calorimetry experiments provide a means for determining the source of overpressure.

Reactive systems in which the increase in pressure is due solely to an increase in vapor pressure are commonly referred to as vapor systems. Vapor systems have the advantage that the latent heat from boiling prevents the temperature from increasing (or tempers the system). As a result, the temperature rise due to the runaway reaction can be controlled by venting. In these cases, the relief system must be designed to accommodate the vapor generated during venting. As such, the relief requirement is based on the temperature rise rate. Adiabatic calorimetry experiments for these types of systems are typically run as closed cell tests. This allows for vapor pressure data to be collected in addition to reaction rate data.

Reactive systems in which the increase in pressure is due solely to the generation of non-condensable gas are commonly referred to as gassy systems. When dealing with gassy reactive systems, it is important to recognize that the reaction temperature rise cannot be controlled by venting. Because the temperature cannot be controlled in these cases, the relief system must be designed to accommodate the peak rate of non-condensable gas generation. Because obtaining the peak pressure rise rate is important, adiabatic calorimetry experiments for these types of systems are commonly run as open cell tests to allow for a large volume of non-condensable gas formation.

Reactive systems in which the increase in pressure is due to vapor and non-condensable gas generation are commonly referred to as hybrid systems. Latent heat is available to remove reaction heat. Due to the generation of both vapor and non-condensable gas, it is important to consider both temperature and pressure rise rates for relief design. Similar to gassy systems, adiabatic calorimetry experiments for these types of systems are often times run as open cell tests. This not only provides a large volume for non-condensable gas formation but also allows for confirmation of tempering.

Flow Regime Considerations

After the determinations of system classification and reaction rates, it is also important to consider vessel flow regime. The flow regime determines the amount of two-phase flow that will occur during venting. This in turn affects the size of the required emergency relief vent. Predictions of the flow regime based on material properties of the vessel contents are not reliable. Prior industrial experience with the chemistry in question or experimental data are required.

Either adiabatic calorimeter previously mentioned can be used for flow regime determination. The ARSST utilizes a Flow Regime Detector (FRED) for distinction between foamy or non-foamy venting. The FRED uses a heater and thermocouple combination in the headspace of the test cell. The heater is programmed to heat the headspace above the test cell to a temperature hotter than the sample temperature. Foamy conditions are detected when the thermocouple in the headspace is quenched by the sample. Specially designed tests for flow regime determination in the VSP2 are called blowdown tests. A blowdown test involves depressurizing the test cell and determining the quantity of material remaining afterwards. The small scale VSP2 test is designed to mimic the dimensionless superficial velocity that would be seen on the large scale during venting. This way, one can make the distinction between churn-turbulent (non-foamy) or bubbly and homogeneous (foamy) flow regimes.

Fauske's Simple Relief Sizing Equation

Fauske has worked to develop a simplified relief sizing methodology for reactive systems that can be used with low phi-factor adiabatic calorimetry data and limited knowledge of material properties. This methodology bases the required vent size on the all vapor or gas vent area by allowing for a modest overpressure. While the vent area is calculated on this basis, it does not imply that two-phase flow will not occur. The following equation can be applied to all classifications of reactive systems venting at or above the critical pressure ratio.

$$A = \frac{1}{0.61 \cdot C_D}\left[\frac{m \cdot c_p \cdot \dot{T}}{\lambda \cdot P}\left(\frac{R \cdot T}{MW_v}\right)^{1/2} + \frac{m \cdot v \cdot \dot{P}}{m_t \cdot P}\left(\frac{MW_g}{R \cdot T}\right)^{1/2}\right]$$

where
A = required vent area (m^2)
C_D = discharge coefficient (-)
m = mass of vessel contents (kg)
c_p = heat capacity of vessel contents (J/kg·K)
$\dot{T}$ = temperature rise rate from calorimetry test (K/s)
λ = latent heat of vessel contents (J/kg)
P = venting pressure (Pa)
R = universal gas constant (8314.47 J/kmol·K)
T = venting temperature (K)
MW_v = molecular weight of vapor (kg/kmol)
v = freeboard volume of test (m^3)
$\dot{P}$ = pressure rise rate from calorimetry test (Pa/s)
m_t = mass of test sample (kg)
MW_g = molecular weight of gas (kg/kmol)

In order to use the aforementioned equation for vapor systems, the overpressure available during venting is required to be at least 40% on an absolute basis. The temperature, pressure, temperature rise rate, and material properties for vapor systems are evaluated at the set pressure of the relief device. It is also important to note that the equation is applicable to non-foamy systems. If a foamy flow regime is expected to prevail, the calculated vent area should be increased by a factor of 2.

For gassy systems, the temperature, pressure, and material properties are evaluated at the maximum allowable accumulation pressure (MAAP). Applicable code permits an overpressure of 10% above the maximum allowable working pressure (MAWP) of the vessel on a gauge basis. The pressure rise rate is the maximum measured during the adiabatic calorimetry test.

For tempered hybrid systems, the temperature, pressure, temperature rise rate, pressure rise rate, and material properties are evaluated at the set pressure of the relief device.

Vapor System Example

3500 kg of a phenol-formaldehyde resin is produced in a reactor with a volume of 5 m^3 and a maximum allowable working pressure of 30 psig. The reaction is run at a temperature of 50°C with a sodium hydroxide catalyst being metered to a mixture of phenol, water, and formaldehyde. The total batch has a final composition as follows:

Component	Concentration (wt.%)
Phenol	31.25
Water	20.50
50% Formaldehyde	44.25
50% Sodium Hydroxide	4.00

Results from a PHA indicate an abnormally fast addition of the catalyst charge would overwhelm the cooling system of the reactor leading to a runaway reaction. It is desired to size a rupture disk with a set pressure of 10 psig based on this upset scenario.

Phenol-formaldehyde systems such as this are known to be foamy vapor systems, making the VSP2 an excellent choice for obtaining runaway reaction data. A closed-cell VSP2 test was run to simulate this upset scenario. The phenol, water, and formaldehyde were loaded to the test cell (in the appropriate ratios) and heated to a temperature of 50°C. The sodium hydroxide catalyst was then injected to the test cell and the reactants were allowed to run away adiabatically. The data from this test that are pertinent to the vent sizing evaluation are provided in Figures 1 and 2.

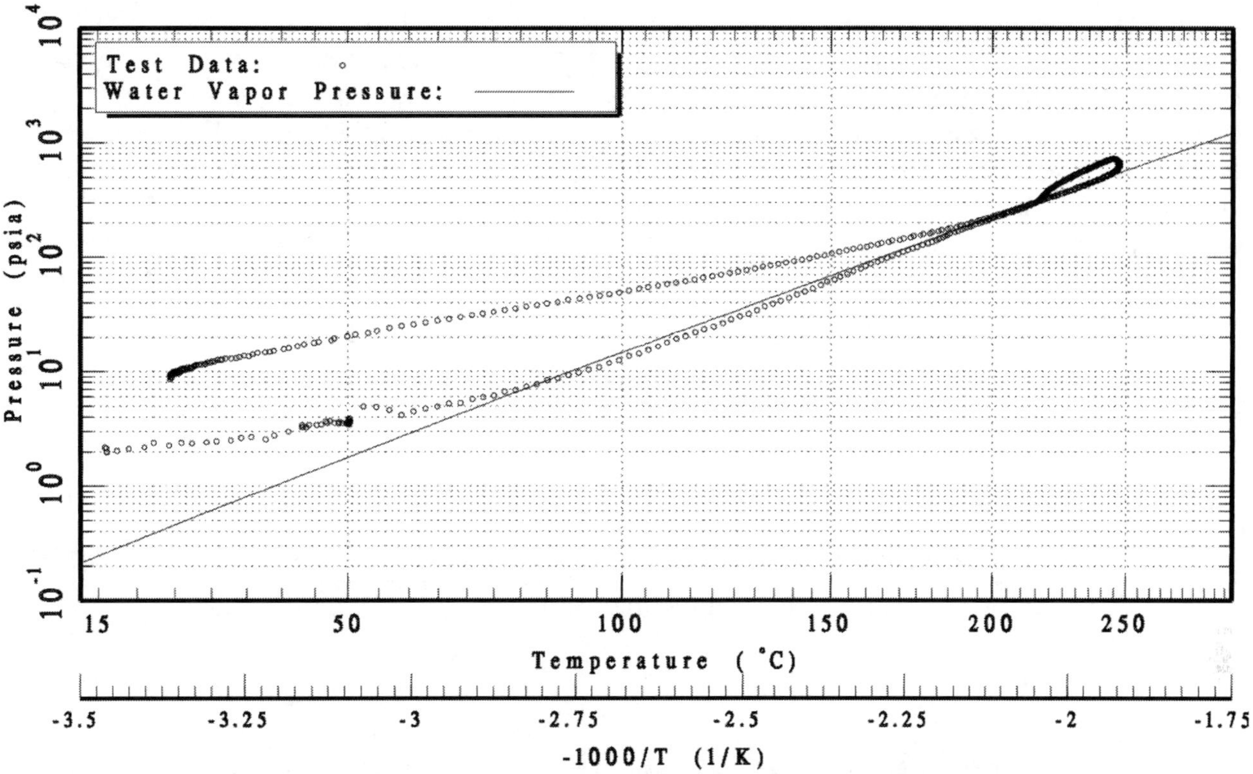

Figure 1 Pressure vs. Temperature for Phenol-Formaldehyde

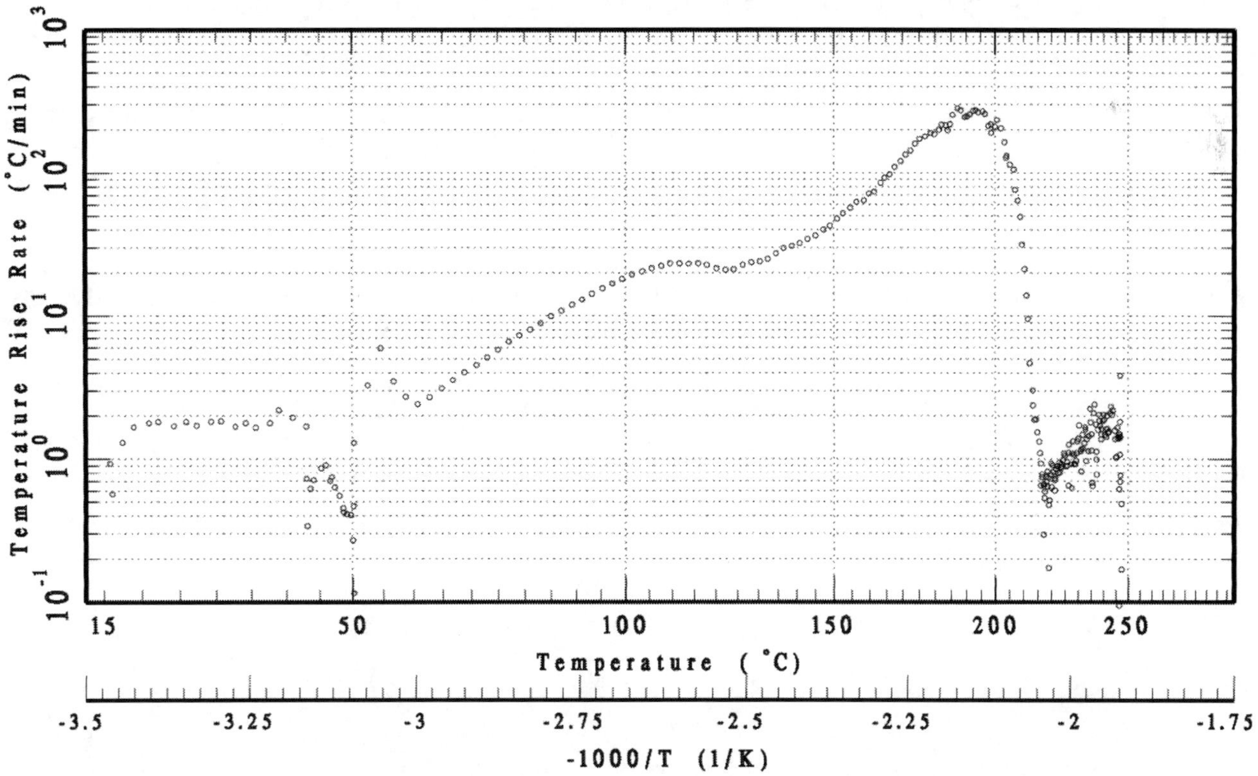

Figure 2 Temperature Rise Rate vs. Temperature for Phenol-Formaldehyde

In this case, the pressure within the test cell closely matches the vapor pressure of water at conditions where relief is expected to occur. Since there is an abundance of water in the batch, this reactive system is expected to be a vapor system during venting. Based on Figure 1, the desired relief set pressure of 10 psig corresponds to a temperature of 115°C. The corresponding temperature rise rate of 23.1°C/min is shown in Figure 2. The vent sizing calculation for this particular upset scenario is shown below. A factor of 2 has been used to account for the expected foamy behavior of the phenol-formaldehyde system. In addition, a discharge coefficient of 0.5 has been chosen to represent reasonably designed outlet piping from the rupture disk.

$$A = \frac{2}{0.61 \cdot C_D}\left[\frac{m \cdot c_p \cdot \dot{T}}{\lambda \cdot P}\left(\frac{R \cdot T}{MW_v}\right)^{1/2}\right]$$

$$A = \frac{2}{0.61 \cdot 0.5}\,\frac{(3500\ \text{kg})(3380\ \text{J/kg}\cdot\text{K})(0.385\ \text{K/s})}{(2.2\times10^6\ \text{J/kg})(170300\ \text{Pa})}\left[\frac{(8314.47\ \text{J/kmol}\cdot\text{K})(388\ \text{K})}{(18.015\ \text{kg/kmol})}\right]^{1/2}$$

$$A = 3.37\times10^{-2}\ \text{m}^2 \text{ or } 52.3\ \text{in}^2$$

Gassy System Example

210 kg of 40% dicumyl peroxide in 2,2,4-trimethyl-1,3-pentanediol diisobutyrate is stored in a 340 L tank. The maximum allowable working pressure of the tank is 80 psig.

Results from a PHA indicate a fire in the surrounding area would elevate the temperature of the tank to cause decomposition of the dicumyl peroxide. The external fire exposure rate was determined to be 0.5°C/min. It is desired to size a rupture disk with a set pressure of 50 psig based on this upset scenario.

Because non-condensable gas is expected to be generated due to decomposition of the dicumyl peroxide, the ARSST is an excellent choice for obtaining runaway reaction data. An open-cell ARSST test was run to simulate this upset scenario. The dicumyl peroxide and diisobutyrate were loaded to the test cell (in the appropriate ratios) and heated at an external rate of 0.5°C/min. In accordance with open cell methodology, the test was performed with an initial nitrogen backpressure equal to the storage tank MAAP of 88 psig. This ensures sufficient reaction rate data are obtained for vent sizing purposes. The data from this test that are pertinent to the vent sizing evaluation are provided in Figures 3 and 4.

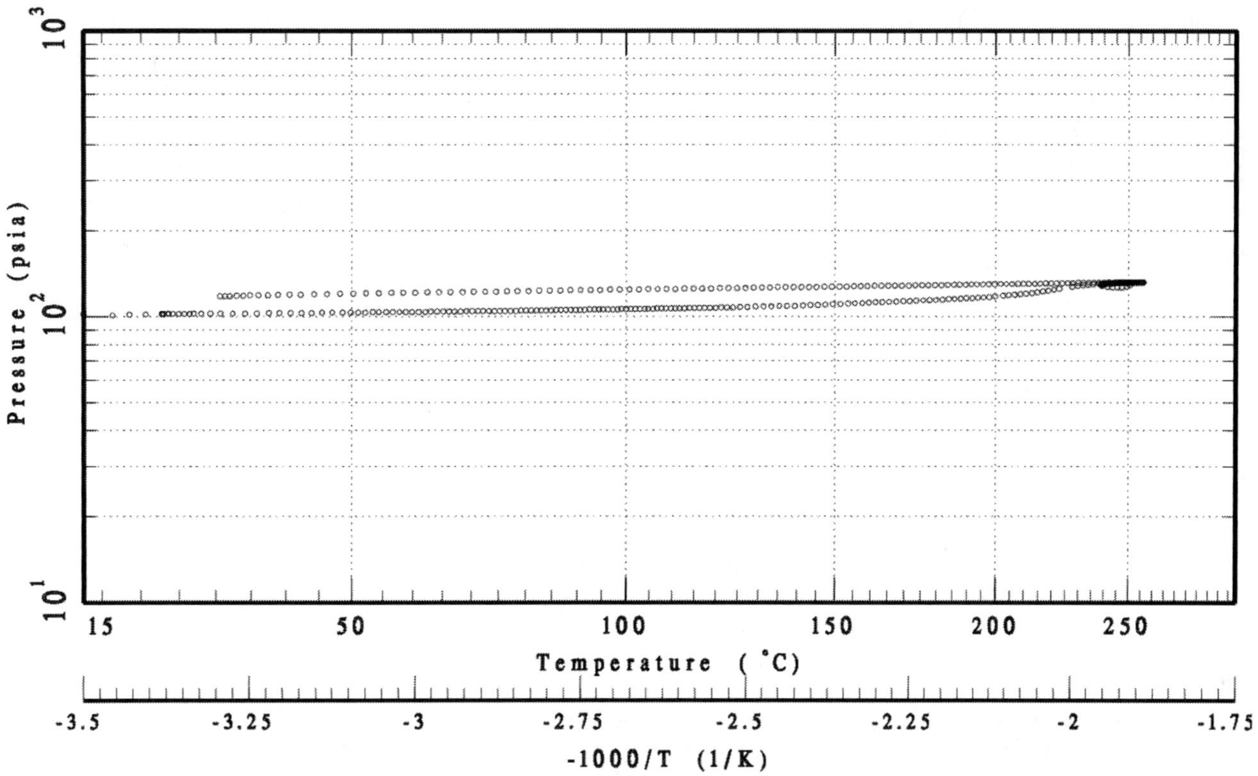

Figure 3 Pressure vs. Temperature for 40% Dicumyl Peroxide in Diisobutyrate

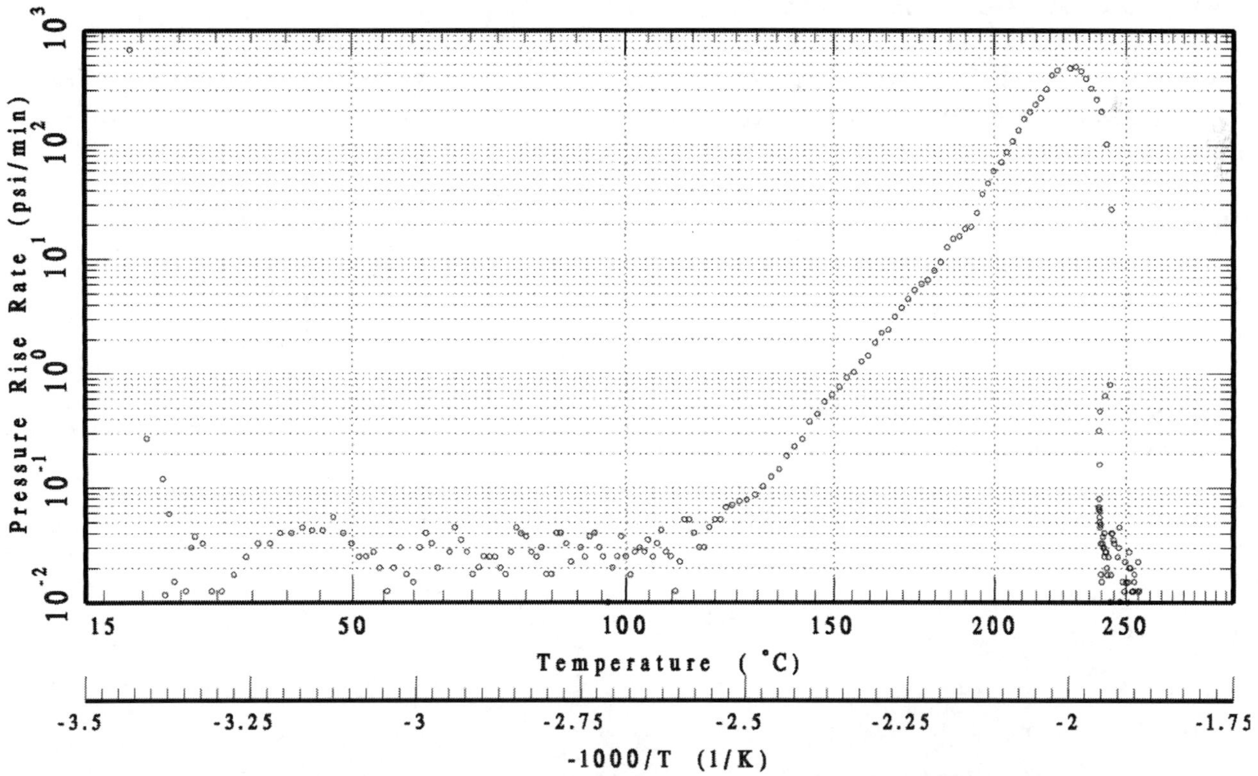

Figure 4 Pressure Rise Rate vs. Temperature for 40% Dicumyl Peroxide in Diisobutyrate

In this case, a significant amount of non-condensable gas was generated as evidenced by the cool down pressure of 103 psig. The peak reaction temperature from the runaway reaction was 250°C. Since the diisobutyrate does not boil until a temperature of 280°C, significant vapor generation will not occur. This reactive system is expected to be a gassy system during venting. Based on Figure 4, the maximum pressure rise rate is 477 psi/min and occurs at a temperature of 230°C. The vent sizing calculation for this particular upset scenario is shown below and based on all gas venting. A discharge coefficient of 0.5 has been chosen to represent reasonably designed outlet piping from the rupture disk.

$$A = \frac{1}{0.61 \cdot C_D}\left[\frac{m \cdot v \cdot \dot{P}}{m_t \cdot P}\left(\frac{MW_g}{R \cdot T}\right)^{1/2}\right]$$

$$A = \frac{1}{0.61 \cdot 0.5}\frac{(210\ \text{kg})(3.5 \times 10^{-4}\ \text{m}^3)(54800\ \text{Pa/s})}{(8 \times 10^{-3}\ \text{kg})(708090\ \text{Pa})}\left[\frac{(44.01\ \text{kg/kmol})}{(8314.47\ \text{J/kmol} \cdot \text{K})(503\ \text{K})}\right]^{1/2}$$

$$A = 7.56 \times 10^{-3}\ \text{m}^2 \text{ or } 11.7\ \text{in}^2$$

Hybrid System Example

2000 kg of 25% hydrogen peroxide is stored in a 700 gallon tank. The maximum allowable working pressure of the tank is 100 psig.

Results from a PHA indicate iron contamination could cause a runaway reaction due to accelerated decomposition rates of the hydrogen peroxide. It is desired to size a rupture disk with a set pressure of 20 psig based on this upset scenario.

Because non-condensable gas is expected to be generated due to decomposition of the hydrogen peroxide, open cell VSP2 testing is an excellent choice for obtaining runaway reaction data. The hydrogen peroxide and iron contaminant were loaded to the test cell (in the appropriate ratios) and subjected to heat-wait-search methodology. In accordance with open cell methodology, the test was performed with an initial nitrogen backpressure equal to the storage tank MAAP of 110 psig. This ensures sufficient reaction rate data are obtained for vent sizing purposes. The data from this test that are pertinent to the vent sizing evaluation are provided in Figures 5 through 7.

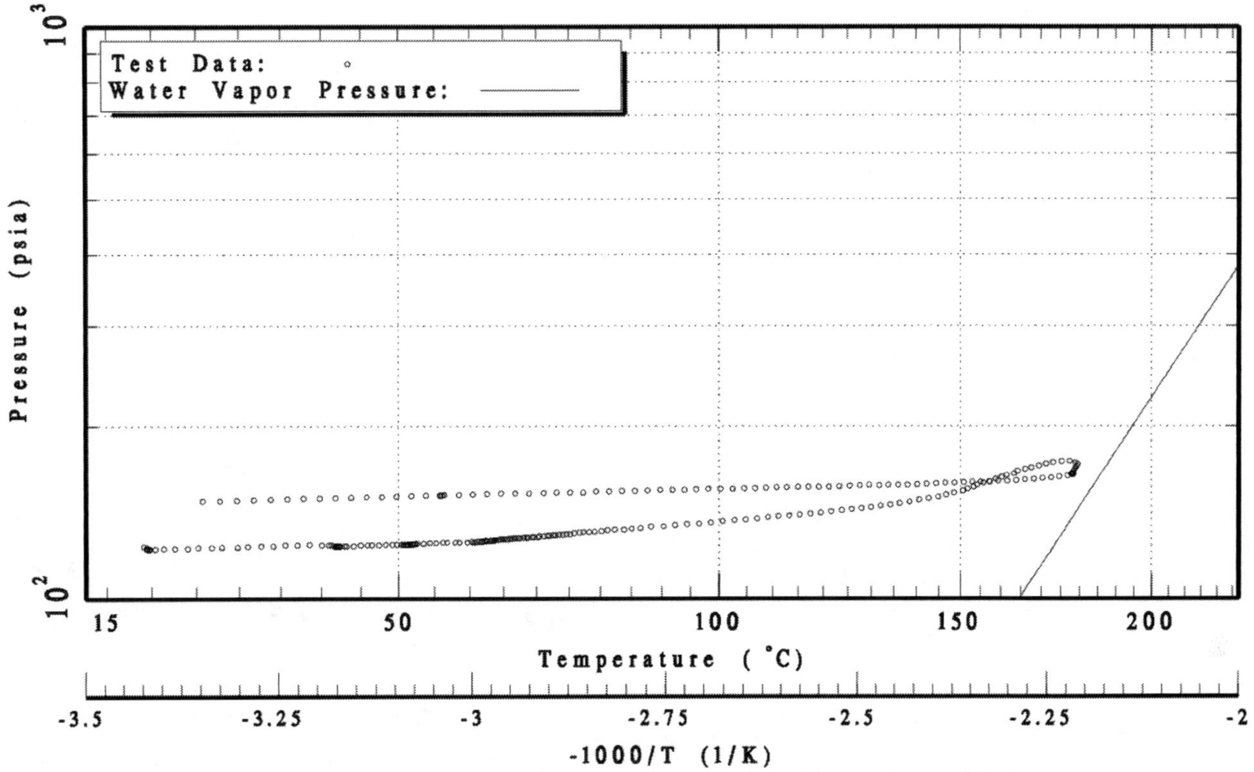

Figure 5 Pressure vs. Temperature for 25% Hydrogen Peroxide System

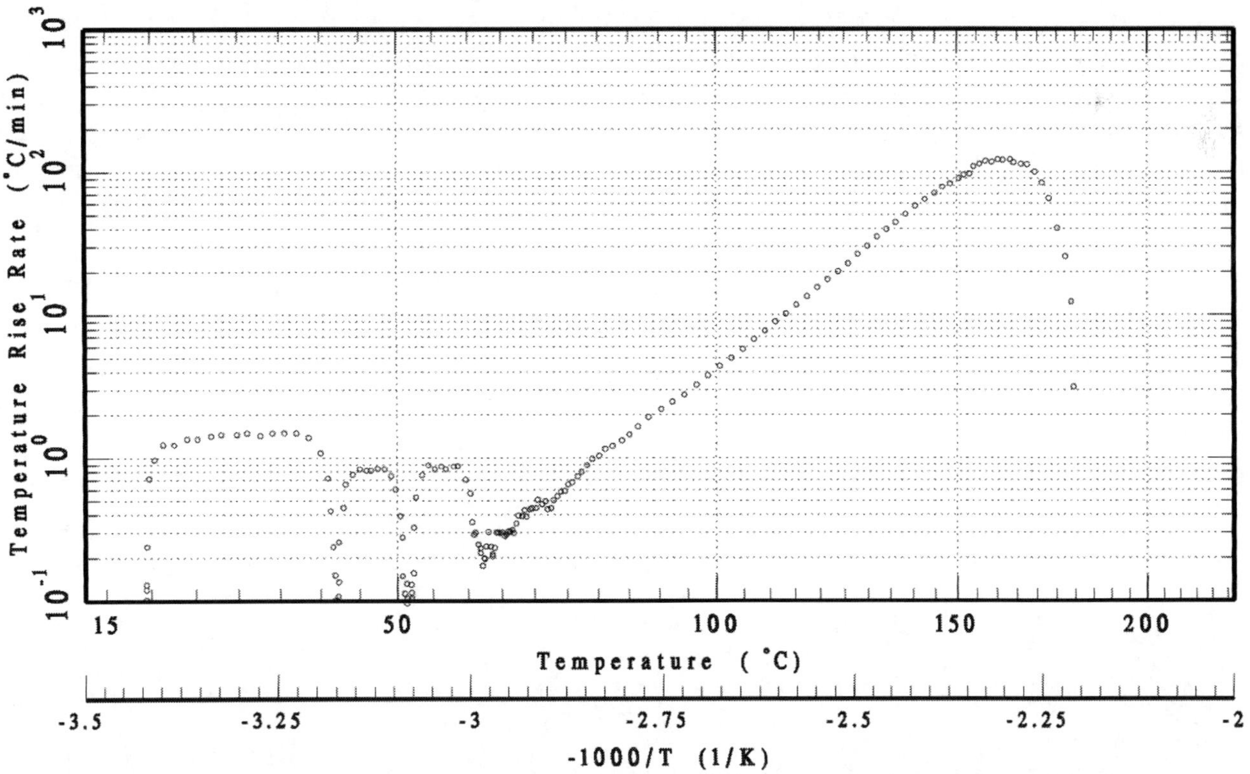

Figure 6 Temperature Rise Rate vs. Temperature for 25% Hydrogen Peroxide System

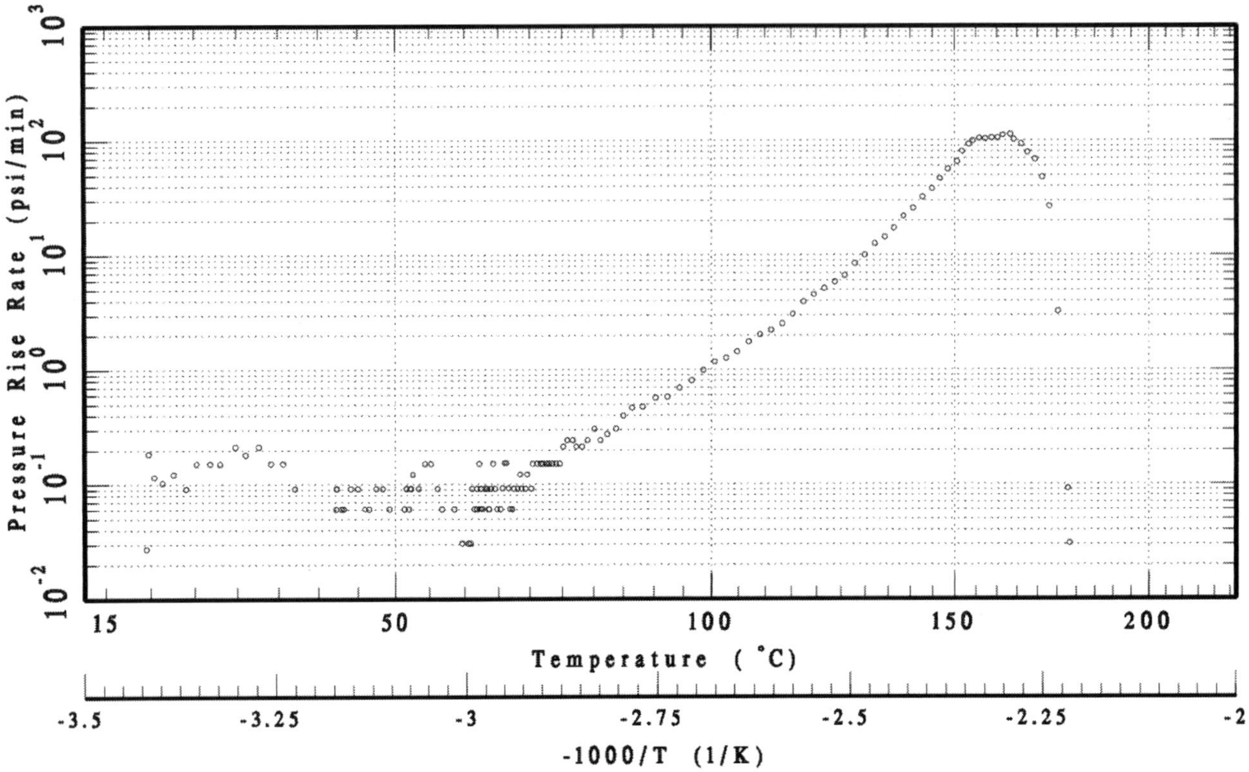

Figure 7 Pressure Rise Rate vs. Temperature for 25% Hydrogen Peroxide System

In this case, a significant amount of non-condensable gas was generated as evidenced by the cool down pressure of 133 psig. The peak temperature during this test was 179°C. This implies that the water will boil during a venting scenario and tempering of the reaction is possible. In order to verify this assumption and determine at what temperature tempering may occur, an open-cell test was run at 20 psig. The data from this test that are pertinent to the vent sizing evaluation are provided in Figure 8.

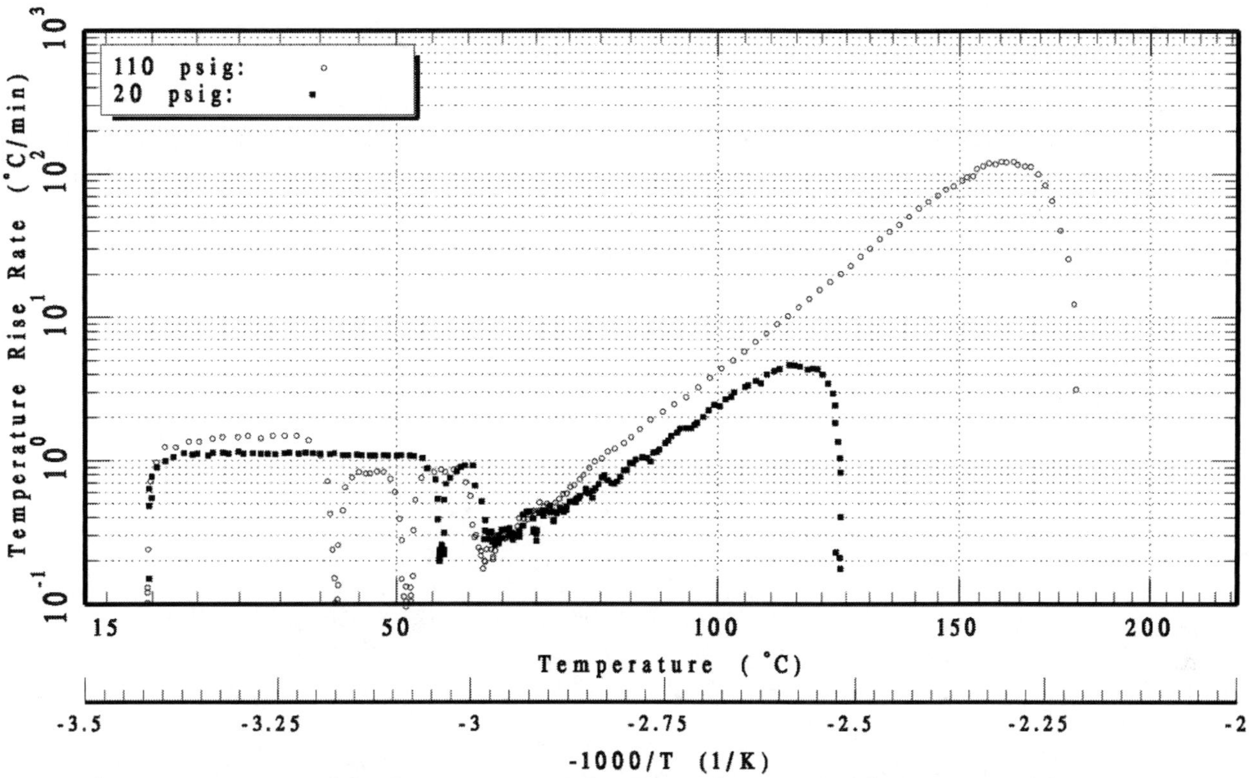

Figure 8 Comparison Temperature Rise Rates for 25% Hydrogen Peroxide System

As evidenced in Figure 8, the low backpressure test exhibited tempering at a temperature of 124°C. This reactive system is thus expected to be a hybrid system during venting. Based on Figures 6 and 7, the reaction rates corresponding to the tempering point are 21°C/min and 6 psi/min. The vent sizing calculation is shown below and based on all gas and vapor venting. A discharge coefficient of 0.5 has been chosen to represent reasonably designed outlet piping from the rupture disk.

$$A = \frac{1}{0.61 \cdot C_D}\left[\frac{m \cdot c_p \cdot \dot{T}}{\lambda \cdot P}\left(\frac{R \cdot T}{MW_v}\right)^{1/2} + \frac{m \cdot v \cdot \dot{P}}{m_t \cdot P}\left(\frac{MW_g}{R \cdot T}\right)^{1/2}\right]$$

$$A = \frac{1}{0.61 \cdot 0.5}\frac{(2000\ \text{kg})(3900\ \text{J/kg} \cdot \text{K})(0.35\ \text{K/s})}{(2.2 \times 10^6\ \text{J/kg})(239248\ \text{Pa})}\left[\frac{(8314.47\ \text{J/kmol} \cdot \text{K})(397\ \text{K})}{(18.015\ \text{kg/kmol})}\right]^{1/2}$$

$$+\frac{1}{0.61 \cdot 0.5}\frac{(2000\ \text{kg})(3.8 \times 10^{-3}\ \text{m}^3)(689\ \text{Pa/s})}{(8.0 \times 10^{-2}\ \text{kg})(239248\ \text{Pa})}\left[\frac{(32.0\ \text{kg/kmol})}{(8314.47\ \text{J/kmol} \cdot \text{K})(397\ \text{K})}\right]^{1/2}$$

$$A = 1.01 \times 10^{-2}\ \text{m}^2 \text{ or } 15.6\ \text{in}^2$$

Summary

A simple comprehensive experimental approach has been provided to size emergency relief devices in reactive service. Once credible reactive upset scenarios have been determined, low phi factor calorimeters can be used to directly simulate the upset conditions and obtain data regarding system classification, reaction rates, and flow regime. Using the simplified vent sizing equation provided by Fauske, required vent sizes can then be determined from these data. The applied methodology is consistent with available experimental data for all three types of reactive systems.

References

Burelbach, J.P., "Advanced Reactive System Screening Tool (ARSST)," Presented at the 28th Annual North American Thermal Analysis Society (NATAS) Conference, Orlando, Florida, October 4-6, 2000.

Askonas, C.F., Burelbach, J.P., and Leung, J.C., "The Versatile VSP2: A Tool for Adiabatic Thermal Analysis and Vent Sizing Applications," Presented at the 28th Annual North American Thermal Analysis Society (NATAS) Conference, Orlando, Florida, October 4-6, 2000.

Fauske, H.K., "Properly Size Vents for Nonreactive and Reactive Chemicals," Chemical Engineering Progress, pp. 17- 29, February, 2000.

Fauske, H.K., Leung, J.C., Fisher, H.G., "Thermal Runaway Reactions in a Low Thermal Inertia Apparatus," Thermochimica Acta, pp. 13-29, 1986.

Fisher, H.G., Forrest, H.S., Grossel, S.S., Huff, J.E., Muller, A.R., Noronha, J.A., Shaw, D.A., Tilley, B.J., Emergency Relief System Design Using DIERS Technology: The Design Institute for Emergency Relief Systems (DIERS) Project Manual, American Institute of Chemical Engineers, New York, NY, 1992.

Choked and New-Choked Real Gas and Two-Phase Flow Analysis of Discharge Piping

Leonid Korelshteyn
Piping Systems Research & Engineering Co (NTP Truboprovod)
Plehanova str., 7, Moscow, 111141, Russia
Korelstein@truboprovod.ru

Prepared for Presentation at
American Institute of Chemical Engineers
2015 Spring Meeting
11th Global Congress on Process Safety
Austin, Texas
April 27-30, 2015

UNPUBLISHED

Choked and New-Choked Real Gas and Two-Phase Flow Analysis of Discharge Piping

Leonid Korelshteyn
Piping Systems Research & Engineering Co (NTP Truboprovod)
Plehanova str., 7, Moscow, 111141, Russia
Korelstein@truboprovod.ru

Keywords: Choked Flow; Real Gas; Two-Phase Gas-Liquid Flow; Discharge Piping; Equations

Abstract

The methods of analysis of real gas and two phase gas-liquid flow in discharge piping of pressure relieve systems with possible multiple choking are discussed. General differential equations for real gas and two phase flow in the pipe in the form of pipe length, temperature, enthalpy and entropy derivatives via pressure for general thermodynamic conditions (including heat exchange with environment) are formulated. Right parts of these equations are regular functions of thermodynamic properties of the fluid and Mach number, so the equations can be effectively solved numerically using standard method.

The proposed approach was implemented in the software for unbranched discharge piping flow using different thermodynamic libraries. Some problems of implementation for mixtures with "narrow" phase envelope (almost azeotropic) are discussed. It found out that using pressure and enthalpy or pressure and entropy as basic thermodynamic parameters in this case is preferable over pressure and temperature to escape loss of accuracy. This also relates to Mach number calculation for two phase flow. For special cases of vapor-liquid flow of one-component fluid, and for non-flashing gas-liquid flow appropriate equations for Mach number are proposed, which take into account real properties of vapor/gas and compressibility of liquid. These equations are generalization of known equations for ideal gas and incompressible liquid mixtures and can be also used to improve accuracy of omega method.

The methods of calculating minor losses for high Mach numbers for elements typical in discharge piping are reviewed.

1. Equations and algorithms for real gas and two-phase flow with high Mach numbers (Introduction)

Preventing choked flow in process piping is usually the good engineering practice. However, there are a number of important cases, when the occurrence of a choked or almost chocked flow is possible and sometimes inevitable. These include, in particular, flow of gases and two-phase gas and liquid mixtures in the pressure relieve systems discharge piping.

This article is focused on equations, algorithms and software implementation for analysis of such flows.

Note: Do not add page numbers. Do not refer to page numbers when referencing different portions of the paper

Calculation of pipelines requires the implementation of appropriate methods and algorithms at all hierarchy levels of analytical models— from analysis of pipeline individual elements (pipes and local resistances) and unbranched pipeline sections to analysis of distribution of flows in a branched pipeline as a whole. Further we will discuss some equations, methods and algorithms used for analysis of this type of flows in "Hydrosystem" and "Safety valve" software, and some lessons based on the experience of software implementation.

2. Special form of flow equations for near-choked flow in pipes

One-dimensional flows of compressible fluids at high Mach numbers, have been well studied for an ideal gas, when for various special cases analytical solutions for the flow equations are possible (see, for example: [1-5]). However, for the real (non-ideal) gas, and even more so for the two-phase gas-liquid flow of non-ideal gas and liquid, only a numerical solution is usually possible. At that, in the case of almost chocked flow, a standard approach based on solving a system of ordinary differential equations along the pipe length by standard numerical methods, may cause some difficulties. This is due to the fact that, on approaching a critical (choked) flow, rate of pressure and other parameters changing of the fluid along the pipe increases rapidly and tends to infinity at the choked flow. At the numerical integration of the equations, this entails the need of smaller and smaller steps in length, with permanent recalculation of the fluid physical properties at every step.

However, if the pipe, heat insulation and environmental parameters can be considered constant in the given area, then all flow parameters depend only on the fluid parameters (pressure and temperature, or specific enthalpy) in particular cross-section of the pipe. This enables rewriting the flow equations as ordinary differential equations for the pressure. In this case, the singularity disappears, and we can apply standard numerical methods. Although this method is quite common, it has been previously applied only to various special cases of gas and two-phase medium flow equations, and the author failed to find in the literature some corresponding equations for the general case of the flow. Similar approach to the problem was offered by Fesenko in the monograph [5], where in paragraph 4.3 he proposed equations for special case of a single-component two-phase mixture, which correspond on the form and structure to the general case equations delivered in this article. The role of isentropic coefficients for real gas flow equations for isentropic flow was shown by Istomin [24, 25]. Similar results for real gas were recently published by Baltadjiev et al [26, 27], but they didn't take into account influence of friction losses and heat exchange. This section of the article fills this gap.

2.1 Real gas

Classical differential equations of real gas flow in a straight pipe can be written in the form (see, for instance [1-4]):

$$\frac{Aw}{u} = AG = W = \text{const}; \qquad \text{[eq. 1]}$$

$$udp + d(\alpha\frac{w^2}{2}) + g\sin\theta dl + \frac{\lambda}{2D}w^2dl = 0; \qquad \text{[eq. 2]}$$

$$dh + d(\alpha\frac{w^2}{2}) + g\sin\theta dl + \frac{q}{W}dl = 0,$$ [eq. 3]

where A = πD2/4 = const — cross-sectional area of the pipe, sq.m; D – inner pipe diameter, m; G = W/A = const — flux, kg/(m2 s); g — free fall acceleration, m/s^2; h — fluid specific enthalpy J/kg; l — distance from the beginning of the pipe, m; p — pressure, Pa; u — fluid specific volume, m^3/kg; W = const —mass fluid flow rate, kg/s; w — fluid flow average speed along the pipe cross-section, m/s; α —kinetic energy coefficient; λ —friction coefficient, determined by known correlations as a function of the Reynolds's number and the relative roughness of the pipe [2, 6, 7]; θ — pipe elevation angle, rad.; q - heat flow from the fluid into the environment per unit of length, J/(m s), which is defined by the relation q = (T – T_{env})/R_Σ, where T and T_{env} — temperature of the fluid and the environment, K; R_Σ — total heat transfer resistance between the fluid and the environment, m K/W, which includes the thermal resistance of heat transfer from the fluid to the pipe wall, the pipe wall resistance, thermal insulation and resistance to heat transfer from the surface of the heat insulating construction (or from the pipe wall of bare (non-insulated) pipes) into the environment. Corresponding calculation formulas are given, for example, in standards [8, 9].

Since we are interested in the almost chocked flow, which is almost always (except for very exotic cases) turbulent, we further assume α = 1.

Given that dh = Tds + udp, from equations 2 and 3 we obtain the equation for calculation of the specific entropy of the fluid.

$$ds = \frac{1}{T}\left(\frac{\lambda}{2D}w^2 - \frac{q}{W}\right)dl$$ [eq. 4]

According to equation 1,

$$d(\alpha w^2/2) = d\left(u^2G^2/2\right) = G^2udu$$ [eq. 5]

Expressing du via dp and ds, we get

$$du = \left(\partial u/\partial p\right)_s dp + \left(\partial u/\partial s\right)_p ds,$$ [eq. 6]

and

$$\left(\partial u/\partial p\right)_s = -\left(u/w_c\right)^2 = -\left(w/w_c \cdot u/w\right)^2 = -M^2/G^2$$ [eq. 7]

where the Mach number $M = w/w_c = G/G_c$; sound speed $w_c = \sqrt{nZRT}$; critical flux $G_c = w_c/u$; Z — compressibility factor; R — gas constant; $n = -\left(\partial \ln p/\partial \ln u\right)_s = -\left(u/p\right)\left(\partial p/\partial u\right)_s$ — isentropic coefficient for u and p.

There are also known thermodynamic equations

$$ds = c_p d\ln T - \left(\partial u/\partial T\right)_p dp$$ [eq. 8]

$$\left(\partial u/\partial s\right)_p = T\left(\partial u/\partial T\right)_p/c_p = uT\beta/c_p,$$ [eq. 9]

where c_p - gas specific heat at constant pressure, J/(kg K);

$\beta = \left(\partial u/\partial T\right)_p/u = T^{-1}\left[1 + \left(\partial \ln Z/\partial LnT\right)_p\right]$— volume expansion coefficient (at constant pressure), 1/K.

Substituting equations 4, 7 and 9 into equation 6, we get

$$du = -\frac{M^2}{G^2}dp + \frac{\beta u}{c_p}\left(\frac{\lambda}{2D}w^2 - \frac{q}{W}\right)dl \qquad \text{[eq. 10]}$$

And combining equation 10 with equations 2 and 5, we get

$$u\left[1-M^2\right]dp + g\sin\theta dl + \left\{\left[1+M^2 n\beta T\frac{ZR}{c_p}\right]\frac{\lambda}{2D}u^2G^2 - M^2\frac{\beta TnZRq}{c_pW}\right\}dl = 0 \qquad \text{[eq. 11]}$$

Or in dimensionless form

$$\frac{d(l/D)}{d\ln p} = -\frac{1-M^2}{\left[1+\left(1-\Lambda_q\right)n\Lambda M^2 + \Lambda_g\sin\theta\right]\frac{\lambda n}{2}M^2}, \qquad \text{[eq. 12]}$$

Here the non-dimensional parameters are:

$$\Lambda = ZR\hat{\beta}/c_p = \gamma^{-1}(\gamma-1)\hat{\beta} = (\partial pu/\partial h)_p = p(\partial u/\partial T)_p/c_p \qquad \text{[eq. 13]}$$

(part of mechanical work in specific enthalpy, when heating at the constant pressure);

$$\hat{\beta} = \beta T = T(\partial u/\partial T)_p/u = 1 + (\partial\ln Z/\partial\ln T)_p \qquad \text{[eq. 14]}$$

(non-dimensional linear expansion coefficient at the constant pressure);

$$\gamma = c_p/\left(c_p - ZR\right),\ \Lambda_g = 2Dg/\left(\lambda w^2\right),\ \Lambda_q = 2Dq/\left(\lambda w^2 W\right). \qquad \text{[eq. 15]}$$

The physical meaning of the expression in square brackets in the denominator of the right-hand side of equation 12 is the following: the first (primary) term corresponds to the friction losses, the second term (in order 1–0.1) takes into consideration the fluid temperature change because of heating, due to internal friction and heat exchange with the environment, and the third term (in order 10^{-2}) takes into consideration influence of the height difference. At that, the dimensionless parameters Λ_g and Λ_q represent relationships of specific values of the potential energy of gravity (Λ_g) and heat (Λ_q), transmitted to the environment, to the energy converted into heat due to hydraulic friction.

Another meaning of Λ coefficient follows from known thermodynamic expression $Tds = c_p dT - T(\partial u/\partial T)_p dp$, which in combination with equation 13 gives

$$\Lambda = (\partial\ln T/\partial\ln p)_s \qquad \text{[eq. 16]}$$

Equation 16 shows that Λ is isentropic coefficient for T and p.

Note that the dimensionless complex $n\Lambda$ based on the known thermodynamic relations

$$n = -k(u/p)(\partial p/\partial u)_T, \qquad \text{[eq. 17]}$$

$$(\partial u/\partial T)_p(\partial T/\partial p)_u(\partial p/\partial u)_T = -1, \qquad \text{[eq. 18]}$$

$$c_p - c_V = T(\partial p/\partial T)_u(\partial u/\partial T)_p = -T\left[(\partial u/\partial T)_p\right]^2(\partial p/\partial u)_T \qquad \text{[eq. 19]}$$

can also be written as

$$n\Lambda = -k(u/p)(\partial p/\partial u)_T\, p(\partial u/\partial T)_p/c_p =$$
$$= -k(u/p)\left[-(c_p - c_V)T^{-1}(\partial u/\partial T)_p^{-2}\right]p(\partial u/\partial T)_p/c_p = \qquad \text{[eq. 20]}$$
$$= k\frac{c_p - c_V}{c_p T(\partial u/\partial T)_p}u = (k-1)/\hat{\beta},$$

where $k = c_p/c_V$, c_V - specific heat at constant volume, J/(kg K). This is exactly the same expression as was obtained in [25].

From equations 16 and definition of n it is easy to see that $n\Lambda$ is also an isentropic coefficient for T and u.

Let us now establish the equations for other thermodynamic parameters of fluid – entropy, enthalpy and temperature.

From equations 4 and 12 after simplifications and taking into account $dh = Tds + udp$

$$\frac{Tds}{pud\ln p} = -\frac{(1-\Lambda_q)(1-M^2)}{\left[1+(1-\Lambda_q)n\Lambda M^2 + \Lambda_g \sin\theta\right]},\quad \frac{dh}{pud\ln p} = 1 - \frac{(1-\Lambda_q)(1-M^2)}{\left[1+(1-\Lambda_q)n\Lambda M^2 + \Lambda_g \sin\theta\right]}$$

[eq. 21]

From equation 8, 14 and 21:

$$\frac{d\ln T}{d\ln p} = \Lambda - \frac{\Lambda(1-\Lambda_q)(1-M^2)}{\hat{\beta}\left[1+(1-\Lambda_q)n\Lambda M^2 + \Lambda_g \sin\theta\right]} \qquad \text{[eq. 22]}$$

Equation 22 has a simple physical meaning: dominant term of the right-hand side describes the isentropic process of real gas, and the additional term — influence of the non-isentropic factors: heat exchange with the environment and irreversible losses due to friction. As we can see, the closer the Mach number M to 1, the closer the flow to isentropic one.

In the general case of Fanno flow (flow without heat exchange with the environment), equations 12, 21, 22 take the form:

$$\frac{d(l/D)}{d\ln p} = -\frac{1-M^2}{\frac{\lambda}{2}nM^2\left[1+n\Lambda M^2 + \Lambda_g \sin\theta\right]} \qquad \text{[eq. 23]}$$

$$\frac{Tds}{pud\ln p} = -\frac{(1-M^2)}{\left[1+n\Lambda M^2 + \Lambda_g \sin\theta\right]},\quad \frac{dh}{pud\ln p} = 1 - \frac{(1-M^2)}{\left[1+n\Lambda M^2 + \Lambda_g \sin\theta\right]} \qquad \text{[eq. 24]}$$

$$\frac{d\ln T}{d\ln p} = \Lambda - \frac{\Lambda}{\hat{\beta}}\frac{1-M^2}{\left[1+n\Lambda M^2 + \Lambda_g \sin\theta\right]} \qquad \text{[eq. 25]}$$

Note that for an ideal gas and a horizontal pipe, when n = γ = k, $\hat{\beta} = 1$ and Λ=(k–1)/k, equation 25 takes a well-known classic form [1]

$$d\ln T/d\ln p = (k-1)M^2/\left[1+(k-1)M^2\right], \qquad \text{[eq. 26]}$$

which shows how with increase of Mach number M the temperature mode transfers from practically isothermal to adiabatic one.

To get equations of Reyleigh flow (frictionless flow with external cooling or heating) let's consider what happens with equations 12, 21, 22, when $\lambda \to 0$. In this case $\Lambda_q \to \infty$, $\Lambda_g \to \infty$, $\Lambda_g/\Lambda_q \to gW/q$, $\Lambda_q\lambda \to 2Dq/(w^2W)$ and the equations take the form

$$\frac{d(l/D)}{d\ln p} = \frac{w^2W}{Dq}\frac{1-M^2}{\left[n\Lambda M^2 - gWq^{-1}\sin\theta\right]nM^2}, \quad \text{[eq. 27]}$$

$$\frac{Tds}{pud\ln p} = -\frac{(1-M^2)}{\left[n\Lambda M^2 - gWq^{-1}\sin\theta\right]}, \quad \frac{dh}{pud\ln p} = 1-\frac{(1-M^2)}{\left[n\Lambda M^2 - gWq^{-1}\sin\theta\right]} \quad \text{[eq. 28]}$$

$$\frac{d\ln T}{d\ln p} = \Lambda - \frac{\Lambda(1-M^2)}{\hat{\beta}\left[n\Lambda M^2 - gWq^{-1}\sin\theta\right]} \quad \text{[eq. 29]}$$

For horizontal pipes equation 29 gives (taking into account that $\Lambda = \gamma^{-1}(\gamma-1)\hat{\beta}$, $n\Lambda = (k-1)/\hat{\beta}$)

$$\frac{d\ln T}{d\ln p} = \frac{kM^2-1}{\hat{\beta}nM^2} \quad \text{[eq. 30]}$$

which for ideal gas (n = k, $\hat{\beta}=1$) takes a known form [1]: $d\ln T/d\ln p = 1-(kM^2)^{-1}$.

Equations 12 and 22 (or 21) can be solved using standard numerical methods. In particular, for solution of the equations 20 or 21, a standard Runge-Kutta method with a constant step on the logarithm of pressure can be used, upon which quadrature formulas (e.g., Simpson formula) can be used to integrate the equations 12 on the same grid.

2.2 Two-phase gas-liquid flow

Let us show that the same approach is applicable to almost chocked two-phase flow as well, for which we can establish similar equations.

To describe a two-phase medium flow in the pipe, we'll use the following mixed model equations: [10–12]:

$$AG = W = \text{const}; \quad \text{[eq. 31]}$$

$$u_m dp + G^2u_m du_m + u_m g\sin\theta\rho_{mv}dl + \frac{\lambda}{2D}G^2u_m^2 dl = 0; \quad \text{[eq. 32]}$$

$$dh_m + G^2u_m du_m + g\sin\theta dl + \frac{q}{W}dl = 0, \quad \text{[eq. 33]}$$

Here and elsewhere, the following system of indexes is used: g and l — values for gas and liquid phases, correspondingly; gl — difference between the values for gas and liquid phases; m — average the two-phase mixture: $f_m = xf_g + (1-x)f_l$, where x — mass void fraction; mv — true mean by volume (e.g. density $\rho_{mv} = \varepsilon\rho_g + (1-\varepsilon)\rho_l$, where $\varepsilon = \left[1+\left(\frac{1-x}{x}\right)\frac{u_l}{u_g}S\right]^{-1}$ — actual volumetric gas content; S — the ratio between velocities of gas and liquid phases).

Equations 31-33 describe a mixed model of equilibrium separated and homogeneous flow, at that in equations 32, 33 an acceleration term is calculated by the homogeneous model, and while in the other terms possible phase slip can be taken into account. According to [12], this approach better describes almost adiabatic flows and flows with the product boiling up, and is commonly used. Application of the this model to the flows with condensation and intensive heat exchange with the environment, as well as to the critical flows over short distances at a very low gas content (when there is a thermodynamic non-equilibrium and delayed boiling) usually produce a conservative estimation of system capacity. HEM model is the special case of this model when S=1.

Calculation in accordance with this model was successfully implemented in the "Hydrosystem" program (for more details, see [13]). There is a variety of methods for two-phase hydraulic friction and holdup calculation with different ranges of applicability (for more details on selecting of methods and their implementation in the "Hydrosystem" program see [14]). Review of appropriate methods of calculation is also given in [15, 16].

A critical flux for this model can be defined from the expression

$$G_c^2\left(\partial u_m/\partial p\right)_{s_m}=-1,\; G_c=\left[-\left(\partial u_m/\partial p\right)_{s_m}\right]^{-1/2} \qquad \text{[eq. 34]}$$

as well as the critical speed and Mach number for two-phase flow:

$$M=G/G_c=G\left[-\left(\partial u_m/\partial p\right)_{s_m}\right]^{1/2};\; w_c=u_m G_c \qquad \text{[eq. 35]}$$

Then, for isentropic coefficient at two-phase flow, we have

$$n=-\left(\partial \ln p/\partial \ln u_m\right)_{s_m}=-\left(u_m/p\right)\left(\partial p/\partial u_m\right)_{s_m}=u_m G^2/\left(pM^2\right),\; w_c=\sqrt{npu_m}, \qquad \text{[eq. 36]}$$

and taking into account that

$$dh_m=Tds_m+u_m dp,\; T^{-1}\left(\partial u_m/\partial s_m\right)_p=\left(\partial u_m/\partial h_m\right)_p=\Lambda/p,\; \Lambda=\left(\partial p u_m/\partial h_m\right)_p, \qquad \text{[eq. 37]}$$

from equations 31-33 we obtain

$$\left[1-M^2\right]dp=$$
$$=-\left[1+M^2 n\Lambda\right]\lambda n M^2\frac{p}{2D}dl-\left[1+M^2 n\Lambda(1-\eta^{-1})\right]g\rho_{mv}\sin\theta dl+M^2\frac{n}{u_m}\Lambda\frac{q}{W}dl \qquad \text{[eq. 38]}$$

$$Tds_m=\lambda G^2\frac{u_m^2}{2D}dl-\frac{q}{W}dl-(1-\eta)g\sin\theta dl \qquad \text{[eq. 39]}$$

where $\eta=u_m\rho_{mv}=\left(x/\rho_g+(1-x)/\rho_l\right)\left(\varepsilon\rho_g+(1-\varepsilon)\rho_l\right)$ — dimensionless coefficient depending on the phase slip. In particular

$$1-\eta=-\frac{x(1-x)(S-1)}{1+(1-x)(S-1)}u_{gl}\rho_{mv},\; 1-\eta^{-1}=\frac{x(1-x)(S-1)}{1+(1-x)(S-1)}\frac{u_{gl}}{u_m} \qquad \text{[eq. 40]}$$

For HEM model $\eta=1$

As a result, in dimensionless form the equation 38 takes form

$$\frac{d(l/D)}{d\ln p}=-\frac{1-M^2}{\left[1+\left(1-\Lambda_q\right)n\Lambda M^2+\left[1+M^2 n\Lambda(1-\eta^{-1})\right]\eta\Lambda_g\sin\theta\right]\lambda\frac{n}{2}M^2} \qquad \text{[eq. 41]}$$

where the dimensionless parameters $\Lambda_g = 2Dg/\left(\lambda u_m^2 G^2\right)$, $\Lambda_q = 2Dq/\left(\lambda u_m^2 G^2 W\right)$ have the same meaning as in a single-phase flow case. From equation 39 and previous expressions it is also easy to get

$$\frac{T}{pu_m}\frac{d(s_m)}{d\ln p} = -\frac{\left(1-\Lambda_q-(1-\eta)\Lambda_g\right)\left(1-M^2\right)}{1+\left(1-\Lambda_q\right)n\Lambda M^2+\left[1+M^2 n\Lambda(1-\eta^{-1})\right]\eta\Lambda_g\sin\theta}. \quad \text{[eq. 42]}$$

$$\frac{1}{pu_m}\frac{d(h_m)}{d\ln p} = 1-\frac{\left(1-\Lambda_q-(1-\eta)\Lambda_g\right)\left(1-M^2\right)}{1+\left(1-\Lambda_q\right)n\Lambda M^2+\left[1+M^2 n\Lambda(1-\eta^{-1})\right]\eta\Lambda_g\sin\theta}. \quad \text{[eq. 43]}$$

Equation 42 again shows that when the flow approaches critical, the thermodynamic mode tends to isentropic.

Now let's get equations for temperature (for non-azeotropic multicomponent mixture) and for void fraction (for one-component two phase flow on saturation line).

For non-azeotropic multicomponent mixture

$$dh_m = c_p dT + \left[\left(u_m - T\left(\frac{\partial u_m}{\partial T}\right)_p\right)\right]dp = c_p T d\ln T + pu_m\left(1-\hat{\beta}\right)d\ln p \quad \text{[eq. 44]}$$

and from the equation 43, we obtain

$$\frac{1}{\Lambda}\frac{d\ln T}{d\ln p} = 1-\frac{\left(1-\Lambda_q-(1-\eta)\Lambda_g\right)\left(1-M^2\right)}{\hat{\beta}\left[1+\left(1-\Lambda_q\right)n\Lambda M^2+\left[1+M^2 n\Lambda(1-\eta^{-1})\right]\eta\Lambda_g\sin\theta\right]}, \quad \text{[eq. 45]}$$

Here $c_p = \left(\partial h_m/\partial T\right)_p$ and $\hat{\beta} = \beta T = \left(\partial \ln u_m/\partial \ln T\right)_p$ are calculated with consideration for the phase composition changes at the temperature variation,

$$\Lambda = \left(\partial pu_m/\partial h_m\right)_p = pu_m T^{-1}\left(\partial \ln u_m/\partial \ln T\right)_p/\left(\partial h_m/\partial T\right)_p = pu_m\hat{\beta}/\left(c_p T\right). \quad \text{[eq. 46]}$$

In the same way as for gas case, it is easy to show also that Λ is isentropic coefficient for T and p:

$$\Lambda = \left(\partial \ln T/\partial \ln p\right)_{s_m}. \quad \text{[eq. 47]}$$

For a single-component vapour-liquid mixture on the saturation line (applying $Tds = c_p dT - T\left(\partial u/\partial T\right)_p dp$ equation for each of phases)

$$dh_m = h_{gl}dx + c_{pm}\frac{dT_{sat}}{dp}dp + \left(u_m - T_{sat}\left(\frac{\partial u_m}{\partial T}\right)_{p,x}\right)dp =$$
$$= h_{gl}dx + \left[pu_m + c_{pm}T_{sat}\frac{pu_{gl}}{h_{gl}} - pT_{sat}\left(\frac{\partial u_m}{\partial T}\right)_{p,x}\right]d\ln p \quad \text{[eq. 48]}$$

And from the equation 43, we obtain

$$\frac{dx}{d\ln p} = \frac{u_m}{u_{gl}}\Lambda\left\{\hat{\beta}_m - \frac{c_{pm}T_{sat}}{pu_m}\Lambda - \frac{\left(1-\Lambda_q-(1-\eta)\Lambda_g\right)\left(1-M^2\right)}{1+\left(1-\Lambda_q\right)n\Lambda M^2+\left[1+M^2 n\Lambda(1-\eta^{-1})\right]\eta\Lambda_g\sin\theta}\right\} \quad \text{[eq. 49]}$$

At that, from to Clapeyron-Clausius equation

$$\Lambda = \left(\partial pu_m/\partial h_m\right)_p = pu_{gl}/h_{gl} = (p/T)T\left(u_{gl}/h_{gl}\right) = d\ln T_{sat}/d\ln p \quad \text{[eq. 50]}$$

and Λ for a single-component vapour-liquid mixture on the saturation line has a meaning of tangent of saturation line in logarithmical coordinates.

2.2.1 "Frozen" flow and generalization of Tangren formula.

For calculation of n and M, the thermodynamic relation can be used

$$(\partial u_m/\partial p)_{s_m} = (\partial u_m/\partial p)_T + (T/c_p)\left[(\partial u_m/\partial T)_p\right]^2 \quad \text{[eq. 51]}$$

Applying equation 51, in combination with equation 46 we get

$$n = -u_m p^{-1}(\partial p/\partial u_m)_{s_m} = -u_m p^{-1}\left[(\partial u_m/\partial p)_T + (T/c_p)\left[(\partial u_m/\partial T)_p\right]^2\right]^{-1} = \left[\hat{\kappa} - \Lambda\hat{\beta}\right]^{-1} \quad \text{[eq. 52]}$$

where $\hat{\kappa} = -(\partial \ln u_m/\partial \ln p)_T$ — non-dimensional coefficient of isothermal compressibility.

In particular, in case of flow without mass-transfer ("frozen" flow), mass void fraction and phase composition are constants, and

$$\hat{\beta} = (\partial \ln u_m/\partial \ln T)_p = \varepsilon_H\hat{\beta}_g + (1-\varepsilon_H)\hat{\beta}_l,\ \hat{\kappa} = -(\partial \ln u_m/\partial \ln p)_T = \varepsilon_H\hat{\kappa}_g + (1-\varepsilon_H)\hat{\kappa}_l, \quad \text{[eq. 53]}$$

$c_{pm} = xc_{pg} + (1-x)c_{pl}$, $\varepsilon_H = xu_g/u_m$ — volumetric void fraction by the homogeneous model.

Note that practically always (except very near the critical point) $\hat{\kappa}_l \ll \hat{\kappa}_g$, so the term with $\hat{\kappa}_l$ in the expression for $\hat{\kappa}$ can be ignored (except the cases, when ε_H is very small). Then the expression for n can be written as:

$$n^{-1} \approx \varepsilon_H\hat{\kappa}_g - \varepsilon_H p x u_g \hat{\beta}_g^2\,\Upsilon/(c_{pm}T), \quad \text{[eq. 54]}$$

where $\Upsilon = \left[1 + (1-\varepsilon_H)\hat{\beta}_l/(\varepsilon_H\hat{\beta}_g)\right]^2$ is close to 1.

Considering that there is a similar relation for gas, $n_g = \left[\hat{\kappa}_g - \Lambda_g\hat{\beta}_g\right]^{-1}$, and that $n_g\Lambda_g = n_g ZR\hat{\beta}_g/c_{pg} = (k_g - 1)/\hat{\beta}_g$, $\hat{\kappa}_g = k_g/n_g$, where $k_g = c_{pg}/c_{Vg}$, after transformations the following formula for n can be obtained:

$$n \approx \frac{1}{\varepsilon_H}\frac{n_g}{k_g}\frac{xc_{pg} + (1-x)c_{pl}}{xc_{pg}\left(1 - \Upsilon + \Upsilon/k_g\right) + (1-x)c_{pl}},\quad \Upsilon = \left[1 + \frac{(1-\varepsilon_H)\hat{\beta}_l}{\varepsilon_H\hat{\beta}_g}\right]^2 \quad \text{[eq. 55]}$$

Equation 55 is a generalization for a mixture of real gas and liquid, of the well-known Tangren's formula for of ideal gas and liquid (see, for instance [17]):

$$n \approx \frac{1}{\varepsilon_H}\frac{xc_{pg} + (1-x)c_{pl}}{xc_{pg}/k_g + (1-x)c_{pl}} \quad \text{[eq. 56]}$$

It is possible also to get exact expression for n^{-1} in the form which clearly shows how its value for mixture corresponds to the values and parameters of gas and liquid phases. From equations 37 and 13 it is easy to get in this case, that

$$\Lambda = \frac{\hat{\beta}}{\varepsilon_H\hat{\beta}_g/\Lambda_g + (1-\varepsilon_H)\hat{\beta}_l/\Lambda_l} \quad \text{[eq. 57]}$$

Substituting this expression to equation 52 and taking into account equations 53, after simplification the expression for n^{-1} takes the form

$$n^{-1} = \varepsilon_{\mathrm{H}} n_g^{-1} + (1-\varepsilon_{\mathrm{H}}) n_l^{-1} + \frac{\varepsilon_{\mathrm{H}}(1-\varepsilon_{\mathrm{H}})(\Lambda_g - \Lambda_l)^2}{\varepsilon_{\mathrm{H}} \Lambda_l/\hat{\beta}_l + (1-\varepsilon_{\mathrm{H}}) \Lambda_g/\hat{\beta}_g} \qquad \text{[eq. 58]}$$

Equation 58 shows that always $n \le \left[\varepsilon_{\mathrm{H}} n_g^{-1} + (1-\varepsilon_{\mathrm{H}}) n_l^{-1}\right]^{-1}$, and $n \to n_l$ when $\varepsilon_{\mathrm{H}} \to 0$, and $n \to n_g$ when $\varepsilon_{\mathrm{H}} \to 1$.

2.2.2 One-component vapor-liquid flow on saturation line and exact omega expression.

Let us get formulas for n (and M) calculation in variables p, x for one-component two-phase flow on saturation line. We have

$$(\partial u_m/\partial p)_{s_m,sat} = (\partial u_m/\partial p)_{x,sat} + (\partial u_m/\partial x)_p (\partial x/\partial p)_{s_m,sat}, \qquad \text{[eq. 59]}$$

where sat in the index means that a differentiable function is taken on the saturation line.

To establish formula for $(\partial x/\partial p)_{s_m,sat}$, note that $dh_m = T ds_m + u_m dp$ and at constant entropy $ds_m = 0$ and $dh_m = u_m dp$. On the other hand, $dh_m = h_{gl} dx + (\partial h_m/\partial p)_{x,sat} dp$ and at constant entropy we obtain the equation $u_m dp = h_{gl} dx + (\partial h_m/\partial p)_{x,sat} dp$ and

$$(\partial x/\partial p)_{s_m,sat} = \left[u_m - (\partial h_m/\partial p)_{x,sat}\right]/h_{gl} \qquad \text{[eq. 60]}$$

As $(\partial u_m/\partial x)_p = u_{gl}$, from equations 59 and 60 we have

$$(\partial u_m/\partial p)_{s_m,sat} = \left[u_m - (\partial h_m/\partial p)_{x,sat}\right] u_{gl}/h_{gl} + (\partial u_m/\partial p)_{x,sat}. \qquad \text{[eq. 61]}$$

Then,

$$(\partial u_m/\partial p)_{x,sat} = \left[(\partial u/\partial p)_T\right]_m + \left[(\partial u/\partial T)_p\right]_m (dT_{sat}/dp), \qquad \text{[eq. 62]}$$

$$(\partial h_m/\partial p)_{x,sat} = \left[(\partial h/\partial p)_T\right]_m + \left[(\partial h/\partial T)_p\right]_m (dT_{sat}/dp) = \\ = u_m - T\left[(\partial u/\partial T)_p\right]_m + c_{pm}\, dT_{sat}/dp \qquad \text{[eq. 63]}$$

And finally from equations 61-63,

$$(\partial u_m/\partial p)_{s_m,sat} = \frac{u_{gl}}{h_{gl}}\left[T\left[\left(\frac{\partial u}{\partial T}\right)_p\right]_m - c_{pm}\frac{dT_{sat}}{dp}\right] + \left[(\partial u/\partial p)_T\right]_m + \left[(\partial u/\partial T)_p\right]_m \frac{dT_{sat}}{dp} \qquad \text{[eq. 64]}$$

Given that $dT_{sat}/dp = T u_{gl}/h_{gl}$, we obtain:

$$(\partial u_m/\partial p)_{s_m,sat} = -c_{pm} T (u_{gl}/h_{gl})^2 + 2T(u_{gl}/h_{gl})\left[(\partial u/\partial T)_p\right]_m + \left[(\partial u/\partial p)_T\right]_m. \qquad \text{[eq. 65]}$$

Here $T\left[\left(\partial u/\partial T\right)_p\right]_m = xu_g\hat{\beta}_g + (1-x)u_l\hat{\beta}_l$, $p\left[\left(\partial u/\partial p\right)_T\right]_m = -xu_g\hat{\kappa}_g - u_l(1-x)\hat{\kappa}_l$ and $n^{-1} = -\left(\partial \ln u_m/\partial \ln p\right)_{s_m,sat} = -\left(p/u_m\right)\left(\partial u_m/\partial p\right)_{s_m,sat} =$

$$= \frac{c_{pm}Tp}{u_m}\left(\frac{u_{gl}}{h_{gl}}\right)^2 - 2\frac{pu_{gl}}{u_m h_{gl}}\left(xu_g\hat{\beta}_g + (1-x)u_l\hat{\beta}_l\right) + \frac{1}{u_m}\left(xu_g\hat{\kappa}_g + u_l(1-x)\hat{\kappa}_l\right) = \quad \text{[eq. 66]}$$

$$= \left(\varepsilon_{\mathrm{H}}\hat{\beta}_g\Lambda_g^{-1} + (1-\varepsilon_{\mathrm{H}})\hat{\beta}_l\Lambda_l^{-1}\right)\Lambda^2 - 2\Lambda\left(\varepsilon_{\mathrm{H}}\hat{\beta}_g + (1-\varepsilon_{\mathrm{H}})\hat{\beta}_l\right) + \left(\varepsilon_{\mathrm{H}}\hat{\kappa}_g + (1-\varepsilon_{\mathrm{H}})\hat{\kappa}_l\right) =$$

$$= \varepsilon_{\mathrm{H}}\hat{\beta}_g\Lambda_g\left(\Lambda/\Lambda_g - 1\right)^2 + \left(1-\varepsilon_{\mathrm{H}}\right)\hat{\beta}_l\Lambda_l\left(\Lambda/\Lambda_l - 1\right)^2 + \varepsilon_{\mathrm{H}}n_g^{-1} + \left(1-\varepsilon_{\mathrm{H}}\right)n_l^{-1}$$

Where dimensionless isothermal compressibility coefficients $\hat{\kappa}_g$ and $\hat{\kappa}_l$ are equal $\hat{\kappa}_g = \left(p/u_g\right)\left(\partial u_g/\partial p\right)_T = k_g/n_g$, $\hat{\kappa}_l = \left(p/u_l\right)\left(\partial u_l/\partial p\right)_T = k_l/n_l$.

Equation 66 gives the general formula for calculation of isentropic coefficient of the single-component two-phase mixture on the saturation line. For ideal gas and liquid phases $\hat{\beta}_l = \hat{\kappa}_l = 0$ and $\hat{\beta}_g = \hat{\kappa}_g = 1$, and equation 66 transforms into the equation for ω parameter in the well-known omega method of calculating critical two-phase flows [18]:

$$\omega = n^{-1} = \frac{c_{pm}Tp}{u_m}\left(\frac{u_{gl}}{h_{gl}}\right)^2 + \varepsilon_{\mathrm{H}}\left(1 - 2p\frac{u_{gl}}{h_{gl}}\right) \quad \text{[eq. 67]}$$

The equation 66 (instead of equation 67) could be used for more correct evaluation of ω parameter in omega method to take into account non-ideal properties for liquid and vapour phases, when all or at least some of the parameters describing non-ideal phases behaviour $(\hat{\beta}_g, \hat{\beta}_l, \hat{\kappa}_g, \hat{\kappa}_l)$ are known. The difference becomes significant at reduced temperatures 0.9 (or reduced pressures 0.5) and higher, where classical omega method becomes inaccurate.

Note, that the last form of equation 66 shows that always $n \le \left[\varepsilon_{\mathrm{H}}n_g^{-1} + \left(1-\varepsilon_{\mathrm{H}}\right)n_l^{-1}\right]^{-1}$ and the value of difference between n and $\left[\varepsilon_{\mathrm{H}}n_g^{-1} + \left(1-\varepsilon_{\mathrm{H}}\right)n_l^{-1}\right]^{-1}$ depends on the difference between isentropic coefficients Λ for the mixture and correspondent coefficients for liquid and vapour phases Λ_l, Λ_g. For liquid Λ_l is usually very small, so there is a huge difference between n and n_l, when $\varepsilon_{\mathrm{H}} \to 0$. This indicates that the model doesn't describe the flow correctly, and it is in fact well known that thermodynamic non-equilibrium effects (boiling delay) occur and should be taken into account for very small values of ε_{H}. But it is interesting, that when $\varepsilon_{\mathrm{H}} \to 1$, $n \to \left[\hat{\beta}_g\Lambda_g\left(\Lambda/\Lambda_g - 1\right)^2 + n_g^{-1}\right]^{-1}$, which also differs (slightly smaller) from n_g. This indicates that some thermodynamic non-equilibrium effects (probably deviation of bulk vapour temperature from remaining liquid phase temperature) also take place when $\varepsilon_{\mathrm{H}} \to 1$.

3. Phase diagrams of the fluid and a selection of 2nd independent thermodynamic parameter for two phase flow

It is possible to integrate equations 41 and 42 (or 43, or 45, or 49) by standard methods (Runge-Kutta method and forth integration using Simpson formula for equation 41 with a constant step on the logarithm of pressure), obtaining a length corresponding to pressure from the beginning or the end of the pipe section and value of the second corresponding thermodynamic parameter, and based on it — all other product properties. As the 2nd thermodynamic parameter, it is possible to select specific enthalpy or entropy, as well as temperature (except for a single-component two-phase or azeotropic mixture), or void fraction (for a single-component two-phase mixture on the saturation line).

However, for the case of two-phase multicomponent mixture, selecting the 2nd independent parameter, one should keep in mind peculiarities of the mixture phase diagram. Based on the experience of practical numerical implementation in "Safety valve" and "Hydrosystem" software, in this case the choice of temperature as an additional thermodynamic parameter is often unsuccessful. For azeotropic-like mixtures (with a narrow two-phase field in coordinates p, T) solution of the system of equations 41, 45 is unstable, and even very small calculation error leads to unreal values of void fraction and other flow parameters. The reason for this is easy to understand, looking at the phase diagrams of the fluid.

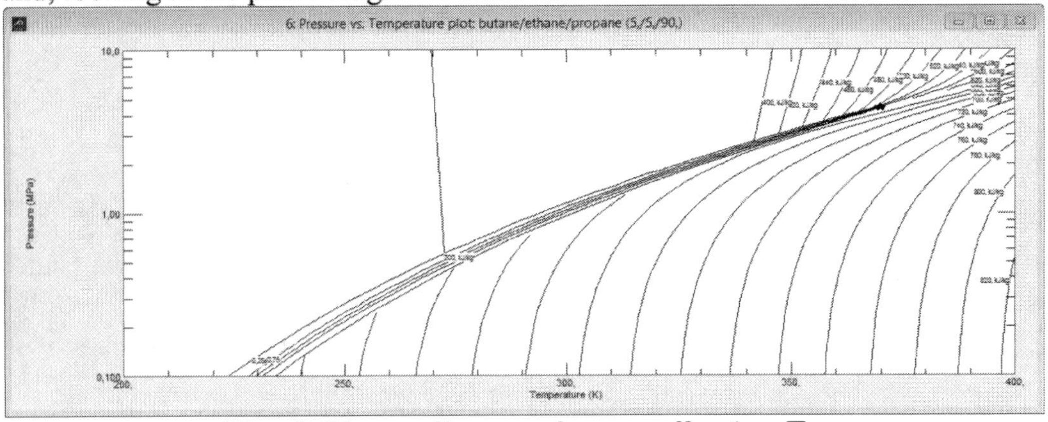

***Fig. 1.* Phase diagram in coordinates T-p.**

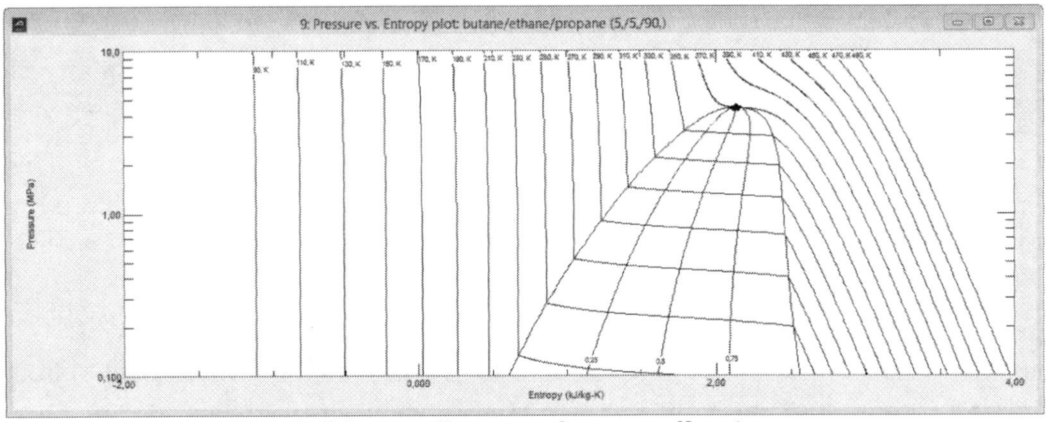

***Fig. 2.* Phase diagram in coordinates s-p.**

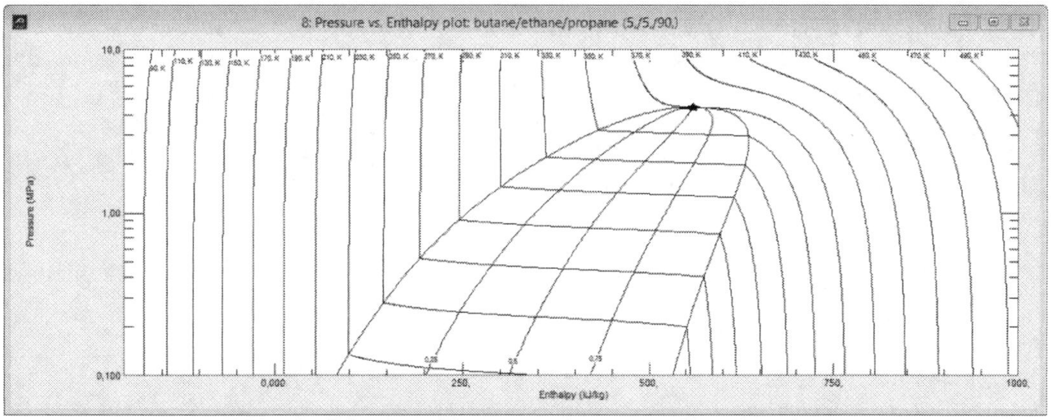

***Fig. 3.* Phase diagram in coordinates h-p.**

Figures 1-3 shows the phase diagram (envelope) of 90% propane with admixtures of ethane and butane in coordinates T-p, s-p и h-p. These phase diagrams are obtained using thermodynamic package NIST REFPROP.

As it is seen from Figure 1, the slightest inaccuracy in the temperature calculation leads to an enormous error for void fraction and other product properties. At the same time (see Figures 2 and 3), while selecting an entropy or enthalpy as the 2nd thermodynamic parameter, the inaccuracy does not lead to any significant errors at calculation.

Selecting of enthalpy in "Hydrosystem", and "Safety valve" programs as an additional thermodynamic parameter for multicomponent two-phase mixtures, provided calculation model to be tolerant to error accumulation, and this way eventually proved to be more effective, even despite additional calculation costs in some of used thermodynamic libraries.

4. Approximate equations of Fanno flow for real gases and two phase flow

While integrating equations 12, 21, 22 for gas or 41, 42, 43, 49 for two-phase flow, there is a problem of effective selection of the integration step, thus it is important to find some way to estimate a possible change in pressure at the specified length of the pipe. Provided that the effect of terms with parameters Λ_g and heat Λ_q is usually insignificant, they may be neglected for rough estimation, and in most cases it is quite enough to consider just a Fanno flow (i.e. adiabatic flow in a horizontal pipe) for the first estimation.

Let us show that for Fanno flow for real of gas and two-phase gas-liquid mixture, it is possible to establish approximate equations, similar to the ones of Fanno flow for an ideal gas.

4.1 Real gas

Let's start from Equation 36, which is valid both for real gas and two phase flow and (as G=const) can be rewritten in the form

$$d\ln u = d\ln n + d\ln p + d\ln M^2 \qquad \text{[eq. 68]}$$

As n almost always (expect in the vicinity of critical point) changes much more slowly than p and M, we can omit the first term of right part of equation 68:

$$d\ln u = d\ln p + d\ln M^2 \qquad \text{[eq. 69]}$$

Next, for Fanno flow the equation 3 takes the form

$$dh + G^2 u du = 0 \qquad \text{[eq. 70]}$$

Dividing by pu and taking into account equation 36, we get

$$\frac{dh}{pu} + nM^2 d\ln u = 0 \qquad \text{[eq. 71]}$$

Substituting into equation 71 expression for $\frac{dh}{pu}$ and $d\ln u$ from equations 24 and 69:

$$\left[1 - \frac{1-M^2}{1+n\Lambda M^2}\right] d\ln p + nM^2\left(d\ln p + d\ln M^2\right) = 0 \qquad \text{[eq. 72]}$$

And finally

$$d\ln p = -\frac{1+n\Lambda M^2}{\left(1+1/n+\Lambda+nM^2\right)M^2} dM^2 \qquad \text{[eq. 73]}$$

From equation 73 and equation 26

$$d\ln T = -\frac{\Lambda}{\hat{\beta}} \frac{\hat{\beta}-1+kM^2}{\left(1+1/n+\Lambda+nM^2\right)M^2} dM^2 \qquad \text{[eq. 74]}$$

From equation 73 and equation 23

$$d(l/D) = \frac{1-M^2}{\left(1+1/n+\Lambda+nM^2\right)\frac{\lambda}{2}nM^4} dM^2 \qquad \text{[eq. 75]}$$

As isentropic coefficients n and Λ, as well as coefficients $\hat{\beta}$ and λ change very slowly for real gas flow, so we can consider them constant and integrate equations 73 – 75 analytically. Integration gives:

$$\frac{p_1}{p_2} = \left(\frac{M_2}{M_1}\right)^{2C_p/A_p} \left(\frac{M_2^2 + A_p/B_p}{M_1^2 + A_p/B_p}\right)^{\left(\frac{D_p}{B_p} - \frac{C_p}{A_p}\right)} \qquad \text{[eq. 76]}$$

$$\frac{T_1}{T_2} = \left(\frac{M_2}{M_1}\right)^{2C_T/A_T} \left(\frac{M_2^2 + A_T/B_T}{M_1^2 + A_T/B_T}\right)^{\left(\frac{D_T}{B_T} - \frac{C_T}{A_T}\right)} \qquad \text{[eq. 77]}$$

$$(L/D)\frac{\lambda n}{2} = \frac{1}{A_p}\left(\frac{1}{M_1^2} - \frac{1}{M_2^2}\right) + \frac{B_p + A_p}{A_p^2} \ln\left[\frac{M_1^2\left(M_2^2 + A_p/B_p\right)}{M_2^2\left(M_1^2 + A_p/B_p\right)}\right] \qquad \text{[eq. 78]}$$

where indexes 1 and 2 correspond to the beginning and the end of pipe, and

$$A_p = 1+1/n+\Lambda,\ B_p = n\Lambda,\ A_T = \hat{\beta}A_p,\ B_T = \hat{\beta}B_p \qquad \text{[eq. 79]}$$

$$C_p = 1,\ D_p = n\Lambda,\ C_T = \Lambda\left(\hat{\beta}-1\right),\ D_T = k\Lambda \qquad \text{[eq. 80]}$$

If we neglect change in the coefficients of these equations of temperature and pressure, then the equations 28-30 can be integrated analytically and give:

Note that equations 76-78 have a form similar to the Fanno flow equations of an ideal gas [1-3, 6], and in fact represent their generalization. Solving equation 78 for specified pipe length (using Newton method for example), from equations 76 and 77 we then can obtain evaluations for pressure and temperature variations within the pipe. Average accuracy of these equations for real gas Fanno flow is very good - within 1%.

4.2 Two-phase gas-liquid flow

All equations of previous section (except 74 and 77) can be also applied to the case of two-phase flow. However, while Λ and λ in two-phase case still change very slowly, the value of coefficient n can change more rapidly, and the equations in this case should be less accurate.

Preliminary calculation tests show that the approximate equations 77, 77 allow in most cases of two phase flow to estimate the change in pressure within accuracy 15-20%, which is quite sufficient for a reasonable choice of the integration step more accurate calculations on the base of exact equations.

5. Calculation of minor losses

Methods for calculating of minor losses on fittings and valves for chocked and near chocked flows often cause discussions between researchers for the gas flow case, and are purely studied for a two-phase flow. In the latter case, however, use of the same correlations as for an ordinary two-phase flow, is within the scope of a common two-phase flow calculation error and can be accepted as a temporary solution. For gas flow, usage of correlations for the incompressible flow may cause an error of 20-30%. Although most of such calculated losses exceed the ones calculated with allowance for fluid compressibility and high Mach numbers, it is desirable, as far as possible, to use more accurate methods.

Fortunately, in the pressure relive discharge pipelines, for which such flows are typical, a range of types of piping components is severely limited, and usually is restricted to bends, diffusers, abrupt expansions, pipe exit and flow separating and merging tees (for branched pipelines).

For pressure losses on bends , in accordance with proposal set forth in [6], equation 5 for straight pipes is used, at that hydraulic resistance coefficient λ is replaced by an equivalent coefficient, with allowance for local losses $\lambda+\zeta D/L$, where local resistance coefficient ζ is calculated by the procedure of incompressible fluid.

For pipe exits the calculation is based on common assumption that there is no net "exit loss" for a fluid exiting from a pipe into a much larger container or vessel. If the fluid is discharged into a space or container of the same fluid, the friction loss is approximately one "kinetic energy head" and the pressure recovery is also the same.

Abrupt expansions are calculated on the based on known approach extending Borda-Carnot equation for the case of compressible flow (see [2, 19]), with correction proposed in [20].

The choked and near-choked gas flow in diffusers is still needed to be studied more thoroughly, though this problem was studied by several researchers – see [7, 21, 22]. As a temporary solution the method proposed in [6] was used.

The special iteration procedure was implemented in the software for diffusers, abrupt expansions and pipe exits analysis, which takes into account the properties of real gas. The program uses ideal gas solution as a preliminary, and then takes into account deviations from ideal gas behavior. Also the program (if necessary) checks for possible chocked flow on these components and calculates such chokes when they take place.

For tee analysis the algorithms proposed in [23] are scheduled to be implemented.

6. Conclusion

The equations and algorithms described above for choked and near choked real gas and two-phase gas-liquid flow, cover the effect of non-ideal behaviour of gas and liquid phases. The important role of isentropic coefficients was shown. In the case of non-condensable gas / non-flashing liquid mixture, and in the case of one-component vapour/liquid mixture, exact expressions obtained for isentropic coefficients of mixture via thermodynamic parameters of the phases. Approximate equations, proposed in section 4, can be successfully used (along with available thermodynamic tables or libraries) by process engineers for evaluation of flow parameters in discharge piping. The equations of section 2 were successfully implemented in “Hydrosystem” and "Safety valve" software, which are now widely used in engineering practice of hundreds of EPC and OO companies to correctly design pressure relief systems.

The author would like to express deep appreciation to his colleagues Elena Yudovina, Tatyana Kuznecova, Alexey Babenko, Sergey Lisin, Valery Lisman, who provide successful software implementation of described algorithms. The author also would like to thank Vyacheslav Spitsyn for his kind and timely assistance in translation of the article.

7. References

[1] Shapiro A. H. The Dynamics and Thermodynamics of Compressible Fluid Flow. V. 1. — New York: The Ronald Press Company, 1953. — 647 p.

[2] Benedict R. P. Fundamentals of Pipe Flow. — New York: A Wiley-InterScience Publication, 1980. — 529 p.

[3] Anderson J. D. Modern Compressible Flow with Historical Perspective. — New York: McGraw-Hill Publishing Company, 1990. — 671 p.

[4] Attou A. Thermodynamique de L'Ecoulement Diphasique Compressible a deux Constituants de Fanno // Revue de L'Institut Francais du Petrole. — 1998. — V. 53. — N 6. — P. 813–837.

[5] Fesenko V.V. Choked two phase flow (in Russian). – Moscow: Atomizdat, 1978, 160 p.

[6] Miller D.S. Internal Flow Systems. 2nd Edition. – BHR Group Limited, 1990. – 396 p.

[7] Idelchik I.E. Handbook of Hydraulic Resistance. 4th Edition. – New York: Begell House, Inc. 2007. – 878 p.

[8] ASTM Standard C680-10 “Standard Practice for Estimate of the Heat Gain or Loss and the Surface Temperatures of Insulated Flat, Cylindrical, and Spherical Systems by Use of Computer Programs”.

[9] ISO 12241–2008 “Thermal Insulation for Building Equipment and Industrial Installations — Calculation Rules”.

[10] Wallis G.B. One-Dimensional Two-Phase Flow. - New York: McGraw-Hill Publishing Company, 1969. – 408 p.

[11] Chisholm D. Two-Phase Flow in Pipelines and Heat Exchangers. – New York: 1983. – 324 p.

[12] Azzopardi B. J. Gas-Liquid Flows. — New York: Begell House, 2006. — p. 331.

[13] Yudovina E.F., Pashenkova E.S., Korelshteyn L.B. Software package "Hydrosystem" and its use for hydraulic calculations of piping systems (in Russian) // In books: Mathematical modelling of piping systems of energy industry. Proc. XII All-Russia Scien. Sem. with intern. participation. "Mathematical models and methods of analysis and optimal synthesis of developing piping and hydraulic systems", September 20-26, 2010 — Irkutsk, ISEM SO RAN, 2010. — p. 475–485.

[14] Babenko A.V., Korelshteyn L.B., Gartman T.N. Mathematical modelling of steady gas-liquid flows in the industrial pipelines. Analysis of branches. (in Russain) // Chemical Engineering and Processing. — 2012. — V. 13. — No. 7. — p. 429-439.

[15] IHS ESDU 01014 "Frictional Pressure Gradient in Adiabatic Flows of Gas-Liquid Mixtures in Horizontal Pipes: Prediction Using Empirical Correlations and Database". — 2007.

[16] Ghajar A. J., Woldesemayat M. A. Comparison of Void Fraction Correlations for Different Flow Patterns in Horizontal and Upward Inclined Pipes // Int. J. Multiphase Flow. — 2007. — V. 33. — P. 347–370.

[17] Morris S. D. Vapour Adiabatic Exponent for Flashing Flow in Nozzles // Chemical Engineering and Processing. — 2000. — V. 39. — P. 275–281.

[18] Leung J.C. Easily Sized Relief Devices and Piping for Two-Phase Flow// Chemical Engineering Progress. – 1996. – V. 92. – N 12 – P. 28-50.

[19] Marjanovic P., Djordjevic V. On the Compressible Flow Losses Through Abrupt Enlargements and Contractions// Transaction of the ASME. – 1994. – V.116 – P.756-762.

[20] Morimune T., Hirayama N., Maeda T. Study of Compressible High Speed Gas Flow in Piping System (2nd Report, Piping Systems with Sudden Enlargement)// Bulletin of the JSME. – 1980. – V.23. – N 186. – P. 2005-2012.

[21] David Japikse, Nicholas C. Baines. Diffuser Design Technology. 2nd Edition. 2000. Concepts ETI. 287 p.

[22] IHS ESDU 90025 «Performance of conical diffusers in subconic compressible flow». – 2007.

[23] J. Perez-Garcia, E. Sanmiguel-Rojas, A. Viedma. New coefficient to characterize energy losses in compressible flow at T-junctions// Applied Mathematical Modeling. – 2010. – V. 34 – N 12– P. 4289-4305.

[24] Istomin V.A. Real Gas Isoentropic Indices: Definitions and Basic Relations// Russian Journal of Physical Chemistry. – 1997. – V. 71 – N 6 – P. 883-888.

[25] Istomin V.A. Real Gas Isentropic Indexes and Their Use in Thermodynamics of Gas and Gas Dynamics. – 1998. – V. 72 – N 3 – P. 334-339.

[26] Baltadjiev N.D., Spakovsky Z.S., Lettieri C. An Investigation of Real Gas Effects in Supercritical CO2 Centrifugal Compressors// Journal of Turbomachinery. – 2015. – V. 137 – N 9.

[27] Baltadjiev N.D. An Investigation of Real Gas Effects in Supercritical CO2 Centrifugal Compressors. Master Thesis. – 2012. – Massachusetts Institute of Technology.

Models for Multi-Phase & Single-Phase Flow in Pressure Relieving System Using Bernoulli Integration

Freeman Self (feself@bechtel.com)
Sanjay Ganjam (sviswana@bechtel.com)
Bechtel Oil, Gas and Chemicals, Inc.

Gerald Jacobs (gerald@virtualmaterials.com)
Virtual Materials Group

Prepared for Presentation at
American Institute of Chemical Engineers
2015 Spring Meeting
11th Global Congress on Process Safety
Austin, Texas
April 27-29, 2015

Session
DIERS T1G00 Practical Methods for Two-Phase Flow

UNPUBLISHED

Models for Multi-Phase & Single-Phase Flow in Pressure Relieving System Using Bernoulli Integration

Freeman Self (feself@bechtel.com)
Sanjay Ganjam (sviswana@bechtel.com)
Bechtel Oil, Gas and Chemicals, Inc.

Gerald Jacobs (gerald@virtualmaterials.com)
Virtual Materials Group

Keywords: pressure relief valve, safety valve, orifice, control valve, Bernoulli equation, numerical integration, homogeneous equilibrium

Abstract

The Bernoulli (mechanical energy) equation is very important in fluid flow, particularly since it describes how mechanical energy is transformed into other forms of energy. But complications occur in compressible flow since the density changes with pressure. The traditional tool is the analytical solution of the Bernoulli equation. However, to characterize density as a function of pressure for compressible fluid, simplifying assumptions are needed that limit the application. As an alternative, it is becoming more prevalent to use integral methods to numerically solve the Bernoulli equation in compressible flow.

This paper will outline the use of numerical integration for several common situations: orifices, relief valves, and control valves. Explanations and derivations will be provided to illustrate how these models can be used for choked / non-choked, single and multi-phase systems.

1. Introduction

The Bernoulli (mechanical energy) equation is very important in fluid flow, particularly since it describes how mechanical energy is transformed into other forms of energy. But complications occur in compressible flow since the density changes with pressure. However, to characterize density as a function of pressure for compressible fluid, simplifying assumptions are needed that limit the application. Among them are the assumption of an ideal gas, constant heat capacity and the characterization of density changes with pressure. Despite these limitations, analytical equations are the traditional method but are sometimes used inappropriately. For example, they are used for high-pressure gases which do not exhibit ideal gas behavior.

As an alternative, it is becoming more prevalent to use numerical integration to solve the Bernoulli equation in compressible flow. Integral methods are of course not new; the power industry has used them for years for single and two-phase flow of steam and water [Benjamin and Miller-1942]. With increasing computational power, excellent thermodynamic and physical property databases, integral methods may provide good solutions. Furthermore, integral methods lend themselves for use in multi-phases models.

2. General Formulation

The general formulation for numerical integration is written as a mass flux [mass/time-area] time the area plus a coefficient.

$$W \text{ [mass/time]} = \{ A_0 K_d \} \{ G \} \qquad 2\text{-}01$$

G = mass flux [mass/time-area] from either numerical integration or analytical solution
A_0 = area based at the throat [area]
K_d = flow coefficient [dimensionless]

When the mass flux is found by numerical integration it is commonly known as the "direct integration" model. Examples of calculating the integration are provided in several publications [Simpson 1991, Darby 2001, Darby 2002, API-520-I-2014].

For two-phase flow equilibrium (phase, heat and momentum) with the properties described as homogeneous, then the model is additionally called "homogeneous equilibrium model" (HEM). The integration method is thus called "homogeneous direct integration" (HDI).

If not at equilibrium, a variety of models have been proposed include "homogeneous non-equilibrium model" and the "frozen equilibrium model". The integration method may be called "homogeneous non-equilibrium direct integration", which may be modified to account for fluid slip and phase non-equilibrium provided the appropriate data is available.

2.1 Mass Flux

The mass flux at the vena contracta is evaluated from the integral starting at the initial pressure. Since the flow from the inlet to the vena contracta is essentially frictionless, and the flow distance is short so heat transfer is minimal, the flow may be considered a constant entropy process. The mass flow is then determined from the area at the vena contracta and mass flux. The integral method is especially beneficial since it rigorously evaluates the mass flux for variable densities. Additionally, if the fluid chokes at the vena contracta, the choke pressure and conditions are determined directly.

2.2 Coefficients

The area and flow coefficient, depending on the application, may be calculated separately or combined into a lumped parameter (e.g. control valves). There are two coefficients that are commonly employed. ($\beta^2 = A_0/A_1$)

$$K_d = C_d / (1 - \beta^4)^{1/2} \qquad \text{2-02}$$

The term C_d is generally labelled "coefficient of discharge" or "coefficient of discharge without velocity of approach factor". When C_d is combined with the "velocity of approach factor", the term K_d is called the "flow coefficient" or "coefficient of discharge with velocity of approach factor". K_d is used to simplify the parameter, or used when it is not convenient to calculate beta ratio (e.g. relief valves or control valves). Since the literature uses an inconsistence mix of parameters labelled either C or K, it is important to understand the definition. For example, relief valve vendors call K_d a "discharge coefficient" by truncating the term "with velocity of approach factor".

The coefficient of discharge C_d is used to adjust the theoretical model to match experimental data. The discharge coefficient will differ depending on the model and its assumptions and the quality of the experimental data. Thus discharge coefficients between different models can only be compared qualitatively. It is defined from mass flow W or mass flux G.

$$C_d = W_{experimental} / W_{Model} = G_{experimental} / G_{Model} \qquad \text{2-03}$$

3. Flow Orifices (square-edge)

3.1 Overview of Flow Orifices

Flow orifices are used for flow measurement and flow restrictions. The application may be complicated due to the variations in configurations, standards, and flow characteristics. This paper will separate the discussion into two main categories: thin orifices and thick orifices. Thin orifices will not choke, while thick orifice will choke depending on the pressure difference.

3.2 Mass Flow Relation (British Standard Units)

$$W\ [\text{mass/time}] = A_0\ [\text{area}]\ K_d\ \ G\ [\text{mass/time-area}] \qquad 3\text{-}01$$

$$W\ [\text{lb/hr}] = A_0\ [\text{in}^2]\ K_d\ 2407\ \rho_2\ [\text{lb/ft}^3] \left(\int_{P2}^{P1} (1/\rho)\, dP \right)^{1/2} \qquad 3\text{-}02$$

$$\text{where } K_d = C_d / (1 - \beta^4)^{1/2}$$
$$\text{and } \beta^2 = (A_0 / A_1) \qquad 3\text{-}03$$

The schematic illustrates the flow streamlines.

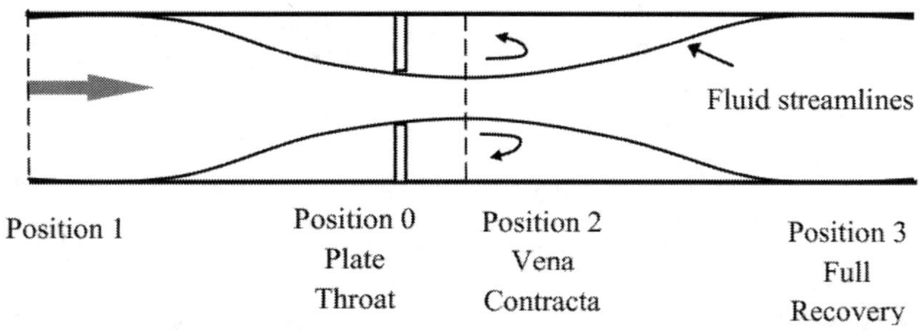

Figure 1

3.3 Discharge Coefficients - Thick Orifices – Choking and Non-Choking Flow

3.3.1. Overview of Thick Orifices

Thick orifices in gas, flashing liquid, and two-phase flow will choke with sufficient pressure drop. Thus they are also known in the trade as "critical orifices". Generally for the orifice to choke, the orifice is required to have a certain thickness relative to the orifice hole. Thick orifices, manufactured with square edges, will choke in compressible flow if the thickness / diameters are approximately greater than 1.0. [Ward-Smith-1979]. Simplistically, choking occurs since the flow remains attached to the lagging edge of the orifice, thus creating a rapid volume expansion at the discharge.

The description of flow coefficients in choking flow are complicated since K_ds appear to be a function the type of phase during choking (gas or flashing liquid), the number of phases (single or two-phase), the quality (vapor percent) in two-phase flow, equilibrium (phase, thermal or slip), and the plate thickness ratios (thickness / diameter).

3.3.2 Flow Coefficient (K_d) for Choking in Thick Orifices – Data Evaluated

Compressible flows in thick orifices under choking conditions were determined by Richardson [Richardson-2006]. The flow coefficient K_d is used since the beta (area ratio) cannot be positively confirmed from the article, although the beta has been roughly estimated as approximately 0.5.

There are four basic regimes are discussed.

(1) Liquids which are not flashing and not choking. These were not part of the experimental data but have been included for reference, with a nominal K_d of approximately 0.6.

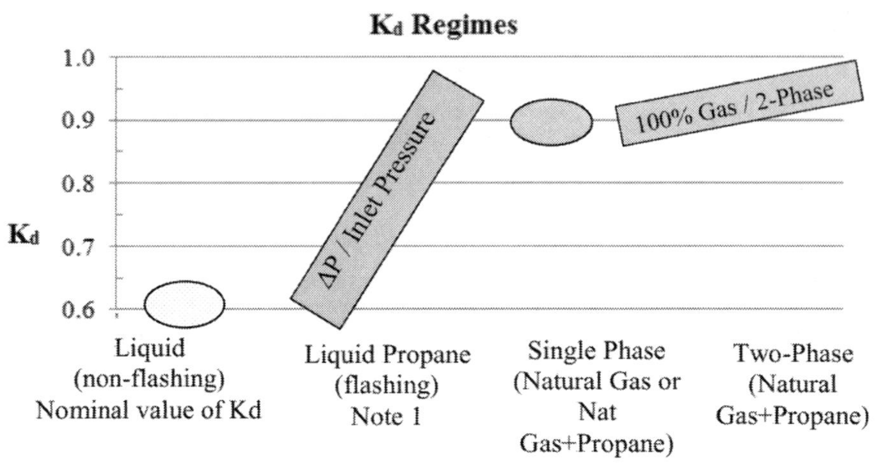

Note 1- ΔP/P1 is inlet pressure minus bubblepoint pressure at inlet temperature divided by inlet pressure

Figure 2 -A

(2) Liquid propane, which is flashing and choking at the bubblepoint at it flows through the orifice. This is discussed in Section 3.3.5.

(3) Single-phase gas. This includes both natural gas and natural gas+propane, which are single-phase into the orifice. This is discussed in Section 3.3.3.

(4) Two-phase gas. This includes only natural gas+propane, which are two-phase into the orifice. This is discussed in Section 3.3.4.

3.3.3 Flow Coefficient (K_d) for Choking in Thick Orifices - Single-Phase Natural Gas with Propane

Several sets of hydrocarbon data were evaluated.

- Natural Gas (100%). The phase inlet to the orifice was gas to the right of the phase envelope and the flows choked before condensing at the phase envelope.

- Natural Gas plus Propane. The phase inlet to the orifice was gas to the right of the phase envelope. However, small amounts of gas were condensed at the choke point (for all points and less than 8% mass).

For single-phase fluids, the flow coefficient K_d was approximately 0.90. The introduction of propane even with slight condensing had little effect on the discharge coefficient. Both direct numerical integration and Richardson integration gave similar results.

These flow coefficients are slightly higher than the air data (Figure 4) for thick orifices.

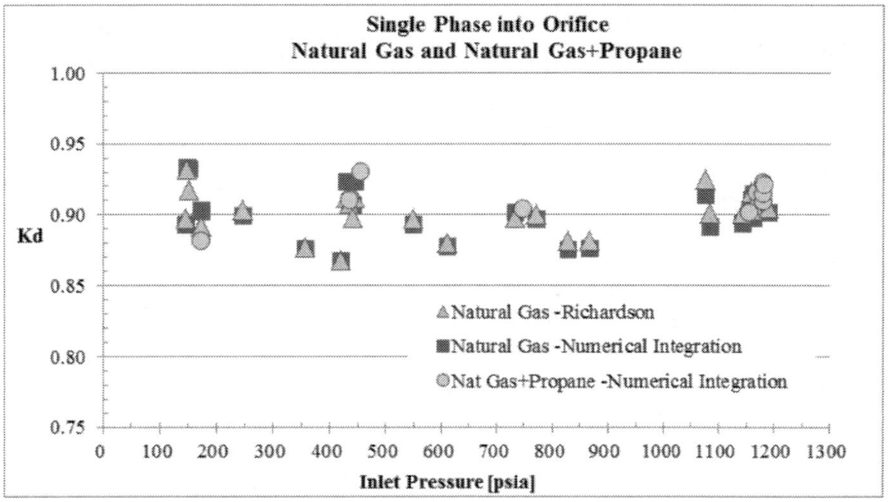

Figure 2 -B

3.3.4 Flow Coefficient (K_d) for Choking in Thick Orifices - Two-Phase Natural Gas+Propane

Natural gas plus propane, in different proportions but approximately 50%-50% were studied for two-phase flow. (Richardson also had a case with condensate added but the results were similar.) The phase inlet to the orifice was two-phase.

Two-phase fluids exhibited slightly higher K_ds approximately 0.95. This may be attributed to slip [Giacchette-2014]. When condensate is added to the mixture, the average Kd remains the same although the range increases. Both direct numerical integration and Richardson integration gave similar results.

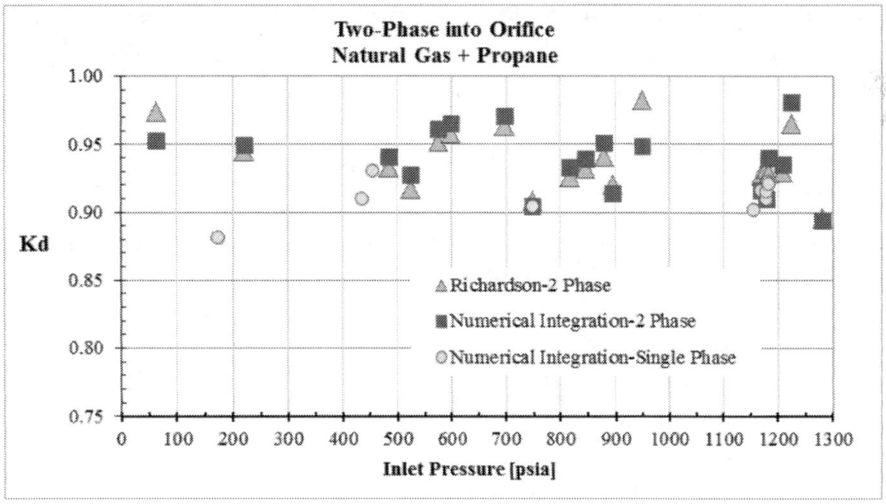

Figure 2 -C

Single phase flow is also shown for comparison; flow coefficients K_d were approximately 0.90.

3.3.5 Flow Coefficient (K_d) for Choking in Thick Orifices - Flashing Propane Liquid

Plotted are data for propane which is liquid inlet to the orifice, and which flashes and chokes sharply at the bubble-point. The bubble-points for all points were approximately 90 psia.

Propane resembled an incompressible liquid at higher pressures resulting in large mass fluxes and K_ds starting at 0.63. At low pressure drops, the propane was slightly more compressible with lower mass fluxes and high K_ds.

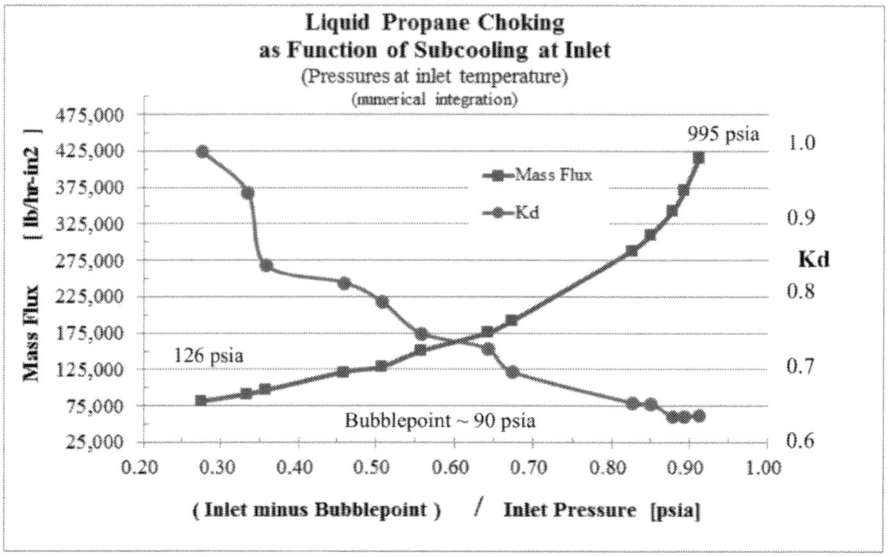

Figure 3

Richardson's coefficients are approximately 0.6 since Richardson modeled flashing propane as a liquid. However, when using the inlet pressures and choke points calculated from numerical integration, the results followed the same pattern as the plot with variable K_ds. (Four points from the data set were not plotted. These had pressures a few pounds above the bubblepoint, choked in the two-phase region, and all had calculated K_ds of approximately 2.6.)

3.3.6 Flow Coefficient (K_d) as Function of Plate Thickness for Choking Flow in Thick Orifices using Air

The previous data are based on thick orifices with plate thickness /diameter ratio of approximately two (2) and beta of zero. However, it has been found the flow coefficient for thick orifices is constant over a range of plate thickness ratios (thickness / diameter) approximately from 1.0 to 7.0 [Ward-Smith-1979]. For thicker plates, friction is important and the K_ds decrease with increasing thickness.

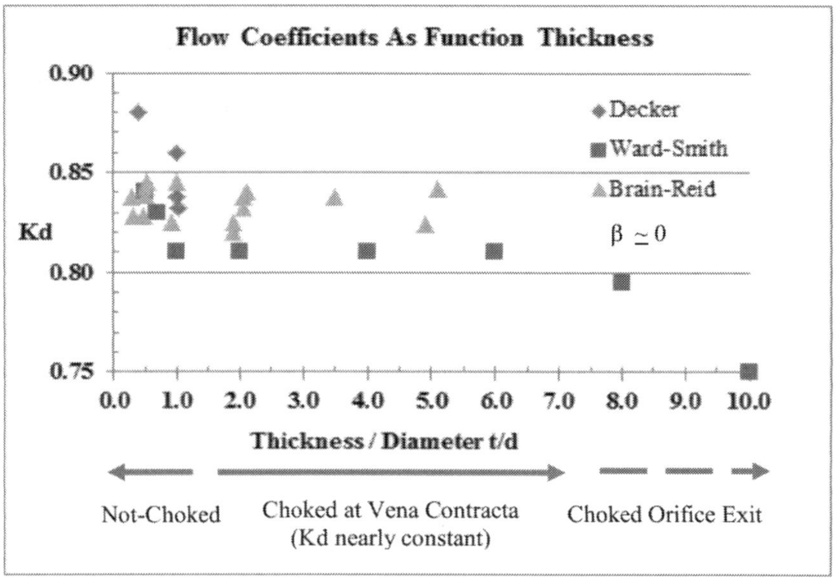

Figure 4

3.4 Discharge Coefficients - Thin Orifices – Not Choking

3.4.1. Overview of Thin Orifices

This section has been included to complete the overview of discharge coefficients. When one thinks of flow orifices as measurement devices, it is generally the thin orifices outlined in standards such as ISO-5167, ASME MFC-3M or BS-1042.

Unlike thick orifices that may choke with sufficient pressure drop, the defining characteristic of a thin orifice is that it does not choke. The mass flux always increases with pressure difference; the mass flux does not ever reach a maximum value regardless of the pressure drop. Simplistically, a thin orifice offers little surface for flow attachment and the fluid "jets" through the orifice without forming a deep vena contracta.

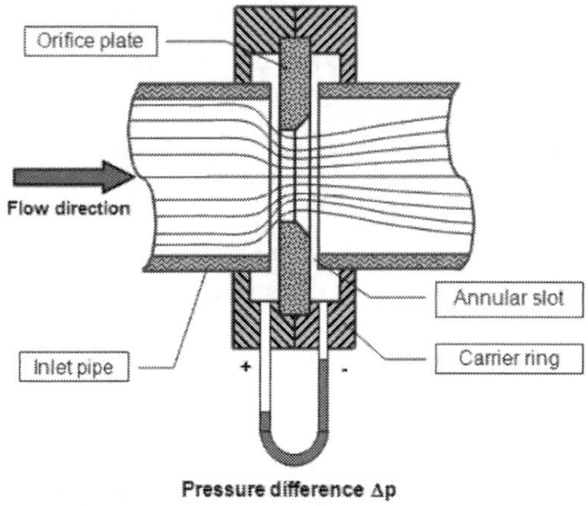

Figure 5

3.4.2. Flow Calculations

Flow through orifices is calculated with analytical Bernoulli models. The discharge coefficient to be used in the Bernoulli models are given by corresponding fitted equations. The standards such as ISO-5167, ASME MFC-3M or BS-1042 provide the equations, which are specific to each type of orifice, and generally provide similar results, although the standards have been frequently modified over the years. In general, limitations include single phase, non-choking, non-pulsating flow, specific location for pressure taps and minimum Reynolds numbers. Refer to each standard for more detailed information.

Engineering handbooks also provide the equations, generally based on one of the above standards. One must be careful since the nomenclature will differ between standards and handbooks. Three correlations are plotted in Figure 6.

- Miller-1996 provides a pre-1996 discussion of the standards. The handbook also provides graphs and equations for discharge coefficients.
- Crane-2009. Several editions are available but with different formulations and Reynolds number dependencies for discharge coefficients.

3.4.3. Use As a "Loss Coefficient" K

Often the pressure drop due to friction in the orifice is required. This may be calculated using the traditional loss coefficient K, which is familiar to engineering applications.

$$\text{Pressure Drop (psi)} = K\,(\,\rho v^2 \,/\, 9278\,)$$

(with density ρ [lb/ft^3], velocity v [ft/sec] and includes gc [as 32.2 lb-ft/sec^2])

Many handbooks present variations on the equations to relate K, the loss coefficient, to C_d the discharge coefficient for thin orifices based on pipe velocity. Following are two examples. ($\beta^2 = A_0/A_1$)

$$K = \left(\frac{[\,1 - \beta^4(1 - C_d^2)]^{0.5}}{C_d\,\beta^2} - 1 \right)^2 \qquad \text{[Crane-2009, Eq 4-2]}$$

$$K \simeq \left(\frac{(\,1 - \beta^4)\,(1 - \beta^2\,)}{C_d^2\,\beta^4} \right) \qquad \text{[Darby-2001, Eq 10-22]}$$

3.4.4. Coefficient of Discharge

To utilize the loss coefficient K, the discharge coefficient is needed. Coefficients of discharge are the same for both vapor and liquid. Shown are discharge coefficients C_d calculated from various sources for a beta of 0.5 [Hollinghead-2012, Miller-1996, Crane-2009]. Discharge coefficients used in the standards are calculated from empirical equations derived from fitted experimental data. They are correlated as a function of Reynolds number, beta (area) ratio and the location of the pressure taps, of which there are numerous variations. Each standard may use slightly different formulations for discharge coefficients and expansion factors.

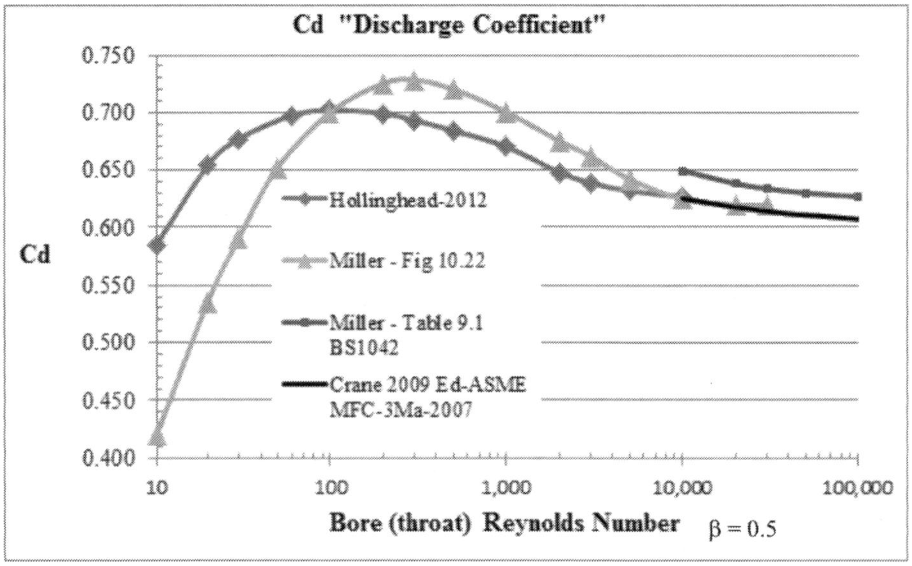

Figure 6

Also shown are the discharge coefficients calculated from a computational fluid dynamic (CFD) model. Discharge coefficients could have been calculated using numerical integration but the results would have been similar. Above Reynolds numbers of 10,000, discharge coefficients become constant. At low Reynolds numbers, discharge coefficients are variable and subject to model differences.

The Table below provides the details for the models.

Reference	Theoretical Model	Discharge Coefficient	Reynolds #	Equation for discharge coefficient
Miller Table 9.1	Analytical Bernoulli with expansion factor for gas flow	Provides the "discharge coefficient" C_d. The nomenclature in Table 9.1 is "C"	Re is based on pipe diameter. Applicable for Re(pipe) greater than 5,000	• Uses standard BS-1042 for corner taps. The table used the term "C" which is "C_d" in this paper. • The Re(pipe-based on pipe diameter) is converted to the Re(core-based on throat diameter) by dividing by Beta.
Miller Figure 10.22	Analytical Bernoulli with expansion factor for gas flow	Provides the "discharge coefficient" C_d. The nomenclature in Figure10.22 is "C"	Re is based on core (throat) diameter. Applicable for Re(core) between 5,000 10 to 100,000	Not provided
Crane -2009 Edition	Based on ASME MFC-3Ma-2007 with analytical Bernoulli with expansion factor	Provides the "flow coefficient" K_d. The nomenclature used is "C"	Re is based on pipe diameter. Applicable for Re(pipe) greater than 5,000	• The flow coefficient Kd has been converted to Cd using the "velocity of approach factor" with beta equal to 0.5 $K_d = C_d / (1 - \beta^4)^{1/2}$ • The Re(pipe-based on pipe diameter) is converted to the Re(core-based on throat diameter) by dividing by Beta.
Hollingshead-2012	Based on computational fluid dynamics	Provides the "discharge coefficient" C_d.	Re is based on core (throat) diameter.	CFD model.

4. Control Valves

4.1 Overview of Control Valve Flow

Control valve design exhibits several unusual features. First, the coefficients of discharge for gases and liquids are not dimensionless but have units of measure. As a result, the area and flow coefficient are lumped together as one parameter. Additionally, although control valve equations are generally treated as orifice equations, the flow is calculated based on the service pressure drop (inlet minus discharge pressure) rather than the pressure drop that produces flow, or the difference between the inlet pressure and the vena contracta. This requires accounting for pressure recovery in the valve outlet body.

4.2 Mass Flow Relation

The traditional mass flow relationship is utilized. Due to the historical development of control valve sizing, the area and flow coefficient are combined into one parameter, an *area-loss coefficient parameter*.

$$W\text{ [mass/time]} = \{ A K_d \} \{ G \} \qquad \text{5-01}$$

$$W\ [\text{lb/hr}] = \left\{ A\ K_d \right\} \left\{ 2407\ \rho_T \left(\int_{P2}^{P1} (1/\rho)\, dP \right)^{1/2} \right\} \quad \text{5-02}$$

The mass flux G is in units of lb/hr-in^2, area is in inch^2 and pressure in psi.

4.3 Flow Coefficient

The *area-flow coefficient parameter* $\{ A\ K_d \}$group is found by equating the ISA control valve equations to traditional orifice equations. For discussion, the original Fisher derivation is easier to physically understand. Refer to Appendix 4 for full derivations including Fisher and ISA-75.01.01. ISA-75.01.01 presents tables for conversion into various units.

4.3.1 Incompressible fluids

$\{ A\ K_d \} = \{ (F_L\, Cv) / 38 \}$, which is a well known relation. The parameter F_L is the liquid recovery factor and used to relate the downstream pressure to the vena contracta pressure. Without this conversion, the value of Cv will overstate the flow rate. The liquid discharge coefficient Cv has British Standard units in gallons / minutes -psi$^{1/2}$. 5-03

4.3.2 Compressible fluids,

- using ISA parameters
  - $\{ A\ K_d \} = 12.873\ [Cv / C\gamma]\ [(X_T\, F\gamma)^{1/2}]$ or rearranged 5-04
  - $\{ A\ K_d \} = [(Cv / 38)]\ [489.174\ (X_T\, F\gamma)^{1/2} / C\gamma]$ 5-05
- Fisher parameter C_g
  - $\{ A\ K_d \} = [Cg / 1100]$ or rearranged 5-06
  - $\{ A\ K_d \} = [Cv * C_1 / 1100]$ or rearranged 5-07
  - $\{ A\ K_d \} = [Cv / 38]\ [C_1 / 28.9]$ 5-08

X_T is the choking pressure ratio. $F\gamma$ is used to adjust heat capacity ratio from air and $C\gamma$ is a consolidation of terms that are functions only of the ideal heat capacity ratio. .

4.4. Compressible Ideal Gas - Low Pressure Drop

The data is for pressure drops less than 100 psi using air and water. [Buresh & Schuder-1964] In this situation, air is similar to an ideal gas, as judged by the criteria in ISA-75.01.01 Section 7.3, so the ISA equations are applicable.

4.4.1 Adjusting for Pressure at the Vena Contracta – Definition of Gas Pressure Adjustement Coefficient F_G

The numerical integration model uses the pressure drop that controls the flow (inlet minus vena contracta). On the other hand, the flow rate is experimentally measured based on the service pressure drop (inlet minus discharge), which must be adjusted for use with numerical integration. A "*gas pressure-adjustment coefficient*" is defined F_G and is formulated similar to the "*liquid-pressure recovery coefficient*" F_L (refer to Appendix 4).

$$F_G^2 = (P_1 - P_2) / (P_1 - Pvc) \qquad 5\text{-}09$$

The parameter F_G is called an "*adjustment*" parameter, since unlike the "*liquid-pressure recovery coefficient*" F_L, the value may be greater than one as well as less than one. A value greater than one large service pressure drops.

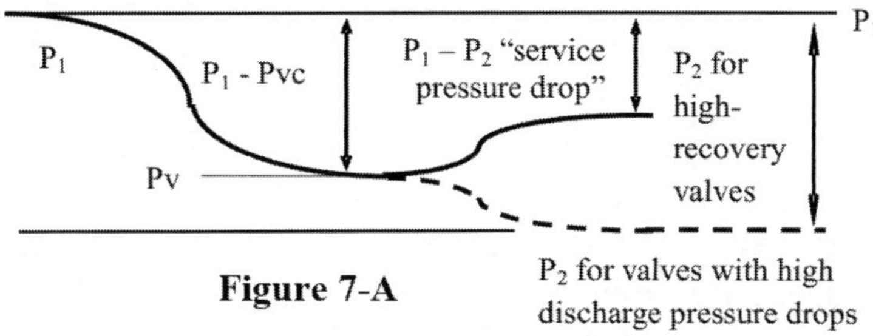

Figure 7-A

- For high-recovery control valves (low C_1 and X_T), The value of F_G is less than one and the choking pressure drop at the vena contracta (P1-Pvc) is greater than the service pressure drop (P_1-P_2).
- For zero-recovery control valves ($C_1 \sim 29$ and $X_T \sim 0.5$), the value of F_G is equal to one and the choking pressure drop at the vena contracta is the same value as the service pressure drop (i.e. $Pvc = P_2$).
- For large-pressure drop control valves ($C_1 \sim 35$ and $X_T \sim 0.8$), the value of F_G is greater than one and the choking pressure drop at the vena contracta (P1-Pvc) is the less than the service pressure drop (P_1-P_2). This phenomenon is due to large pressure drops at the discharge of the control valve. Double-ported globe valves are examples.

To compare numerical integration at flow pressure drop (P1 –Pvc), with experimental data and ISA-75.01.01 equations, the service pressure drop must be converted to the flow pressure drop.

- For the experimental data, the service pressure drop is adjusted using F_G.

$$(P_1\text{-}Pvc)^{0.5} = (P_1\text{-}P_2)^{0.5} / F_G \quad \text{where } P_1\text{-}Pvc \text{ is used in direct integration}$$

- Similarly, using the ISA-75.01.01 equations, the sizing pressure drop is adjusted to provide the proper mass flow rate.

$(P_1\text{-}P_{vc}) / P_1 = F_G^2 (P_1\text{-}P_2) / P_1$ where P_1-Pvc is used in direct integration

4.4.2 Value of F_G

To determine the value of *gas pressure-adjustment coefficient* F_G, the ($A\ K_G$) *area-flow coefficient parameter* is assumed constant. This is a good assumption if the Reynolds number, based on the upstream flow area, is greater than 100,000.

- At choke pressure, the *gas area-flow coefficient parameter* is:
  $A * K_{G\text{-choke}} = Cg / 1100 = (Cv / 38) * (C1 / 28.9)$

- At low pressure drops, the *gas area-flow coefficient parameter* is:
  $A * K_{G\text{-low pressure}} = (Cv / 38) (F_G)$

- Equating yields an expression for the *gas pressure-adjustment coefficient* F_G. F_G, similar to C_1 and X_T, indicates the degree of pressure recovery.

$F_G = C1 / 28.9$ (Fisher parameters)

or

$F_G = (490 / C\gamma)(F\gamma\ X_T)^{0.5}$ (ISA-75-01.01)

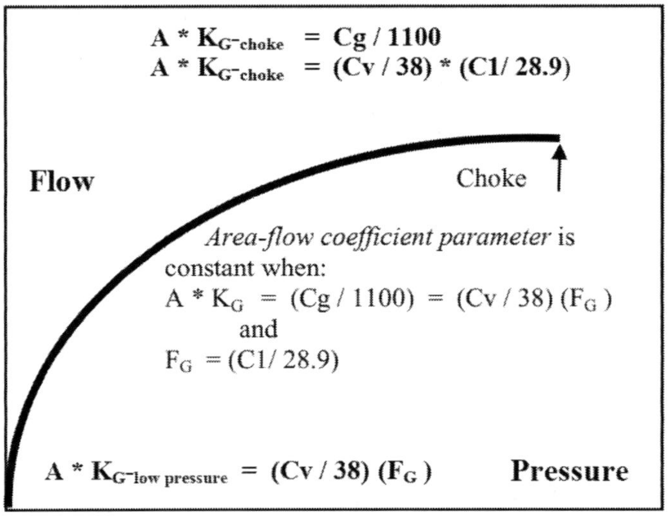

Figure 7-B

4.4.3 Comparison of Fisher Air Data

Figures 8 through 10 present six control valves with flow data (labelled "Fisher") compared with ISA-75.01.01 (labelled "ISA) predictions and numerical direct integration (labelled HDI). The data matches quite well for all six control valves. The control valves include both high-recovery and low recovery valves. The pressure drop is (P_1 – Pcv).

4.4.4 Choke Pressures

The choke pressures from the data and ISA predictions are different from numerical integration. The integration choke points are calculated directly from the integral. The ISA-75.01.01 choked points are determined by the equation and are not expected to exactly match the actual choke point. The data choke points are taken from the graph in the article where the flow rate is a maximum, and are not expected to be accurately read.

4.5 Non-Ideal Gas

The ISA-75.01.01 gas equations were developed from the ideal gas Bernoulli equation. The ISA criterion states the equations are valid when the specific heat ratio ranges between 1.08 and 1.65. For specific heat ratios above 1.65, numerical integration should provide reasonable accuracy, since the mass flux is determined using actual densities and thus is appropriate for high-pressure gases.

In contrast, Marc Riveland [Riveland-2012 and Riveland-2014] has explored non-ideal gas flow using the ISA concept. He has proposed a non-linear modification to the factor $F\gamma$ to account for non-ideal gas behavior so that the ISA-75.01.01 equations may be used directly. His modification agrees well with an analytical model that uses the isentropic exponent ratio rather than the specific heat ratio.

4.6 Incompressible Liquid

As expected, numerical integration matches the ISA-75.01.01 equations well. Since the match is so close, no plot is provided

4.7 Flashing Liquid

Completely incompressible liquids do not choke since the velocity of a sound wave in the liquid is higher than the velocity of the liquid flow. However, if a liquid containing gas flashes, then the mixture density decreases, such that choking may occur. This phenomenon is often called off-gassing or out-gassing. An example in amine system is the flashing

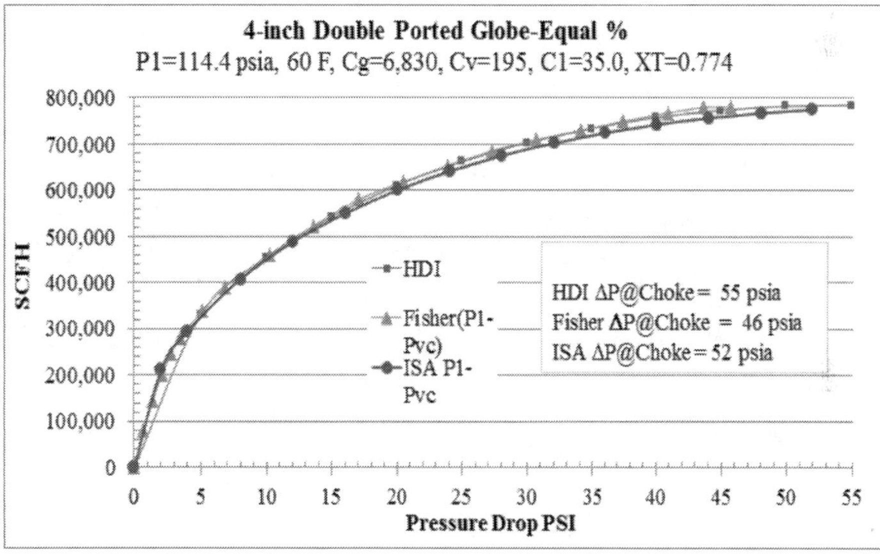

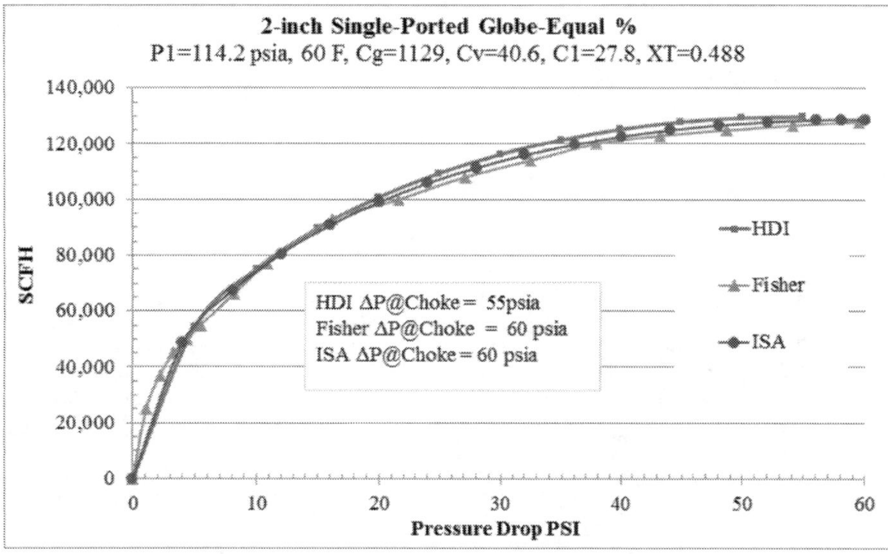

Figure 8

of the rich amine from the bottom of the absorber through the liquid level control valve.

The calculation of the pressure at which the flows choke, due to flashing gas, is important for the control valve sizing. The traditional method is to determine the choking pressure using the ISA-75.01.01 equations from Clause 6.2., Equation 3

$\Delta P_{choked} = (P_1 - P_{vc})$ where P1 is the upstream pressure and Pvc is the pressure at the vena contracta.

Equation 3 calculates the choking pressure at the vena contracta, using the liquid critical pressure recovery factor F_F which is formulated as a function of the pure-component vapor pressure and thermodynamic critical pressure.

$P_{vc} = F_F P_v$ where Pvc is the choking pressure at the vena contracta, Pv is the vapor pressure of the liquid at temperature, and F_F is the liquid critical pressure recovery factor.

Interestingly the empirical equation was derived from pure component data, by determining the choke pressure by numerically integrating the Bernoulli equation.

However, this correlation does not predict accurately the choking pressure for multicomponent fluids [Riveland-2007]. A simple and satisfactory solution for a multi-component fluid is to

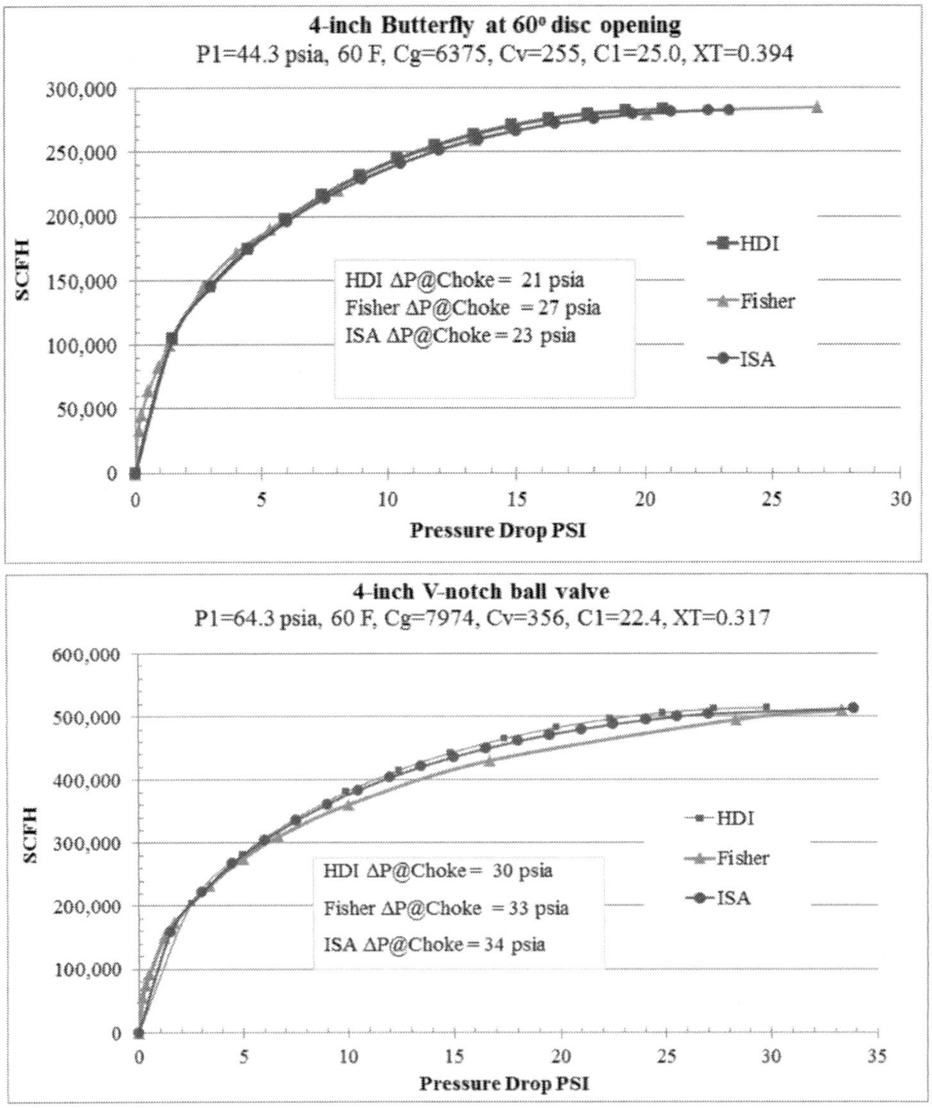

Figure 9

determine the choking pressure at the vena contracta (Pvc) via numerical integration. If the backpressure of the control valve is lower than the choking pressure, then the fluid chokes. The choking pressure at the vena contracta (Pvc) can then be used to find the choking pressure drop in Equation 3.

4.8 Two-Phase Flow

Two-phase flow is not covered by IS-75.01.01. However, numerical integration provides a platform for predicting two-phase flow. But, there are several issues to address.

- A common description of mixed density is $1/\rho_T = x/\rho_G + (1-x)/\rho_L$ where x is the vapor mass fraction and ρ_T, ρ_G, amd ρ_L are the densities for mixed, gas and liquid respectively
- Equilibrium and non-equilibrium may be incorporated into numerical integration, provided non-equilibrium data is available
- A flow coefficient relationship is needed. Based on typical values for a variety of control valves, the flow coefficient for gas (K_{d-G}) and the flow coefficient for liquids (K_{d-L}) may be obtained from parameters in Table D-2, ISA-75.01.01. High-recovery valves such as venturi valves have ratios approaching 1.0. Low-recovery valves have high ratios. Refer to Figure 11.

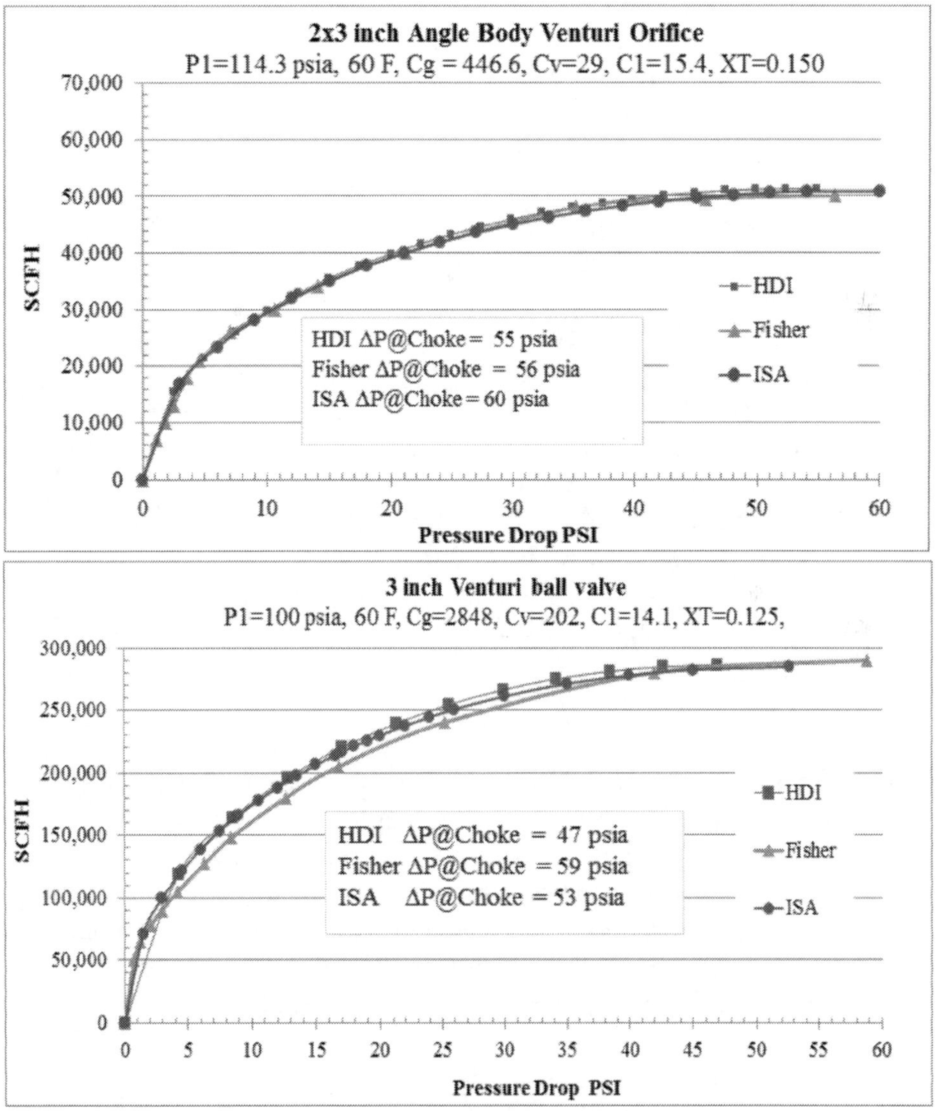

Figure 10

- Liquids: $\{ A * K_{d\text{-}L} \} = (Cv / 38) (F_L)$
- Gases: $\{ A * K_{d\text{-}G} \} = [Cv / 38] [C_1 / 28.9]$
- Then the ratio $(K_{d\text{-}G} / K_{d\text{-}L}) = C_1 / (28.9 F_L)$

4.9 Multi-Stage Control Valves

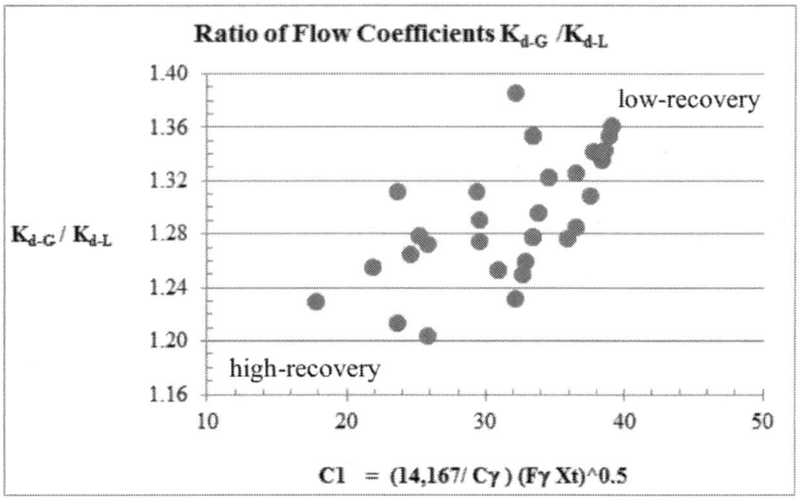

Figure 11

Multi-stage control valves have trims with several stages separated by a gap. The trim may have single or multiple flow passages. The valves are used to create large pressure drops and to reduce noise by lowering the velocities in the valve. Since friction is high, flow coefficients need to be reformulated with a friction component. ISA-75.01.01 (Annex B) provides a method for compressible flow through multistage control valves consistent with the equations for single-stage valves. These valves will not be considered at this time.

5. Pressure Relief Valves

5.1 Overview of Relief Valve Flow

The use of numerical integration has been most often applied to relief valves, rather than orifices and control valves. The integral method has proven successful for relief valves under a full range of fluids and conditions, including flashing liquids, high-pressure gases, dense-phase fluids and multi-phase fluids for both choking and non-choking conditions. Additionally the equation may be easily arranged to account for fluid slip between gas and liquid phases (i.e. the velocity of the phases are not equal) and phase non-equilibrium if appropriate data are available.

5.2 Mass Flow Relation

The mass flow relation will be written in terms of mass flux G [mass/time-area] and area A.

$$W\ [\text{lb/hr}] = A_0\ K_d \left\{ 2407\ \rho_T \left(\int_{P2}^{P1} (1/\rho)\, dP \right)^{1/2} \right\} \qquad 4\text{-}01$$

5.3 Discharge Coefficient

The discharge coefficients for industrial relief valves may be obtained either from the National Board Pressure Relief Device Certifications NB-18 ("Redbook"), or API Standard 526 "Flanged Steel Pressure Relief Valves", which provides a standard framework for valve physical size and discharge coefficients. Relief valve vendors have configured their designs such that either NB-18 or API-526 discharge coefficients and areas can be utilized. To meet the requirements of ASME Boiler & Pressure Vessel Code, the parameters have been selected such that the following relationship is applicable. As such, it should be emphasized that parameters from NB-18 should not be comingled with parameters from API-526.

$$\{ Kd * Area \}\text{-based on NB-18} < \{ Kd * Area \}\text{-based on API-526}$$

API-526 was developed to facilitate engineering projects and it is commonly used since the parameters are almost all standard regardless of the vendor.

- For gases, K_d is 0.975 for many major vendors based on API Standard 526 "Flanged Steel Pressure Relief Valves".
- For liquids, K_d varies between vendors but is on the order of 0.75 based on API-526 values.
- For multi-phase flow, there are several methods to obtain the mixed discharge coefficient.
    - Average the liquid and gas K_ds based on the volume-weighted average of the gas and liquid flows
    - Average with the volume-weighted flows, where each phase is calculated with the respective Kd
    - Kds are chosen depending on whether the flow chokes

5.4. Incompressible Fluids

Numerical integration and analytical equations provide similar relief sizing results for constant density fluids. For fluids that may be have variable densities, such as those above the peak of a phase envelope (typically called critical or dense-phase fluids), numerical methods will provide more accurate representation of the mass flux than an incompressible analytical method.

5.5. Compressible Gas - Ideal and Non-Ideal Gas

API Standard 520-Part I, section 5.6.1 suggests a criterion when it is appropriate to assume that the gas is ideal and relief valve sizing results are acceptable.

- If the real compressibility factor Z is between 0.8 and 1.1 (generally low pressures and high temperatures), then ideal gas behavior may be assumed.

• If the real compressibility factor Z is less than 0.8 or greater than 1.1, then the behavior may not be ideal and the use of analytical equations may not be appropriate. In this case, numerical integration may be used.

5.6. Two-Phase Flow

Two-phase flow in relief valves provided the interest for development of methods other than ideal-gas or incompressible liquids. The subject is well covered in the literature, and the intent is to provide only a short summary.

Although there are have been many models proposed, there are two primary methods covered by international standards.

- API Standard 520-Part I (2014) discusses numerical integration and the Omega method, which is an analytical development for two-phase flow. Other references for the Omega methods include [Leung-2004, Diener-2004]. Since the Omega method is analytical method, it will not be considered in this paper.
- ISO-4126-Part 10 (2010) discusses only the Omega method.

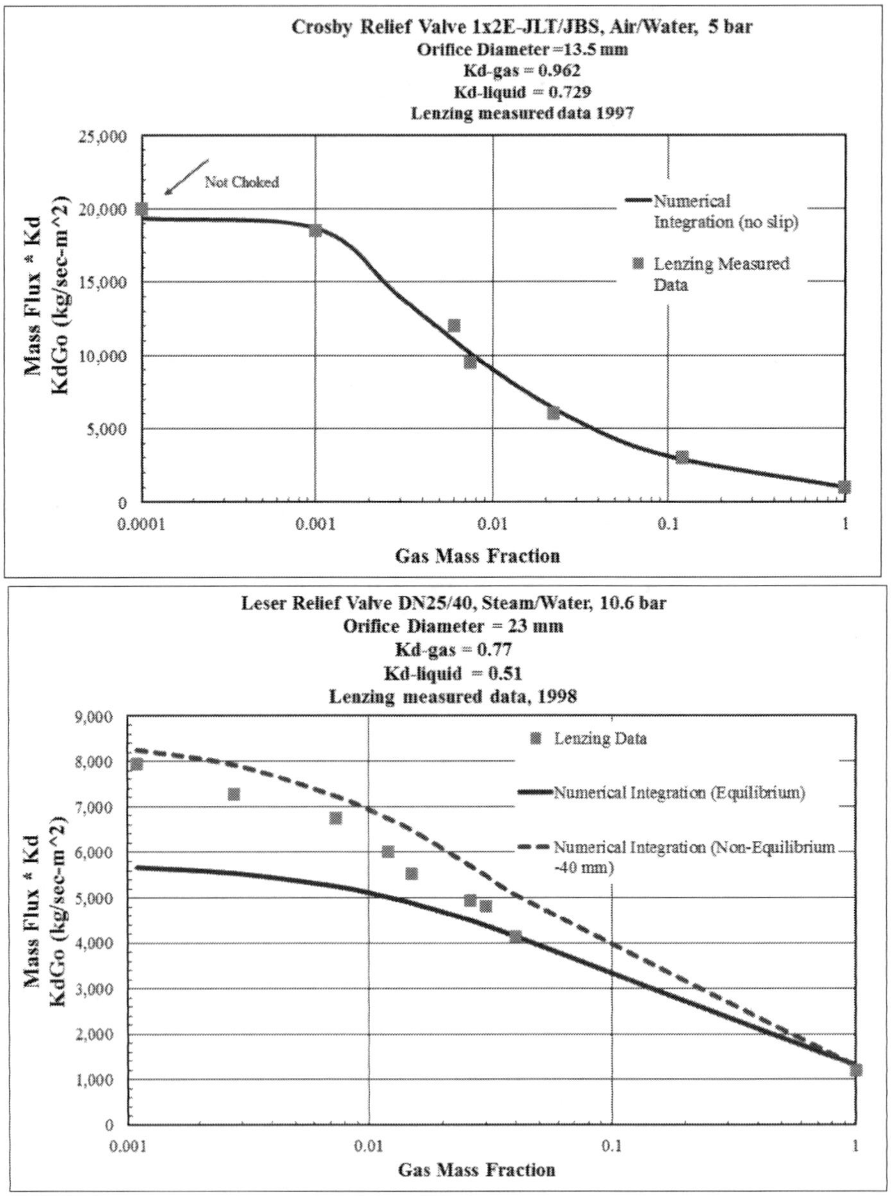

Figure 12

The summary of the numerical integration method is based on several articles by Ron Darby [Darby-2004, Darby-2005]. The references for the experimentally measured data are provided in the articles. The two examples show that numerical integration provides an adequate model of two-phase flow.

The air-water data may be modeled assuming equilbrium. However, the data for team and water shows that a non-equilibrium model provides a better fit.

The discharge coefficients are chosen depending on whether the flow chokes. [CCPS-1998]. This method accommodates compressible gas, multi-phase flow or choking liquids.

- If the flow chokes, only the inlet nozzle flow of the relief valves is involved and "gas" discharge coefficient is used
- If the flow does not choke, both the inlet nozzle and body downstream of the vena contracta is involved and the "liquid" discharge coefficient is used.

7. Conclusions

We demonstrate the use of numerical integration for flow through orifice, control valves and relief valves.

• Orifices. Single-phase hydrocarbon flows through choking thick orifices exhibit flow coefficients K_d expected for gases. On the other hand, two-phase flow into the orifice of natural gas plus propane exhibited slightly higher flow coefficients, possibly due to non-equilibrium effects of the propane and methane. The experimental data for liquid propane, which chokes at the propane bubble point, displays variable flow coefficients, depending on the choking pressure drop. At high pressure drops, the flow coefficient is similar to a liquid; at low pressure drops, the flow coefficients resemble gas flow.

• Control Valves. Numerical integration is an elegant and accurate method to predict ideal gas flows. Since the mass flux is calculated from actual densities, numerical integration is also appropriate for non-ideal gases. Additionally, the method offers a conceptual basis for two-phase flow.

• Relief Valves. Numerical integration has proven successful for relief valves under a full range of fluids and conditions, including flashing liquids, high-pressure gases, dense-phase fluids and multi-phase fluids for both choking and non-choking conditions. Additionally the equation may be easily arranged to account for non-equilibrium conditions if appropriate data are available.

8. Nomenclature

A area [length^2]
{ A K_d } area-flow coefficient parameter for control valves
C "discharge coefficient" used by some publications; same as C_d
$C\gamma$ consolidation of terms that are functions only of the ideal heat capacity ratio k
$C_C = A_2 / A_0$ contraction coefficient $= (A_2/A_1) / \beta^2$ since $(A_0 / A_1) = \beta^2$.
$Cc^2 = 1/\alpha_2$ when velocity profile at position 2 is defined as $\alpha_2 = (A_0/A_2)^2$

C_d "discharge coefficient" or "discharge coefficient without velocity of approach factor" (W experimental / W model)
Cv flow coefficient for control valve [volume/time-pressure^0.5]
Cg apparent gas discharge coefficient for choked gas flow
C_1 parameter representing pressure drop, defined by $Cg = C_1\ Cv$
C_2 consolidation of terms that are functions only of the heat capacity ratio and used to adjust data taken on air for different values of k.
D_H hydraulic diameter
F_L liquid pressure-recovery coefficient for control valves
F_G gas pressure-adjustment coefficient for control valves
$F\gamma$ adjustment for ideal gas heat capacity ratio = $\gamma / 1.4$
f Fanning friction factor
gc constant for British Standard Units 32.174 ft-lb/sec^2
K "loss coefficient" or "resistance coefficient" or "K factor" or "pressure loss coefficient" or "friction loss factor", depending on publication
K_d "flow coefficient" or "discharge coefficient with velocity of approach factor", or "meter coefficient' or "discharge coefficient" for relief valves, depending on publication
L length of pipe
M molecular weight
P pressure [(mass*length / time^2) / length^2] or [mass / time^2 – length]
R the gas constant
T temperature
v velocity [length/time]
W mass flow rate [mass/time]
x vapor mass fraction
X_T critical pressure ratio for gas at choking
Y "expansion factor" which accounts for density changes for a gas
Z compressibility factor

α velocity profile adjustment
β area ratio $(A_0/A_1)^{0.5}$
ρ density [mass/volume]
γ ideal gas heat capacity ratio

Subscripts
0 throat position
1 upstream position
2 vena contracta position 3 vena contracta position
3 downstream position
T mixed phase
G gas density
L liquid density
γ pertains to heat capacity ratio

9. References

9.1 Articles

• M.W. Benjamin and J.G. Miller, "The flow of a Flashing Mixture of Water and Steam Through Pipes", Transactions of the American Society of Mechanical Engineers, Volume 64, October 1942

• J. F. Buresh and C.B. Schuder, "Development of a Universal Gas Sizing Equation for Control Valves", ISA Transactions, Vol 3, 1964, pp 322-328

• R. Darby, *Chemical Engineering Fluid Mechanics*, 2nd Ed, CRC Press, 20001.

• R. Darby, "On Two-Phase Frozen & Flashing Flows In Safety Relief Valves", JLPPI, Vol 17, 2004, pp 255-259

• R. Darby, F. Self, and V. Edwards, "Properly Size Pressure Relief Valves For Two Phase Flow", Chemical Engineering Magazine, June 2002, pp 68-74

• R. Darby, F. Self, and V. Edwards, "Methodology For Sizing Relief Devices For Two Phase (Liquid/Gas) Flow", AIChE Spring National Meeting, Session T5a12 - Pressure Relief Systems, April 23-27, 2001

• R. Diener, and J. Schmidt, "Sizing of Throttling Device for Gas/Liquid Two-Phase Flow, Part 1: Safety Valves, Process Safety Progress, Vol 23(4), March 2004, pp335-344,

• G. Giacchette, M. Leporini, B. Marchetti, and A. Terenzi, "Numerical Study of Choked Two-Phase Flow of Hydrocarbons Fluids Through Orifices", Journal of Loss Prevention in the Process Industries, Vol 27 (2014), pp 13-20

• C. L. Hollingshead, "Discharge Coefficient Performance of Venturi, Standard Concentric Orifice Plate, V-Cone, and Wedge Flow Meters at Small Reynolds Numbers", Graduate Thesis Paper 869, Utah State University, May 2011 <http://digitalcommons.usu.edu/etd/869>

• J. C. Leung, J.C., "A theory on the discharge coefficient for safety relief valve", JLPPI, Vol 17, 2004, pp 301-313

• S. M. Richardson, G. Saville, S. A. Fisher, A. J. Meredith and M. J. Dix, "Experimental Determination Of Two-Phase Flow Rates Of Hydrocarbons Through Restrictions", Trans IChemE, Part B, Process Safety and Environmental Protection, 2006, Volume 84(B1) January 2006, pp 40–53

• M. Riveland, "Use of Control Valve Sizing Equations with Simulation Based Process Data", Presented at ISA Automation Week 2012.

• M. Riveland, "The Impact of non-Ideal Compressible Fluid Behavior on Process Control Valve Sizing and Selection", Presented at Valve World Conference 2014.

• M. Riveland and V. Mezzano, "Critical Service Control Valve Sizing and Selection for Hydrotreating", presented at Emerson Global Users Exchange, 2007

• L. L. Simpson, "Estimate Two-Phase Flow in Safety Devices", Chemical Engineering Magazine, August 1991, pp 98-102.

• A.J. Ward-Smith, "Critical Flowmetering: The Characteristics of Cylindrical Nozzles with Sharp Upstream Edges", International Journal Heat and Fluid Flow, Vol 1(3), September 1979, pp 123-132

9.2 Books

- CCPS. *Guidelines for Pressure Relief and Effluent Handling Systems*, Center for Chemical Process Safety (CCPS), American Institute of Chemical Engineers, New York, March 1998
- Crane, "Flow of Fluids" TP-410, 2009 Edition, 2013
- Richard W. Miller, *Flow Measurement Engineering Handbook*, 3rd Edition, McGraw-Hill, 1996
- National Board of Boiler Inspectors, *Pressure Relief Device Certifications*, (Http:/www.nationalboard.org/Redbook/Redbook.html)

9.3 Standards

- API Standard 520-I (9th Edition, July 2014), "Sizing, Selection, and Installation of Pressure-Relieving Devices-Part I-Sizing & Selection", Section 5.10 and Annex C,
- API Standard 526 (6th Edition, July 2009), "Flanged Steel Pressure Relief Valves"
- ASME MFC-3M-2004 "Measurement of Fluid flow in Pipes Using Orifices, Nozzles and Venturi".
- BS-1042-1-1.2-1989 "Measurement of fluid flow in closed conduits."
- IEC-60534-2-3 (2nd Edition march 2007), "Industrial-process control valves – Part 2-3: Flow capacity – Test procedures"
- IEC-60534-2-1 (2nd Edition march 2011), "Industrial-process control valves – Part 2-1: Flow capacity – Sizing equations for fluid flow under installed conditions"
- ISA-75.01.01 (1st Edition August 2012) (60534-2-1 MOD), "Industrial-Process Control Valves - Part 2-1: Flow capacity – Sizing equations for fluid flow under installed conditions"
- ISO-4126.10.01 (1st Edition-Oct 2010), "Safety devices for protection against excessive pressure - Part 10: Sizing of safety valves for gas/liquid two-phase flow"
- ISO 5167-2:2003 "Measurement of fluid flow by means of pressure differential devices inserted in circular cross-section conduits running full - Part 2: Orifice Plates, 2003

Appendix 1 – Bernoulli Equation – General Form - Details of Derivations

The mechanical energy steady-state balance illustrates how energy is transformed due to kinetic energy, potential energy, process work (free energy), mechanical work and friction loss. The balance is shown as flowing from Point 1 to Point 2; the signs of the terms reflect this convention. Please note that sign convention differs between authors so careful attention should be exercised. The terms are shown as energy per mass {[(length/time)^2-mass] / mass}.

Kinetic energy –work to change the velocity of the fluid	Process work due to pressure changes	Potential energy due to elevation	Work done by fluid to surrounding system (e.g. shaft of a machine)	Friction loss or the irreversible conversion of friction into thermal energy for turbulent flow in straight pipe	Friction loss or the irreversible conversion of friction into thermal energy
↓	↓	↓	↓	↓	↓

$$\Delta \tfrac{1}{2} v^2 + \int_{P2}^{P1} (1/\rho)\, dP + g\,\Delta h + W - \Sigma\left(\tfrac{1}{2} v^2 f\, 4L/D_H\right) - \Sigma\left(\tfrac{1}{2} v^2 K\right) = 0$$

A1-01

The mechanical energy steady-state balance will account for errors due to the velocity profiles, friction estimates, and others, by introduction of three adjustable parameters.

- α are used to adjust for velocity profiles. For example, in turbulent flow with a flat velocity profile $\alpha = 1$.
- C_d is the "discharge coefficient", or actually correlating parameter which accounts for deviations in assumptions. The discharge coefficient is used to calibrate the equation to flow data and is the ratio of the experimentally measured flowrate (or mass flux) to that calculated from the model. Thus the discharge coefficient will differ depending on the model and assumptions and the quality of the experimental data. Discharge coefficients between different models can only be compared qualitatively. Finally, the placement of the discharge coefficient within the equation is based on convenience and is not theoretically established.

Expanding above equation and adding the parameters, the following general equation is the starting point for the individual situations.

$$\frac{\tfrac{1}{2} v_1^2}{\alpha_1} - \frac{\tfrac{1}{2} v_2^2}{\alpha_2} + C_d^2 \int_{P2}^{P1} (1/\rho)\, dP + g\,(h_1 - h_2) - \Sigma\left(\tfrac{1}{2} v^2 f\, 4L/D_H\right) - \Sigma\left(\tfrac{1}{2} v^2 K\right) = 0$$

A1-02

Appendix 2 – Flow Orifices (square-edge) - Details of Derivations

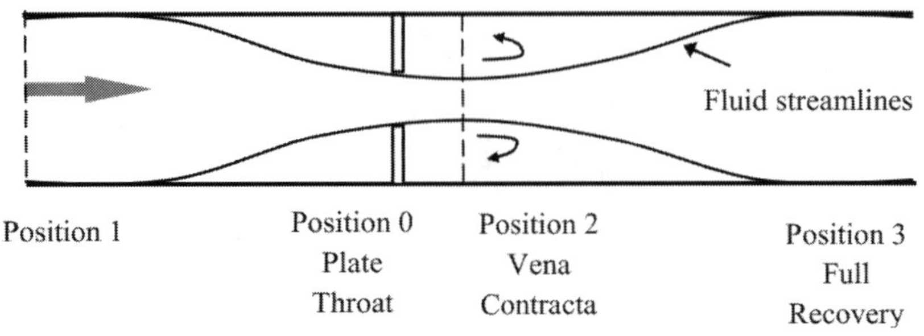

Figure 1

1. The general equation will now be modified to develop the basic equations for numerical integration. The order of the integral and signs must be carefully considered since different authors integrate in different orders. The relief valve is physically short in length so that wall friction as a function of straight pipe length is neglected. However other sources of friction are included. In this paper, the friction formulation K will be related to the inlet velocity. The K will take a different form if related to discharge velocity. If comparing to other forms in the literature, some authors assume the kinetic energy adjustment for the shape of the velocity profiles (alpha) is a multiplier, rather than a divisor.

(e.g. $(\frac{1}{2} v^2 \alpha)$ rather than $(\frac{1}{2} v^2 / \alpha)$)

$$\frac{\frac{1}{2} v_1^2}{\alpha_1} - \frac{\frac{1}{2} v_2^2}{\alpha_2} + C_d^2 \int_{P2}^{P1} (1/\rho)\, dP + \overset{\nearrow 0}{g\,(h_1 - h_2)} - \overset{\nearrow 0}{\Sigma\,(\tfrac{1}{2} v^2 f\, 4L/D_H)} - \Sigma\,(\tfrac{1}{2} v^2 K) = 0 \qquad \text{A2-01}$$

2. The equation will be written at the vena contracta (position 2) since that is the position where choking occurs. Note the change in the signs.

$$+ \frac{\frac{1}{2} v_2^2}{\alpha_2} -- \frac{\frac{1}{2} v_1^2}{\alpha_1} + \Sigma\,(\tfrac{1}{2} v_1^2 K) = C_d^2 \int_{P2}^{P1} (1/\rho) \qquad \text{A2-02}$$

3. The Bernoulli equation explains how energy is transformed. And together with the continuity equation, it can establish the mass flow. The kinetic energy term will be used since it contains the velocity term.

4. The mass flow is the same regardless $W = W_1 = W_2 = W_0$

$$W = \rho_1 v_1 A_1 = \rho_2 v_2 A_2 \quad \text{or} \quad v_1^2 = v_2^2 \, [\rho_2 A_2 / \rho_1 A_1]^2 \qquad \text{A2-03}$$

$$v_2^2 = W^2 / [\, \rho_2^2 A_2^2 \,] \qquad \text{A2-04}$$

5. We can rearrange into a typical form by rearranging terms and taking square root. The following is a common form of the equation.

$$W = \frac{(\alpha_2)^{0.5}\,\rho_2 A_2 C_d \left(2\int_{P2}^{P1} (1/\rho)\,dP\right)^{1/2}}{\{1 - ([\alpha_2/\alpha_1][\rho_2 A_2/\rho_1 A_1]^2) + \Sigma([K\alpha_2][\rho_2 A_2/\rho_1 A_1]^2)\}^{1/2}} \qquad \text{A2-05}$$

6. The general equation can be made more specific for orifices or venturis.

- As usual, a Beta ratio is defined using A_1 and A_0 as $(A_0 / A_1) = \beta^2$.
- Since the throat is a known experimental parameter, it is convenient to write the equations for the throat (position 0). Thus we need to define a relationship between A_0 and A_2. Typically, a contraction coefficient is defined as follows. The relationship to the velocity profile is discussed later.

  $A_2 = A_0\, C_C$ or
  $C_C = A_2 / A_0$

- For venturi meters, $A_2 \sim A_0$ and $C_C \sim 1$
- For orifices, the vena contracta (A_2) is narrower than the orifice throat (A_0). We can then combine the ratio of (A_2/A_1) as: $(A_2/A_1) = \beta^2 C_C$. This will be substituted in the above.
- Consolidate terms in the dominator and considering signs

$$W = \frac{(\alpha_2)^{0.5}\,\rho_2 A_2 C_d \left(2\int_{P2}^{P1} (1/\rho)\,dP\right)^{1/2}}{\{1 - ([Cc\,\beta^2 (\rho_2/\rho_1)]^2\,[(\alpha_2/\alpha_1) - (\alpha_2 K)])\}^{1/2}} \qquad \text{A2-06}$$

7. One last expression will complete the general equation, which still includes a friction term. This form includes the Beta ratio in the denominator.

- At the discharge, the velocity profile for the orifice may be given by $\alpha_2 = (A_0/A_2)^2$. This is an important assumption since it provides a method to quantify alpha α. Further, by assuming a profile for α_2, then α_2 may be eliminated and we only need a value for α_1. As noted earlier for venture meters, $A_0 \sim A_2$ and the velocity profile results in $\alpha_2 = 1.0$.
- Note that since the contraction coefficient is defined as $Cc = A_2 / A_0$
- With the above, then we can write an identity is $Cc^2 = 1/\alpha_2$.

$$W = \frac{\rho_2 A_0 C_d \left(2 \int_{P2}^{P1} (1/\rho)\, dP \right)^{1/2}}{\{ 1 - ([\beta^2 (\rho_2 / \rho_1)]^2 \ [(1/\alpha_1) - (K)]) \}^{1/2}} \qquad \text{A2-07}$$

9. The value of $\alpha_1 = 1.0$, since the flow is turbulent at the inlet. If we assume that friction is considered sufficiently small to be included in the coefficient of discharge, then $K \simeq 0$. Also $\beta^2 = (A_0 / A_1)$

$$W = \frac{\rho_2 A_0 C_d \left(2 \int_{P2}^{P1} (1/\rho)\, dP \right)^{1/2}}{\{ 1 - [\beta^4 (\rho_2 / \rho_1)^2] \}^{1/2}} \qquad \text{A2-08}$$

$$G = \frac{W}{\dfrac{A_0 C_d}{\{ 1 - [\beta^4 (\rho_2 / \rho_1)]^2 \ ] \}^{1/2}}} = \rho_2 \left(2 \int_{P2}^{P1} (1/\rho)\, dP \right)^{1/2} \qquad \text{A2-09}$$

10. The discharge coefficient term may be rearranged depending on whether the fluid is a vapor or liquid or two-phase.

$$K_d = C_d \ / \ (1 - [\beta^4 (\rho_2 / \rho_1)^2])^{1/2} \qquad \text{A2-10}$$

If the differences in densities between the inlet and discharge are sufficiently small, the error is included in the coefficient of discharge. In this form, the term discharge coefficient C_d is sometimes called "coefficient of discharge without velocity of approach factor". [Crane-2009, Miller-1996] Similarly, the flow coefficient K_d is called the "coefficient of discharge with velocity of approach factor".

$$K_d = C_d / (1 - \beta^4)^{1/2} \qquad \text{A2-11}$$

Appendix 3 – Control Valves - Details of Derivations

An expression for the term representing area and discharge coefficient is needed in terms of parameters utilized by the control valve industry.

App 3.1 Area and Flow Coefficient For Choking Gas

An expression for the area-gas discharge coefficient parameter for choking will be found by equating the analytical Bernoulli equation to the choking control valve gas equation.

• <u>Bernoulli Equation for Choking Gas Flow</u>.

$W = A * K_{GAS} * C\gamma * P1 * (M/TZ)^{1/2}$ A3-01

Where:

- K_{GAS} is the gas flow coefficient
- $C\gamma$ is a consolidation of terms that are functions only of the heat capacity ratio γ=Cp/Cv. (For ideal gas, γ=Cp/ [Cp – R] where R is the gas constant)
- $C\gamma = 520 * [\ \gamma * \{ [2/(\gamma+1)]^{((\gamma+1)/(\gamma-1))} \}\ ]^{0.5}$ (e.g. $C\gamma$ is 355.9 for air.)

• <u>Fisher Equation for Gas Flow</u>. The historical Fisher equation will be used since it is the traditional form. The equivalent ISA-75.01.01 is provided in a later section. From air data Fisher has defined the apparent gas discharge coefficient for choked gas flow as Cg, which is determined experimentally.

$Q = Cg\ C_2\ P1\ [520 / (T\ SG)]^{1/2}$ (refer to Buresh & Schuder eq 13) A3-02

$W = 0.3234 * Cg * C_2 * P1 * (M/TZ)^{1/2}$

Where:

- SG is defined as M / 28.965 with the molecular weight of air being 28.965
- Q [scf/h] = W [lb/hr] * 379.49 [scf/mole] / M [lb/mol] at 14.696 psia & 60°F.
- $C_2 = [\ 2.0665\ ] * [\ \{ \gamma/(\gamma+1) \} * \{ [2/(\gamma+1)]^{(2/(\gamma-1))} \}\ ]^{0.5}$ where C_2 is a consolidation of terms that are functions only of the heat capacity ratio and used to adjust data taken on air for different values of k. (e.g. C_2 is 1.00 for air)

• <u>Area and Discharge Coefficient for Choking Gas Flow {A K_{GAS} }</u> – Fisher Formulation. When equated both equations

$\{ A\ K_{GAS} \} = Cg / 1100 = (Cv/38)(C_1/28.95)$ A3-03

Where:

- K_{GAS} is the gas flow coefficient
- $(C\gamma / C_2) = 355.9$ for all values of γ.
- $C_1 = Cg/Cv$ or $Cg = C_1 * Cv$ by definition

App 3.2 Area and Flow Coefficient For Incompressible Liquid

The expression for incompressible flow is found by equating Bernoulli equation with ISA-75.01.01 expression.

- Bernoulli equation. Analytical Bernoulli equivalent in terms of area and liquid discharge coefficient

$Q_{GPM} = A * K_{LIQUID}\ 38\ ((P_1 - P_{VC}) / G)^{0.5}$ A3-04

where P_1 is the upstream pressure and P_{VC} is the pressure at the vena contracta.

- ISA-75.01.01 Expression. For the liquid discharge coefficient Cv is defined.

$Q_{GPM} = Cv ((P_1 - P_2) / G)^{½}$

$Q_{GPM} = (F_L * Cv) ((P_1 - P_{VC}) / G)^{0.5}$ A3-05

where

- $(P_1 - P_2)$ is the "service pressure drop"
- P_1 is the upstream pressure
- P_2 is the downstream pressure located after the vena contracta. P_2 will reflect any pressure recovery that occurs in the downstream valve body after the vena contracta.
- An expression is needed to convert from downstream pressure to the vena contracta pressure. Without this conversion, the value of Cv will overstate the flow rate. The "liquid pressure recovery coefficient" F_L is used.

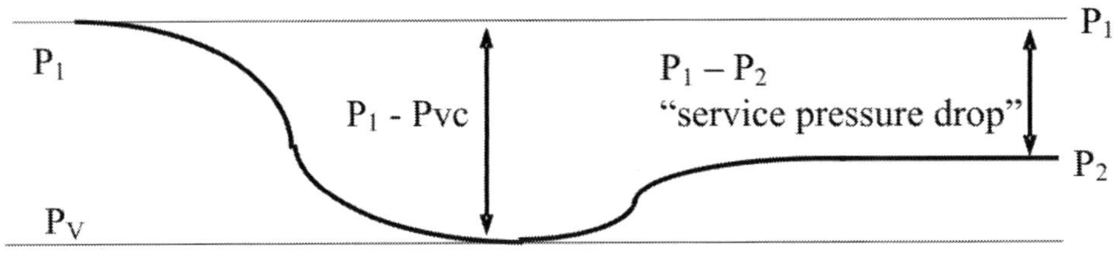

$F_L^2 = (P_1 - P_2) / (P_1 - Pvc)$

- Area and Flow Coefficient for a Liquid {A*K_{LIQUID}} When equated equations:

{ A * K_{LIQUID} } = (F_L * Cv) / 38 A3-06

App 3.3 Appendix- ISA Form for {A*K_d} for Compressible Fluid

Similarly, an expression may be derived using ISA-75.01.01 parameters.

- Bernoulli. $W = A * K_{GAS} * C\gamma * P1 * (M/TZ)^{½}$

- ISA 75.01.01-2012-Equation 06

$W\ (lb/hr) = 19.3 * Cv * Y * (X_{size})^{0.5} * P1 * (M/T_1Z_1)^{0.5}$

- Expansion Factor Y

$Y = 1.0 - (X_{size} / 3 * X_{choked})$

Y is the "expansion factor" which accounts for density changes for a gas.

- At lower pressures, Y has a maximum value 1.0 at zero pressure drops.
- At choking, the value has been established at 0.667. The ISA parameter X_T [IEC 60534-2-3] is experimentally measured to match 0.667. The value of 0.667 is approximately the value of Y that one may find for a typical flow.

• Pressure Ratio. The substitutions are:

- If Not Choked

$X_{size} = \Delta P / P_1 = (P_1 - P_2) / P_1$ and Y if found from above equation

- If Choked

$X_{size} = X_{choked} = F\gamma\, X_T = F\gamma\, (P_1 - P_{Choke}) / P_1$ and $Y = 0.667$

• Area and Flow Coefficient for Compressible Fluid {A K_{GAS} } – ISA Formulation
Equate equations:

$$\{A\, K_{GAS}\} = 12.873\, (Cv / C\gamma)(F\gamma\, X_T)^{0.5} \qquad \text{A3-07}$$

Where:

- $F\gamma = \gamma / 1.4$ and γ is the ratio of heat capacities Cp and Cv. The term $F\gamma$ is use to adjust the gas for heat capacity ratios other than 1.4. (Note air γ=1.40)

App 3.4 Relationship Between C1 and XT

Equating A3-03 and A3-07,

$$C_1 = (14{,}167 / C\gamma)(F\gamma\, X_T)^{0.5} \qquad \text{A3-08}$$

$$C_1 = (39.807 / C_2)(F\gamma\, X_T)^{0.5}$$

$$(F\gamma\, X_T) = (C_1 * C_2)^2 \; / \; 1{,}584.6 \qquad \text{A3-09}$$

$$(F\gamma\, X_T) = (C_1 * C\gamma)^2 \; / \; (14{,}167)^2$$

Since

- $(C\gamma / C_2) = 355.9$ for all values of γ.
- $Cg = C_1\, Cv$

Appendix 4 – Pressure Relief Valves - Details of Derivations

1. The general equation will now be simplified for pressure relief valves.

• For relief valves, the coefficient of discharge is labelled K_d. Generally K_d is called the "flow coefficient" or "coefficient with velocity of approach factor", but for relief valves it is termed "discharge coefficient". For relief valve sizing, the term $[\beta^4 (\rho_2/\rho_1)^2] = [(A_0/A_1)^2 (\rho_2/\rho_1)^2]$ is considered sufficiently small so that it is lumped in the relief valve coefficient of discharge. ($\beta^2 = (A_0/A_1)$)

$$K_d = C_d / \left(1 - [\beta^4 (\rho_2/\rho_1)^2]\right)^{0.5} \qquad \text{A2-10}$$

2. Substituting into the orifice equation A2-08, in British Standard Units, G in units of lb/hour-inch^2

$$W\ [\text{lb/hr}] = \left\{ A_0 K_d \right\} \left\{ 2407\ \rho_2 \left(\int_{P2}^{P1} (1/\rho)\, dP \right) \right\}^{1/2} \qquad \text{A4-01}$$

If multi-phase, then the density at the vena contracta may be expressed as an homogenous average density. For flow through a relief valve, homogeneous flow is a reasonable assumption.

The density used is ρ_T = average density lb_m/ft^3 at point of maximum mass flux density (vena contracta) from the homogenous vapor-liquid average

$$1/\rho_T = x/\rho_G + (1-x)/\rho_L \qquad \text{A4-02}$$

11TH GLOBAL CONGRESS ON PROCESS SAFETY

How to Size a Rupture Disk Vent-line for Two-Phase Gas/Liquid Flow based on Current Engineering Practices

Prof. Dr. Jürgen Schmidt
Center of Safety Excellence
Joseph-von-Fraunhofer Str. 9
76327 Pfinztal (near Karlsruhe) / Germany
juergen.schmidt@cse-institut.de
www.cse-institut.de

CSE CENTER OF SAFETY EXCELLENCE

Prepared for Presentation at
American Institute of Chemical Engineers
2015 Spring Meeting
11th Global Congress on Process Safety
Austin, Texas
April 27-29, 2015

How to Size a Rupture Disk Vent-line for Two-Phase Gas/Liquid Flow based on Current Engineering Practices

Prof. Dr. Jürgen Schmidt
Center of Safety Excellence
Joseph-von-Fraunhofer Str. 9
76327 Pfinztal (near Karlsruhe) / Germany
juergen.schmidt@cse-institut.de
www.cse-institut.de

Keywords: rupture disk, two-phase flow, engineering practice, sizing, safety device

Abstract

Emergency vent lines are classically sized by one-dimensional, steady state numerical simulations of the pressure drop in all piping elements. Typically, one or more cross sections with critical flow conditions occur. Due to the complex nature of a two-phase flashing flow and the uncertainty of the sizing models, large deviation may arise when sizing rupture disk vent-lines. The current state of knowledge in sizing vent-line systems represented by the HNE-CSE model is outlined and exemplified by 4 industrial rupture disk vent line systems. The mass flow rate and the pressure profile highly deviate from simple sizing procedures based on homogenous equilibrium flow with the assumption of 60 % flow contraction within the rupture disk. The simple sizing procedure may lead to large underestimation of piping forces, malfunction of retention systems and undesired environmental hazards.

1. Introduction

In the chemical, pharmaceutical and petrochemical industries, vessels and pipes are protected against overpressure by safety relief devices, typically a rupture disk or a safety valve. Especially for rupture disks, the dischargeable mass flow rates are predominantly dependent on the flow resistance in the inlet and outlet lines. Rupture disk devices and vent-lines have to be considered as a combined relief unit.

In contrast to a safety valve relief line, the inlet and outlet line of a rupture disk vent-line can have a significantly smaller diameter, which may be important in case of re-installation of existing vent-lines. Critical flow conditions can be established at multiple locations, in the rupture disk or immediately behind an area change in the inlet and/or outlet line. During the relief of a vessel, the location(s) with critical flow may change. Hence, to size a rupture disk device under typical plant operating conditions, extensive fluid dynamic knowledge is necessary. Current sizing procedures often highly underestimate the narrowest flow area of a rupture disk device and the mass flow rates released through rupture disk vent-lines, [1, 2, 3]. A precise sizing procedure, especially for two-phase flashing flow, is lacking. Neither a test facility is standardized nor are any measurements published.

During the last twenty years a tremendous effort has been made in sizing relief lines for flashing two-phase flow by considering phase slip between gases and liquids (mechanical non-equilibrium) and boiling delay as the thermodynamic non-equilibrium effects. Both increased significantly the precision of sizing safety devices. To harmonize the methods for sizing safety valves and rupture disk vent-lines [4, 5], the HNE-DS model for flashing gas/liquid-two-phase flow through throttling devices, recommended in ISO 4126-10 [5], is extended to the flow through pipes and rupture disk devices.

2. Hazard Prevention from an Industrial Emergency Relief

Most of the chemical, pharmaceutical and petrochemical plants are protected with classic end-of-pipe technology, a safety valve or a rupture disk connected to a disposal system or discharging a gas to atmosphere. Cyclone separators [6, 7] or catch drums [8] are typically installed in the chemical industry to catch the liquid phase of a two-phase gas/liquid stream in case of an emergency relief.

Safety valves are installed mostly because of their reclosing nature. In contrast, the advantage of rupture disks is the almost full opening of a relief line, see Figure 1. This leads to smaller diameter inlet and outlet lines for the same mass flow rate to be discharged. In addition, high quality rupture disks have a high safety integrity [9]. The reliability of mechanical devices under plant conditions is currently under consideration in Europe.

Figure 1: Bursted rupture disk, Rembe GmbH, Brilon

In the chemical and pharmaceutical industries, rupture disk devices are usually not installed directly on a reactor. Instead, they are placed at a location where inspection and service can effectively be afforded. Typically, a relatively short inlet line connects the reactor with the rupture disk device, which is mounted to an outlet line and further on to a gathering relief pipe outside the plant building. The gathering line leads into a separator and the gas is released over the roof into the ambient or in a quench, washer or flare [10]. To avoid any unintentional rupture of the disk in case of product release or due to reactor cleaning, vacuum supports are used in front of the rupture disks.

In the petrochemical industry, long and often large diameter rupture disk vent-lines are used to discharge gas into a flare system.

The mass flow rates of liquid and gas discharged through an emergency relief vent-line are highly dependent on the worst case scenario considered and the overestimation of the safety device due to the uncertainty of the sizing procedure. Based on the transient mass flow rate through the emergency relief line and the pressure profile,

- The forces acting on the vent-line are estimated
- Separator and drums are sized
- Scrubber and flares are proven to be additionally suitable for emergency relief rates
- The accumulation of explosive gases is computed
- The pressure resistance of plant buildings and measuring stations is checked
- The toxic hazard potential of discharged gases at the fence of a chemical plant is compared to the accepted exposure limits, i.e. in Germany typically the ERPG 2 value [11, 12].

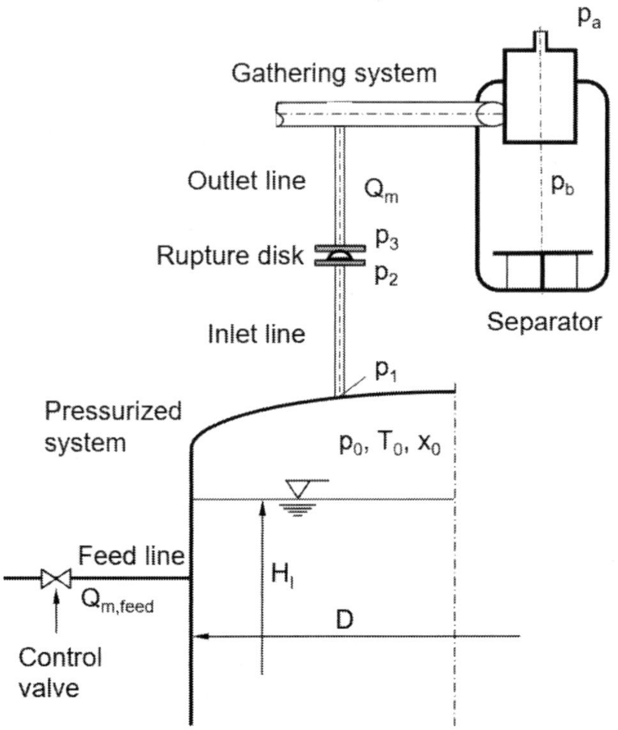

Figure 2: Industrial rupture disk vent-line system

Overestimation of emergency relief lines may lead to tremendous extra costs, malfunction of downstream process equipment, higher gas and liquid discharge rates and, subsequently, additional environmental problems.

3. Steps to size a rupture disk

5 steps are necessary to size a rupture disk vent-line: (1) the sizing case must be defined based on a risk analysis (HAZOP study) to identify the worst credible case out of all reasonable possible failures. (2) The flow regime at the entrance of the rupture disk vent-line shall be estimated by means of the level swell in the pressurized system and (3) the minimum flow rate to be discharged must be determined. It is generally calculated by an energy and mass balance around the pressurized system. The following two steps belong to the sizing of the vent-line system. Prior to a final calculation, the geometry and size of the rupture disk vent-line has to be estimated (4) and in the final step, the estimation must be validated by a precise and detailed pressure drop calculation (5).

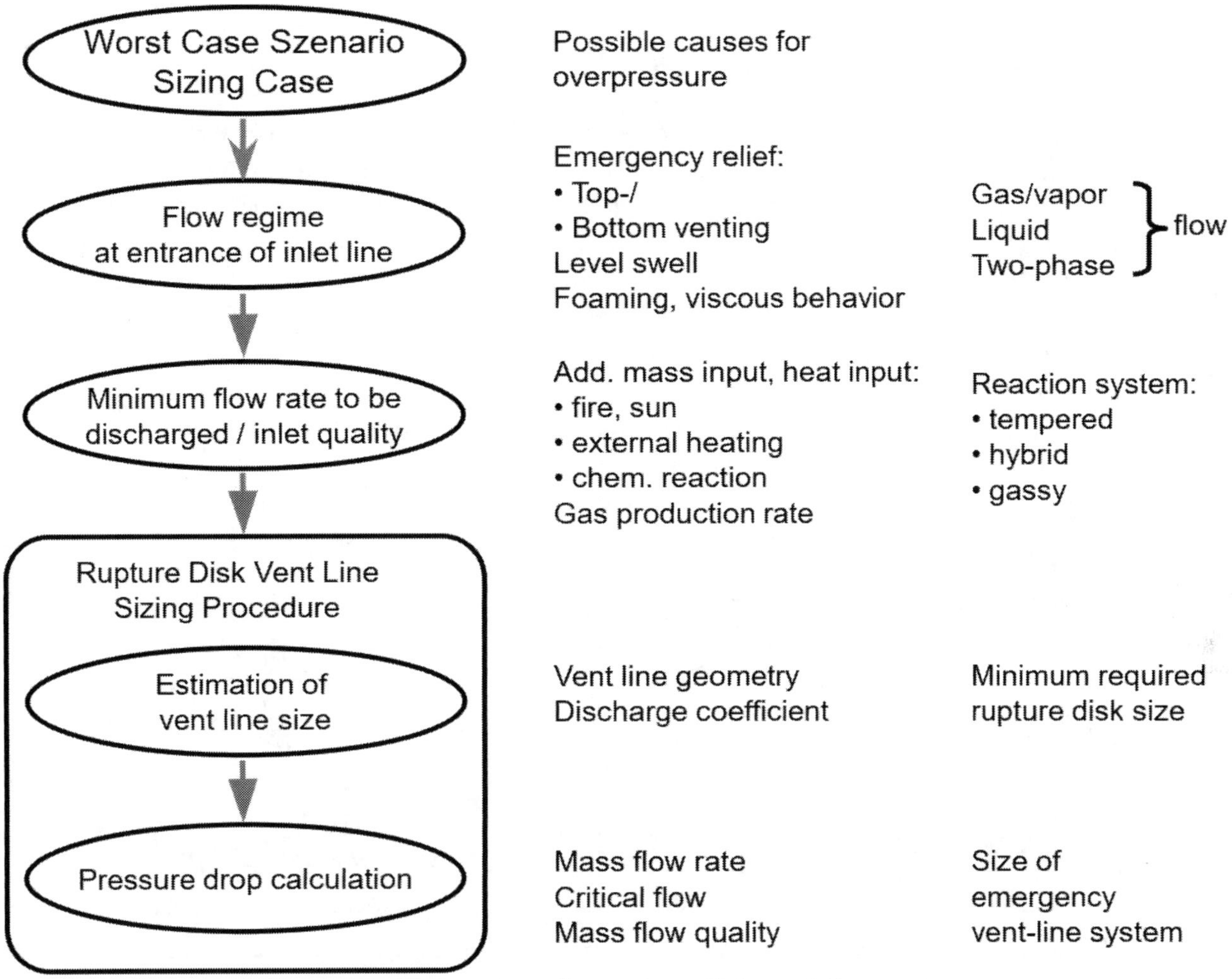

Figure 3: Sizing steps for rupture disk vent-line systems

In the following, all steps are described in detail.

3.1 Risk Analysis (Worst Case Scenario)

Reasonably conceivable deviations from normal plant operation shall be identified and evaluated regarding their hazardous risk potential in a Process Hazard Assessment and Safety Evaluation (PHASE) [13]. Several well established procedures such as HAZOP (Hazard and Operability Study) [14], PHA (Preliminary Hazard Analysis) [15], What-if-Analysis or the more quantitative methods including Fault-Tree-Analysis [16] and Event-Tree-Analysis [17] are used in practice. These methods are often supplemented by checklists [18] with several levels of detail. All those procedures provide the means to assess the causes for a pressure increase. These causes can be related to changes in mass and energy transfer to or from the pressurized system, or a deviation from the normal reaction system, e.g., when a thermal runaway reaction occurs. Schmidt et al. [19] summarized the most common causes for a pressure rise. Some of these may occur simultaneously or immediately one after another. Subsequently, one or more independent process deviations may have to be taken into account as possible sizing cases for the rupture disk vent-line.

3.2 Flow Conditions at the Entrance of the Inlet Line

The size of a rupture disk vent-line is mainly determined by the flow condition at the entrance of the inlet line – liquid, gas or a two-phase gas/liquid flow. If the liquid hold-up in a reactor exceeds a threshold value, the content swells up to the reactor lid. Many experiments with several fluids, mostly under laboratory conditions, have been evaluated to find a lower boundary for the limiting filling threshold in a pressurized system at sizing conditions, which avoids a two-phase release of the vessel inventory (solid line in Figure 4, [20,21]). Above the curve, single-phase gas flow cannot be ensured, two-phase flow is most likely.

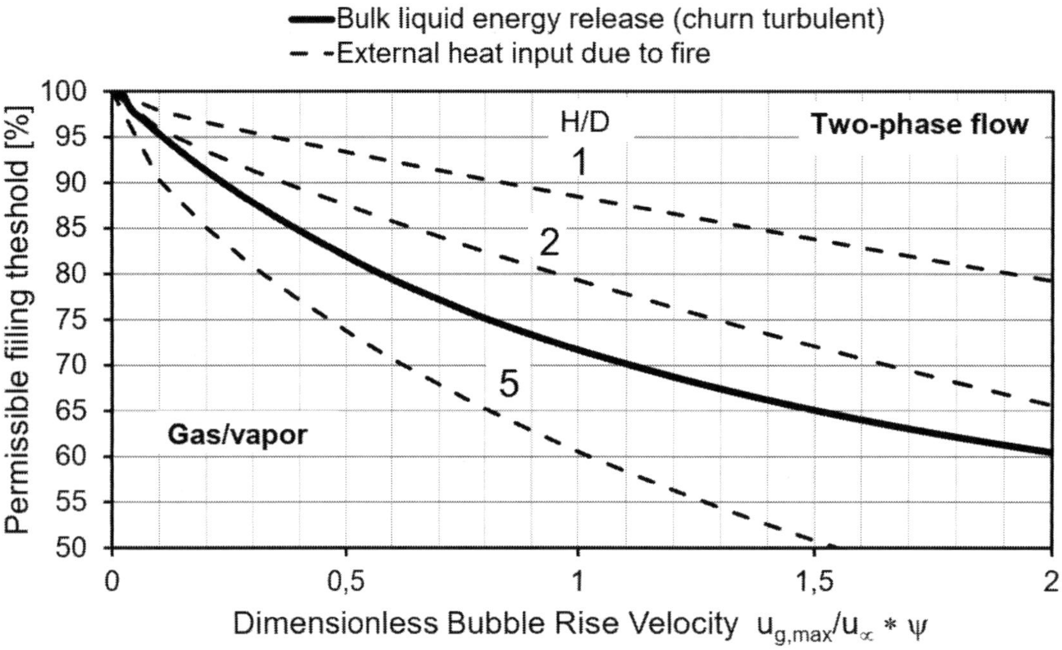

Figure 4: Maximum permissible filling threshold at sizing conditions for external heat input due to fire and a bulk liquid heat release (e.g. chemical reaction) in dependence of the dimensionless bubble rise velocity and the reactor height to diameter ratio [22]

In Figure 4 the permissible filling thresholds for reactor inventories with a viscosity of a non-foaming liquid at sizing conditions of the rupture disk less than 100 mPas are given for heat releases due to a chemical reaction (solid line) and for heat inputs by fire through the cylindrical wall of a vertical vessel (dashed lines). The threshold is given depending on the dimensionless bubble rise velocity, i.e. the ratio of the maximum gas volumetric flux (superficial gas velocity) above the liquid level at sizing conditions $u_{g,\max} = u_g(Q_{in,\max})$ and the rise velocity of a single bubble in a liquid plume u_∞ (churn turbulent flow):

$$u_g = \frac{\dot{Q}_{in}}{\Delta h_{v0}} \cdot \frac{v_{g0}}{\frac{\pi}{4} \cdot D^2} \qquad (1)$$

$$u_{\infty} = 1.53 \cdot \sqrt{v_{l0}} \cdot \left[\mu_{l0} \cdot g \cdot \left(\frac{1}{v_{l0}} - \frac{1}{v_{g0}} \right) \right]^{0.25} \quad (2)$$

Herein, $k_{\infty} = 1.53$ for non-foaming, low viscous liquids ($\mu_{l0} < 0.1 Pas$) and $k_{\infty} = 1.18$ for other fluids. For typical solvents used in industry, the rise velocities of single bubbles u_{∞} according to equation 2 at boiling conditions are in the range of 14 to 18 cm/s [22].

The dimensionless bubble rise velocity $u_{g,\max} / u_{\infty}$ at maximum heat input $\dot{Q}_{in}$ is weighted by the factor Ψ in Figure 4 to compare the level swell due to heat input at bulk energy release, e.g. a chemical reaction, and the energy input due to external heating, e.g. by a fire. The weighing factor Ψ in case of a homogeneously distributed bulk heated reactor (e.g. chemical reaction) is defined as $\Psi = \Psi_{bulk} = 1$. In case of fire, Ψ is the ratio of the cross sectional area of the reactor related to the wetted area of the cylindrical part of the reactor between the bottom of the reactor and a maximum height of fire of 7.5 m:

$$\Psi = \Psi_{fire} = \frac{\frac{\pi}{4} D^2}{\pi \cdot D \cdot H_{wet}} = \frac{D}{4 \cdot \left(H_{fire} - \Delta H_{bottom} \right)}; \quad \text{with} \quad H_{fire} \leq 7.5m \quad (3)$$

The level swell is calculated by the model of Fisher et al. [23] valid for vertical cylindrical vessels not under fired (dashed lines in Figure 4). Often a reactor skirt avoids any accumulation of flammable liquids below the reactor. If the reactor is installed at a height of more than 7.5 m above the ground, no heat input due to fire is assumed.

Heat transfer due to external heating (e.g. with steam) is done at lower energy rates compared to a fire. Bubbles are formed mainly at the wall. As a consequence, the level swell process no longer occurs evenly distributed over the reactor cross section. Instead, large eddies establish in the reactor promoting the mixing and condensation of bubbles. The swell up process is less pronounced in this case. Further information on level swell calculations is given in [5].

3.3 Mass Flow Rate to be Discharged and Inlet Mass Flow Quality

The energy transferred into a pressurized system at sizing conditions leading to an impermissible pressure or temperature (worst credible scenario) must be released through the emergency relief vent-line. Therefore, an energy balance around the pressurized system would yield the energy to be released. The energy is transported by a mass flow rate – the mass flow rate required to be discharged $Q_{m,out}$. Generally, this mass flow rate can be calculated from the heat release rate $\dot{Q}_{in}$ divided by the heat of vaporization of the reaction mixture at sizing conditions of the safety device Δh_{v_0} and the pressure relief factor ξ:

$$Q_{m,out} = \frac{\dot{Q}_{in}}{\Delta h_{v0} \cdot \xi} \quad (4)$$

The heat release rate $\dot{Q}_{in}$ has to be calculated depending on the mode of heat input, see Table 1, i.e. external heat transfer through the wall or a bulk heat release in the liquid phase.

Less conservative results do arise for low viscous (< 100 mPas), non-foaming liquids if the vapor liquid disengagement is considered prior to the inlet of the safety device, see [19, 24].

Table 1: Equations for the heat input into the reactor

Way of heat input	Heat rate into the reactor
External heating	$\dot{Q}_{in} = U \cdot A_{wet} \cdot \left(T_{heat} - T_b\left(p_0\right)\right)$
Fire	$\dot{Q}_{in} = 43200 \cdot F \cdot A_{fire}^{0.82}$ $F = 1$ no fire insulation $F < 1$ fire insulation of reactor (cf. Tab. 6, ISO 23251 [25])
Chemical reaction (tempered reaction)	$\dot{Q}_{in} = \frac{M_0 \cdot cp_{l0}}{2}\left(\left.\frac{dT}{dt}\right\|_0 + \left.\frac{dT}{dt}\right\|_{max}\right)$

Table 2: Equations for the calculation of the pressure relief factor for reactors with contents partly vaporizing at sizing conditions (tempered systems.)

Composition of the relief mass flow rate	Way of heat input	Pressure relief factor ξ
vapor	External heating	$\xi = 1$
	Chemical reaction	Iterative solution: $\frac{cp_{l0} \cdot \Delta T_{over}}{\Delta h_{v0}} - \xi + \frac{v_0}{\left(v_{g0} - v_{l0}\right)}\left\{1 - ln\left(\frac{v_0}{\left(v_{g0} - v_{l0}\right) \cdot \xi}\right)\right\} = 0$ with $\Delta T_{over} = T_b\left(p_{max}\right) - T_b\left(p_0\right)$;
Two-phase vapor liquid flow	External heating fire	Iterative solution: $\frac{1}{\xi}\left\{\frac{v_0}{\left(v_{g0} - v_{l0}\right)} - \xi - \frac{cp_{l0} \cdot \Delta T_{over}}{\Delta h_{v0}}\right\} - ln\left(\frac{v_0}{\left(v_{g0} - v_{l0}\right) \cdot \xi}\right) = 0$ with $\Delta T_{over} = T_{max} - T_b\left(p_0\right)$; $T_{max} = T_b\left(p_{max}\right) \cup T_{heat}$
	Chemical reaction	$\xi = \left(\sqrt{\frac{v_0}{v_{g0} - v_{l0}}} + \sqrt{\frac{cp_{l0} \cdot \Delta T_{over}}{\Delta h_{v0}}}\right)^2$ with $\Delta T_{over} = T_b\left(p_{max}\right) - T_b\left(p_0\right)$

The temperature difference ΔT_{over} in Table 2 corresponds to the increase in temperature within the reactor during the pressure relief. This can be the maximum temperature of the heat transfer medium for an externally heated reactor; at a chemical reaction, the temperature can rise up to the boiling temperature at the maximum pressure during a runaway. The temperature difference ΔT_{over} influences substantially the size of the relief cross-section of the safety device [26]. In principle, the set pressure of the safety device should be adjusted as low as possible in case of exothermic chemical reactions in vaporizing systems so that the speed of reaction and, hence, the heat release into the mixture at sizing conditions of the safety device is small.

All equations in Table 2 are given for tempered reaction systems. If the reaction produces gases or the system is not vaporizing, a classic vent sizing according to ISO 4126-10 [5] is recommended to account for the gas release rate of the reaction.

In addition to the mass flow rate to be discharged, the void fraction ε_0 or quality $\dot{x}_0$ at the entrance into the relief vent-line is needed. Both depend on the reactor volume V_R and the mass within the reactor M_0. The void fraction under assumption of a homogeneous mixed reactor $\varepsilon_{0,hom}$ is:

$$\varepsilon_0 = \varepsilon_{0,hom} = \frac{V_R - M_{l0} \cdot v_{l0}}{V_R} \qquad (5)$$

The mass flow quality at the entrance of the relief vent-line under stagnation conditions reads [23, 3]:

$$\dot{x}_0 = \frac{\varepsilon_0 \cdot v_{l0}}{\varepsilon_0 \cdot v_{l0} + (1 - \varepsilon_0) \cdot v_{g0}} \qquad (6)$$

3.4 Estimation of Vent-line Size

The rupture disk device and the vent-line have to be considered as a fluid dynamically connected system, i.e. in contrast to a safety valve, the rupture disk cannot be sized separately. Instead, sizing is an iterative procedure starting with the geometry of the rupture disk vent-line under consideration (existing plant) or an estimated size of a vent-line. The size of the rupture disk vent-line should be as low as possible to avoid e.g. undesired level swell inducing two-phase flow, extra piping forces or an overestimation of downstream retention systems. But it must be large enough to ensure at least a discharge of the mass flow rate required to be discharged out of the pressurized system ($Q_{m,RD} \geq Q_{m,out}$).

A good first estimate for the required area of the relief vent-line and the minimum required size of the rupture disk A_{RD} is to neglect the resistance of the inlet and outlet piping and, additionally, to assume a narrowest flow cross section in the rupture disk equal to 61% of the rupture disk net flow area [27]:

$$A_{RD} \cong \frac{Q_{m,out}}{C_{VC} \cdot \sigma_{RD} \cdot \dot{m}_{nozzle}} \quad \text{with} \quad C_{VC} = 0.61 \quad \text{and} \quad \sigma_{RD} = \left(\frac{d_{RD}}{d_{pipe}} \right)^2 \qquad (7)$$

$\dot{m}_{nozzle}$ is the mass flux through a frictionless nozzle without heat transfer to the wall and σ_{RD} is the ratio of the net flow area of a burst rupture disk to the area of the disk holder, which shall be equal to the pipe inner diameter. The net flow area of the rupture disk is defined as an equivalent area of an orifice with the same flow cross section as the rupture disk and is typically between 0.6 (disk with a permanent vacuum support) almost 1.

3.5 Pressure Drop Calculation

Based on the first estimate for geometry and size of the rupture disk vent-line, the mass flow rate and pressure drop shall be calculated and iterated, until the vent-line systems fits the sizing requirements. This final step of the calculation procedure is outlined in detail in the next chapter.

4. Mass Flow Rate and Pressure Profile in Vent-Line Systems

In industry, sizing of emergency relief vent-lines is mostly done by means of a one-dimensional steady state numerical simulation of the pressure drop through all pipe elements. Because of the complex nature of the flow a simulation can only be performed with high uncertainties.

4.1 Phenomenological description of the flow through rupture disk vent-line systems

A top vented emergency relief of a pressurized system most often leads to a compressible discharge of steam or a compressible gas/liquid two-phase flow. The flow stream contracts and therefore accelerates from stagnation conditions in the pressurized system to the entry of the vent-line system, before the stream reattaches the wall. The initial pressure is partly converted into kinetic energy and internal friction losses. From the entrance, the fluid stream follows the piping system and the pressure is further decreased due to wall friction. In front of the rupture disk device, the flow starts to contract up to a vena contracta – the narrowest flow cross section behind the rupture disk device. Further downstream, the fluid stream expands again until it reattaches the pipe wall, see schematic in Figure 5. In the narrowest flow cross section, the fluid reaches a local pressure minimum. After the rupture disk device, the stream follows the piping system until it reaches the end of the vent-line. Herein, the pressure drops further due to wall friction until the exit pressure is reached. The pressure decrease along the relief line leads to a reduced density of the gas phase and, in case of a saturated gas/liquid two-phase flow, it causes the mass flow quality to increase. Any decrease in density is accompanied by an increase in velocity.

Any flow contraction and any diameter enlargement is a potential location where a critical pressure ratio may be reached and this will limit the dischargeable mass flow rate. Pipe inlets, bends, T-junctions, pipe enlargements, the end of the relief line and the rupture disk device are typical sources for critical flow conditions in vent-lines to occur.

The dischargeable mass flow rate of a compressible fluid through a rupture disk vent-line system depends on the resistance of the piping geometry between the pressurized system and the first cross section where critical flow condition were established (choking area). This choking area determines the maximum flow rate for given inlet conditions. Rather seldom is subcritical or an incompressible flow through the relief line. In case of critical compressible flow, the flow resistance of the complete vent-line system is relevant for the calculation of the mass flow rate to be discharged.

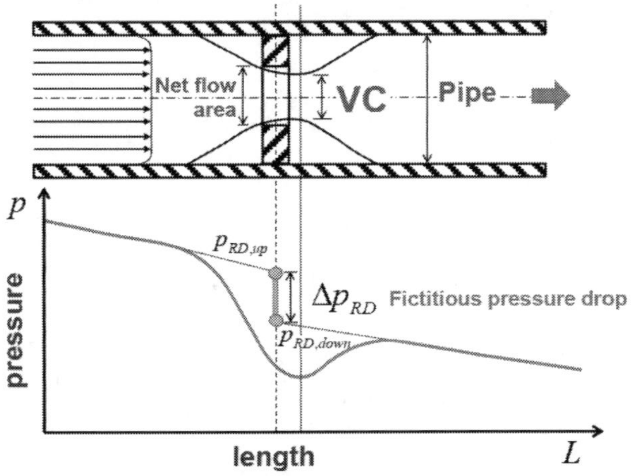

Figure 5: Flow stream in a rupture disk vent line

In a safety relief valve, the choking area is by definition the seat area of the safety device itself. In contrast, choking conditions in rupture disk vent-lines may occur in any of the local narrowest flow cross sections of the complete vent-line system.

4.2 Industrial Sizing Codes

The two-phase discharge of reactors and vessels through rupture disk vent-lines typically encountered in industry including a complex piping system and critical flow conditions is not yet investigated in detail. Almost no experimental data is available and very few standards are published. Neither API 520 [1] nor ISO 4126-10 [5] are applicable for complex rupture disk vent-line systems. AD2000 [3], ASME Pressure Vessel Code [2] and ISO 4126-6 [28] are limited to single-phase flow based on resistance coefficients measured in subcritical almost incompressible fluid flow. Test facilities are not yet standardized for compressible flow, [29, 4]. The resistance coefficients K_R typically measured to characterize rupture disk devices are not applicable for compressible fluids, see [19].

According to the current sizing convention, the mass flow rate to be discharged and the pressure profile in the piping system is determined by the fluid acceleration, gravity effects and losses. The pressure losses are assumed to be equal to the sum of all resistances of individual pipe sections with the same diameter, fittings like bends, T-junctions, pipe enlargements and the rupture disk. The flow is often simulated by a one-dimensional, steady state frictional flow in an adiabatic tube. All flow parameters are averaged in time and over the cross sectional area. Interfacial and wall friction is considered by empirical mathematical models, which have been derived mainly from subcritical single-phase flow experiments.

In general, critical flow conditions are not considered in pipe inlets, bends, T-junctions and other fittings. Most often this is additionally true for rupture disks, although they may cause a distinct flow contraction. Pipe contractions and bends are not or only rarely installed in relief lines. Piping codes often determine critical flow conditions only in pipe enlargements, valve seat areas and at the pipe ends. Non-equilibrium effects like slip and boiling delay are most often not considered at all.

The mass flow rate through complete industrial emergency relief vent-lines is not yet validated by tests.

Figure 6 gives a calculated example of a two-phase steam/water-flow through a rupture disk vent-line flowing into a gathering system applying a typical industrial sizing code. Critical flow conditions are predicted in the rupture disk device (considered as symmetrical contraction followed by an enlargement), at the end of the outlet line and at the end of the gathering system. The first cross section with critical flow conditions limits the mass flow rate (choking conditions).

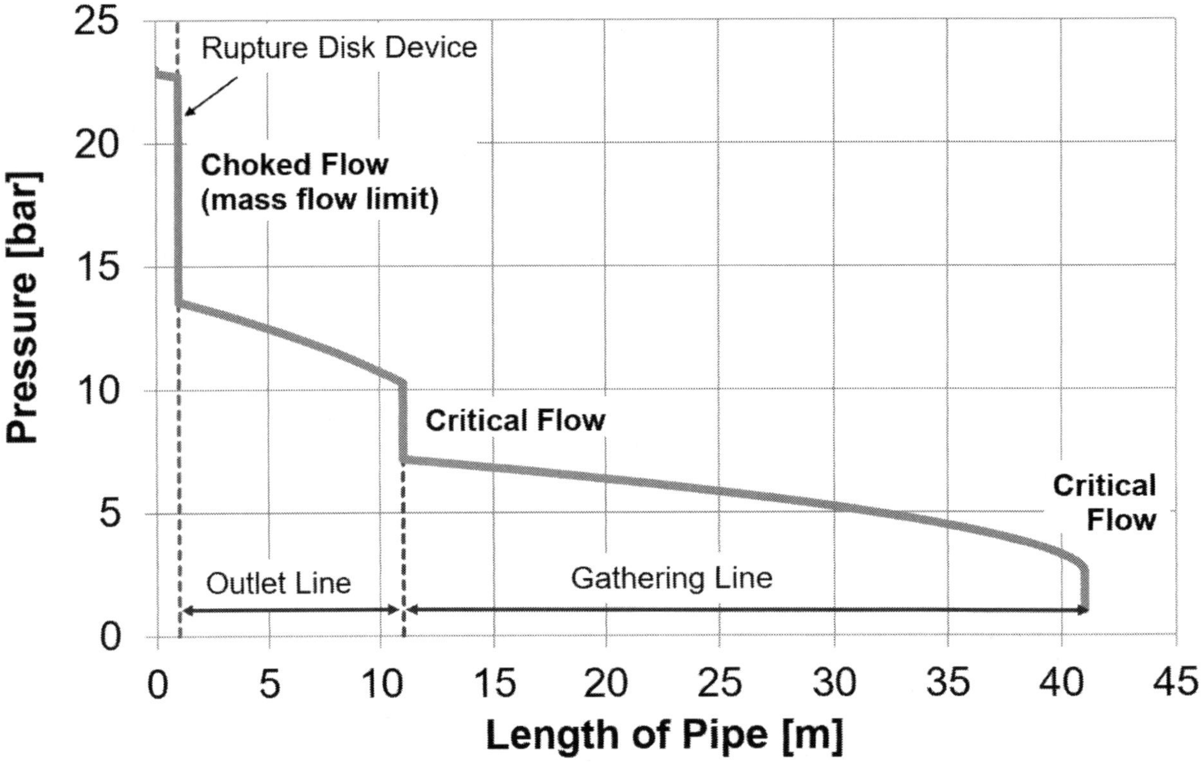

Figure 6: Pressure profile in a typical rupture disk vent-line system in the chemical industry during an emergency relief of a two-phase steam/liquid flow.

During the last 10 years industrial sizing codes have significantly improved. Property data handling and the thermodynamics of gases and liquids (single- and multi-component) have been implemented consistently. Nevertheless, sizing of pressure relief vent-lines is not yet sufficient, neither in most of the in-house nor in commercially available codes. Sizing codes in practice have several disadvantages, which may prevent a conservative sizing result for a rupture disk vent-line system:

1. Empirical models for resistance calculations of fittings can be arbitrarily chosen and connected to each other. Irrespective of whether physical model simplifications are contradicting or whether the conditions applied are far beyond any validated application range, the user gets a meaningless result without warning.

   As examples: a critical mass flow rate model developed for initially subcooled conditions may be applied in saturated two-phase flow conditions; pressure drop and critical mass flow rate of a fitting with a high viscous two-phase fluid can be calculated based on a resistance coefficient measured in a subcritical incompressible flow. Even sophisticated engineers do not have the chance to overcome those failures.

2. Sizing codes do not include critical flow detection in all of the flow cross sections, where critical flow may potentially be establish.

3. Sizing codes include inconsistencies at critical flow conditions. E.g., the pressure drop across a pipe contraction is calculated based on an empirical model validated for subcritical flow conditions. To account for critical flow conditions within the narrowest flow cross section one of several critical flow models may be chosen and applied to the vena contracta area. The empirically determined pressure drop is then extrapolated by means of the critical flow rate. Any of the critical flow models itself consists of a pressure drop calculation between inlet and vena contracta, which is not consistent to the pressure drop extrapolated by the empirical model [30]. Most of the sizing code users are not aware of these inconsistencies.

4. Conservative results are generally only possible, if each model needed for a certain task – e.g. calculation of mass flow rate or pressure profile – is consistent to the sizing methodology. The choice of models and sub-models is key important for the overall calculation results and its conservatism. The assumption of homogeneous equilibrium two-phase flow may lead to an underestimated mass flow rate but to a conservative minimum required flow area. The same mass flow rate may lead to a highly underestimated pressure profile in the vent-line system. The choice of another, conservative mass flow rate model is needed to calculate the pressure drop.

5. Resistance and discharge coefficients are typically measured under defined laboratory conditions with steam, air or water and are most often not applicable in flows above 50% of the reduced pressure of a substance (pressure divided by the thermodynamic critical pressure) respectively 90% of the reduced temperature. At the thermodynamic critical point the densities of steam and liquids are equal and, hence, the coefficients must coincide. Surprisingly many applications in industry are within this range.

The current trend for industrial sizing codes is a transient flow calculation. Besides the discrepancies mentioned above, additional problems occur: opening characteristics of safety devices have to be considered. For a rupture disk device this can be easily estimated because of its non-reclosing nature. But the opening characteristic of a reclosing safety valve is affected by the flow in the inlet and outlet pipe, which may cause a fully different calculated opening behavior compared to experimental results.

Further experimental evidence is strongly needed to validate the sizing convention for safety relief vent-lines. This is especially true for rupture disk vent-line systems typically encountered in practice.

A first step towards a conservative sizing procedure would be a set of consistent flow models for the complete piping system. The flow through any industrial throttling device may be approximated based on an isentropic nozzle flow model and an appropriate flash calculation. Wall and interfacial friction, slip, mechanical and thermodynamic non-equilibria etc. may be empirically encountered. Such a model is applicable in single-phase gas and liquid flow as well as in two-phase flashing or non-flashing flow. It is useable for safety valves, control valves, orifices, nozzles, pipe contractions and enlargements [24, 31, 32, 33] and in its latest development for pipes, bends and other resistances [19]; it may be applied for subcritical and in critical flow conditions. The model is called HNE-CSE (**H**omogeneous **N**on-**E**quilibrium **C**onsistent set of **S**izing **E**quations). In the following, the HNE-CSE model is derived and applied for sizing rupture disk devices.

4.3 The HNE-CSE model for sizing emergency vent-line systems

The flow through a relief vent-line may be estimated by a one-dimensional, steady state frictional flow in an adiabatic tube, where interfacial and wall friction, acceleration due to a change in cross-sectional area and density, as well as the change in potential energy due to gravity is considered. All flow parameters are averaged in time and space, Figure 7.

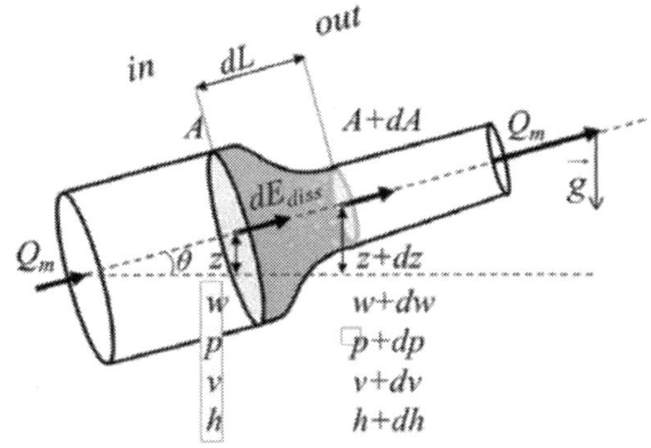

Figure 7: Fluid element representing the flow in a relief vent-line

By means of the first and second law of the thermodynamics and the continuity equation, an equation for the mass flow rate through throttling devices, pipes and pipefittings is obtained by integrating the system between the inlet location – here represented by subscript *in* – and the outlet location – represented by subscript *out*. In dimensionless form it yields the General Sizing Equation, see [19]:

$$C=\frac{1}{v_{out}^{*}}\cdot\sqrt{\frac{-\int_{\eta_{in}}^{\eta_{out}}v^{*}\cdot d\eta-\frac{g\cdot\sin\theta\cdot L}{v_{in,t}\cdot p_{in,t}}}{1-\Gamma\cdot\left(1-K_{R,in}\right)}}=\frac{Q_{m}}{A_{out}\cdot\sqrt{2\frac{p_{in,t}}{v_{in,t}}}}\tag{8}$$

$v_{in,t}$ and $p_{in,t}$ are the inlet specific volume and inlet pressure at stagnation inlet conditions. $K_{R,in}$ is the resistance coefficient regarding wall and interfacial frictional effects related to the static inlet condition and the factor Γ represents the acceleration between inlet and outlet:

$$K_{R,in}=\frac{2}{w_{in}^{2}}\int dE_{diss}\qquad\text{and}\qquad\Gamma=\frac{w_{in}^{2}}{w_{out}^{2}}=\left(\frac{A_{in}^{2}}{A_{out}^{2}}\right)\left(\frac{v_{out}^{2}}{v_{in}^{2}}\right)\tag{9}$$

The dimensionless pressure ratio $\eta=p/p_{in,t}$ and the dimensionless specific volume $v^{*}=v/v_{in,t}$ refer to the inlet stagnation conditions.

The flow coefficient C is the actual mass flow rate Q_m related to the maximum feasible mass flow rate $Q_{m,\max}$, calculated for a frictionless $K_{R,in}=0$, incompressible flow $v=v_{in}=v_{out}$ through a nozzle from an infinitely large reservoir $w=w_{in,t}=w_{in}=0$ with a certain pressure at the inlet $p_{in,t}$ and vacuum conditions at the outlet $p_{out}=0$.

An analytical solution of the General Sizing Equation can be derived for a homogeneous mixture of gas and liquid at mechanical (no slip) and thermodynamic (no boiling delay) equilibrium conditions, see [19]. Following the derivation of the ω-methods [26, 31, 32, 33] and applying the two-phase equation of state $v^{*}=\omega\cdot(1/\eta-1)+1$ yields, see [19]:

$$C=\frac{\eta_{out}}{\left[\omega\cdot\left(1-\eta_{out}\right)+\eta_{out}\right]}\cdot\sqrt{\frac{\left(\eta_{in}-\eta_{out}\right)\cdot(1-\omega)+\omega\cdot\ln\left(\frac{\eta_{in}}{\eta_{out}}\right)-\frac{g\cdot\sin\theta\cdot L}{p_{in,t}\cdot v_{in,t}}}{\left[1-\Gamma^{2}\cdot\left(1-K_{R,in}\right)\right]}}\tag{10}$$

where:

$$\omega=\frac{1}{\kappa_{in,t}}\frac{\dot{x}_{in,t}\cdot v_{g_{in,t}}}{v_{in,t}}+\frac{cp_{l_{in,t}}\cdot T_{in,t}\cdot p_{in,t}}{v_{in,t}}\cdot\left[\frac{v_{g_{in,t}}-v_{l_{in,t}}}{\Delta h_{v_{in,t}}^{2}}\right]^{2}\cdot N\tag{11}$$

$$N=\left(\dot{x}_{in,t}+cp_{l_{in,t}}\cdot T_{in,t}\cdot p_{in,t}\cdot\left(\frac{v_{g_{in,t}}-v_{l_{in,t}}}{\Delta h_{v_{in,t}}^{2}}\right)\cdot\ln\left(\frac{\eta_{in}}{\eta_{out}}\right)\right)^{a}\tag{12}$$

$$\Gamma = \left(\frac{A_{out}}{A_{in}}\right) \cdot \left[\frac{\omega \cdot (1-\eta_{in}) + \eta_{in}}{\omega \cdot (1-\eta_{out}) + \eta_{out}} \cdot \frac{\eta_{out}}{\eta_{in}}\right] \quad (13)$$

This set of equations is referred to as the **H**omogeneous **N**on-**E**quilibrium **C**onsistent set of **S**izing **E**quations (HNE-CSE). The HNE-CSE model defines the dimensionless mass flow rate for a throttling device, a pipe or a pipefitting dependent on the compressibility coefficient ω, the non-equilibrium factor N and the acceleration ratio Γ.

The outlet pressure ratio η_{out} of a throttling device, a pipe or a pipefitting may equal the back pressure ratio $\eta_{out} = \eta_b$ or the critical pressure ratio $\eta_{out} = \eta_{crit}$, where the dimensionless mass flow rate C reaches it maximum value ($dC/d\eta = 0$). The critical pressure ratio shall be iteratively calculated:

$$\left(\omega - \frac{\eta_{crit}^2}{2 \cdot C(\eta_{crit})^2}\right) \cdot \frac{C(\eta_{crit})}{\eta_{crit}} \cdot \frac{1}{\left[\eta_{crit} - \omega \cdot (\eta_{crit} - 1)\right]\left[1 - \Gamma(\eta_{crit})^2 \cdot (1 - K_{R,in})\right]} = 0 \quad (14)$$

By means of the critical pressure ratio $\eta_{out} = \eta_{crit}$ in the HNE-CSE model, equations 10-14, the critical mass flow rate $C_{crit} = C(\eta_{crit})$ in a throttling device or pipefitting with a flow contraction or at the end of a constant diameter straight pipe section can be calculated. In case of non-flashing gas/liquid two-phase flow it is recommended to correct the dimensionless mass flow rate with the mechanical non-equilibrium factor derived in [31; 33].

The HNE-CSE model is an extension of the HNE-DS model including not only non-equilibrium effects but also the kinetic energy change at the entry and the flow resistance of a throttling device, pipe or pipefitting. The HNE-CSE model reduces to well-known sizing equations at boundary conditions, e.g. (1) the HNE-DS model in case of a plenum flow into an ideal frictionless nozzle ($\Gamma_{nozzle} = 0, K_{R,in} = 0, \theta = 0$) or (2) the original ω-method if additionally $N = 1$ is set, or (3) into incompressible pipe flow ($\Gamma_{pipe} = 1$) respectively compressible flow in a pipe ($\Gamma_{pipe} = \left(v_{in}^* / v_{out}^*\right)^2$).

The HNE-CSE model may also be applied to multi-component flashing systems whose saturation temperature range does not exceed 100 K. It may lead to uncertain results if the sizing conditions are close to the thermodynamic critical conditions of a fluid. Here, the state of knowledge is to solve the General Sizing Equation even if non-equilibrium effects are not considered.

In addition to the mass flow rate and outlet pressure, the mass flow quality in the narrowest flow cross section and at the end of an element is needed in case of flashing two-phase flow. It can be deduced from the energy balance between the inlet of the element and the flow cross section considered. The enthalpy at the inlet of the element h_{in} is transformed in to specific kinetic energy due to acceleration of the fluid Δe_{kin} (change in area and/or density) and in to specific potential energy Δe_{pot}:

$$\dot{x}_{out} = \frac{h_{out} - h_{l,in}}{\Delta h_v} = \frac{h_{in} + \Delta e_{kin} + \Delta e_{pot} - h_{l,in}}{\Delta h_v} \tag{15}$$

Without any area and density change (incompressible flow with same inlet and outlet diameter) the change of state would be isenthalpic. In practice, it is often favorable to calculate the flash based on fictitious stagnation conditions. Stagnation conditions can easily be calculated by the HNE-CSE model assuming an isentropic diffusor flow from a current condition to a plenum state with zero velocity ($A_{out}/A_{in} = 0$, $w_{out} = 0$, $\Gamma = 0$, $K_{R,in} = 0$).

4.4 Application of the HNE-CSE model

The HNE-CSE model is a consistent set of equations to calculate the mass flow rate and pressure profile in vent-line systems with single-phase or multi-phase flashing or non-flashing flow. The vent-line system may include almost any type of throttling device (e.g. rupture disk, safety valve, control valve, ball valve, orifice, diffusor, nozzle, pipe contraction, pipe enlargement, etc.), straight pipe elements and pipefittings (bends, T-junction, etc.).

The total vent-line system must be stream-downwards divided into pipe elements like the pipe entrance, throttling devices, pipefittings, the pipe exit and constant diameter straight pipe sections. For each pipe element the following inlet parameter shall be known: pressure, temperature, mass flow quality, inlet diameter. Additionally, the narrowest geometrical flow cross section and the outlet diameter of each element are needed. Based on these parameters, the mass flow rate to be discharged shall be guessed. Then, the HNE-CSE model shall be applied to each pipe element in eight steps:

1. Estimate the narrowest flow cross section (vena contracta) if it may differ from the inlet or outlet diameter, see chapter 3.4.
2. Compute the inlet stagnation conditions by means of the HNE-CSE isentropic diffusor model.
3. Calculate the dimensionless critical pressure and critical mass flow rate and check that the guessed mass flow rate is equal or less.
4. If a flow contraction occurs inside the pipe element: calculate the pressure in the narrowest flow cross section iteratively by assuming an isentropic flow from the inlet into that section. Define all other fluid conditions.
5. Determine the flow resistance coefficient either from the device manufacturer or from literature, see chapter 4.5. The two-phase resistance coefficient may be deduced from a two-phase multiplier and the resistance coefficient for single-phase gas and liquid flow.
6. Calculate the outlet pressure of the pipe element.
7. Calculate the mass flow quality at the outlet of the pipe element
8. Determine all other outlet conditions

The procedure shall be applied repeatedly throughout all pipe elements until the exit of the vent-line is reached. The mass flow rate shall be varied until the calculated exit pressure is equal to the given back pressure or the calculated critical pressure, in case critical flow is reached in one of the pipe elements.

4.5 Resistance coefficient

The resistance coefficients $K_{R,in}$ for piping elements like bends, contraction, enlargements and tee-junctions are given e.g. in [34]. For single-phase gas flow in pipes it is defined as $K_{R,in} = \lambda(\mathrm{Re}) \cdot L/d$. Especially in two-phase gas/liquid flow, resistance coefficients for pipes and fittings are often historically defined by means of a two-phase multiplier Φ. The multiplier corrects the resistance calculated based on single-phase flow assuming the same mass flow rate of either only gas or only liquid flowing through the pipe or pipefitting. The most general form of a two-phase multiplier is defined by means of a logarithmic average of both the pressure drop in single-phase gas flow $\Delta p_{g,in}$ and liquid flow $\Delta p_{l,in}$:

$$\Phi^2 = \frac{\Delta p_{2ph}}{\Delta p_{1ph,\log}} = \frac{(\eta_{in} - \eta_{out})}{\Lambda} \tag{16}$$

$$\text{with} \quad \Lambda = \frac{\Delta p_{1ph,\log}}{p_{in,t}} = \gamma_g \frac{\Delta p_g}{p_{in,t}} + \gamma_{gl} \sqrt{\frac{\Delta p_g}{p_{in,t}} \cdot \frac{\Delta p_l}{p_{in,t}}} + \gamma_l \frac{\Delta p_l}{p_{in,t}} \tag{17}$$

$$\text{and} \quad \frac{\Delta p_g}{p_{in,t}} = K_{Rg} \cdot \frac{\dot{x}_{in}^{\,2} C^2}{2 v_{g_{in}} p_{in}} \eta_{in} \quad \text{and} \quad \frac{\Delta p_l}{p_{in,t}} = K_{Rl} \cdot \frac{(1-\dot{x}_{in})^2 C^2}{2 v_{l_{in}} p_{in}} \eta_{in} \tag{18}$$

where the values for gas, liquid and both phases for γ are model "switch" constants with values of either 0 or 1 depending on a certain model for a two-phase multiplier.

The application of a two-phase multiplier within the HNE-CSE model shall lead to the same pressure drop result. Generally, frictional and acceleration effects are taken into consideration in the two-phase multiplier, but not gravitational pressure changes. Hence, considering the dimensionless outlet pressure $\eta_{out} = \eta_{in} - \Phi^2 \cdot \Lambda$ the resistance coefficient $K_{R,in}$ is defined by:

$$K_{R,in} = 1 - \frac{1}{\Gamma^2} \left[1 - \frac{\Phi^2 \cdot \Lambda \cdot (1-\omega) - \omega \cdot \ln\left(1 - \Phi^2 \frac{\Lambda}{\eta_{in}}\right)}{C^2 \cdot \left[\omega \cdot \left(\frac{1}{\eta_{in} - \Phi^2 \cdot \Lambda} - 1\right) + 1\right]^2} \right] \tag{19}$$

Any definition of a two-phase multiplier may be used in conjunction with the HNE-CSE model, e.g. for straight pipes the Lockhart-Martinelli [35] approximation, the model of Chisholm [36,37] or Friedel [38], which may be more appropriate in certain two-phase flow conditions [39, 40].

Table 3: Two-phase multiplier for frictional pressure drop in straight pipes

	γ_g	γ_l	γ_{gl}	Φ
Lockhardt Martinelli [35]	0	1	0	$\Phi^2 = 1 + \frac{21}{X} + \frac{1}{X^2}$ with: $X = \sqrt{\left(\frac{dp}{dz}\right)_l \Big/ \left(\frac{dp}{dz}\right)_g}$
Chisholm [36, 37]	0	1	0	$\Phi^2 = 1 + \left(\Gamma_{Chis}^{\ 2} - 1\right) \cdot \left[B\left(\dot{x}_{in}(1-\dot{x}_{in})\right)^{0.9} + \dot{x}_{in}^{\ 1.8} \right]$
				B ; $\Gamma_{Chis} = \left(\frac{v_g}{v_l}\right)^{0.5} \left(\frac{\mu_g}{\mu_l}\right)^{0.1}$; $\dot m$
				$4{,}8$; $\le 9{,}5$; $\dot m \le 500$
				$\frac{2400}{\dot m}$; $\le 9{,}5$; $500 < \dot m \le 1900$
				$\frac{55}{\sqrt{\dot m}}$; $\le 9{,}5$; ≥ 1900
				$\frac{520}{\Gamma_{Chis}\sqrt{\dot m}}$; $9{,}5 < \Gamma_{Chis} < 28$; ≤ 600
				$\frac{21}{\Gamma_{Chis}}$; $9{,}5 < \Gamma_{Chis} < 28$; > 600
				$\frac{15000}{\Gamma_{Chis}^{\ 2}\sqrt{\dot m}}$; ≥ 28 ; > 600
Friedel [38]	0	1	0	$\Phi = E + \frac{3.43 \cdot F_{Fr} \cdot H_{Fr}}{Fr_l^{0.047} We_l^{0.0334}}$ with: $Fr_l = \frac{\dot m^2 v_l^2}{g \cdot d}$; $E = (1-\dot x)^2 + \dot x^2 \frac{v_g \lambda_g}{v_l \lambda_l}$; $F_{Fr} = \dot x^{0.685}(1-\dot x)^{0.24}$; $H_{Fr} = \left(\frac{v_g}{v_l}\right)^{0.8} \left(\frac{\mu_g}{\mu_l}\right)^{0.22} \left(1 - \frac{\mu_g}{\mu_l}\right)^{0.89}$; $We_l = \frac{\dot m^2 d \cdot v_l}{\sigma}$

The two-phase multiplier is an average value for the pipe section considered. In flashing liquids or at large density changes due to the pressure drop, e.g. in long pipes, the pipe length shall be splitted into smaller sections to solve for pressure losses and changes in mass flow quality. The HNE-CSE model is used repeatedly until the end of the pipe section is reached.

For rupture disk devices the resistance factor $K_{R-RD,in}$ is often given by manufactures in accordance with ASME PTC25 [4]. It represents the velocity head loss in incompressible liquid or low Mach number gas flow. But in compressible gas flow, and especially in two-phase flow, this resistance factor is too uncertain. The Reynolds analogy as argument for a constant resistance factor either under measured laboratory conditions or under plant operation conditions is not valid here. In the rupture disk, the flow contraction and re-attachment to the pipe wall is affected by the compressibility of the gas and the change in mass flow quality. This is especially true for high velocity flows, which are typically encountered at emergency vent-lines. The resistance factor in these types of flow is not a constant value; it depends on geometrical parameters, flow conditions and fluid properties. Further measurements and an extended test procedure are necessary [19].

As a best estimate, the resistance of a rupture disk device may be depicted from orifice flow measurements. Even more important is the narrowest flow cross section in this device, because it may limit the mass flow rate to be discharged.

4.6 Narrowest flow area in Throttling Devices

The minimum flow area in throttling devices and pipefittings must be known for the HNE-CSE model. In a nozzle this represents the throat area, whereas in safety and control valves it is often defined as the geometrically narrowest flow area, e.g. the seat area. For rupture disk devices, it highly depends on the net flow area of the open device. The net flow area ratio slightly differs with the rupture disk nominal size and depends on the specific disk type (e.g. vacuum support), disk material, burst pressure, and the disk shape due to the burst process. The disturbance of the flow in a bursted rupture disk device may differ between devices of very low and large burst pressures, because the bursted flow area must not necessarily have the same shape after opening. There are no experimental data available for rupture disk devices to determine the narrowest flow cross section – it must be estimated.

Schmidt [41, 42, 43] has measured the pressure profile along the centerline in sudden pipe contractions with single phase gas, liquid and two-phase non-flashing flow, Figure 8. He showed the location of the vena contracta to be 0.7 diameters downstream of the contraction. Although in single-phase liquid and in single-phase gas flow a distinct pressure minimum was measured, in two-phase flow a vena contracta was not detectable. High speed photos in a sudden pipe contraction made of visible Plexiglas showed a transient flow behavior where liquid was first collected in front of the contraction before it was spread with lots of droplets and larger liquid plugs into the downstream pipe. The droplet formation took place at the edge of the pipe contraction and hindered the gas flow to form a stream contraction. In these experiments, the vena contracta coefficient for two-phase flow was almost one.

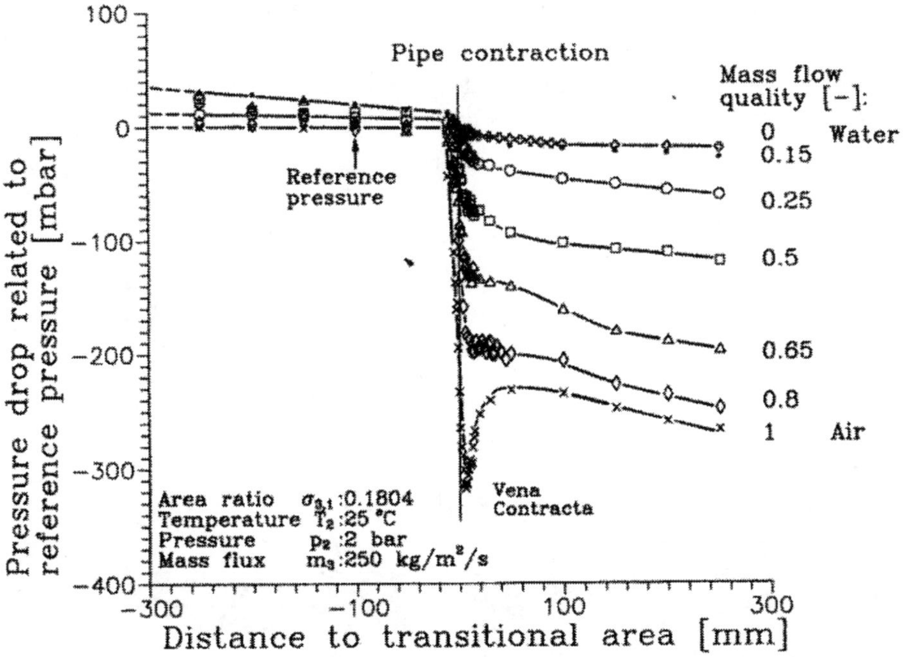

Figure 8: Static pressure profile in the centerline of a sudden pipe contraction measured in an air/water two-phase flow in a sudden pipe contraction with an area ratio of 0.1804 (inner pipe diameter inlet line 24 mm and outlet line 56.2 mm) [41, 43]

A transition from single-phase liquid and single-phase gas flow with measureable contraction areas to two-phase flow with almost no contraction must establish at very low mass flow qualities and at very high mass flow qualities. Shannak [44] got similar results in air/water flow through an orifice.

As a first estimate, the results of Schmidt [41] for two-phase flow and the well-known contraction coefficients of Benedict et. al. [45] for single-phase flow may be used to size a rupture disk device [19]. The mass flow rate through the device is limited by the area of the vena contracta and the backpressure downstream of the disk. In case of a single-phase gas flow the critical mass flow rate is not constant, but still depends on the backpressure. By tendency, the vena contracta enlarges with a decrease in backpressure. Smaller recirculation zones establish. There is evidence by measurements that the vena contracta in gas flow slightly increases under critical flow conditions up to a pressure ratio of 1:32. Thereafter, no further change in mass flow rate was detected.

Further investigations are necessary to simulate the size of the vena contracta in rupture disk devices and also in pipefittings like elbows and T-junctions. Especially in two-phase gas/liquid flow there are no data available. ASME VIII Div.1 UG131 [2] recommends a vena contracta coefficient of 0.62 for single-phase gas and liquid flow as conservative value. In practice, the same value is often used for two-phase gas/liquid flow.

5. Industrial example for sizing rupture disk vent-line systems

To explain the uncertainties of the sizing procedure for rupture disk vent-lines 4 examples are depicted which cover the typical range for the chemical, pharmaceutical and parts of the petrochemical industry.

Table 4: Initial sizing parameter for calculation examples

Example No		1	2	3	4
Fluid		Water	Water	Water	Water
Burst pressure	p_{burst} [barg]	2	2	50	50
Opening pressure	p_0 [bar]	3.2	3.2	56	56
Initial temperature	T_0 [°C]	135.7 saturated	135.7 saturated	271.3 saturated	271.3 saturated
Inlet mass flow quality	x_0 [%]	0	0	0	0
Vent-line inner diameter	d_0 mm]	50	50	50	50
Inlet pipe length	L_{12} [m]	1	1	1	1
Outlet pipe length	L_{34} [m]	1	10	1	10
Pipe roughness	ε [mm]	0.07	0.07	0.07	0.07
Inlet resistance	$K_{R,entry}$	0.584	0.584	0.584	0.584
Device resistance	$K_{R,RD}$	0.7	0.7	0.7	0.7
Backpressure	p_a [bar]	1.0132	1.0132	1.0132	1.0132
Model		HEM	HNE-CSE	HEM	HNE-CSE
Vena Contracta coefficient	C_{CV}	0.6	0.95	0.6	0.95

For simplicity, initially saturated water was used as medium to be discharged, but the overall results are also representative for other fluids, other vent diameters, mass flow rates and initial mass flow qualities. Two burst pressures, 2 and 50 barg, are depicted for comparison. The full opening of the rupture disk devices is assumed at pressures 10% above the burst gauge pressure.

Figure 9 shows a schematic of the rupture disk vent-line systems, each with a sharp entrance (resistance factor $K_{R,in}$), an inlet pipe of 1m length up to a rupture disk type KUB (resistance factor $K_{R,RD}$) of Rembe GmbH, Brilon, and a simple, vertically upward outlet pipe of either 1m or 10 m length. The rupture disk vent-line system has an inner diameter of 50 mm.

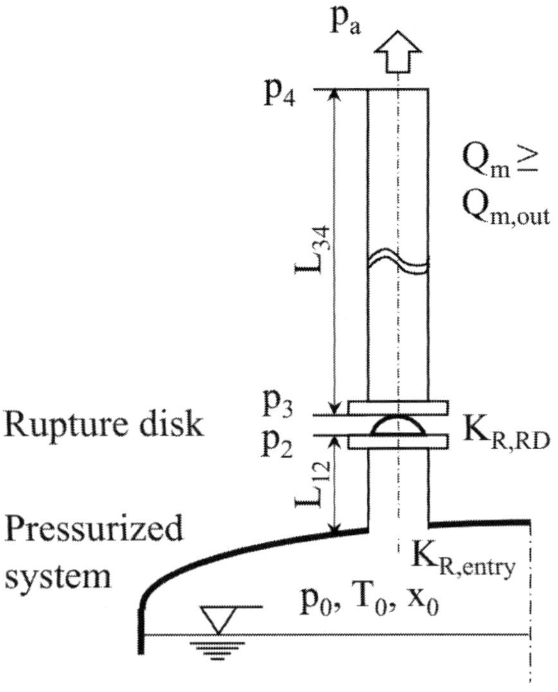

Figure 9: Rupture disk vent-line

The size of the rupture disk device was calculated by means of two models: the classical homogeneous equilibrium model (HEM) with a vena contracta coefficient assumed for the rupture disk of 0.6

and the HNE-CSE model with a coefficient of 0.95, see chapter 4.6. The results are shown in Figure 10 for a burst pressure of 50 barg and in Figure 11 for a burst pressure of 2 barg.

Applying the HEM model with a vena contracta coefficient of 0.6 to size the rupture disk vent line with a burst pressure of the disk of 50 barg results in a mass flow rate of 55158 kg/h irrespective of whether the length of the outlet pipe is 1 or 10 m. The pressure drop in the inlet pipe is very small. In the vena contracta of the rupture disk device, critical flow is computed. Downstream of the device pressures of 30 bar (abs) respectively 18 bar (abs) establish depending on the upcoming pressure drop in the outlet pipe. At the end of the rupture disk vent-line system, a second cross section with critical flow conditions is predicted.
By means of the HNE-CSE model with a contraction coefficient of 0.95 the mass flow rates differ significantly: 99487 kg/h in a vent-line with a total length of 11 m (10 m outlet pipe) and 137500 kg/h at 2 m length (1 m outlet pipe). The calculated mass flow rate is roughly 200% to 250% greater of that calculated by means of the HEM model with a lower contraction coefficient. In both, the short rupture disk vent-line and the long vent-line, critical flow conditions are reached only at the end of the vent-line system and not within the rupture disk device. Hence, the size of the vent-line limits the overall mass flow rate. Due to different mass flow rates and local mass flow qualities, the pressure profiles are very different.

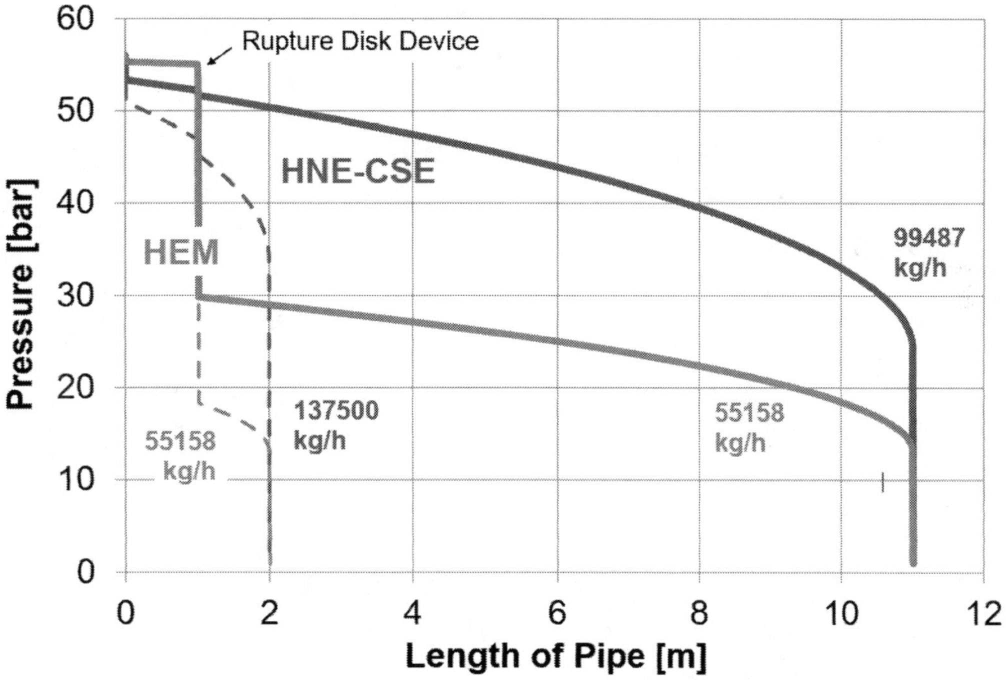

Figure 10: Pressure profile and mass flow rate in rupture disk vent-lines at 50 barg burst pressure discharging initially saturated water

A similar result is given in Figure 11 for a burst pressure of the rupture disk device of 2 barg. Here, critical flow conditions with a mass flow rate of 6176 kg/h are computed by means of the HEM model with a vena contracta coefficient of 0.6. At the end of the vent-line system the ambient pressure is reached. In contrast, the HNE-CSE model with a vena contracta coefficient of 0.95 predicts a critical mass flow rate only at the end of the system. The mass flow rates are again much higher compared to the homogeneous approach.

The results presented here show the difficulty in sizing a rupture disk vent-line for two-phase flashing flows. The homogeneous equilibrium flow assumption in combination with a vena contracta coefficient of 0.6 leads to highly underestimated flow rates through the rupture disk device, which are not representative and not conservative for the pressure profile in the vent-line system, pipe forces, a retention system or disposal calculations to evaluate toxic effects of hazard substances. Applying the HNE-CSE model gives to the current sizing knowledge much more reasonable results, but the results are equally not validated by tests.

The calculated profiles of the pressure and mass flow quality in the inlet line up to the mass flow limiting cross section determine the critical mass flow rate through the system. In short inlet lines with a flashing two-phase flow the HEM model may not be a conservative choice, because the mass flow quality is highly overestimated.

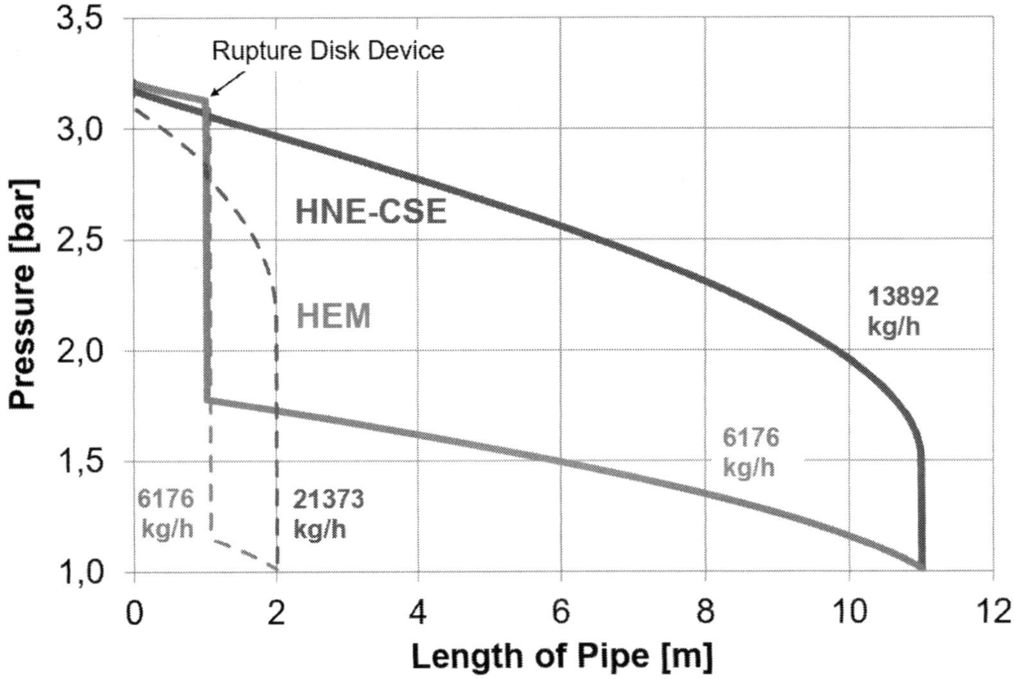

Figure 11: Pressure profile and mass flow rate in rupture disk vent-lines at 2 barg burst pressure discharging initially saturated water

6. Summary

Emergency relief vent-line systems in the chemical, pharmaceutical and petrochemical industry typically consist of a complex piping system with throttling devices (rupture disk, safety valve etc.), pipefittings and many straight pipe sections. For sizing such systems a deep understanding of the fluid dynamic processes, detailed property data (especially mixture property data) and consistent models for all pipe elements are necessary. Several commercial codes are inconsistent mostly because the models applied are validated under subcritical flow conditions – often in non-flashing flow - but extrapolated up to critical flow conditions for flashing gas/liquid single- and multi-component mixtures. There exist many self-defined "good engineering practices" but almost no experimental validation for complete vent-line systems and no standardized procedure for sizing.

Based on the first and second law of thermodynamics the HNE-CSE model is presented as a consistent set of fluid dynamic equations for steady state one-dimensional sizing of emergency relief vent-lines. Existing resistance coefficient models and two-phase multiplier models may be introduced into the set of equations to account for pressure losses within the throttling devices, pipe or pipefittings. The HNE-CSE model is an extension of the HNE-DS model for safety valves and control valves. The potential kinetic energy, gravitational effects and the pressure loss in any pipe element are additionally considered.

By means of the HNE-CSE model, rupture disk vent-line systems may be sized in accordance to the current state of knowledge. Nevertheless, the sizing procedure is mainly a convention and not validated by tests.
It is highly necessary to perform experiments for two-phase non-flashing and flashing flows. In addition, there is a strong need to extend the test procedure of ASME PTC25 for measuring the resistance factor of rupture disk devices to compressible fluid flow and to determine coefficients or contraction coefficients for this type of devices.

7. Literature

1. API RP-520, "Sizing, Selection, and Installation of Pressure-Relieving Devices in Refineries. Part 1 — Sizing and Selection", 8th edition, 2010.
2. ASME, "Boiler and Pressure Vessel Code", The American Society of Mechanical Engineers, Section VIII Pressure Vessels, Division 1, UG 131, 2013.
3. AD2000 Merkblatt A1: Sicherheitseinrichtungen gegen Drucküberschreitungen – Berstsicherungen. Beuth Verlag, Berlin, 2006.
4. ASME-PTC25, "Pressure Relief Devices - Performance Test Codes (ASME PTC 25-2008)" THE AMERICAN SOCIETY OF MECHANICAL ENGINEERS, New York, 2009.
5. ISO 4126-10, "Safety devices for protection against excessive pressure – sizing of safety valves for gas/liquid two-phase flow". DIN Deutsches Institut für Normung e.V., Beuth Verlag GmbH, Berlin, 2010.
6. Schmidt, J.; Stoessel, F., "Sicherheitskonzept zur Absicherung von Mehrzweckanlagen", 10. Fachtagung Anlagensicherheit-, Arbeits- und Umweltsicherheit am 4./5. November 2010 in *Köthen* (Hochschule Anhalt (FH)), 2010.
7. Schmidt, J., „Sicherheits- und Rückhalteeinrichtungen für Mehrzweckanlagen Teil 2: Rückhalteeinrichtungen", Chem. Ing. Technik, July 2011.
8. Center for Chemical Process Safety, "Guidelines for Pressure Relief and Effluent Handling Systems", Wiley, 1998.
9. Lottermann, J.: Risikominderung - SIL äquivalente Kenngrößen für Produkte der Explosionsdruckentlastung, CIT Plus, 12, pp 27-29, 2014.

10. Friedel, L., Schmidt, J., "Design of long safety valve vent line systems for gas relief", J. Loss Prevention Process Ind. Vol. 6, No 5, 1993.
11. Emergency response planning guideline values, https://www.aiha.org/get-involved/AIHAGuidelineFoundation/EmergencyResponsePlanningGuidelines/Documents/2013ERPGValues.pdf
12. Kommission für Anlagensicherheit, „TRAS 410 - Technische Regeln Anlagensicherheit der Kommission für Anlagensicherheit, Erkennen und Beherrschen exothermer chemischer Reaktionen", Fassung 10/2012.
13. Barton J., Rogers R., (ed.), "Chemical Reaction Hazards", Second Edition, IChemE, Rugby, 1997.
14. Knowlton E., "Creative Checklist Hazard and Operability Studies, Chemical Manufacturers Association", Process Safety Management Workshop, Arlington, Virginia, 1985.
15. Department of Defense, "Military Standard System Safety Program Requirements", MIL-STD-882B, Washington, D. C., 1984.
16. Fussel J. B., "Fault Tree Analysis: Concepts and Techniques", Generic Techniques in System Reliability Assessment, 1976.
17. Lees F. P., "Loss Prevention in the Process Industries", Butterworths, 1980.
18. CCPS —"Guidelines for Hazard Evaluation Procedures", AIChE, New York, 1985.
19. Schmidt J., Claramunt S., "Sizing of Rupture Disks for Two-Phase Gas/Liquid Flow According to HNES-Model", Proc. 17th Annual International Symposium Mary Kay O'Connor Process Safety Center, College Station TX, pp.166-206, 2014.
20. Friedel L., Wehmeier G., Westphal F., "Pressure relief system design for vapor or two-phase flow? Process Safety and Environmental Protection", Trans IChemE Vol. 72, Part B, pp. 142-148, August 1994.
21. Fisher, H.G., "Emergency Relief System Design Using DIERS Technology", The Design Institute for Emergency Relief Systems of the AIChE, New York, 1992.
22. Schmidt, J., "Auslegung von Sicherheitsventilen für Mehrzweckanlagen nach ISO 4126-10", Chemie Ingenieur Technik, 83 (6), 1–18, 2011.
23. Fisher, H. G., and Forrest, H.S., "Protection of storage tanks from two-phase flow due to fire exposure", Process Safety Progress, 14(3), 183–199, 1995.
24. Schmidt J., Sizing of safety valves for multi-purpose plants according to ISO 4126-10, Journal of Loss Prevention in the Process Industries, 25, 181-19, 2011.
25. ISO 2325, "Petroleum, petrochemical and natural gas industries — Pressure-relieving and depressuring systems", 2006.
26. Leung, J.C., "Simplified vent sizing equations for emergency relief requirements in reactors and storage vessels", AICHE Journal, 32(10), 1622-1634, 1986.
27. Fauske H., "Sizing Rupture Disks (RDs) for two-phase flow", Fauske&Associates, Inc. (n.d.), Burr Ridge, Illinois.
28. EN ISO 4126-6, "Safety devices for protection against excessive pressure. Part6: Application, selection and installation of bursting disc safety devices", ISO, 2013.
29. ISO/DIS 4126-11, "Safety devices for protection against excessive pressure. Part 11: Performance Testing", ISO, 2014.
30. Private communication to Robert Pack, BASF SE on 12th February 2015
31. Diener, R.; Schmidt, J., "Sizing of throttling devices for gas liquid two-phase flow, Part 2: control valves, orifices and nozzles", Process Safety Progress, 24 (1), 29-37, 2005.

32. Schmidt, J., "Sizing of Nozzles, Venturis, Orifices, Control and Safety Valves for Initially Sub-cooled Gas/Liquid Two-Phase Flow – the HNE-DS method", Forsch. Ing. Wesen. Vol. 71, 47-54, 2007.
33. Diener, R.; Schmidt, J., "Sizing of throttling devices for gas liquid two-phase flow, Part 1: safety valves", Process Safety Progress, 23(4), 335-344, 2004.
34. Miller, D.S., "Internal flow systems", 2nd completely revised edition, 1996.
35. Lockhart R.W., "Martinelli R.C.: proposed correlation of data for isothermal two-phase two-component flow". Chem. Engng. Progress, 45(1), p. 39, 1949.
36. Chisholm D., "A theoretical basis for the Lockhart-Martinelli Correlation for Two Phase Flow", International Journal of Heat and Mass Transfer, vol. 10, pp.1767-1778, 1967.
37. Chisholm D., "Pressure Gradients due to Friction during the flow of evaporating two-phase mixtures in smooth tubes and channels", International Journal of Heat and Mass Transfer, 16(2), pp.347-358, 1973.
38. Mayinger F., "Two-Phase Flow Phenomena with Depressurization-Consequences for the Design and Layout of Safety and Pressure Relief Valves", Chem. Eng. Process, vol. 23, p.1-11, 1988.
39. Leung J. C., "Pipe discharge flow calcualtions", DIERS Users Group Meeting, Houston Texas, October 2009.
40. Leung J.C., "The Discharge of Two-phase Fashing Flow in a Horizontal Duct" AIChE Journal, 33(3), p. 524-527, 1987.
41. Schmidt J., "Berechnung und Messung der Druckänderung über scharfkantige plötzliche Rohrerweiterungen und -verengungen bei Gas/Dampf Flüssigkeitsströmung", Fortschritt-Berichte VDI. Reihe 7: Strömungstechnik, Bochum, 1992.
42. Schmidt, J., Friedel, L., "Two-phase pressure drop across sudden contractions in duct areas", European Two Phase Flow Group Meeting, Hannover, June 7.-10., 1993.
43. Schmidt. J., Friedel, L. "Two phase pressure drop across sudden contractions in duct areas", Int. J. Multiphase Flow, 23(2), pp 283-299, 1997.
44. Shannak B., "Experimental and theoretical investigation of gas–liquid flow pressure drop across rupture disks", Nuclear Engineering and Design, 240(6), p. 1458–1467, 2010.
45. Benedict R.P., "Fundamentals of gas dynamics", John Wiley & Sons Australia, 1983.

8. Nomenclature

Variable	Definition	Unit
a	Exponent of the non-equilibrium factor N	—
A	Flow cross section of a certain pipe geometry	m2
A_{fire}	Wetted wall area of a reactor up to a height of 7.5 m from the bottom with heat transfer due to fire at a maximum liquid hold-up	m2
A_{in}	Flow cross section at inlet conditions	m2
A_{out}	Flow cross section at outlet conditions	m2

A_{pipe}	Cross sectional area of the pipe	m
A_{RD}	Nominal area of the rupture disk	m2
A_{wet}	Heat transfer area of the wetted wall from a reactor at a maximum liquid hold-up	m2
B	Dimensionless factor at Friedel two-phase flow correlation	—
C	Flow coefficient	—
C_{crit}	Critical flow coefficient	—
C_{VC}	Contraction coefficient of the flow: ratio between the narrowest flow cross section in the rupture disk and the effective rupture disk flow area	—
cp_{l_0}	Specific heat capacity of the liquid reactor content at sizing conditions	J/(kg K)
$cp_{l_{in,t}}$	Specific heat capacity of the liquid at inlet stagnation conditions	J/(kg K)
D	Reactor inner diameter	m
d	Pipe element diameter	m
d_0	Vent-line inner diameter	mm
d_{pipe}	Diameter of the relief pipe	m
d_{RD}	Equivalent diameter of the rupture disk net flow area	m
d_{VC}	Vena contracta diameter	m
E	Dimensionless factor at Friedel two-phase flow correlation	—
E_{diss}	Specific dissipation energy	J/kg
Δe_{kin}	Variation in specific kinetic energy	J/kg
Δe_{pot}	Variation in specific potential energy	J/kg
F	Factor to judge the fire insulation of a reactor	—
F_{Fr}	Dimensionless factor at Friedel two-phase flow correlation	—
Fr_l	Dimensionless factor at Friedel two-phase flow correlation	—
g	Gravity acceleration	m/s2
h	Specific enthalpy of the fluid	J/kg
H_{fire}	Height of fire in case of external heating	m
H_{Fr}	Dimensionless factor at Friedel two-phase flow correlation	—
h_{in}	Specific enthalpy of the fluid at the inlet of a pipe element	J/kg
$h_{l,in}$	Specific enthalpy of the liquid at the inlet of a pipe element	J/kg
h_{out}	Specific enthalpy of the fluid at the outlet of a pipe element	J/kg
H_{wet}	Heat of the wetted wall from a reactor	m
ΔH_{bottom}	Distance from the reactor to the ground	m
Δh_{v_0}	Heat of vaporization at sizing conditions	J/kg
$\Delta h_{v_{in,t}}$	Heat of vaporization at inlet stagnation conditions	J/kg
k_{∞}	Factor of the rising velocity of a single bubble in a liquid plume	—

K_R	Flow resistance coefficient	—
$K_{R-pipe,in}$	Flow resistance coefficient of the pipeline referred to inlet conditions.	—
K_{Rg}	Flow resistance coefficient for all-gas content	—
$K_{R,\text{in}}$	Flow resistance coefficient referred to inlet conditions of an element.	—
$K_{R,\text{entry}}$	Flow resistance coefficient referred to inlet conditions of the rupture disk vent-line.	—
K_{Rl}	Flow resistance coefficient for all-liquid content	—
$K_{R,RD}$	Flow resistance coefficient of a rupture disk	—
L	Pipeline length	m
L_{12}	Inlet vent line pipe length	m
L_{34}	Outlet vent line pipe length	m
M_0	Maximum mass in the reactor at sizing conditions	kg
M_{l_0}	Liquid mass in the reactor at sizing conditions	kg
$\dot{m}$	Mass flux	kg/ (m² s)
$\dot{m}_{nozzle}$	Mass flux through an ideal nozzle	kg/(m² s)
N	Non-equilibrium factor	—
p	Pressure	Pa
p_0	Pressure in the reactor at sizing condition	Pa
p_1	Pressure at the inlet pipe entrance	Pa
p_2	Pressure at the inlet pipe exit	Pa
p_3	Pressure at the outlet pipe entrance	Pa
p_a	Ambiance pressure	Pa
p_b	Back pressure	Pa
p_{burst}	Burst pressure	bar g
$p_{in,t}$	Pressure at inlet stagnation conditions	Pa
$p_{\max}$	Maximum pressure in the reactor during a pressure relief	Pa
p_{out}	Pressure at outlet conditions	Pa
dp / dz	Pressure gradient	Pa/m
Δp_{2ph}	Pressure loss in two-phase flow	Pa
$\Delta p_{1ph,\log}$	Logarithmic average of pressure drop in single phase gas and single phase liquid and the same flow area	Pa
Δp_g	Pressure drop when only single phase gas flows through the pipe element	Pa
Δp_l	Pressure drop when only single phase gas flows through the pipe element	Pa
$\dot{Q}_{in}$	Maximum heat input into the reactor	J/s

Q_m	Mass flow rate	kg/s
$Q_{m,\max}$	Calculated maximum feasible mass flow rate	kg/s
$Q_{m,\mathrm{out}}$	Mass flow rate required to be discharged from a reactor	kg/s
$Q_{m,RD}$	Mass flow rate through a rupture disk	kg/s
Re	Reynolds number	-
T	Temperature	K
T_0	Temperature at sizing conditions	K
$T_b(\mathrm{p})$	Boiling temperature at pressure p	K
T_{heat}	Maximum temperature of the heat transfer media	K
$T_{in,\mathrm{t}}$	Temperature at inlet stagnation conditions	K
$T_{\max}$	Temperature at maximum pressure in the reactor	K
dT / dt	Temperature rising velocity of a chemical reaction	K/s
ΔT_{over}	Temperature increase within the reactor after opening of the safety valve until the maximum pressure is reached	K
u_∞	Rising velocity of a single bubble in a liquid plume	m/s
u_g	Superficial gas velocity above the liquid level at relieving conditions	m/s
U	Maximum heat transfer coefficient for a cleaned heat exchanger	W/(m2 K)
v	Specific volume of the fluid	m3/kg
v_0	Specific volume of the two-phase mixture at sizing conditions according to ISO 4126-10 (homogeneous level swell)	m3/kg
v_{in}	Specific volume of the fluid at inlet conditions	m3/kg
$v_{in,\mathrm{t}}$	Specific volume of the fluid at inlet stagnation conditions	m3/kg
v_{g_0}	Specific volume of the gas/vapor at sizing conditions	m3/kg
$v_{g_{in}}$	Specific volume of the gas at inlet conditions	m3/kg
$v_{g_{in,t}}$	Specific volume of the gas at inlet stagnation conditions	m3/kg
v_{l_0}	Specific volume of the liquid at sizing conditions	m3/kg
$v_{l_{in}}$	Specific volume of the liquid at inlet conditions	m3/kg
$v_{l_{in,t}}$	Specific volume of the liquid at stagnation inlet conditions	m3/kg
v_{out}	Specific volume of the fluid at outlet conditions	m3/kg
$v_{out,\mathrm{t}}$	Specific volume of the fluid at stagnation outlet conditions	m3/kg
v^*	Dimensionless specific volume	—
v^*_{in}	Dimensionless specific volume at inlet conditions	—
v^*_{out}	Dimensionless specific volume at outlet conditions	—
V_R	Reactor volume	m3
w	Two-phase flow velocity	m/s

w_{in}	Two-phase flow velocity at inlet conditions	m/s
$w_{in,t}$	Two-phase flow velocity at inlet stagnation conditions	m/s
w_{out}	Two-phase flow velocity at outlet conditions	m/s
We_l	Liquid Weber factor	$(m^2\ s)/kg$
$\dot{x}_0$	Mass flow quality at sizing conditions	—
$\dot{x}_{in}$	Mass flow quality at static inlet conditions	—
$\dot{x}_{in,t}$	Mass flow quality at inlet stagnation conditions	—
$\dot{x}_{out}$	Mass flow quality at static outlet conditions	—
X	Martinelli parameter	—
z	Coordinate in the direction of gravity	m
γ_g	Constant for the gas in two-phase multiplier approximation	—
γ_l	Constant for the liquid in the two-phase multiplier approximation	—
γ_{gl}	Constant in the two-phase multiplier approximation.	—
Γ	Dimensionless acceleration ratio	—
Γ_{Chis}	Dimensionless parameter at Chisholm two-phase flow correlation	—
Γ_{nozzle}	Dimensionless acceleration ratio of an ideal nozzle	—
Γ_{pipe}	Dimensionless acceleration ratio of a pipe	—
ε	Pipe roughness	mm
ε_0	Void fraction at the entrance of the relief vent-line (sizing conditions)	—
$\varepsilon_{0,hom}$	Void fraction at the entrance of the relief vent-line for a homogeneous mixed reactor	—
η	Pressure ratio	—
η_{crit}	Ratio of the critical pressure to the sizing pressure	—
η_{in}	Inlet pressure ratio (ratio between the inlet pressure and the inlet pressure at stagnation conditions)	—
η_{out}	Outlet pressure ratio (ratio between the outlet pressure and the inlet pressure at stagnation conditions)	—
θ	Pipeline inclination	rad
κ	Isentropic coefficient	—
λ	Friction factor	—
Λ	Dimensionless static pressure ratio	—
μ_{l0}	Liquid surface tension at sizing conditions	kg/s2
ξ	Pressure relief factor	—
σ_{RD}	Ratio of the net flow area of a burst rupture disk to the area of the inlet pipe of the disk.	—
Φ	Two-phase multiplier	—

Ψ	Weighting factor for level swell comparison due to heat input at bulk energy release and due to external heating	—
Ψ_{bulk}	Weighting factor for level swell comparison due to heat input at bulk energy release.	—
Ψ_{fire}	Weighting factor for level swell comparison due to external heating	—
ω	Compressibility factor	—

11TH GLOBAL CONGRESS ON PROCESS SAFETY

Modification of the DIERS Fire Exposure Test Methodology

Peter Ralbovsky
NETZSCH Instruments NA, LLC
129 Middlesex Turnpike, Burlington MA 01803
peter.ralbovsky@netzsch.com

Simon Chippett
Consultant, NETZSCH Instruments NA, LLC
129 Middlesex Turnpike, Burlington MA 01803
simon.chippett@gmail.com

Prepared for Presentation at
American Institute of Chemical Engineers
2015 Spring Meeting
11th Global Congress on Process Safety
Austin, Texas
April 27-29, 2015

UNPUBLISHED

Modification of the DIERS Fire Exposure Test Methodology

Simon Chippett
Consultant, NETZSCH Instruments NA, LLC
129 Middlesex Turnpike, Burlington MA 01803
simon.chippett@gmail.com

Peter Ralbovsky
NETZSCH Instruments NA, LLC
129 Middlesex Turnpike, Burlington MA 01803
peter.ralbovsky@netzsch.com

Keywords: Fire Exposure, Adiabatic Calorimetry, Vent Sizing

Abstract

There is a need to be able to obtain proper runaway data for reactions where fire exposure is a credible scenario. Where the engineering design has been refined throughout the years, the testing methodology has mainly remained the same. New techniques have been developed for measuring thermal runaway which can be applied directly to the fire exposure case resulting in much more accurate data and, subsequently, better ERS designs as well as other engineering controls. In the DIERS literature database, through the work of the DIERS Fire Exposure Committee, a detailed methodology for both the testing and ERS design for the Fire Exposure Case can be found [1]. Although this original paper is an excellent summary and tutorial of the method, the treatment of thermal inertia and the proposed instrumentation method can be improved. This paper will build upon the initial work, focusing on the deviations and reasons behind the changes, and describe a more robust testing approach to obtain proper fire exposure data.

1. Introduction

Fire exposure is a common consideration when dealing with tanks of flammable or combustible liquids in chemical industry. Codes like NFPA-30 (National Fire Protection Association) and standards like API-520/521 are generally followed to determine the heat load due to an external fire. The fire heat flux is calculated based on many parameters like tank geometry, insulation, material properties etc. The fire exposure scenario can be simulated using an accelerated rate calorimeter, and the data can be used to determine the required vent size for tanks.

External heat ramp method can be used for this test, but there are several issues with this method. This method assumes that the heat capacity of the sample is known and does not change which is often not the case. There is no actual measurement of heat that is added to the sample. Plus, a constant external heat rate does not comply with code which requires constant heat flux. So a better way to do the fire exposure test is to use an internal heater based on the patented VariPhi technology. The fire heat flux calculated for plant equipment can be scaled to the test cell conditions based on the process material weight and sample mass in test cell. This heat flux can be added to the sample directly by the internal heater. The VariPhi technology will add power to compensate to Φ=1. From this single test, more accurate fire exposure tests can be conducted. The heat of reaction and kinetics can be determined as well.

2. Instrumentation

The calorimeter used for these experiments was the NETZSCH ARC® 254. (The ARC® 254 is an accelerating rate calorimeter [2, 6] designed specifically for adiabatic operation. The high tracking rate (up to 200 K/min) provides more reliable data and a wider range of applications. VariPhi is a patented technology (See Chippett [4]) on the ARC 254. The basis of this technology is an additional controlled variable DC heater (Figure 1). With this option it is possible to define the thermal inertia in order to allow simulation of a real-world thermal environment by compensating for heat lost from the sample to the vessel. By operating in different modes such as isothermal or scanning, endothermic and exothermic transitions can be quantified, pressure data can be measured, and the heat capacity of the sample can be determined. In conducting a fire exposure test, a constant power is added to sample by the internal heater and the software will compensate the heat loss to the vessel to make Φ=1.

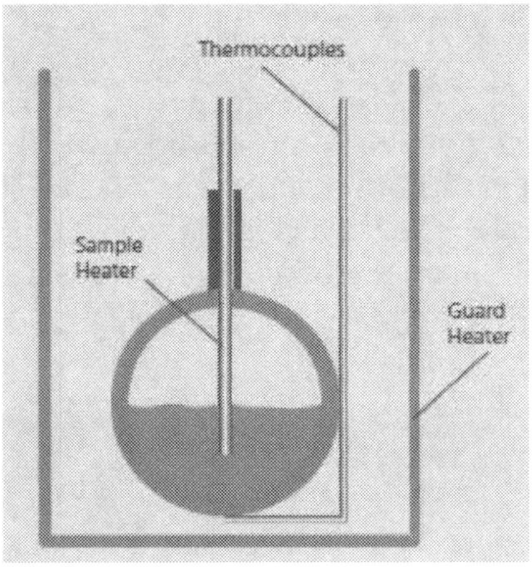

Figure 1. Diagram of *VariPhi*™

3. Experimental Results & Discussion

3.1 Correcting Heater Ramp Rate for Thermal Inertia

The determination of the heat-up rate of the plant vessel or storage tank is well described in Das [1] and how to apply the API rules. The point of the "fire-exposure" experiment is to try to replicate, as best as possible, the heat-up conditions to which the chemicals in the plant are exposed during the fire scenario. Using good engineering practices, a heat-up rate of the chemicals within the plant vessel is calculated. The thermal inertia of the plant vessel is calculated. See Yin [2] for additional details on calculating thermal inertia for plant vessels.

The method then corrects the heat-up rate by Equation 1 for thermal inertia.

Heating Rate Adjusted = ***Calculated Heating Rate x*** $(\phi_{test}/\phi_{plant})$ [Eq. 1]

This correction would lead to the wrong results and needs to be avoided. Thermal inertia relates only to the correction of adiabatic test data. It is used to compensate for the heat "lost" to the sample vessel during a reaction. Thermal inertia affects temperature and pressure as well as their respective rates. In a normal adiabatic test the sample is reacting and producing heat, some of this energy is used in self-heating and some of it is used to heat the sample container. In this case the energy is flowing from the sample and into the container. Using thermal inertia to correct for the exothermic temperature and pressure curves for adiabatic calorimeter is well documented, starting with Townsend's [3] early paper describing Accelerating Rate Calorimetry (ARC®) data correction.

The fire exposure scenario is not an adiabatic process as extra heat is added to the plant vessel due to the external fire. In this case, the plant vessel is hotter than the chemicals inside so the energy is flowing from the wall to the chemical. This is also true for the laboratory test. If the test is set-up to ramp the heaters at a specified rate then the heaters will heat the sample vessel walls which in turn will heat the sample. The mass and thermal conductivity of the sample vessel will have some effect on how quickly the sample achieves the desired rate but given enough time sample heating will achieve a steady state. For example, in Figure 2 it takes roughly 25 minutes for the sample to reach the 2K/min heating rate of the heaters. In comparison, the VariPhi method achieves the desired sample heating in less than 2 minutes.

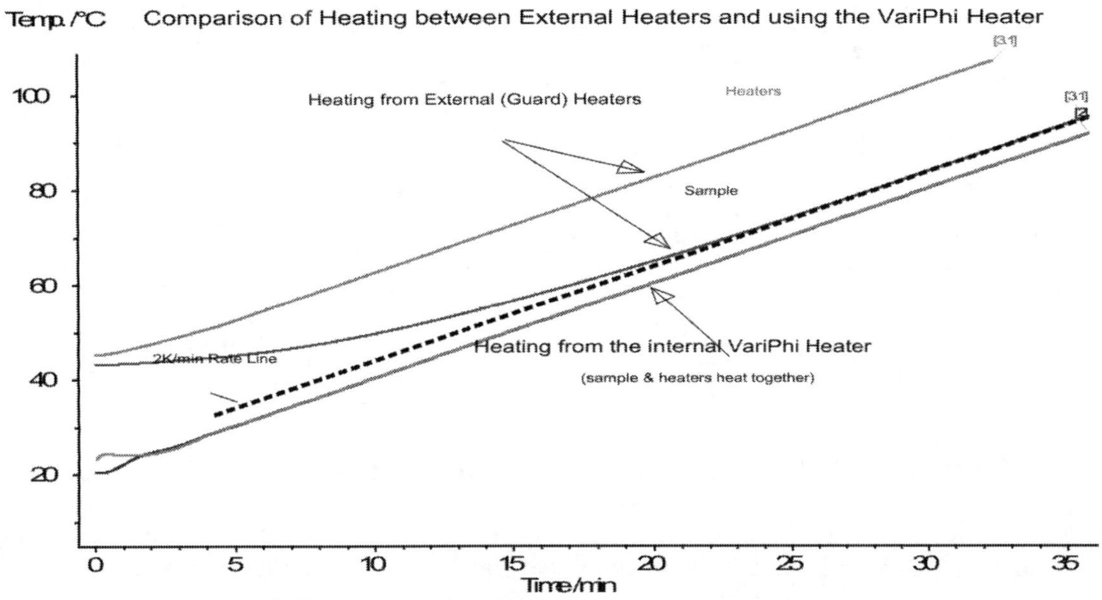

Figure 2. Comparison of Methods during Initial Heating

In some instances the delay in achieving the desired heating may be irrelevant; however there are other cases, with accelerated or inhibited reactions, where this delay could significantly alter the results.

3.2 Correcting Test Results for Thermal Inertia?

Once the sample has reached the desired test ramp rate, the heat energy supplied by the heaters will be more or less consistent, dictated by the established gap between heater and sample temperatures as shown between roughly 25 minutes to 50 minutes in Figure 3. This is a test of 10% DTBP in toluene (Φ=1.5) ramp at 2 K/min. As the test progresses, the sample may begin to self-heat. This self-heating will be initially masked by the ramping process, but eventually the sample temperature rate will begin to increase due to the heat from the guard heaters and the reaction. As this rate increases, the gap between the heater and sample temperature will change. As the gap becomes smaller there is less heat being contributed by the heaters. In this particular test, the sample temperature exceeds the heater temperature just after 70 minutes, almost at the end of the reaction, at which point the sample loses heat to the calorimeter. Soon, the reaction is over and the heater overtakes the sample once again. In this test there is no "adiabatic" data therefore there is no possibility to correct this data for thermal inertia. This data cannot be used for determination of reaction heat or for kinetic analysis.

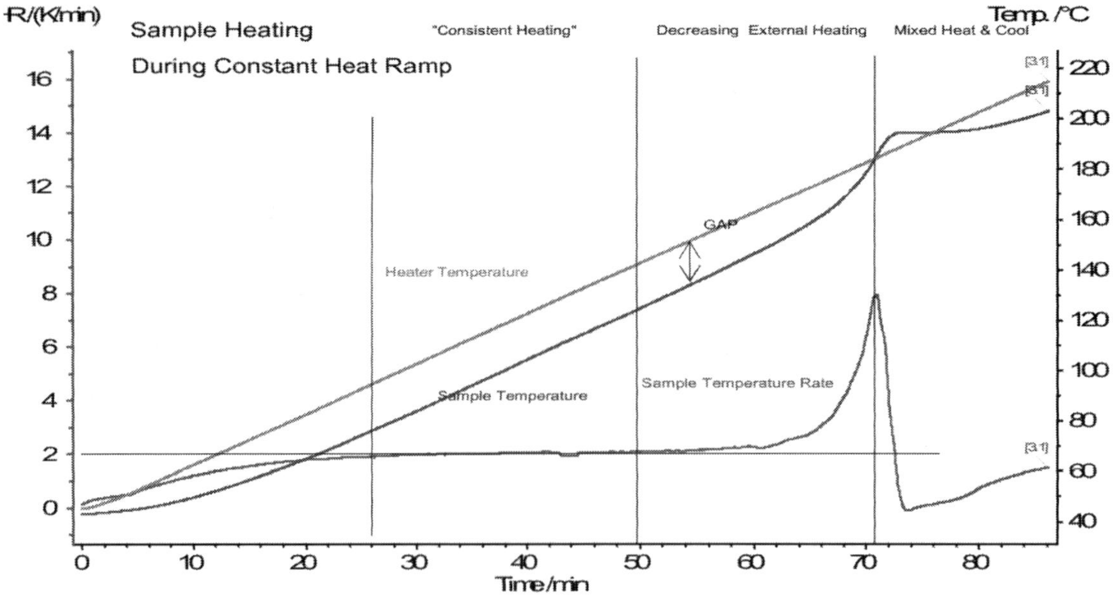

Figure 3. "Mind the Gap" - Heating Changes during Constant Rate

The instrument could have been set to track the exotherm once the temperature of the sample exceeded the heater temperature. This is certainly an option in some adiabatic calorimeters but in this case there would have been little reaction to analyze. Some calorimeters may also have trouble tracking at rates higher than 10 K/min and in this particular situation slower calorimeters would not be able to transition quickly enough from the constant heating rate to tracking mode to obtain any amount of good adiabatic data. It is important to recognize that the low phi data is needed for proper vent sizing calculations. This method, which does not provide adiabatic, Φ-correctable data, would really require testing to be done at or near the thermal inertia of the plant vessel. This is not to say that the method, given the correct set of heating rates and decomposition kinetics, could not ever result in usable adiabatic data, it just can't consistently provide such results.

The changing heat input also undermines the goal of the test: to simulate a heat load (fire) on the process vessel. Unfortunately, the heat load reduces as the reaction rate increases which leads to results that would underestimate the peak temperatures and pressures values and rates.

3.3 Using the VariPhi™ Heater for Constant Rate Heating

The VariPhi can be programmed to keep the heating rate constant from the start of the test. Figure 2 shows the heat-up results from two identical sample tests of 10% DTBP. In one test the guard heaters are used to heat the sample, in the other the VariPhi is used. Given that the VariPhi heater delivers the heat directly to the sample, the heating is more consistent throughout the initial portion of the test, prior to any sample self-heating. In the first few minutes of the test the PID control is adjusting to fix the heating rate at 2K/min. After that, the heating rate is maintained at 2K/min with the VariPhi outputting about 750 mW.

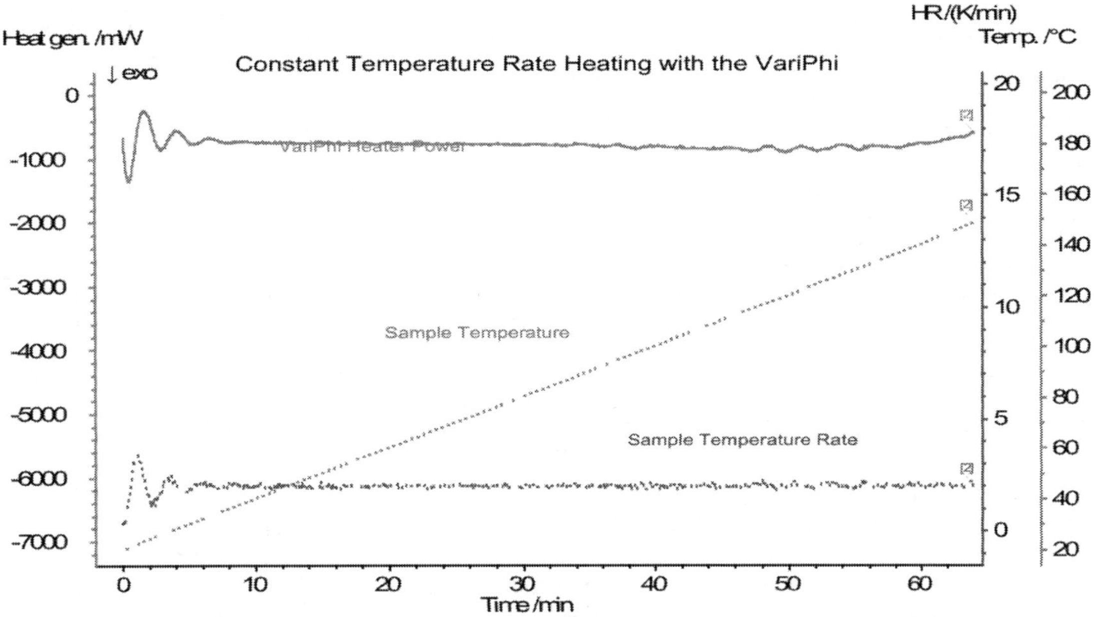

Figure 4. Heating with the VariPhi at a Constant Rate

In this mode the heating rate is kept constant at 2K/min by adjusting the VariPhi heater power. As the sample begins to self-heat, Figure 5 (point 1), the VariPhi heater will turn down to maintain the heating rate. Once the sample temperature rate exceeds 2K/min, Figure 5 (point 2), the VariPhi heater starts to compensate for the thermal inertia. The compensation can be set in the software to any value (from 1 up to the test condition). In this test it was set to compensate to a $\Phi = 1$. As the self-heat rate increases, the amount of heat absorbed by the sample vessel also increases, so during the exotherm the VariPhi power and the self-heat rate are nearly mirror images of each other.

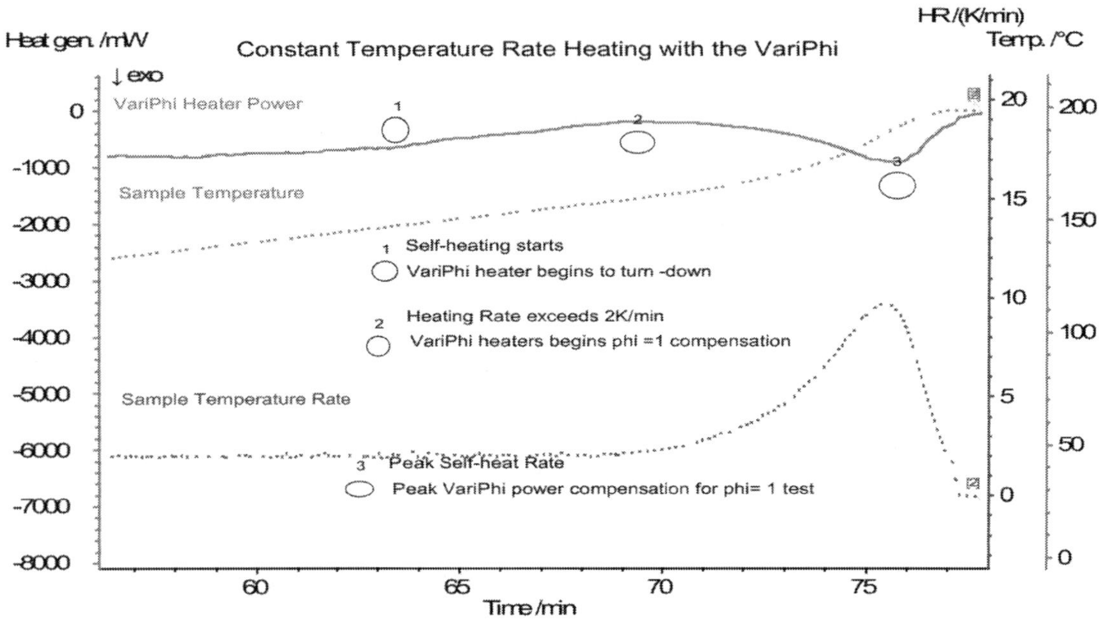

Figure 5. VariPhi Φ-compensation in Constant Rate Mode

Running tests using the constant heating rate in VariPhi mode has some advantages over the previous technique. The heating power delivered is more consistent, at least up to the point where the sample begins to react. The exotherm data is already Φ-corrected and adiabatic so the peak temperatures and pressures can be used directly for vent sizing and/or kinetic analysis. However, even this method will underestimate the actual temperature and pressure rise just as the original method does.

3.4 Using the VariPhi™ Heater for Constant Power/ Φ Compensation Heating

Ideally, for the fire exposure test method to better simulate the actual event, the external power should be delivered as a constant power source and not a constant rate. In the Das [1] fire exposure procedure, the determination of Q, the fire heat flux, is discussed at length. It is from Q that the test heating rate is determined. However, we can scale the Q for the process by dividing it by the ratio of the plant reactant mass to the sample test mass.

$$Q_{test} = Q/(\text{Mass of reactant in plant vessel/mass of the test sample}) \quad \text{[Eq. 3]}$$

The value of Q_{test} can be entered directly into the test set-up. The test must also be conducted at a thermal inertia of 1 (or at the plant vessel thermal inertia). Therefore, the VariPhi heater has to add BOTH the constant value determined from the fire scenario, $Q_{test,}$ as well as $Q_{compensation}$, the heat "lost" to the sample vessel during the test. The determination of the heat required to compensate for the thermal inertia is given in Equation 4. The derivation and limitations of this method are detail by Chippett [4]

$$\mathbf{dQ_{compensation}/dt = 0.5\,\lambda\,M_b\,C_b\,(dT_w/dt + dT_s/dt)} \quad \text{[Eq. 4]}$$

where,

λ is the fraction of the sample container to be compensated (**Φ = 1, λ=1**)
M_b mass of the container
C_b heat capacity of the container
T_w container wall temperature
T_s interior sample temperature

In the earlier tests conducted at constant rate, approximately 750 mW of power was required to keep the sample at a rate of 2K/min. For ease of comparison, a constant rate phi=1 test was conducted was again with a 10% solution of DTBP.

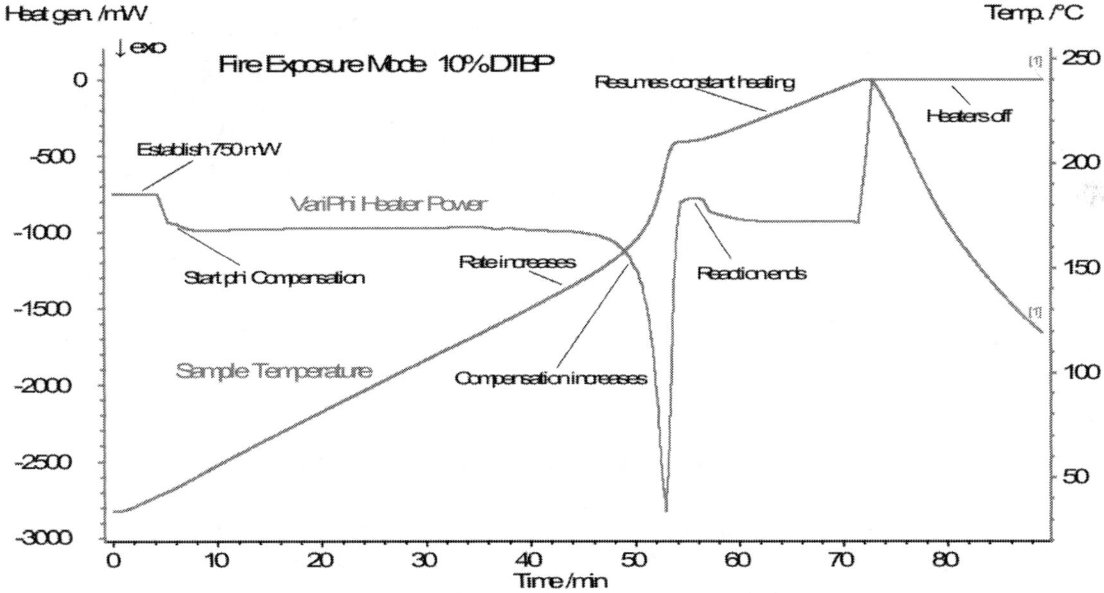

Figure 6. VariPhi Fire Exposure Mode (750mW, Φ =1) 10% DTBP

Figure 6 displays the VariPhi heater power and the sample temperature as a function of time. Initially, the VariPhi heater supplies a constant 750mW of power and this establishes an initial heating rate. Some of the 750mW that is being supplied is going to heat the sample vessel. Based on the heating rate and the temperature and the sample vessel thermal properties (which are automatically entered during test set-up) the software adds the additional power lost to the sample vessel. This in turn increases the rate slightly and the heater power is further adjusted. As long as there are no endotherms or exotherms, the sample will heat-up at nearly a constant rate based on mass and heat capacity. Once the sample begins to react, the self-heat rate will increase and this will trigger more compensation power to be added. So, during this time it is applying the 750mW plus the compensation power. Generally, the heater power curve will follow the self-heat rate plot until the reaction has come to an end. After that, the adjusted 750 mW of constant power will continue until the test ends.

Assuming that the original Q is a good estimate of the heat flux for the scenario and the headspace ratio was similar, the actual temperature and pressure curves measured during this test would match exactly the full scale process. This would be free of caveats associated with thermal inertia correction. Fire exposure testing can produce very high self-heat rates so the reaction must not exceed the heat tracking capability of the calorimeter. Figure 7 is a comparison of the test method results discussed in this paper.

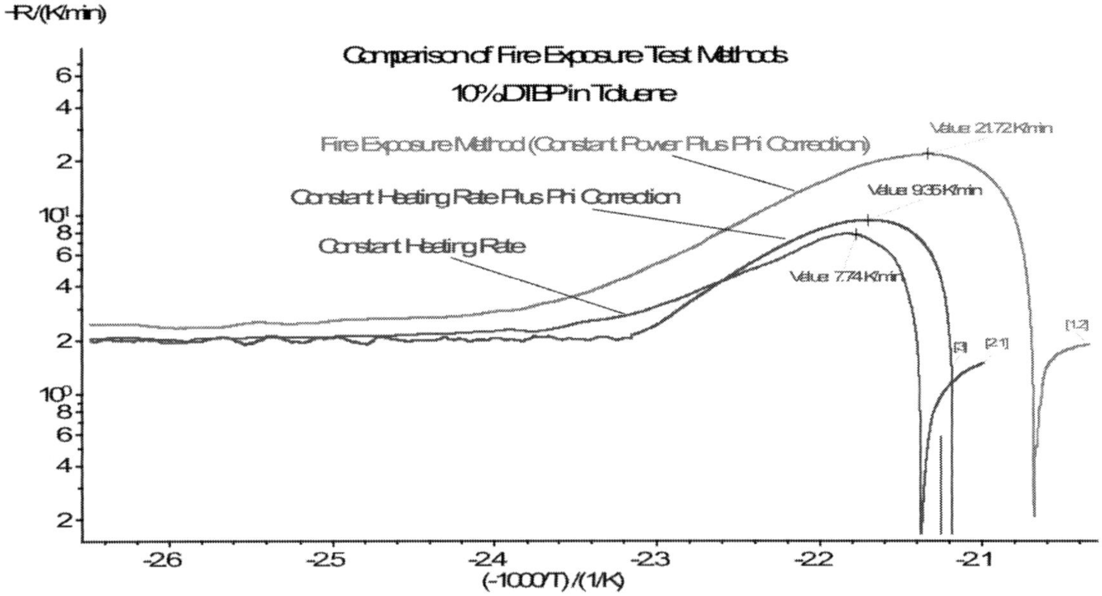

Figure 7. Comparison of Fire Exposure Test Methods

There is another advantage of using the VariPhi method. Since we know at any given time how much heat is being supplied by the internal VariPhi heater and we have a good measure of the sample heating rate, we can calculate the sample power from the following equation.

$$\mathbf{dW/dt = \{P_0\ (dT_s/dt)/(dT_0/dt)\} - 0.5\ M_bC_b\ \{(dT_s/dt) - (dT_w/dt)\} - P} \qquad \text{[Eq 5]}$$

where (dT_s/dt) is the measured sample heat rate and (dT_0/dt) is the base line heating rate where the heater power is P_o. The equation to determine the sample power does not require any knowledge of the heat capacity of the sample (reactants or products). The resulting sample power curve shown in Figure 8 can be analyzed using techniques similar to those for DSC data.

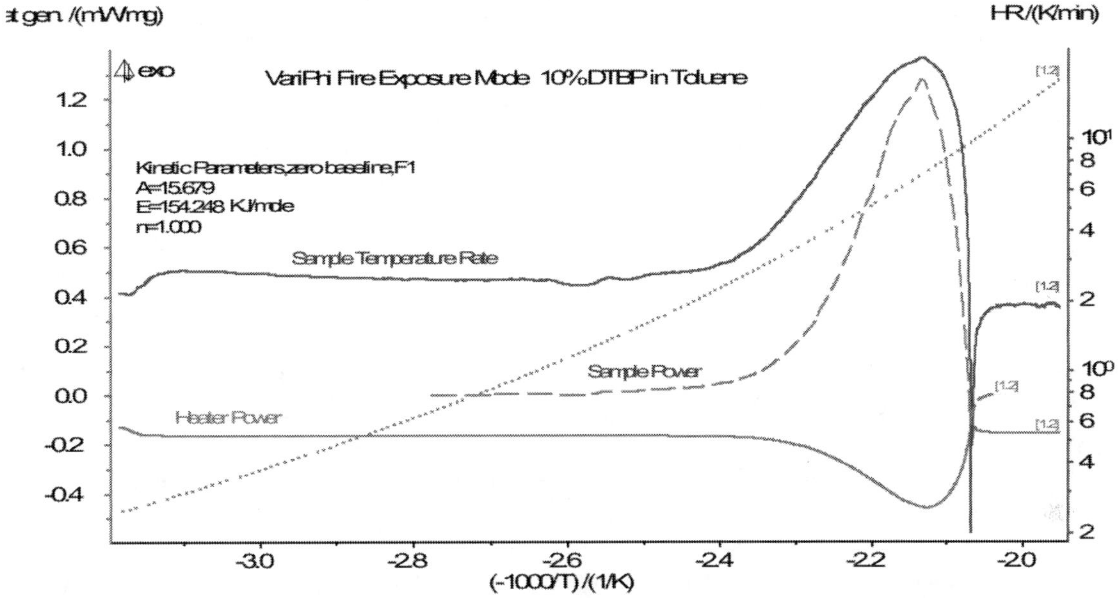

Figure 8. Sample Power curve can be used for kinetic analysis

4. Conclusion

The method described by Das [1] clearly explains and demonstrates a suitable method for the development of a sound fire exposure test. In this paper we note one change as it relates to adjusting the heating rate based on the difference between the test thermal inertia and the plant thermal inertia. As described earlier, such adjustments are unnecessary and should be avoided.

Prior to the capability of the VariPhi to run constant power and thermally compensated tests, the engineer was limited to running such tests in low Φ equipment at a constant rate. Such testing requires large sample sizes and does not do a very good job of replicating the actual fire scenario.

Through exploring several different approaches it is clear that running the VariPhi at constant power, with bomb compensation, provides test data that is more representative of the fire exposure scenario and the test can be used directly for vent sizing. No further adjustments are required.

The original method of constant heating rate underestimates the actual temperature and pressure rise resulting in under-sizing of ERS design for the fire exposure case.

Furthermore, since the contribution of the sample power output can be calculated from the heater power and heating rate curves, the test data can also be used to develop kinetic and thermodynamic data which is not possible with the original constant rate method.

5. References

[1] Das, D "A Comparison of Fire Heat Flux Based on Fundamental Principles of Heat Transfer with Heat Flux calculated by NFPA-30.

[2] Yin, F., "Adiabatic Calorimeter: Fundamentals and Application in Thermal Hazard Evaluation", *International Symposium on Runaway Reactions and Pressure Relief Design*, American Institute of Chemical Engineers, 58-94, 1995

[3] Townsend, D.I., Tou, J.T., Thermal Hazard Evaluation by an Accelerating Rate Calorimeter, *Thermochemica Acta*, 37, 1-30, 1980.

[4] Chippett, S., A New, Low Thermal Inertia, Scanning Calorimeter, CCPS International Conference, Scottsdale, AZ, 2003

[5] Chippett, S., Ralbovsky, P., Improved Performance of an Adiabatic Calorimeter, Presentation at PITTCON 2004, Chicago

[6] Ralbovsky, P., Chippett, S., Brady-Myerov A. "Design Criteria for Reactive Hazards Screening Tools" *International Symposium on Runaway Reactions and Pressure Relief Design*, American Institute of Chemical Engineers, 1998

Guidance for sizing relief devices installed below liquid level

Rahul Raman
Provenance Consulting
301 W. 6th Street Suite 200 Borger Texas 79007
Rahul.raman@provenanceconsulting.com

Prepared for Presentation at
American Institute of Chemical Engineers
2015 Spring Meeting
11th Global Congress on Process Safety
Austin, Texas
April 27-29, 2015

Guidance for sizing relief devices installed below liquid level

Rahul Raman
Provenance Consulting
301 W. 6th Street Suite 200 Borger Texas 79007
Rahul.raman@provenanceconsulting.com

Keywords: Emergency relief system, two-phase flow, bottom venting, DIERS technology, pressure relief valve, external fire,

Abstract

API 520 Part I & API 521 provide guidance for designing emergency relief devices for the external pool fire scenario for pressure vessels built to ASME Code Section VIII, Div.1. Current API standards provide vapor generation and/or two-phase flow sizing models assuming that the relief valve is installed in the top of the vessel. ASME Code Section VIII, Div.1, UG-135, provides guidance in the installation of the relief device. However, the current standards do not limit the location of relief devices to only the top of the vessel. Certain pressure vessels, such as filters or vertical heat exchanger shells may have relief devices installed at a location below the liquid level for maintenance access, space limitations, or piping constraints. The use of traditional sizing methods can grossly undersize the valve as the relief device will need to pass fluids at a rate equal to the volumetric rate of vapor generated by the fire. Existing literature provides limited guidance on sizing and disposal requirements. This paper provides general guidance, sizing method strategies, mitigation, and design of relief devices used in such services.

1. Introduction

During an external pool fire, the vessel is engulfed in flames impinging on the vessel. The contents are considered to be saturated and if the inlet piping to the relief valve is below the liquid level, then the contents relieved will not be in the vapor phase. The vapors generated will push the liquid through the relief path. As the liquid level drops and falls below the outlet nozzle to the relief valve, the flow can change from two phase flow to all vapor flow. However, these systems are typically sized for saturated liquid displacement and may not take the effects of two phase flow into consideration. Firstly, qualify a bottom venting case by looking at the system in detail. Having a relief valve mounted on the bottom of the vessel but having the relief path inlet above the liquid level will not qualify as a bottom venting case. Secondly, understand the state of existing methodologies and best practices that can be used for sizing. Finally, use DIERS technology to predict vapor disengagement and orifice sizing.

2. Current Industry Practice and Guidance

2.1 Current state of RAGAGEP (Recognized and Generally Accepted Good Engineering Practice) plus Industry Practice

The state of RAGAGEP and current industry practices are summarized below, and some of the best methodologies and techniques are highlighted. This paper gives general equations and industry-recognized methods, but should not be considered as an all-inclusive guide. The reader is always advised to use a conservative approach for emergency relief system (ERS) design that considers a broad base of industry practice and company guidelines, and which complies with local, state, and federal regulations.

2.1.1 API (American Petroleum Institute)

In an external fire scenario when the inlet piping to the PRV is below a liquid level, the orifice needs to be sized for displacement of fluid that is equivalent to the volumetric vapor generation rate based on API 521 [1]. This assumes that the vessel is isolated during an external fire scenario to simplify the analysis. A further simplification is to consider that the vessel contents are at saturated conditions during a bottom-venting release that results in two-phase flow. API 520 part I provides guidance regarding sizing of the orifice for two-phase flow, namely the omega method and the direct integration method, and is considered to be industry-recognized best practice [2]. The amount of vapor generated increases the system pressure commensurately and pushes the saturated liquid through the PRV inlet piping.

2.1.2 DIERS (Design Institute for Emergency Relief Systems)

The vapor disengagement methodology provided by DIERS is used extensively to understand the flow behavior inside a vessel assuming that the relief path is above the liquid level (i.e., on the top of the vessel) [3]. In cases where the relief valve is below the liquid level, the flow behavior contrasts to conventional DIERS technology as it is not the same as a vessel with a relief device installed in the vapor space. The DIERS disengagement methodology along with the coupling equation can be used for straight pipe with a negligible void fraction, but oftentimes the piping to the relief valve is complex and the methodology becomes unviable. Some important parameters that can be calculated are quality (vapor fraction) at the vent entrance, (start) onset of two-phase flow, and disengagement void fraction (stop) at which the flow turns from two-phase flow to all vapor flow [4]. This paper is intended to describe a particular case and give simple guidance. For more generalized expressions, a review of the omega method by Leung is recommended [5]. An excerpt of required DIERS technology is described below.

2.1.2.1 Perfect Nozzle Mass Flux

The perfect nozzle mass flux equation reported previously by Leung for saturated liquid with a critical flow through the nozzle is given by [4],

$$G_{PN} = \eta_c \left(\frac{g_c P_o}{\omega_0 v_{fo}}\right)^{0.5} \quad \text{[Eq. 1]}$$

Where the omega parameter, ω_0, is obtained from the flash calculation given by,

$$\omega_0 = \left(\frac{\left(\frac{v_9}{v_{fo}}\right) - 1}{\left(\frac{P_0}{P_9}\right) - 1} \right) \text{ [Eq. 2]}$$

Where, $P_9 = P_0 \times 0.9$

The dynamic pressure ratio at critical flow condition is given by,

$$\eta_c = 0.6055 + 0.1356\,(\ln \omega_0) - 0.0131 \ln(\omega_0)^2 \text{ [Eq. 3]}$$

The above equations together provide the ideal nozzle critical flow mass flux for saturated liquid through the orifice.

2.1.2.2 Disengagement equations

Liquid disengagement equations are presented below for churn-turbulent and bubbly flow models. The model is able to predict the phase through the orifice as a function of Ψ, dimensionless superficial velocity and α, average void fraction [3] [4]. Figure 1 is a model graph showing disengagement curves for different flow regimes and vessel orientations. Two-phase flow typically occurs for data points below the curve with all vapor flow above the curve.

Table 1. Flow regime constant C_o range for different flow regimes

C_o range	Flow Regime
1.18 ~ 1.53	Churn-Turbulent
1.01 ~ 1.18	Bubbly
1.01	Homogeneous

Right Circular Cylindrical Vessel with Fire Exposure on Vertical Sides Only (Churn-turbulent)

$$\alpha_{CT} = \frac{1}{3}\left[\frac{\psi_{CT}}{2 + C_0\psi_{CT}}\right] \text{ [Eq. 4]}$$

Right Circular Cylinder Vessel with Fire Exposure on Vertical Sides and Bottom of Vessel (Churn-turbulent)

$$\alpha_{CT} = \left[\frac{\psi_{CT}}{2 + C_0\psi_{CT}}\right] \text{ [Eq. 5]}$$

Typically, a bubbly flow disengagement curve is very steep and hence a more simplified formulae is given below,

$$\alpha_{BUB} = 0.5 - (0.25 - \psi_{BUB})^{0.5} for\ \psi_{BUB} \leq 0.25 \text{ [Eq. 6]}$$

$$\alpha_{BUB} = 0.8\ for\ \psi_{BUB} > 0.25 \text{ [Eq. 7]}$$

The original bubbly flow disengagement approximation published in the DIERS project manual is given by [3],

$$\psi_{BUB} = \frac{\alpha_{BUB}(1-\alpha_{BUB})^2}{(1-\alpha_{BUB}^3)(1-C_0\alpha_{BUB})} \text{ [Eq. 8]}$$

The generated vapors entrain the liquid through the vent during a relief event. This forms a convergent flow pattern where the two phase mixture rushes towards the relief valve inlet. The entrainment of liquid in the vessel is a function of Ψ, dimensionless superficial velocity, A_{CR}, cross sectional area of the vessel, and H_V, height of the vessel [4].

$$\alpha_{ENT} = 0.2849 \frac{(A_{CR}\psi_{CT})^{0.5}\left(\frac{\rho_{go}}{\rho_{fo}}\right)^{0.25}}{H_v} \text{ [Eq. 9]}$$

$$\alpha_{ENT} = 0.2502 \frac{(A_{CR}\psi_{BUB})^{0.5}\left(\frac{\rho_{go}}{\rho_{fo}}\right)^{0.25}}{H_v} \text{ [Eq. 9]}$$

$$\alpha_D = \alpha_{ENT} + (1-\alpha_{ENT})\alpha_{CT\ or\ BUB} \text{ [Eq. 10]}$$

$$\alpha_D \le \alpha_{vessel} \text{ All vapor flow predicted [Eq. 11]}$$

$$\alpha_D > \alpha_{vessel} \text{ Two phase flow predicted [Eq. 12]}$$

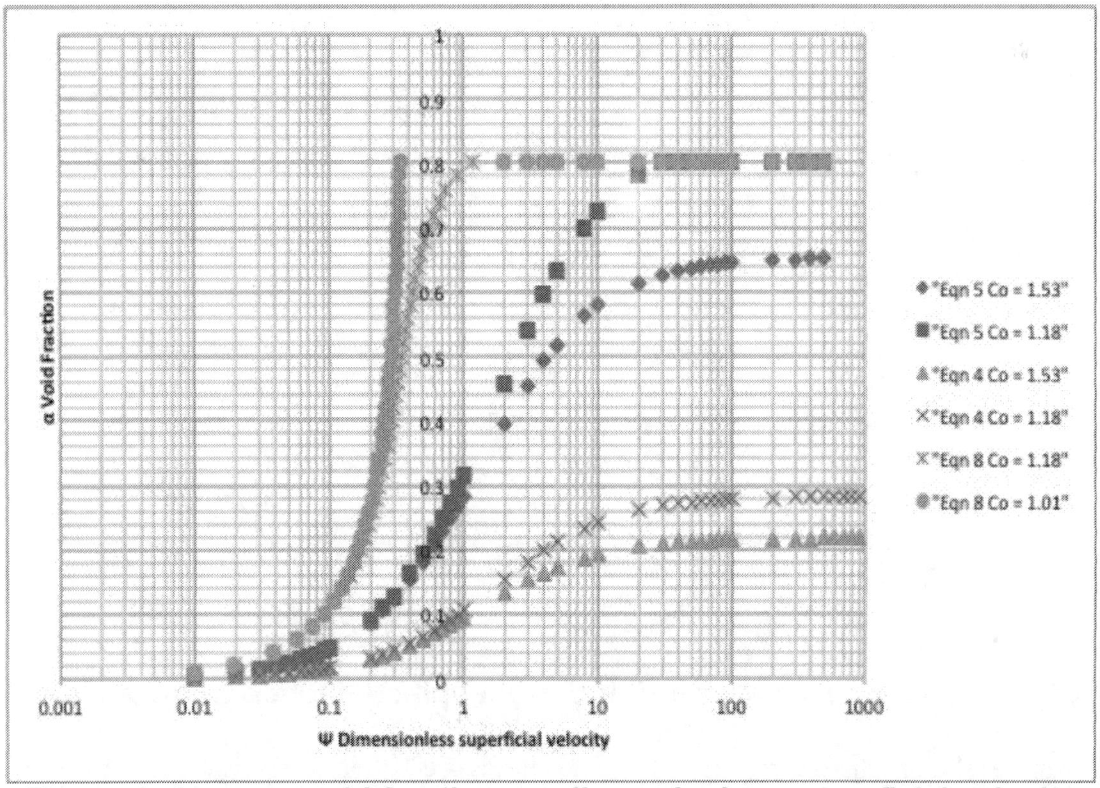

Figure 1. Average void fraction vs. dimensionless superficial velocity

2.2 Common Bottom Venting Examples

When a vessel is exposed to external fire, it is sometimes not apparent for less-experienced engineers to identify a bottom-venting case. This section provides some examples and illustrations to show the different types of bottom venting cases that are possible.

2.2.1 Filters

A widely used unit operation in operating facilities is filtration. Filters are often small, mostly-liquid-filled vessels, and when subjected to an external fire have a low heat up time due to small volume of fluid. A relief device typically protects filters for blocked outlet and external fire. The inlet or outlet piping for filters may be from the bottom, but this does not become sufficient basis to be considered as bottom venting. A more detailed qualitative analysis should be done; vessel internals and the vendor drawing need to be consulted. For example, if the PRV is installed in the inlet pipe of the filter and the internals show that the inlet pipe is welded to a tube sheet then it might be considered as top venting. See figure 1 below for a schematic illustration. For such cases the user can model it as a relief valve installed on the top of the vessel. There are three basic flow regime models that are extensively used in DIERS: Churn-Turbulent, Bubbly, and Homogeneous flow [3]. The flow regimes are listed in relative order: the orifice area is smallest for the churn-turbulent flow model and largest for the homogenous model. It is up to the user to decide which flow model to use, based on the thermo-physical properties of the fluid. Furthermore, it is important to note that the flow path internal to the vessel may need to overcome additional resistance. For example, the fluid may need to pass through a filter cake, filter elements, and a tube sheet to enter the relief piping on the top of the vessel. One suggested method is to use a reduced coefficient C_0 value or change the flow regime to a more conservative model to account for higher superficial velocities through the vessel. This is left to the discretion of the user. Table 1 above shows the typical value of C_0 for the different types of flow regimes. See Figure 1 for the effects of C_0 value. Note that a lower C_0 results in two-phase flow at lower void fractions.

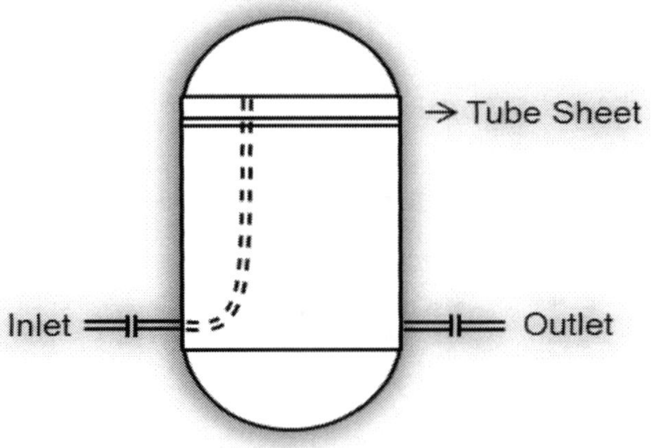

Figure 2. An example of a filter that has inlet pipe attached to the tube sheet

If a PRV is installed in the outlet line of the filter in Figure 1 then the flow will be two-phase and the system must not be sized for all vapor flow.

2.2.2 Vertical heat exchangers

Pressure relief valve design for vertical heat exchangers may be subject to bottom venting cases. The geometry and installation of the vertical heat exchangers do not typically allow for PRV installation directly on top of the vessel. The conventional disengagement methodologies can be used for sizing as long as the PRV is installed close to the swelled liquid void fraction, i.e., close to the top of the exchanger. No further modification may be required for completing the evaluation of such systems. However, due to installation, space, maintenance, and other constraints, the PRV may be installed below the normal liquid level or swelled liquid level, which results in liquid displacement. See Figure 3 for some common examples. Let us consider a liquid filled vessel for both cases. The PRV must be adequately sized to relieve swelled subcooled liquid, saturated liquid, two-phase flow, and perhaps all vapor flow for Case 3B.

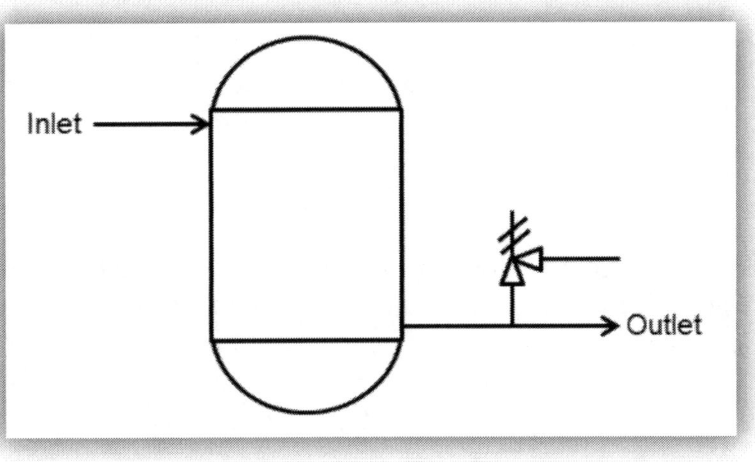

Figure 3A

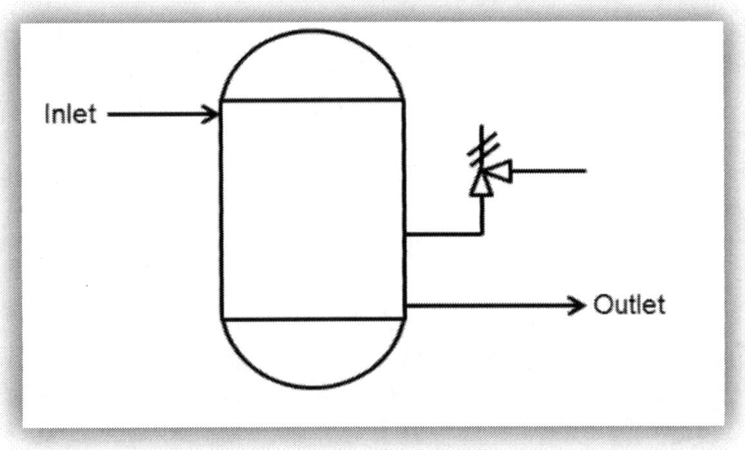

Figure 3B

Figure 3. An example of vertical heat exchangers

2.2.2.1 Case 3A

The PRV is located in the bottom of the vessel. It is exaggerated in Figure 3A to be near the bottom of the shell. Due to the location of PRV the flow through the orifice is going to be two phase flow. Disengagement and vapor flow will occur once all liquid is evaporated or discharged through the PRV and the shell is filled with vapor. The location of the vessel outlet nozzle may be further up the shell, and in such circumstances Case 3A should be evaluated similarly to Case 3B. The volumetric flow rate due to the vaporization of the liquid is to be calculated and Eq. 1 can be used to find the mass flux, and Eq. 13 should be used for calculating the orifice area. W is the maximum required mass flow rate obtained from Eq. 14 using the saturated liquid density and the volumetric vapor generation rate. This section described a simple sizing equation and a more rigorous analysis is detailed in section 2.3.

$$A_{req} = 0.04 * \frac{W}{G_{PN} \prod K} \text{ [Eq. 13]}$$

$$W = m^* * \frac{v_{go}}{v_{fo}} \text{ [Eq. 14]}$$

The product of all correction factors is represented by $\prod K$ that includes, coefficient of discharge K_d, viscosity correction factor K_v, combination capacity factor K_c, and others. Pressure attenuation in the piping system follows an isenthalpic path.

2.2.2.2 Case 3B

Pressure relief valves are also mounted in the side nozzles of vertical heat exchangers as shown in Figure 3 Case 3B. The relief valve will relieve saturated liquid or low-quality two-phase flow until the liquid level drops in the vessel and reaches the nozzle. Below this level, it can be evaluated as a relief valve mounted on the top of the heat exchanger. Therefore it is important to note that in such a system, there will initially be liquid swell that discharges subcooled liquid through the relief valve outlet. Upon further increase in temperature and pressure, saturated liquid or low-quality two-phase flow is discharged through the relief valve resulting in decrease of level inside the exchanger. Once the level drops, there will be two-phase flow of increasing quality and eventually disengagement. Figure 4 highlights the flow and disengagement model. An iterative calculation can provide the maximum flow through the pipe that equals the flow through the orifice. The stagnation enthalpy is conserved in the piping system and an iterative calculation is required, which can be solved by starting with an initial guess. Successive iterations yield a flow rate, temperature, pressure, and vapor quality that is used for sizing the orifice using the ideal nozzle mass flux equation. The iteration ends when the flow rates and conditions are the same for the pipe and nozzle model.

$$W = m^* * \frac{v_{go}}{v_{fgo}} \text{ [Eq. 15]}$$

$$v_{fgo} = (1 - x) * v_{fo} + x * v_{go} \text{ [Eq. 16]}$$

The flow through the pipe is isenthalpic and the pressure, temperature, and quality obtained at the vent entrance can be used in Eq. 13, 15, 16, with an appropriate mass flux equation (e.g., omega method or direct integration method) to obtain the orifice area.

2.3 Bottom venting model

In Figure 3 Case 3B, when the level drops to the relief valve nozzle outlet, the disengagement equations plus the coupling equations can be used to predict the orifice area. Once the level reaches the nozzle height it can be sized using the conventional DIERS methodology with the relief valve on the top of the vessel. Flow may be two-phase depending on the disengagement criteria established in Eq. 10, 11, & 12. Note that when the disengagement criteria is calculated, the nozzle height should be considered as the top of the vessel, that is, α_{vessel} = 0.0001 (a zero value is not used as it leads to a trivial solution).

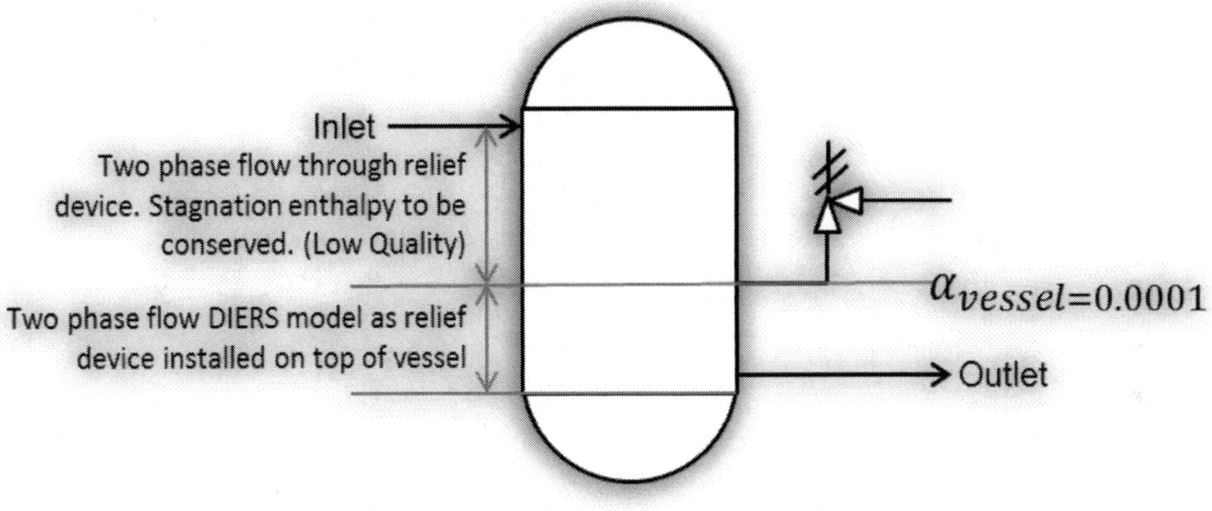

Figure 4. Disengagement model for relief valves installed below the liquid level

When fire is impinging on a vertical heat exchanger, assuming that the entire shell is within the wetted surface area requirements of API 521, the bottom half can be neglected and Eq. 4 can be used for representing the disengagement model. Note that the existing curve for Eq. 4 in Figure 1 shows that curve flattens at about Ψ = 20. This is a conservative disengagement model and the author believes a less conservative model can be used for such cases. The basis for this assumption is that firstly, the convergent flow through the relief valve is going to be less pronounced because of the overhead space above the vessel nozzle. Secondly, disengagement will be more pronounced because of the location of the inlet pipe to the relief valve plus ample space for vapor and liquid droplets to break free. There are currently no models that are available to describe this type of disengagement. Therefore the user can use the Eq. 4 to model the relief valve orifice area until further models are presented. The most conservative orifice area must be selected and used as the design basis for the ERS design.

3. Conclusion

It is suggested to use the preceding guide to model the relief device for pressure relief valves installed below a liquid level. The disengagement model currently available may be the upper bound and a less conservative disengagement model may be required to analyze such system. A liquid trim relief valve is suggested for such applications as liquid displacement is more pronounced. The method allows the designer to determine the vent size for the different flow conditions, firstly when the liquid level is above the PRV nozzle, and then when the liquid level is below the PRV inlet piping nozzle.

4. References

[1] API 521, Pressure-relieving and Depressuring Systems, Sixth edition, American Petroleum Institute, Washington, DC, 2014

[2] API 520, Sizing, Selection, and Installation of Pressure-relieving Devices, Part I - Sizing and Selection, Ninth edition, American Petroleum Institute, Washington, DC, 2014

[3] H.G.Fisher et al., Emergency Relief System Design Using DIERS Technology: The Design Institute of Emergency Relief Systems (DIERS) Project Manual, American Institute of Chemical Engineers, New York, NY, 1993

[4] Harry S. Forrest, Emergency Relief System Design for Fire Exposure With Consideration of Multiphase Flow, International Symposium on Runaway Reactions, and Pressure Relief Design, Center for Chemical Process Safety, American Institute of Chemical Engineers, New York, NY, 1995

[5] Joseph C. Leung, Easily Size Relief Devices and Piping for Two-Phase Flow, Chemical Engineering Progress, Volume 92, Journal Issue: 12, Pages: 28-50, American Institute of Chemical Engineers, New York, NY, 1996

Mechanical Integrity Considerations in LNG Depressurization

A Sensitivity Analysis

Daniel Nguyen‡, Jordi Dunjó, and Hana Gilliam
ioMosaic Corporation
93 Stiles Road, Salem, New Hampshire 03079

‡Presenter (nguyen@iomosaic.com)

Prepared for Presentation at
American Institute of Chemical Engineers
2015 Spring Meeting
11th Global Congress on Process Safety
Austin, Texas
April 27-29, 2015

UNPUBLISHED

Abstract

In a typical LNG installation, a rapid depressurization can cause cryogenic temperatures in both upstream and downstream connected process equipment and piping. This phenomenon, sometimes referred to as auto-refrigeration, can compromise the equipment's mechanical integrity and pose a risk of material embrittlement. As vessel metal walls are exposed to temperatures below the minimum design metal temperature (MDMT), permanent damage is possible. The potential for brittle failure is even more pronounced for a non-fire scenario. The level of severity depends on the initial pressure, initial temperature, content inventory, depressurizing rate, fluid composition, surrounding conditions, and heat transfer mechanisms.

Emergency depressurizing valves must therefore be sized to ensure a reasonable compromise between the impact of pressure and temperature. This paper examines the effects of different liquid levels, depressurizing valve sizes, vessel wall thicknesses, thermal insulations, and fluid compositions. The primary objective is to identify and illustrate the key factors that influence the mechanical integrity of a typical LNG installation, particularly at the mid to lower end of methane fluid compositions, and their impacts on carbon steel.

Key words: LNG, Depressurization Systems, Mechanical Integrity, Embrittlement, Vessel Wall Dynamics, Vessel Wall Segmentation Approach.

Introduction

Emergency depressurizing systems (EDPs) are designed to reduce pressure by expelling the fluids and/or inventory from the protected equipment, thereby reducing the risk of equipment failure. Typical scenarios considered for emergency depressurization are external fire, uncontrolled reactions, and process vessel leaks.

In the event of a pool fire or jet flame impingement, not only do the system contents experience a rise in temperature and pressure, but the temperature of the system's walls rises as well. As the temperature of the metal increases, its mechanical strength decreases. Since the portion of the vessel filled with liquid predictably absorbs most of the heat, the main area of concern would be the unwetted or dry wall exposed to fire. As heating continues, the tensile strength is further reduced. Eventually, the wall metal temperature will reach the vessel's ultimate tensile strength, causing equipment failure.

Much literature[1,3,4] has been devoted to addressing system depressurization for fire scenarios; this paper focuses on the scenario of a vessel leak, alternatively referred to as a non-fired or cold depressurization. Due to expansion cooling and condensation of light ends, rapid depressurization can cause cryogenic temperatures in both upstream and downstream connected process equipment and piping. Depressurization systems in typical LNG installations often experience this phenomenon, known as auto-refrigeration, which may compromise the equipment's mechanical integrity and pose a risk of material embrittlement. As system walls are exposed to temperatures below the minimum design metal temperature (MDMT), permanent damage is possible. The potential for brittle failure is even more pronounced for a non-fire scenario. Its severity depends on the initial pressure, temperature, content inventory, depressurizing rate, fluid composition, surrounding conditions, and heat transfer mechanisms.

Emergency depressurizing valves (EDPV) must therefore be sized to ensure a reasonable compromise between the impact of pressure and temperature. For example, in the event of fire or abnormal heating, a larger EDPV size would generally lower pressure faster, likely meeting the established pressure reducing criteria (e.g.,100 psi or 50% of the initial pressure) in the specified amount of time (often at 15 minutes). However, a larger EDPV size would also generate a higher depressurizing rate, exacerbating the cryogenic effect, which ultimately leads to a lower metal temperature. Likewise, a larger liquid inventory may provide a better liquid contact for heat absorption and sustain the vessel integrity for longer in the event of abnormal heating. However, the larger liquid inventory may also expose the metal wall (at the bottom in contact with the cryogenic liquid) to the lower temperature longer, creating a higher probability of embrittlement failure.

This paper examines the impact of different liquid levels, depressurizing valve sizes, vessel thicknesses, and fluid compositions. The primary objective is to identify and illustrate the key factors that that influence the mechanical integrity of a typical LNG installation, particularly at the mid to lower end of methane fluid compositions and their impacts on carbon steel.

Basis for Sensitivity Analysis

The system chosen for this paper's sensitivity analysis study entails a horizontal cylindrical vessel with 2:1 Elliptical heads with Carbon Steel SA-516-G70. Unless stated otherwise, the basis and mixture composition considered in all vessel wall dynamics analyses are summarized in Table 1 and Table 2.

Table 1: System Specifications and Conditions

Parameter	Value
Length [straight side for cylinders] [m]	6.0
Inside Vessel Diameter [m]	3.0
Shell and Head Thickness [m]	0.0127
Total Vessel Surface Area [m^2]	76.1
Total Vessel Volume [m^3]	49.5
Initial Vessel Wall Temperature [°C]	25
Ambient Temperature [°C]	25
Relative Humidity [%]	70
Wind Speed at reference height [$m \cdot s^{-1}$]	2
Wind speed reference height [m]	10

Table 2: Mixture Composition

Compound	Mole Fraction
METHANE	0.413
ETHANE	0.165
PROPANE	0.165
ISOBUTANE	0.257

Five (5) main cases were analyzed, examining the sensitivity of vessel wall thicknesses, initial liquid levels, EDPV sizes, insulation thicknesses, and fluid compositions. Table 3 provides a summary of the parameters considered in this study.

Table 3: Summary of Sensitivity Analysis Cases

Case	Description
Vessel Wall Thickness	Four (4) wall thicknesses (0.01", 1.5", 3.0", and 20"), considering 10% initial liquid level and an EDPV size of 4"
Initial Liquid Level	Three (3) initial liquid levels (10%, 50%, and 90%), considering all EDPV sizes
EDPV Size	Four (4) EDPV sizes (0.5", 1.0", 2.0", and 4.0"), considering all initial liquid levels
Insulation Thickness	Four (4) insulation thicknesses (0", 2", 7", and 14"), considering all initial liquid levels and an EDPV size of 4"
Fluid Compositions	Six (6) methane mole fractions (0.05, 0.10, 0.20, 0.30, 0.40, and 0.5) at 10% initial liquid level and 22.5 barg using an EDPV size of 2"

Note that evaluating the various initial liquid levels and EDPV sizes results in a total of twelve (12) dynamic simulation runs. Likewise, combining the insulation thicknesses at all initial liquid levels equates to another twelve (12) runs. Adding the four (4) runs from the vessel wall thickness case (at 10% initial level and an EDPV size of 4") and six (6) methane fluid compositions (at 10% filled level and 22.5 barg using an 2" EDPV size), a total of thirty four (34) dynamic simulation runs were conducted in this study.

Except for the methane fluid composition cases, all simulations considered the same mixture composition and same initial fluid temperature. However, to retain saturated conditions at the desired temperature for each selected initial liquid level case, the initial pressure was modified accordingly, as shown in Table 4.

Table 4: Initial Saturated Conditions

Initial Liquid Level [%]	Initial Pressure [barg]	Initial Fluid Temperature [°C]
10	22.5	0.00
50	50.0	0.00
90	62.8	0.00

Table 5 summarizes the Foam Glass properties while Table 6 provides the methane compositions at their associated saturated temperatures used in this study.

Table 5: Foam Glass Insulation Properties

Temperature [°C]	Density* [kg·m^{-3}]	Specific heat capacity* [kJ·kg^{-1}·K^{-1}]	Thermal Conductivity [W·m^{-1}·°C^{-1}]
-75	300	351.5	0.172
-50	300	351.5	0.180
-25	300	351.5	0.188
-10	300	351.5	0.197
0	300	351.5	0.201
10	300	351.5	0.205
25	300	351.5	0.209
50	300	351.5	0.222
75	300	351.5	0.234
100	300	351.5	0.247

*Both insulation density and specific heat capacity are considered not temperature dependent

Table 6: Methane Compositions and Temperatures at 22.5 barg and 10% Filled Level

Item	Mole Fraction					
Methane	0.050	0.100	0.200	0.300	0.400	0.500
Ethane	0.165	0.165	0.165	0.165	0.165	0.165
Propane	0.165	0.165	0.165	0.165	0.165	0.165
Isobutane	0.620	0.570	0.470	0.370	0.270	0.170
Temperature (°C)	78.77	70.71	52.76	31.38	3.73	-26.12

Emergency depressurizing valves are often sized to ensure a reasonable compromise between the impact of pressure and temperature. A larger EDPV size (resulting in a higher depressurizing rate) generally reduces the system pressure in a shorter amount of time. However, this would also create a greater expansion cooling effect, resulting in a lower fluid temperature, which may compromise the mechanical integrity of the vessel.

Consistent with the depressurizing rate guidelines from API 521[2], this analysis examines the following two key criteria:

- The studied system is depressurized to 6.89 barg (100 psig) or 50% of the initial pressure within 15 minutes of initiating depressurization

- Vessel wall metal temperature does not fall below the minimum design metal temperature. This criterion provides an assurance against brittle failure due to expansion cooling or auto refrigeration.

Based on the established criteria, the main variables of interest considered for this analysis are the temperature and pressure during the entire process of depressurization. Our focus is thus designed to answer the following questions:

- What is the lowest wall temperature (the bottom wall segment with direct contact of liquid) reached during depressurization?
- What is the pressure after 15 minutes of depressurization?
- At what time does the system pressure reach 50% of the initial pressure?
- When does the system pressure reach 6.89 barg (100 psig)?
- Whether and when the wall temperature reaches the minimum design metal temperature of carbon steel (-29 °C)?

Study Approach

The sensitivity analysis study was performed using the SuperChems™ component of ioMosaic's Process Safety Office™. The program consists of detailed models for simulating the fluid dynamics of vessels containing multi-phase fluids. To account for detailed vessel wall and fluid heat transfer dynamics, the simulated equipment is segmented into multiple zones, as shown in Figure 1. Detailed heat transfer to/from the surroundings and between the zones is dynamically accounted for. There is no limit on the number of zones a user can specify. The ability to subdivide a vessel into multiple segments allows users to closely examine the dynamics of the fluids and vessel wall thermal effects. Other valuable applications of the segmentation approach include the modeling of external fire, localized heating, and jet flame impingement (which is out of the scope of this paper).

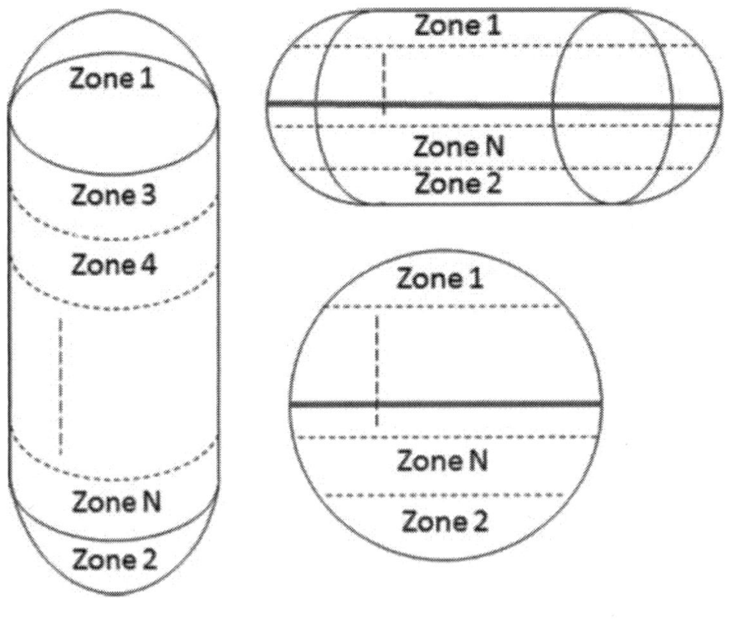

Figure 1. Vessel Segmentation Scheme

Figure 3 illustrates the added value of the vessel segmentation procedure via the vessel wall dynamics analysis. The illustrated simulation considers all parameters highlighted in Table 1 and Table 2 using a 50% initial liquid level and an EDPV size of 2 inches. The initial liquid height is estimated to be 1.5 meters. Based on the segmentation identified (with 10 wall segments as shown in Figure 2), the model confirms that segments 0 to 5 are in direct contact with the liquid. Wall segments 6 to 10 consist of the upper portion in contact with the vapor space of the vessel, where the metal walls are dry or unwetted.

Vessel Wall Heat Transfer Segment	Bottom Segment Elevation, m	Top Segment Elevation,	Segment Metal Volume, m3	Segment Inner Surface Area, m2	Segment Outer Surface Area, m2	Wall-Fluid Conduction Length, m	Segment-Segment Bottom Contact Area, m2	Segment-Segment Top Contact Area, m2	Segment-Segment Conduction Length, m
Bottom Segment 0 -> 1	-0.013	0.121	0.108	8.372	8.927	0.013	0.000	0.460	0.472
Segment 1 -> 2	0.121	0.417	0.108	8.918	8.828	0.012	0.460	0.334	0.273
Segment 2 -> 3	0.417	0.757	0.108	7.529	7.531	0.014	0.334	0.301	0.341
Segment 3 -> 4	0.757	1.126	0.108	6.854	6.880	0.016	0.301	0.292	0.366
Segment 4 -> 5	1.126	1.500	0.108	6.357	6.390	0.017	0.292	0.289	0.374
Segment 5 -> 6	1.500	1.874	0.108	6.357	6.390	0.017	0.289	0.292	0.374
Segment 6 -> 7	1.874	2.243	0.108	6.854	6.880	0.016	0.292	0.301	0.366
Segment 7 -> 8	2.243	2.583	0.108	7.529	7.531	0.014	0.301	0.334	0.341
Segment 8 -> 9	2.583	2.879	0.108	8.918	8.828	0.012	0.334	0.460	0.273
Top Segment 9 -> 10	2.879	3.013	0.108	8.372	8.927	0.013	0.460	0.000	0.472

Figure 2. Detailed Vessel Segmentation

The minimum design metal temperature (MDMT) considered in this study is -29 °C, a typical criteria used for carbon steel material of construction. Figure 3 illustrates that the wall temperature history of segments in direct contact with the liquid are expected to reach the MDMT in less than 20 minutes. However, the top segments (i.e., 6 to 10) are not expected to reach embrittlement temperatures due to the low wall/vapor heat transfer coefficient (5-100 $Wm^{-2}{}^{o}C^{-1}$). This particular case has been selected to highlight the middle portion (Segment 5) of the vessel, which is in contact with both liquid and vapor vessel spaces. At approximately 400 seconds, Segment 5 is still in direct contact with the liquid. As the system is further depressurized, the liquid is reduced and Segment 5 transitions to more contact with the vapor space.

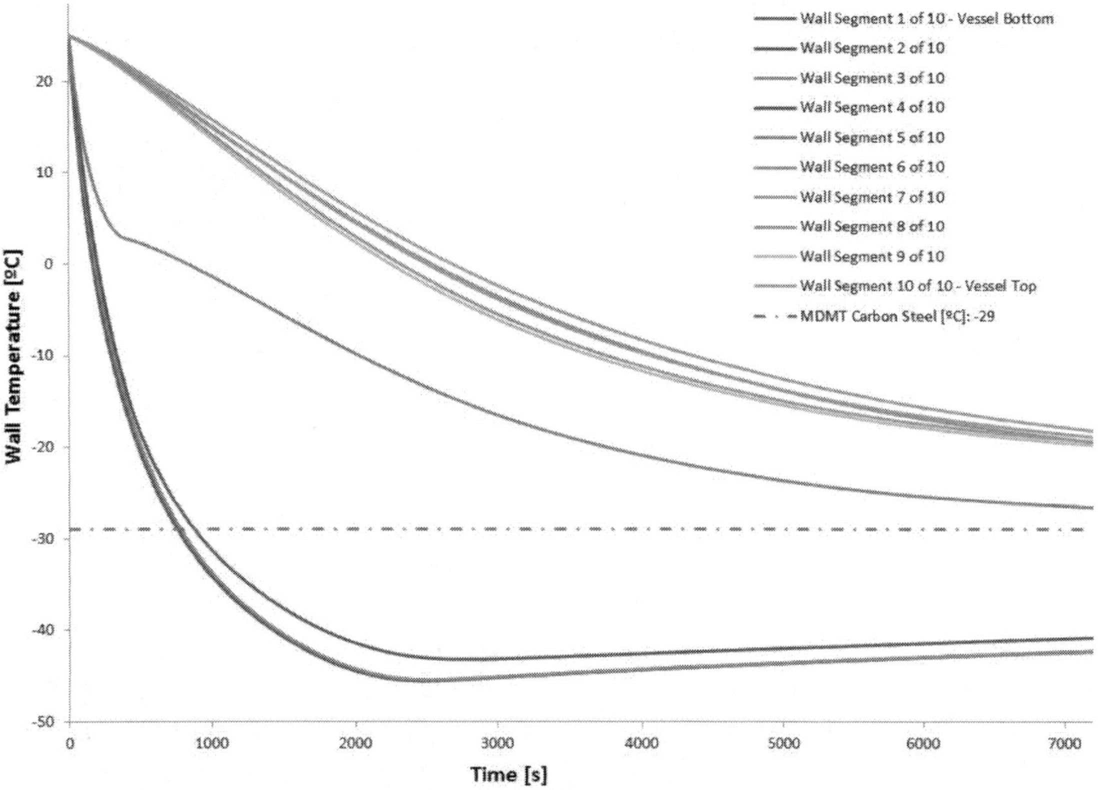

Figure 3. Wall Segment Temperature Profile (50% Initial Liquid Level, 2" EDPV)

Notice that as depressurization progresses, the rate of temperature decrease for Segment 5 slows, tilting the slope sideward and segregating from the other segments still in direct contact with liquid. This simple illustration highlights the significance and dynamic interface of liquid contact impact on the metal walls. It implies that the high value of wall/liquid heat transfer coefficient (100-10,000 $Wm^{-2\circ}C^{-1}$) accelerates the cryogenic fluid heat transfer to the metal wall, posing a higher risk of embrittlement failure.

Analysis of the Results

The following sections illustrate how the selected parameters considered in this study influence the mechanical integrity of the vessel during the depressurization. Special attention has been focused on the bottommost wall segments of the vessel, which are expected to be in direct contact with the liquid during a cold depressurizing scenario.

Impact of Vessel Thickness

This sensitivity analysis examines the impact of four (4) different vessel thicknesses (0.01", 1.50", 3.00", and 20.0"), evaluated using a 10% initial liquid level and an EDPV size of 4 inches. Figures 4 through 5 illustrate the results from the vessel wall dynamic analyses performed.

Table 6 illustrates the estimated metal mass according to the vessel thicknesses selected.

Table 6: Vessel Shell and Heads Mass as a function of Vessel Thickness

Thickness [inches]	Metal Mass [kg]
0.01	150.69
1.50	22,603
3.00	45,207
20.0	301,382

Source: SuperChems Expert™

Taking into account both the mass of the vessel shell and heads and the specific heat of the carbon steel, the wall temperature profiles during a depressurization process are shown. As illustrated in Figure 4, the lowest wall temperature predicted for the vessel thickness of 0.01 inches is -48.0 °C, which occurs approximately 5 minutes after depressurization begins. For vessel thicknesses of 1.5 inches or higher, the MDMT is not expected to be reached. Comparing Figure 4 (bottommost segment in direct contact with liquid) and Figure 5 (topmost segment in contact with the vessel vapor space) indicates that there is a higher risk of a vessel failure at the bottom of the vessel during a cold depressurization. On the contrary, the conclusion is the opposite for the case of a vessel under fire exposure. In the event of fire, the direct contact of fire with the vapor space (top segments) is expected to rapidly increase the metal temperature. The metal strength may be reduced to a value lower than the internal vessel stress. As a result, the top section of the vessel is at a higher risk of failure.

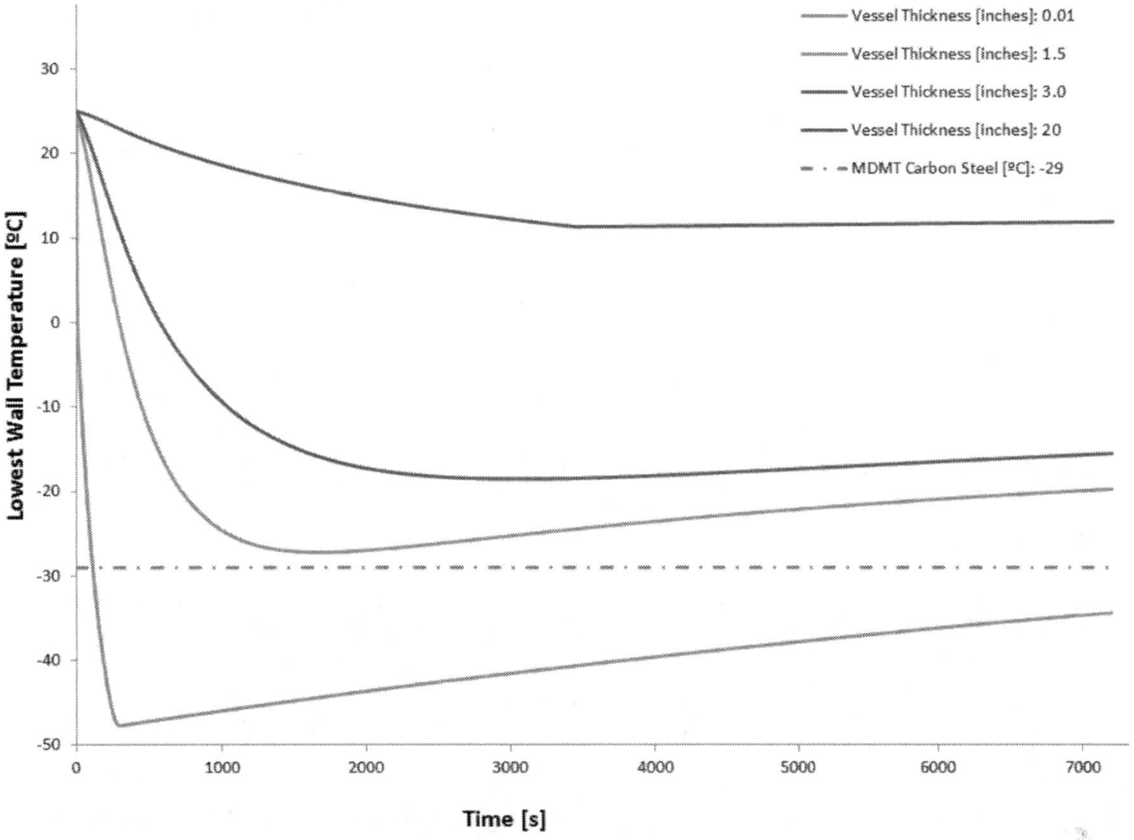

Figure 4. Lowest Wall Temperature History (Bottom Wall Segment; 10% liquid level; EDPV size 4")

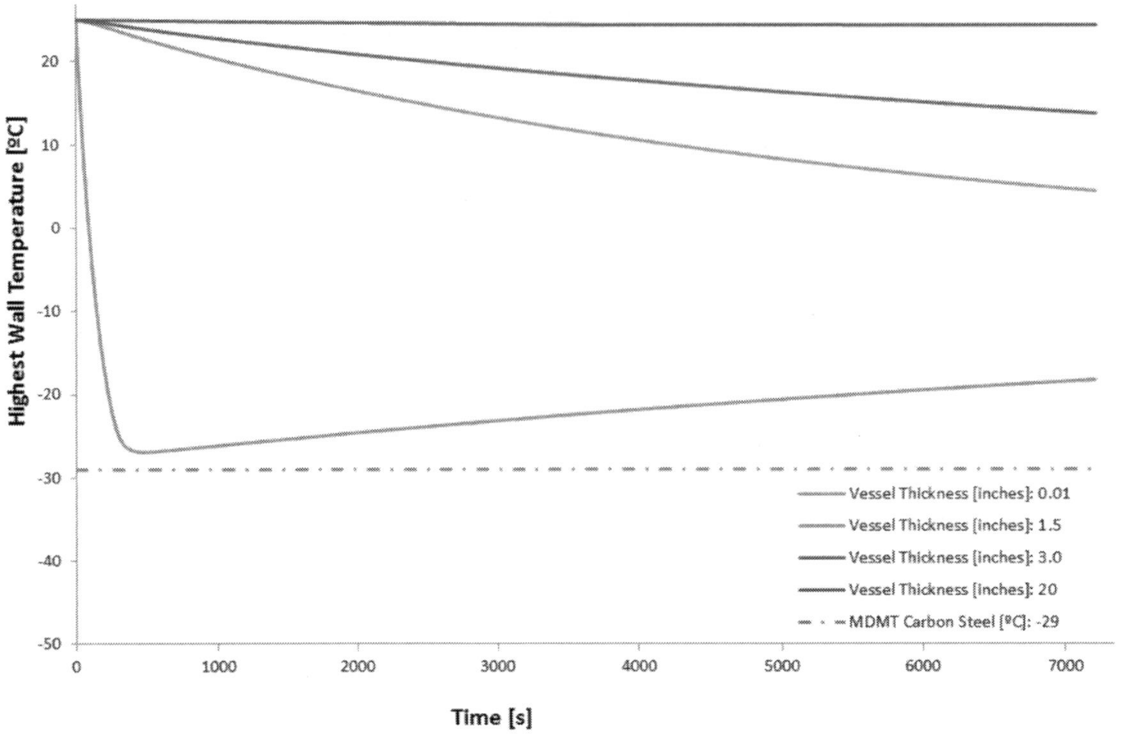

Figure 5. Highest Wall Temperature History (Top Wall Segment; 10% liquid level; EDPV size 4")

Impact of Liquid Level and EDPV Size

This sensitivity analysis examines the impact of three (3) different initial liquid levels (10%, 50%, and 90%), evaluated at four (4) different EDPV sizes (0.5", 1.0", 2.0", and 4.0"). Together, there are a total of twelve (12) simulation runs.

Figures 6 through 11 illustrate both the pressure and wall temperature history results from the identified cases. Note that the temperature profiles are based on the bottommost segment of the vessel, assessing the metal wall in direct contact with liquid.

The lowest initial liquid level (10%) simulations with all EDPV sizes are shown in Figures 6 and 7. The diagrams indicate that the 0.5" EDPV size does not meet any pressure depleting criteria, although it maintains the temperature above the MDMT. In contrast, the pressure criteria are met for the EDPV sizes of 1", 2", and 4", but the metal temperatures fall below the MDMT, posing a risk of brittle failure.

The same situation is found for the remaining cases with 50% and 90% initial liquid levels, illustrated in Figures 8 through 11. Accordingly, the size of the EDPV influences how quickly the vapors are expelled from the system, and thus, how rapidly the system pressure is reduced. While a larger EDPV size depressurizes the system faster and meets the pressure reducing criteria, it also accelerates the cryogenic liquid heat transfer to the metal wall. According to the results, all cases highlight that an EDPV size of 4" will entail a wall temperature lower than the vessel MDMT. Note also that the rapid depressurization prolongs the contact time of the liquid portion of the fluid with the wall, confirming the impact of high heat transfer mechanisms in the liquid phase. This implies that higher liquid levels may pose a greater risk of material embrittlement. Conversely, a reduced initial liquid level results in a shorter contact time between the metal wall and the cryogenic liquid, providing a lower probability of brittle failure.

Figure 12 provides another perspective on the trend of using different EDPV sizes for a given system (using 10% filled level). It shows that larger EDPV sizes produce greater differences between the fluid and the metal wall temperatures. In other words, the vessel fluid temperature is predictably colder using a larger EDPV size, creating a larger driving force between the cryogenic fluid and the metal wall. Notice also that the temperature difference for all EDPV sizes ultimately converges at 2-3 °C after about 30 minutes, indicating steady state with minimal heat transfer driving force.

Note that these cases consider the default vessel thickness of 0.5 inch as outlined in the study basis. They are presented to demonstrate the needed compromise between the pressure reduction and wall temperature criteria. As a result of this analysis, a few actions could be recommended depending on the cycle lifetime of the system. If it is a new design, a greater vessel thickness should be explored. Changing the vessel's material of construction from carbon steel to stainless steel is also a viable alternative. For an existing system under operation, an optimized EDPV size may be required to achieve adequate depressurization while minimizing the auto-refrigeration effect. Alternatively, a combination of EDPV sizes with different activation times could be considered. Note also that an emergency depressurization system is often designed to handle both cold and fire depressurizations. An optimized scheme should therefore be established to address both situations.

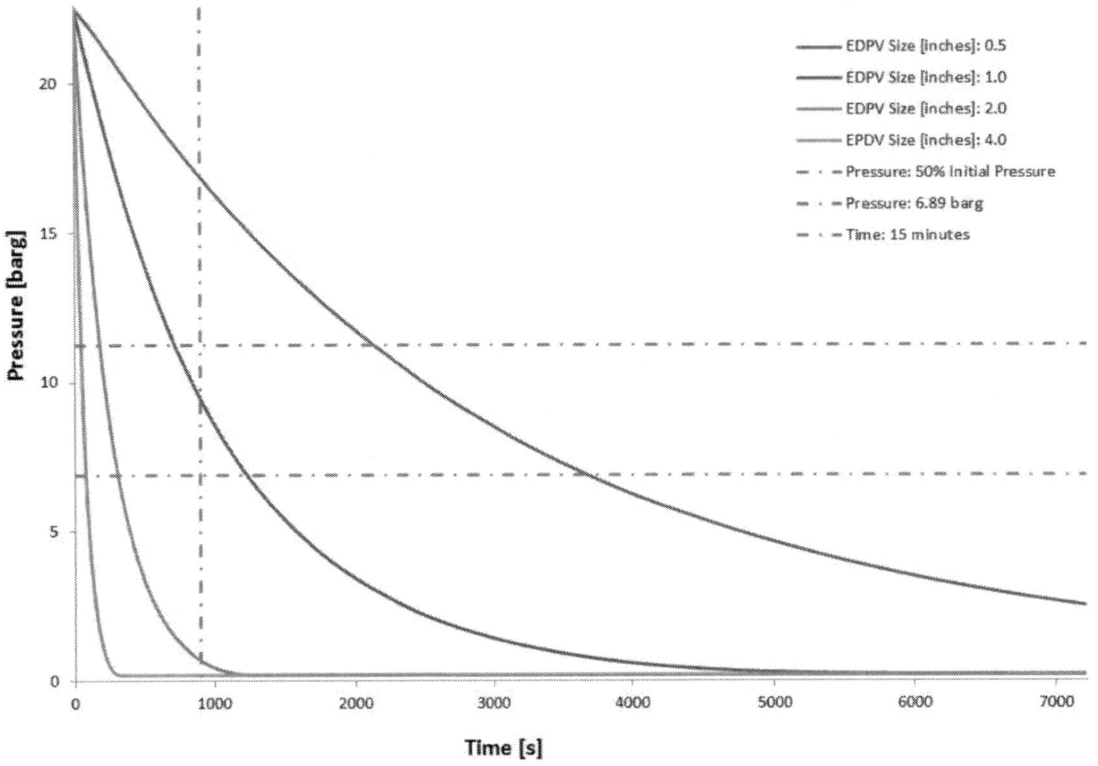

Figure 6: Pressure History (EDPV Size Sensitivity Analysis at 10% Liquid Level)

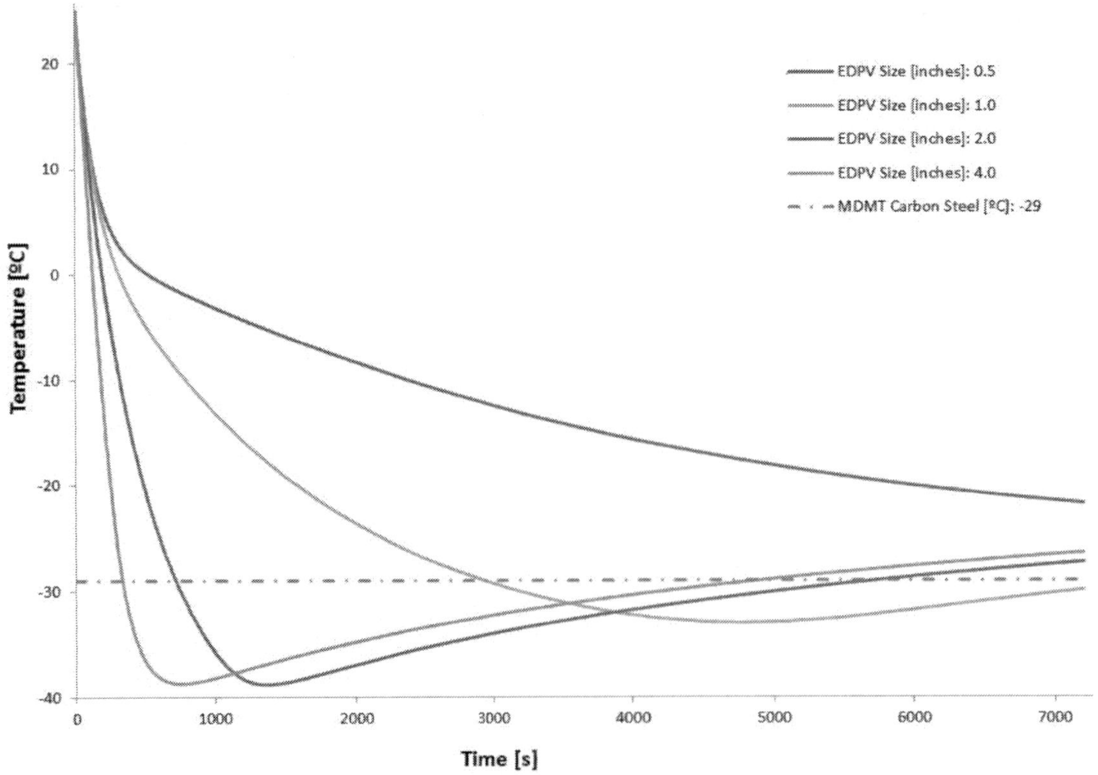

Figure 7: Wall Temperature History (EDPV Size Sensitivity Analysis at 10% Liquid Level)

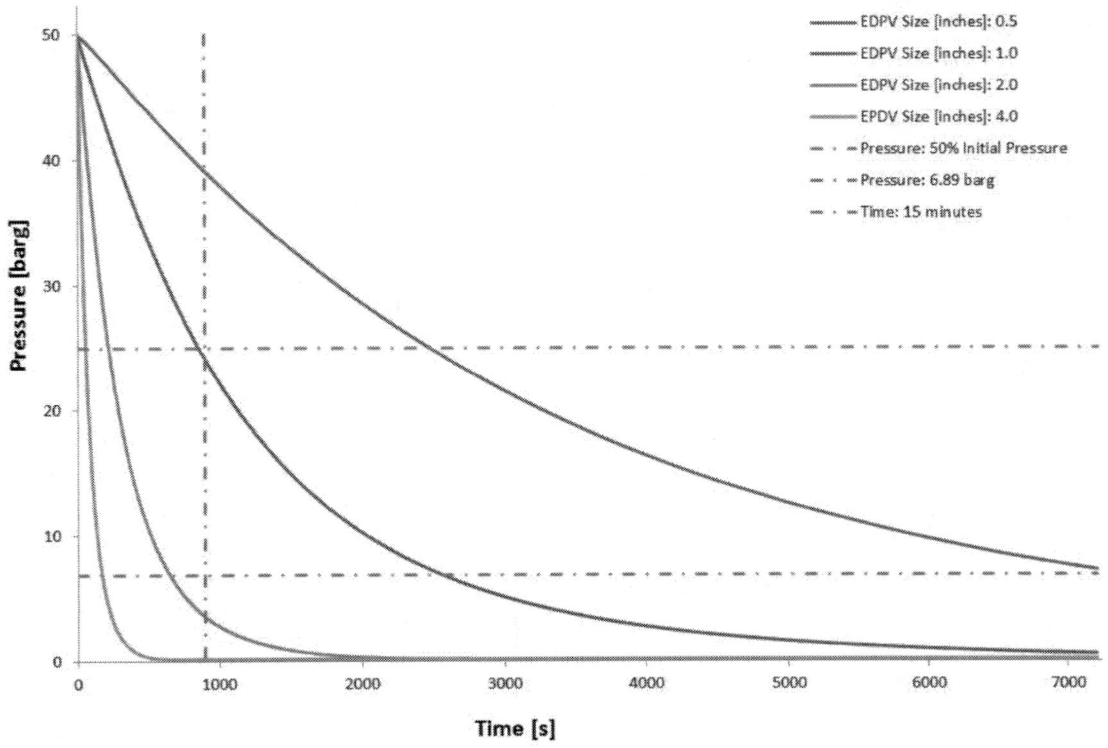

Figure 8: Pressure History (EDPV Size Sensitivity Analysis at 50% Liquid Level)

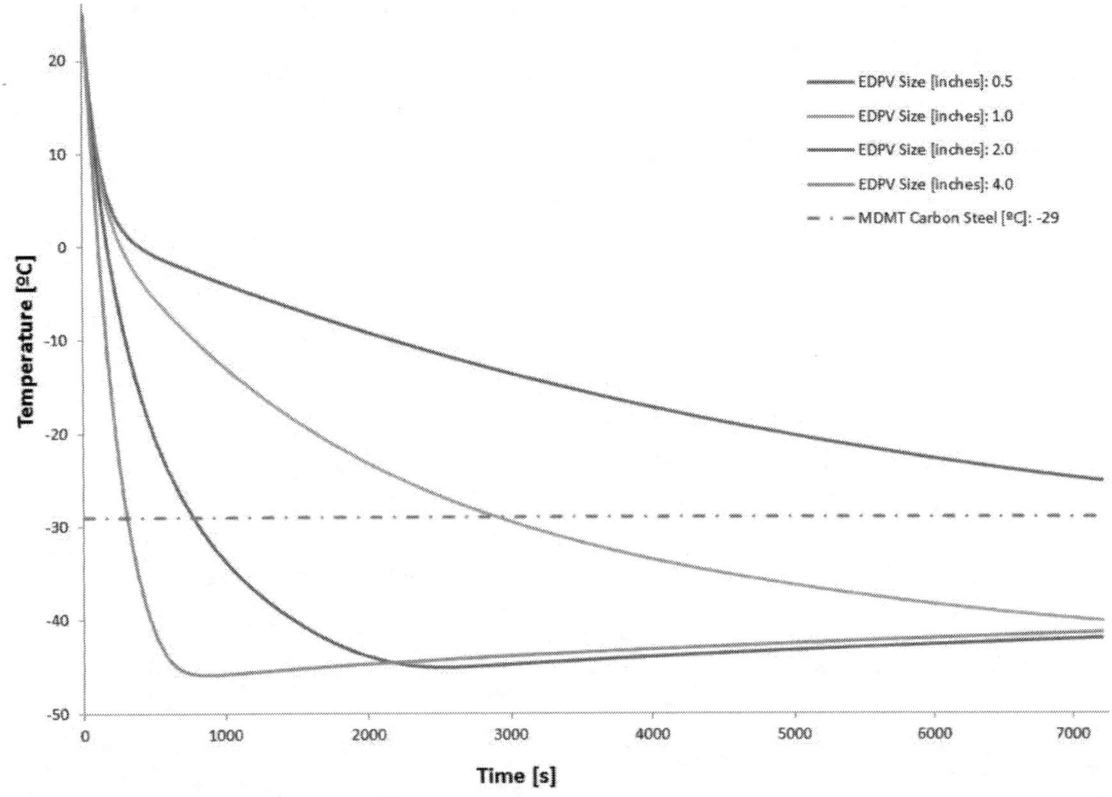

Figure 9: Wall Temperature History (EDPV Size Sensitivity Analysis at 50% Liquid Level)

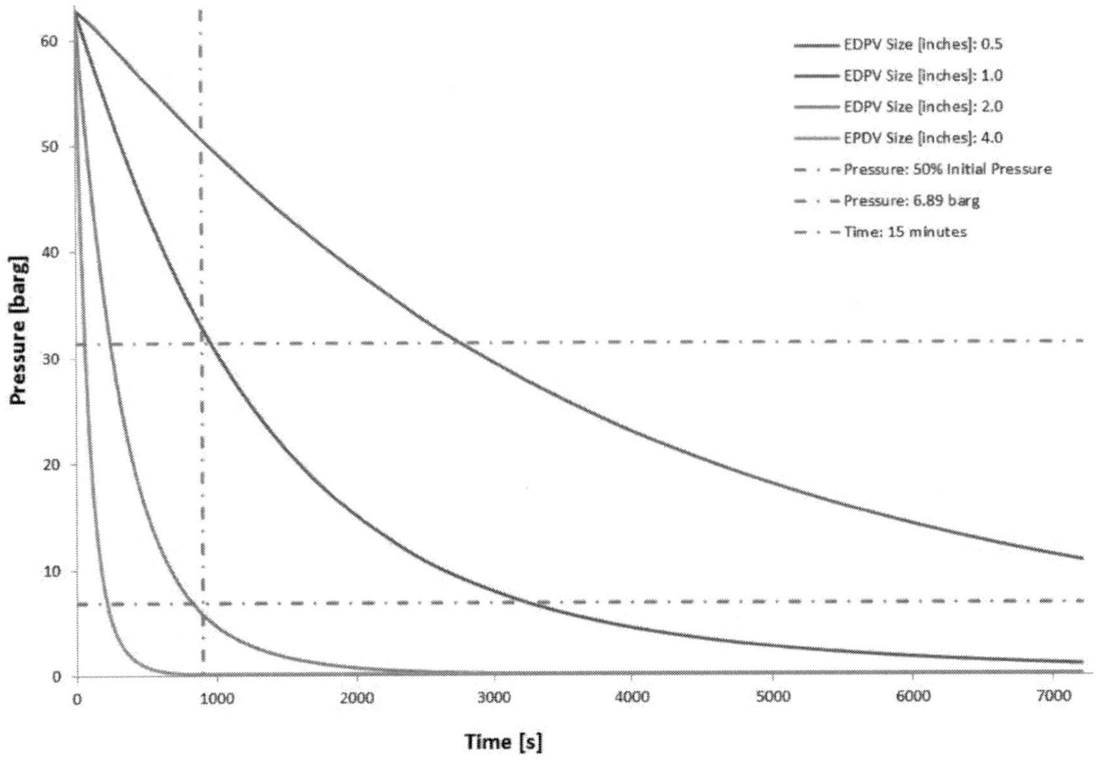

Figure 10: Pressure History (EDPV Size Sensitivity Analysis at 90% Liquid Level)

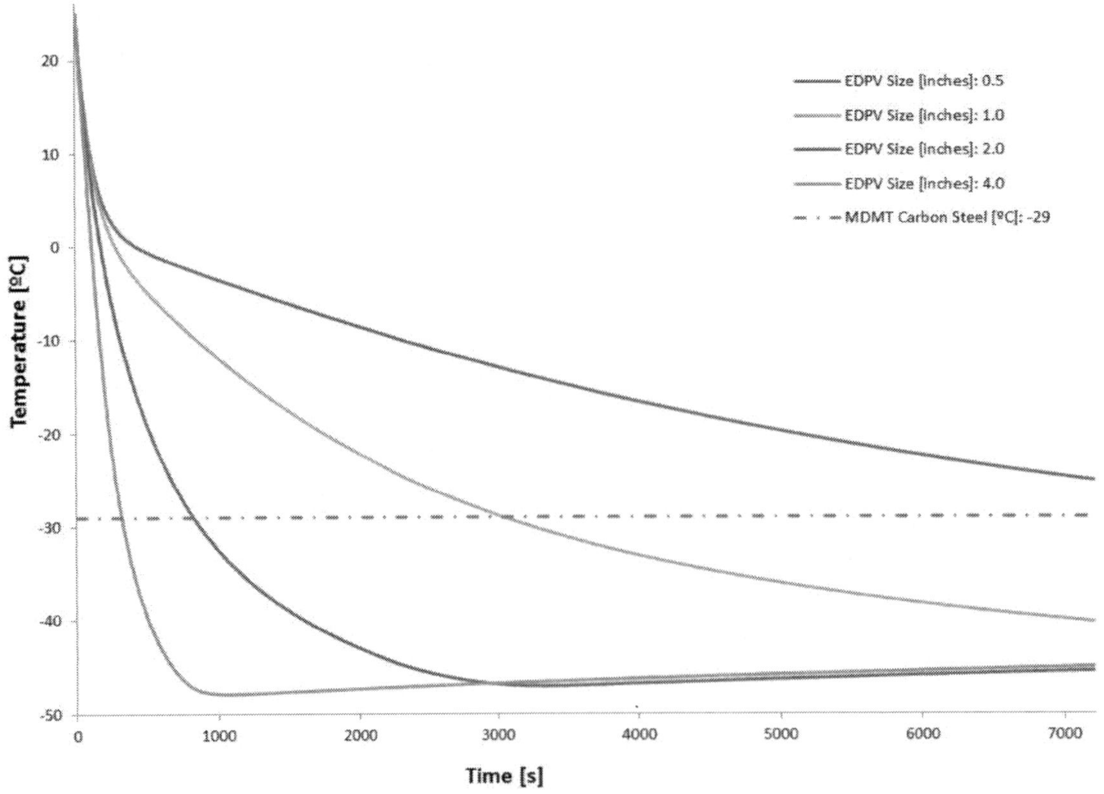

Figure 11: Wall Temperature History (EDPV Size Sensitivity Analysis at 90% Liquid Level)

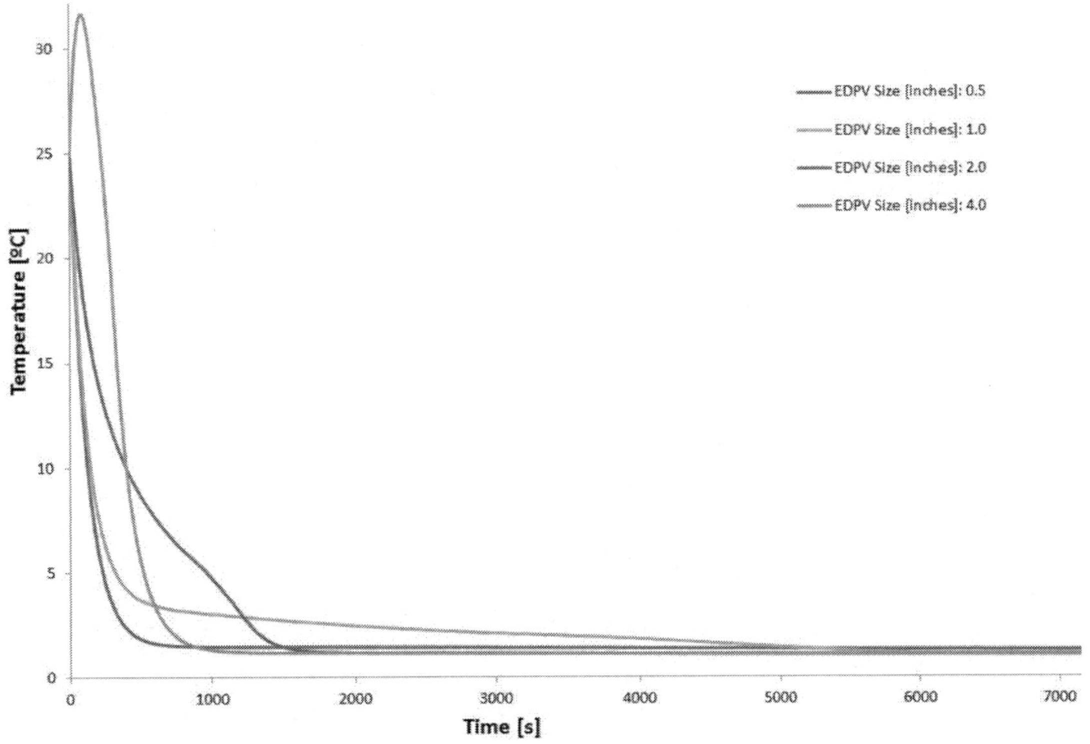

Figure 12: Fluid and Wall Differential Temperature History (10% Liquid Level)

Impact of Insulation Thickness

This sensitivity analysis examines the impact of four (4) different thermal insulation thicknesses (0”, 2”, 7”, and 14”), evaluated at three (3) different initial liquid levels (10%, 50%, and 90%).

As shown in Figures 13 through 15, the presence of insulation in cold depressurization inhibits effective heat transfer between the vessel wall and the surrounding. The impact of insulation is, however, not significant. Notice the metal temperature difference between the case of no insulation and one with 2”. In all cases, the difference is less than a fifth of a degree. The results also show that the insulation thickness impact is further reduced at higher initial liquid levels.

In the event of a non-fired depressurization, it can be concluded that no insulation is generally preferred; this is the opposite of an external fire scenario. In the event of fire, insulation has been found to be one of the few effective mitigation[2] measures, preventing heat transfer from the fire to the vessel wall. Since insulation has minimal impact on a cold depressurization and provides effective mitigation for an external fire, it should be seriously considered when sizing an EDPV for both scenarios.

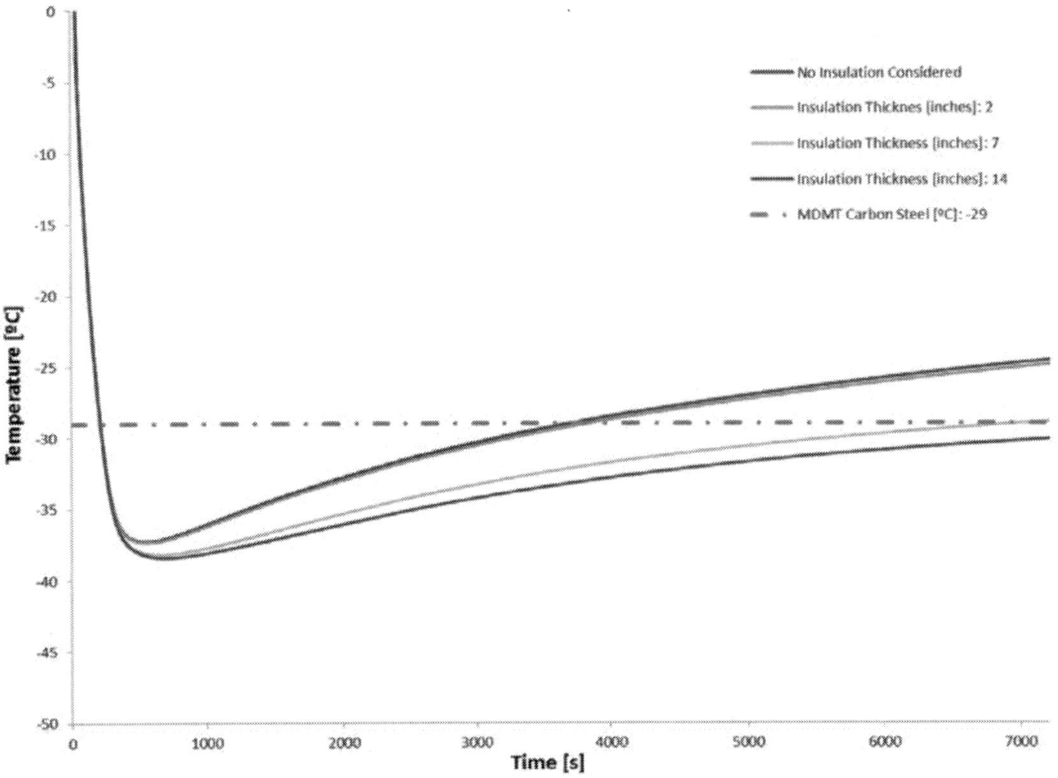

Figure 13. Wall Temperature History (Insulation Impact; 10% liquid level; EDPV size 4")

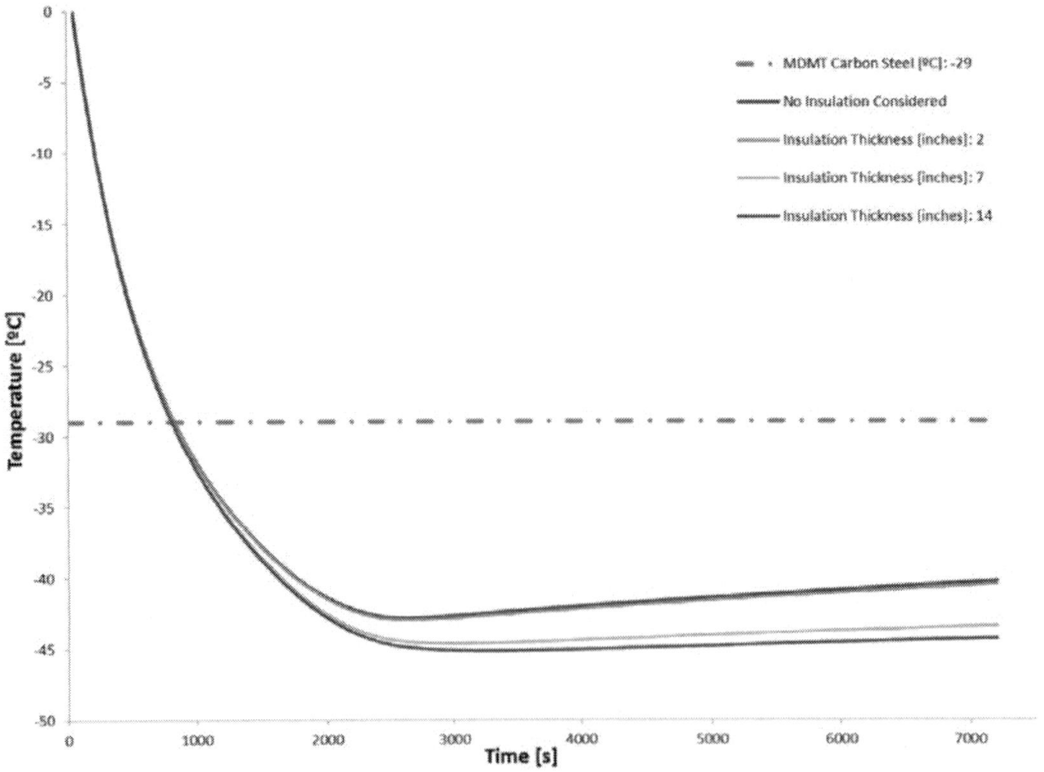

Figure 14. Wall Temperature History (Insulation Impact; 50% liquid level; EDPV size 2")

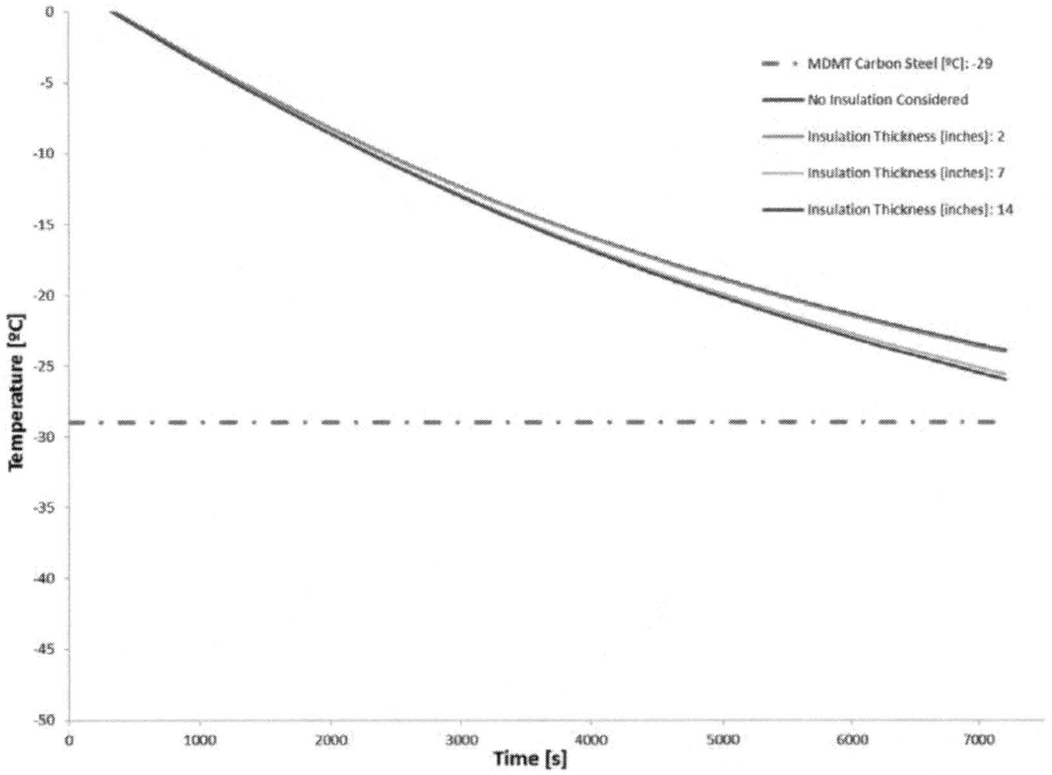

Figure 15. Wall Temperature History (Insulation Impact; 90% liquid level; EDPV size 0.5")

Impact of Fluid Compositions

This sensitivity analysis examines the impact of six (6) different methane fluid compositions (0.05, 0.10, 0.20, 0.30, 0.4, and 0.50), starting at 22.5 barg, a 10% initial liquid level, and with a 2" EDPV.

Figures 16 and17 show both the pressure and wall temperature histories at different methane compositions. Note that the temperature profiles are based on the bottommost segment of the vessel, assessing the metal wall in direct contact with liquid. The results indicate that while a richer methane content would help depressurize more rapidly, it also accelerates the cryogenic liquid heat transfer to the metal wall. At a methane mole fraction of 0.4 or higher, the metal wall temperature falls below -29°C, lower than the normal MDMT desired for carbon steel. This implies that a higher methane content may pose a greater risk of material embrittlement.

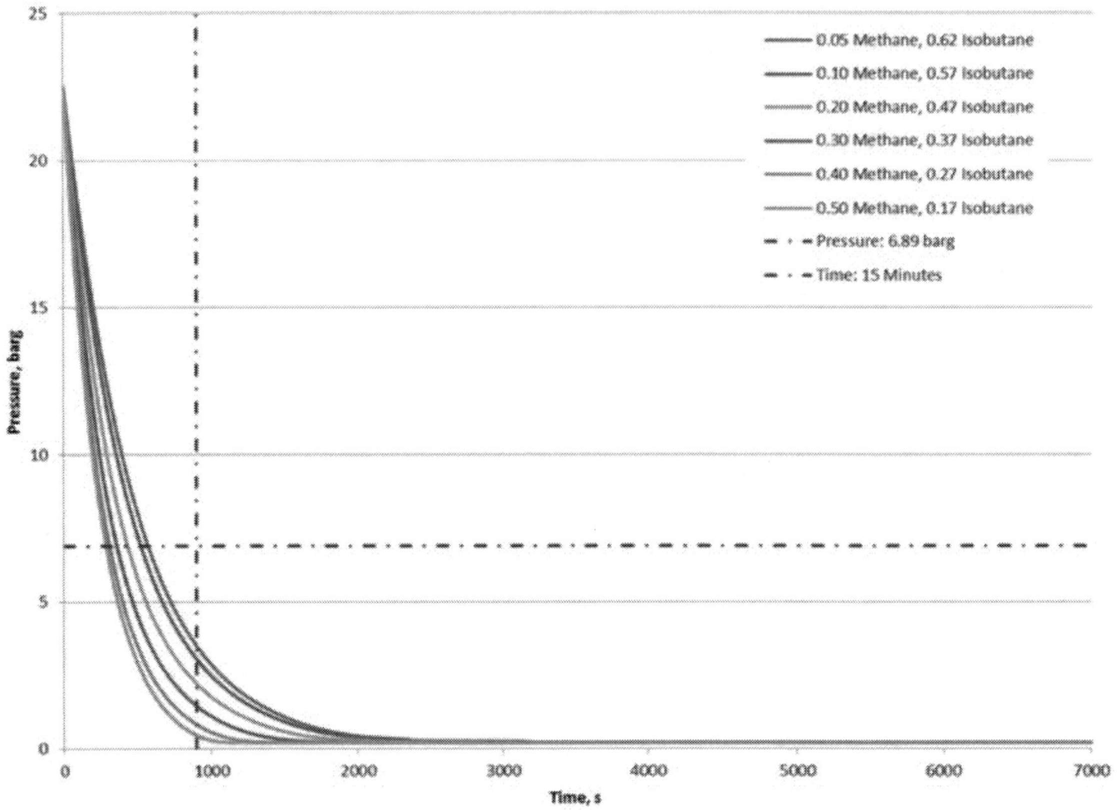

Figure 16. Pressure History (Methane Composition Impact; 10% liquid level; EDPV size 2")

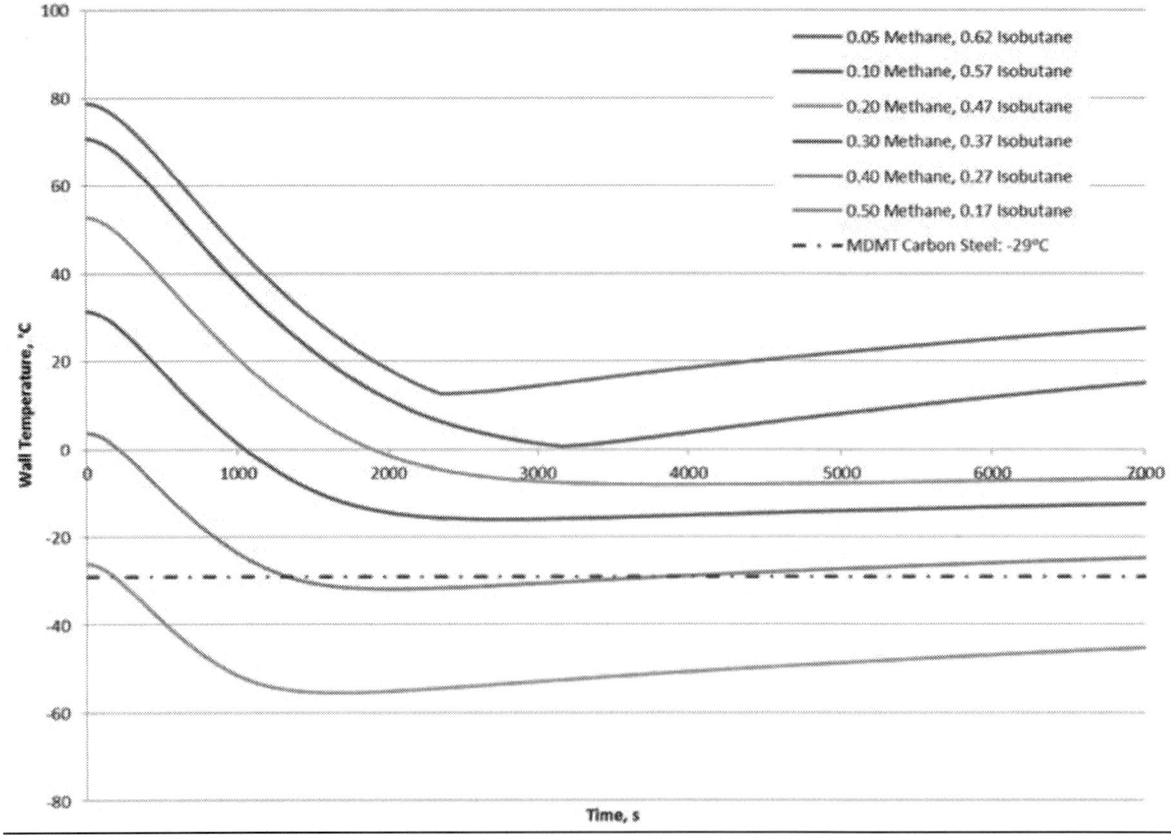

Figure 17. Wall Temperature History (Methane Composition Impact; 10% liquid level; EDPV size 2")

Conclusions

The vessel segmentation used in this paper demonstrates a reliable approach for modeling detailed vessel wall dynamics. Due to the direct contact between the metal wall and the cryogenic fluid, the probability of embrittlement failure is higher at the bottom portion of the vessel during a cold LNG depressurization. A lower initial filled level would typically lessen the risk of material embrittlement. A larger depressurizing rate, on the other hand, would exacerbate the cryogenic effect, causing a lower metal temperature. For that reason, while a sufficiently large EDPV size is required for depressurization within the specified pressure and time criteria, an oversized EDPV is generally not advised.

In the event of a non-fired depressurization, no thermal insulation is generally more preferred. This is the opposite of an external fire scenario, where insulation[2] has been found to be one of the few effective mitigations. Since insulation has minimal impact on a cold depressurization and provides an effective mitigation measure for an external fire, it should be seriously considered when sizing for both scenarios.

The sensitivity analysis on the methane fluid compositions indicates that while a richer methane content would help depressurize more rapidly, it also accelerates the cryogenic liquid heat transfer to the metal wall. This implies that a higher methane content may pose a greater risk of material embrittlement.

The thickness of the vessel can significantly influence wall temperatures. A thinner vessel thickness generally results in lower wall temperatures, exposing the equipment to a higher risk of brittle failure.

During the project phase, material of construction should be carefully selected, accounting for the vapor pressure of the fluid mixture whenever a cold depressurization is expected.

References

[1] Melhem, A. G., “Properly Calculate Vessel and Piping Wall Temperatures During Depressurizing and Relief,” published online 2014; DOI 10.1002/prs. 11689

[2] American Petroleum Institute, “Pressure Relieving and Depressurizing Systems”, ANSI/API Standard 521, Sixth Edition, 2013

[3] Salater, P., et al., “Size Depressurization and Relief Devices for Pressurized Segments Exposed to Fire,” Chemical Engineering Progress, September 2002

[4] Overa, S., et al., “Determination of Temperatures and Flare Rates During Depressurization and Fire,” Presented at 72nd Annual GPA Convention, San Antonio, Texas, March 15-17, 1993

11TH GLOBAL CONGRESS ON PROCESS SAFETY

Statistical Review of Runaway Reaction Kinetics

Enio Kumpinsky, Ph.D.
Ashland Inc.
P.O. Box 2219
Columbus, OH 43216
ekumpinsky@ashland.com

Prepared for Presentation at
American Institute of Chemical Engineers
2015 Spring Meeting
11th Global Congress on Process Safety
Austin, Texas
April 27-29, 2015

UNPUBLISHED

Statistical Review of Runaway Reaction Kinetics

Enio Kumpinsky, Ph.D.
Ashland Inc.
P.O. Box 2219
Columbus, OH 43216
ekumpinsky@ashland.com

Keywords: Collinearity, Correlation, Confidence, Prediction, Reparametrization

Abstract

The regressed self-heating rate of vinyl acetate polymerization runaway reaction was analyzed by means of statistical tools. The data come from a VSP2 experiment that was performed for a DIERS round-robin test. It is shown that all predictors are collinear, i.e., there is a linear dependence between the predictors, meaning that the model is inflated. Consequently, the parameters may be distorted, many of which do not make physical sense. It is shown how to reduce collinearity to a manageable level and how to perform reparametrization to lessen interdependency between the parameters. The outcome is the shortest and less correlated form of the self-heating rate equation, which can be used for process-safety evaluations.

1. Introduction

Loss of temperature control is a typical scenario to be considered when a process involves a potentially hazardous exothermic reaction. The study of runaway reactions in adiabatic calorimeters is a standard protocol to quantify the hazards (Townsend and Tou, 1980). Once the experimental data are collected, it is necessary to study them carefully to ensure that key process vulnerabilities are thoroughly explored.

An essential feature of a good analysis of experimental data from adiabatic calorimetry is the development of a kinetic model for the self-heating rate as a function of temperature. For dynamic methods, such as simulations with SuperChems (Melhem, 2014), a detailed kinetic model is needed in order to account for the formation and consumption of chemical species with time. Analytical techniques, such as the Omega Method (Leung, 1986, 1995), are based on one or two data points for vent sizing. However, corrections for thermal inertia may be necessary, and they involve the use of the activation energy, which comes from a kinetic model (Fisher et al., 1992). Other applications, such as the determination of self-accelerating decomposition temperature (SADT) for thermally unstable materials, are also based on kinetic modeling.

The most generic form of the self-heating rate equation, as it appears in SuperChems, takes the following form:

$$\frac{dT}{dt} = AT^d \exp\left(-\frac{E}{RT} + \frac{B}{T^{\frac{1}{3}}} + \frac{C}{T^{\frac{2}{3}}}\right)N \qquad [1]$$

Note that SuperChems (see Melhem, 2014) does not endorse the simultaneous use of all the available kinetic parameters, but for demonstration purposes, all of them will be employed in this study. Typically, predictors are the independent variables, i.e., T, $1/T$, $(1/T)^{1/3}$ and $(1/T)^{2/3}$. Parameters are estimated by regression analysis, i.e., A, d, E/R, B and C. When working with SuperChems or other dynamic simulators, N in Eq. [1] is typically represented by the product of reactant concentrations raised to appropriate powers, whose sum defines the order of reaction. When doing regression in Microsoft Excel for other applications, such as the Omega Method, N can be represented by

$$N = \left(T_f - T\right)^n \left(T - T_o\right)^m \qquad [2]$$

The symbols mean the following:

A	pre-exponential factor
B	temperature exponent
C	temperature exponent
d	temperature exponent
dT/dt	self-heating rate, the model response
E/R	activation energy divided by the universal gas constant
m	rate of reaction
n	rate of reaction
N	rate multiplier that depends on concentration.
t	time
T	instant temperature
T_f	final temperature
T_o	onset temperature

In Eq. [2], N represents concentration in terms of thermal conversion. The power m is zero when the experiment follows a standard Arrhenius shape. For a quick-start reaction, as in free-radical initiation, m is larger than zero, while in SuperChems it is accounted for by forward multipliers (Melhem, 2004).

As seen in Eqs. [1] and [2], all terms depend on temperature. Accordingly, there might be associations between the predictors. This interdependency between the predictors is called collinearity. If two terms are strongly collinear, it means that it suffices to have only one of them in the regressed equation. Otherwise, the coefficients of both terms may become corrupted, losing physical meaning. Such a model is said to be inflated, because it has redundant terms. To address this issue, the worst cases of collinearity must be identified and removed. However, collinearity cannot be totally eliminated from this type of modeling.

It is important to evaluate the residuals once the worst cases of collinearity are removed. A residual is the difference between an experimental point and its respective regressed value. It is said that the workmanship of a piece of furniture can be assessed by the uniformity of the wood dust. Likewise, the quality of a regression model can be assessed by the residuals. Normally and randomly distributed residuals indicate good regression.

Given the structure of the self-heating rate equation, there may be interdependency between the coefficients of the model, such as activation energy, pre-exponential factor and others. Strong correlations cause model degeneration. A technique called reparametrization can be used to mitigate correlations and therefore increase the quality of the model.

Regardless of collinearity between predictors, the quality of the regression model is dependent on the worth of the experimental data. This can be quantified in terms of regression confidence and predictive intervals. By performing multiple experiments in the future, the mean response of observations at a given temperature is expected to fall within the confidence interval obtained with the original experiment. Similarly, a single future experiment should yield self-heating rates falling within the predictive interval of the original experiment. With high-quality experimental data, the confidence and predictive intervals are expected to be in narrow bands, centered on the regressed line. An increase in dispersion of the experimental data translates into larger confidence and predictive intervals.

The following topics will be covered:

- Effect of collinearity (linear dependence of regression predictors)
- Variance inflation factors
- Development of confidence and predictive intervals for the kinetic equation
- Residual analysis
- Reduction of dependency between some coefficients through reparametrization

The data used in this development come from an adiabatic experiment in the VSP2, Vent Sizing Package 2, which was performed at Ashland Inc. for a DIERS round-robin test (DIERS = Design Institute for Emergency Relief Systems). The test was done initially under vacuum with 70 g of material in a VSP2 test cell of the type VT-2SS4, which has a volume of about 110 ml. The original formula, by mass, was 60% vinyl acetate monomer (VAM), 39% toluene and 1% dilauroyl peroxide. Vinyl acetate was inhibited with 15 PPM of monomethyl ether hydroquinone (MEHQ).

The experimental output is shown in Figure 1. It is a reverse Arrhenius plot from the onset of self-heating to the end of this exotherm. A second exotherm due to thermal polymerization is realized at temperatures above 200°C, but it is not part of this study. As a typical free-radical polymerization, the reaction begins quickly once the inhibitor is depleted. The quick onset is represented by the squared data points in Figure 1, developed with SigmaPlot 13. For simplicity, without any negative effect on the conclusions of this work and use in process-safety studies, those initial points were disregarded from the analysis, so that m could be set to zero in Eq. [2].

Figure 1. VAM Solution DIERS Round Robin in the VSP2, self-heating rate vs. temperature

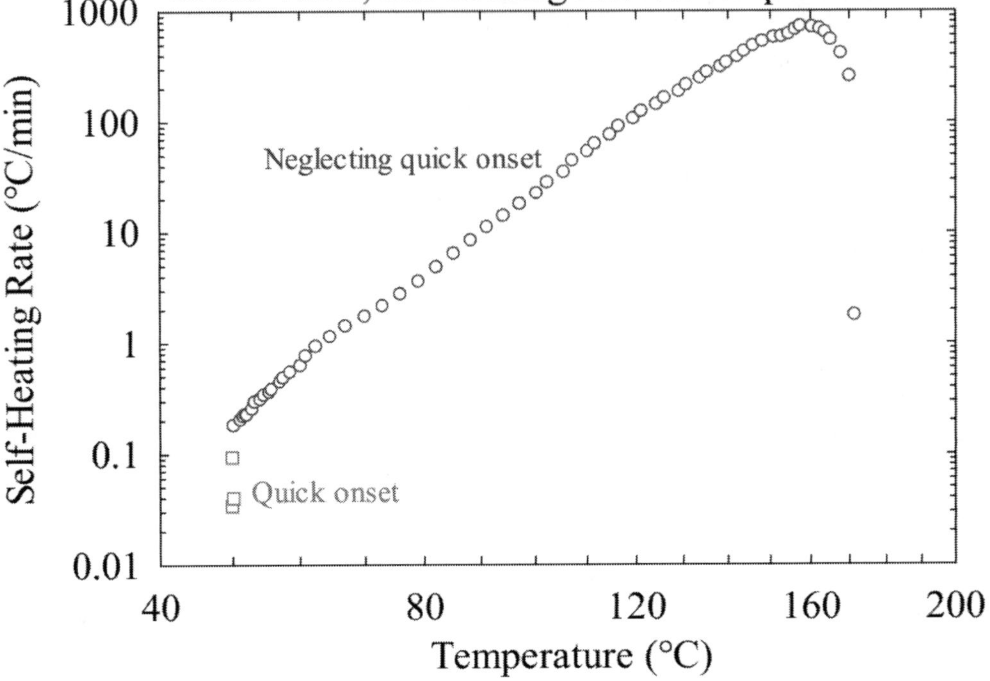

2. Discussions

2.1 Collinearity

With the assumptions described in the Background section, the combined Eqs. [1] and [2] can be set in logarithmic form for $m = 0$:

$$\ln\left(\frac{dT}{dt}\right) = \ln A + d\ln T + n\ln\left(T_f - T\right) - \frac{E}{RT} + \frac{B}{T^{\frac{1}{3}}} + \frac{C}{T^{\frac{2}{3}}} \qquad [3]$$

To clarify, the multiplier of $\ln T$, d, represents an exponent, not a differential. The experimental data, except for the quick-onset points, were entered in Minitab 17 statistical software, to be regressed by the least-squares method (Montgomery et al., 2012). The first six points out of 63 are shown in Figure 2.

Figure 2. A fraction of the data used in regression analysis

DIERS Round Robin ***

↓	C1	C2	C3	C4	C5	C6	C7	C8	C9
	T (°C)	dT/dt (°C/min)	T (K)	ℓn(T)	ℓn(Tf-T)	1/T	1/T^(1/3)	1/T^(2/3)	ℓn(dT/dt)
1	50.30	0.142	323.46	5.77908	4.79827	0.0030916	0.145678	0.0212220	-1.94982
2	50.27	0.181	323.43	5.77898	4.79852	0.0030919	0.145682	0.0212233	-1.70981
3	51.26	0.204	324.42	5.78204	4.79032	0.0030824	0.145534	0.0211801	-1.59160
4	51.63	0.219	324.79	5.78318	4.78724	0.0030789	0.145479	0.0211640	-1.52005
5	51.99	0.228	325.15	5.78429	4.78424	0.0030755	0.145425	0.0211484	-1.47753
6	52.13	0.226	325.29	5.78472	4.78307	0.0030742	0.145404	0.0211423	-1.48766

The first step in a statistical analysis is to define the hypotheses, which come in a pair. There are several ways to define the hypotheses for multiple regression, but they must hold the same meaning. One way is the following. Null Hypothesis: H_o: No predictor has any effect on dT/dt. Alternative Hypothesis: H_a: At least one predictor has an effect on dT/dt. Another way of stating them is the following. Null Hypothesis: $\ln A = d = n = E/R = B = C = 0$. Alternative Hypothesis: At least one among $\ln A$, d, n, E/R, B and C is different from zero.

The significance level adopted for this regression was the typical $\alpha = 0.05$ (confidence level of 95%). This value is assigned. The regression analysis shows P-values, which characterize the probability that the null hypothesis is correct. These values are calculated. Predictors with a P-value smaller than α are statistically significant.

Equation [4] represents the results of multiple regression analysis in Minitab 17 by the least-squares method. All the regressed coefficients had a P-value of 0. The coefficient of determination, R^2, was 99.97%, a very good number.

$$\ln\left(\frac{dT}{dt}\right) = -14700.8 + 53.105\ \ln(T) + \frac{6047170}{T} - \frac{2403600}{T^{\frac{2}{3}}} + \frac{321272}{T^{\frac{1}{3}}} + 0.578468\ \ln(T_f - T) \qquad [4]$$

T_f was 171.6°C in all cases, or the equivalent 444.76 K. Eq. [4] can be set back in the original form, as seen in Eq. [5]:

$$\frac{dT}{dt} = T^{53.105} \exp\left(-14701 + \frac{6047170}{T} + \frac{321272}{T^{\frac{1}{3}}} - \frac{2403600}{T^{\frac{2}{3}}}\right)(T_f - T)^{0.578} \qquad [5]$$

Upon examining Eq. [5], it can be verified that the pre-exponential factor, $A = \exp(-14701)$, is extremely small. The exponent of T, $d = 53.105$, is too high. The reaction order $n = 0.578$, is a mathematically reasonable value. However, the activation energy/universal gas constant, $E/R = -6047170$ K, is excessive and with the wrong sign. The coefficients of $(1/T)^{1/3}$, $B = 321272$ and of $(1/T)^{2/3}$, C = -2403597, are exceedingly high. Therefore, just considering $R^2 = 99.97\%$ and all P-values equal to zero is not sufficient to assess the quality of a regression model. Collinearity has to be considered.

An indication of a high degree of collinearity came from the warning issued by Minitab: * ERROR * X-matrix is nearly singular. This means that the determinant of the matrix is almost zero. Should the determinant be zero, the least-squares matrix would be singular, thus indicating that at least one of the predictors would be an exact linear function of one or more of the other predictors.

An alternative method to least squares is partial least squares (see Geladi and Kowalski, 1986). The partial least squares method is valuable when the predictors are highly collinear as seen in Eq. [5]. The application of this method in Minitab 17 leads to the regression shown in Eq. [6], with $R^2 = 99.95\%$.

$$\frac{dT}{dt} = T^{-145705} \exp\left(1664799 - \frac{54500222}{T} - \frac{9450694}{T^{\frac{1}{3}}} + \frac{34044147}{T^{\frac{2}{3}}}\right)\left(T_f - T\right)^{0.573} \quad [6]$$

As seen in Eqs. [5] and [6], regressions by least squares and partial least squares lead to the same qualitative results, i.e., misaligned parameters with a high coefficient of determination. A high value of the latter does not mean much without assessing collinearity. Even for partial least squares the collinearity is excessive, a clear indication that equation [3] should not be used with all of its parameters.

The subject of collinearity can be explored in more detail in two additional ways, going back to the least-squares method. First, by checking correlations among predictors. Second, by examining a matrix plot of the predictors.

Table 1 shows two numbers for each cell. The top number is the Pearson coefficient (Montgomery et al., 2012). The lower number is the P-value, or the probability that the null hypothesis should be favored. Null Hypothesis H_o: The pair of predictors is not correlated. Alternative Hypothesis H_a: The pair of predictors is correlated. The Pearson coefficient is -1 for a perfectly negative correlation and +1 for a perfectly positive correlation. As can be seen in Table 1, all P-values are zero, meaning that all predictors are correlated. The strength of the correlation is measured by the Pearson coefficient. Those equal to 1 or -1 are very strong. When such strong correlations exist, generally it means that only one of the two predictors should be part of the regression.

Table 1. Pearson coefficients and P-values for each pair of predictors

Predictors	$\ln(T)$	$\ln(T_f-T)$	$1/T$	$(1/T)^{1/3}$
$\ln(T_f-T)$	-0.630 0.000			
$1/T$	-0.999 0.000	0.615 0.000		
$(1/T)^{1/3}$	-1.000 0.000	0.625 0.000	1.000 0.000	
$(1/T)^{2/3}$	-1.000 0.000	0.620 0.000	1.000 0.000	1.000 0.000

Figure 3 is a matrix plot of all predictors of Eq. [3]. For instance, the graph in row 3 of column 1 represents the relationship between $\ln(T)$ and $1/T$. As can be seen, many predictors can be linearly expressed in terms of the other predictors, thus confirming the results in Table 1.

The conclusion, then, is that all predictors are correlated, but for some pairs collinearity is higher than for other pairs. One component of a highly collinear pair of predictors must be eliminated. The first attempt is to eliminate two among the highly correlated exponent predictors $1/T$, $(1/T)^{1/3}$ and $(1/T)^{2/3}$ from the model. The most common practice for this type of chemical reaction is to work with $1/T$ to obtain the activation energy, so B and C are removed from the model. The regressed formula without predictors $(1/T)^{1/3}$ and $(1/T)^{2/3}$ is shown in Eq. [7]. It has a coefficient of determination R^2 of 99.91%, barely a change from 99.97%, which was obtained from the full model, Eq. [5].

$$\frac{dT}{dt} = T^{1.423} \exp(21.192 - 10140/T)(T_f - T)^{0.616} \qquad [7]$$

The numerical values in Eq. [7] are much more reasonable than those in Eq. [5]. However, this simplification may not be sufficient, as it does not address the high correlation between predictors $\ln(T)$ and $1/T$ (see Table 1 and Figure 3). It does not matter if a logarithmic transform has been applied to this pair. T is still highly correlated with $\exp(1/T)$ or $\exp(-1/T)$. Hence, the next step is to eliminate $\ln(T)$ from the regression. The resulting expression is shown in Eq. [8]. The coefficient of determination R^2 is 99.89%. Again, hardly a decrease from the previous two cases.

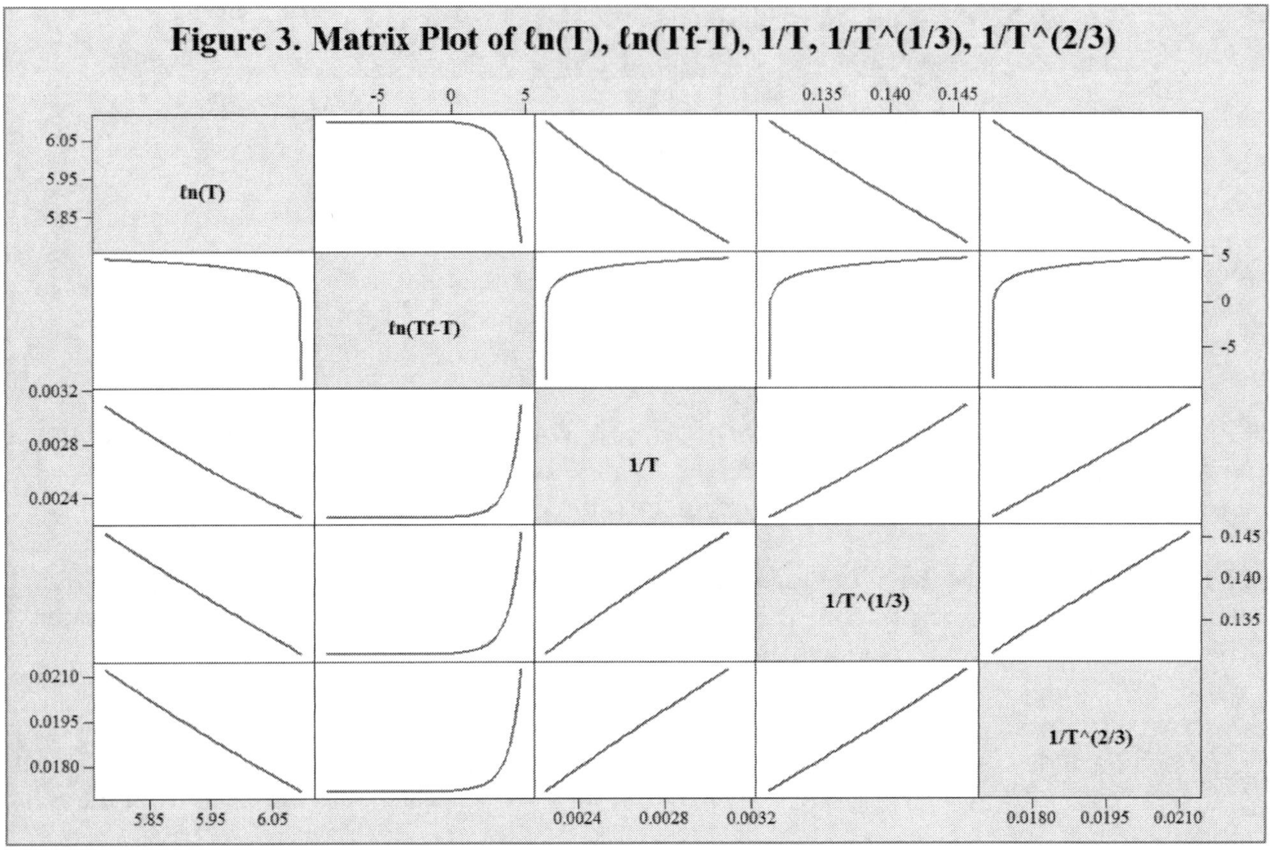

$$\frac{dT}{dt} = \exp\left(33.302 - 12252/T\right)\left(T_f - T\right)^{0.627} \qquad [8]$$

Hence, using more predictors in a regression model is not necessarily better. In fact, it is worse due to an increase in collinearity. Additionally, the elimination of three collinear predictors had a negligible effect on the coefficient of determination R^2:

$R^2 = 99.97\%$ with T^d, $(1/T)^{1/3}$ and $(1/T)^{2/3}$ in the model, Eq. [5]
$R^2 = 99.91\%$ with T^d in the model, Eq. [7]
$R^2 = 99.89\%$ with the standard kinetic equation, Eq. [8]

Several calculations depend on the activation energy, such as corrections for thermal inertia and calculations of self-accelerating decomposition temperature (SADT). As shown, the activation energy may be distorted in inflated models. This can be seen by comparing E/R in Eqs. [5], [7] and [8], i.e., -6047170 K, 10140 K and 12252 K.

Note also in Eqs. [5], [7] and [8] that the order of reaction did not change significantly with the elimination of collinear predictors: 0.578 in Eq. [5], 0.616 in Eq. [7] and 0.627 in Eq. [8]. This can be explained by looking at the second row and second column in Figure 3. All graphs involving $\ln(T_f\text{-}T)$ have a strong curvature. Collinearity still exists between $\ln(T_f\text{-}T)$ and the other predictors, but it is relatively mild compared to the other cases. This matter can be further elucidated by variance inflation factor.

If partial least squares are applied to an equation in the same form of Eq. [8], the following is obtained:

$$\frac{dT}{dt} = \exp\left(33.385 - 12288 / T\right)\left(T_f - T\right)^{0.629} \qquad [9]$$

Eq. [9], obtained by partial least squares regression using Minitab 17, with $R^2 = 99.87\%$, is very similar to Eq. [9], with $R^2 = 99.89\%$.

2.2 Variance inflation factor (VIF)

When performing multiple regression analysis for self-heating rate, or for multiple regression in general, it is best to have no good way to express one predictor as a function of the other predictors. Ideally, there should be only randomness. The difficulty with the modeling of thermal kinetics is that all predictors depend on temperature. Therefore, some level of interdependence is expected. The best situation, then, is to have the weakest possible relationships among the predictors.

It is desirable to quantify the extent of multicollinearity among the predictors. Multicollinearity exists when two or more predictors are highly correlated in multiple regression. Variance inflation factor (*VIF*) is a measure of multicollinearity strength. *VIF* for the kth predictor is defined as follows:

$$VIF_k = \frac{1}{1 - R_k^2} \qquad [10]$$

In Eq. [10], R_k^2 is the coefficient of determination obtained when regressing the kth predictor as a function of the remaining predictors:

$$R_k^2 = 1 - \frac{\sum_{i=1}^{N}\left(y_{k,i} - \hat{y}_{k,i}\right)^2}{\sum_{i=1}^{N}\left(y_{k,i} - \overline{y}_k\right)^2} = 1 - \frac{\text{Sum of residuals squared}}{\text{Sum of squared deviations from the mean}} \qquad [11]$$

In Eq. [11], $y_{k,i}$ is the actual value of predictor k in the ith row of data, $\hat{y}_{k,i}$ is its corresponding regression value as a function of the other predictors and $\overline{y}_k$ is the mean value of predictor k, i.e., the sum of all $y_{k,i}$ values divided by N, the number of rows of data. The levels of correlation acceptance are summarized in Table 2.

Table 2. Variance inflation factors and levels of correlation acceptance

VIF Value	Correlation level
VIF = 1	No correlation, acceptable
1<VIF<5	Moderately correlated, acceptable
5<VIF<10	Highly correlated. Multicollinearity should be investigated.
VIF > 10	Unacceptably correlated, poorly estimated regression coefficients

Variance inflation factors are displayed in Table 3 for the predictors of Eqs. [7] and [8].

Table 3. Variance inflation factors for the predictors of regressed dT/dt

Equation	Predictor Regression	VIF
	$1/T = f[\ln(T_f - T), \ell n(T)]$	277.17
[7]	$\ln(T_f - T) = f[1/T, \ell n(T)]$	2.32
	$\ln(T) = f[\ln(T_f - T), 1/T]$	261.00
[8]	$1/T = f[\ln(T_f - T)]$	1.60
	$\ln(T_f - T) = f[1/T]$	1.60

Table 3 shows that estimates of two coefficients of Eq. [7], d and E/R, are poorly evaluated. Indeed, two of the predictor regressions have variance inflation factors well above the threshold of 10. As mentioned earlier, the reaction order, n, is not very different in Eqs. [7] and [8]. In fact, these are not even too far removed from n in Eq. [5] with all terms included. The visual observations on the curvatures of the graphs involving $\ln(T_f - T)$ in Figure 3 can now be numerically corroborated by low variance inflation factors in Table 3.

Given these statistical facts, a very good justification would be needed in order to use a model with more terms than those in Eq. [8] for this particular chemical reaction. What follows, then, is a study of Eq. [8] using different techniques.

2.3 Confidence and predictive intervals

If multiple experiments are performed in the future, 95% of the mean responses at a given temperature are expected to fall within the confidence interval, if the same experimental conditions are applicable. On the other hand, 95% of single future experiments are projected to yield self-heating rates falling within the predictive interval. The predictive interval is wider than the confidence interval, because of the higher uncertainty of a single value compared to the mean.

Figure 4, generated with SigmaPlot 13, contains six sets of data: the experimental data shown with circles, the regressed Eq. [8] as a solid line, the upper and lower confidence limits and the upper and lower predictive limits as dashed lines.

A comparison between experimental data and regression line indicates that clusters of points are sequentially distributed slightly above and slightly below the regression line. This means that an argument could be made for a minor case of lack-of-fit. This can be seen more clearly under residual analysis. A systematic random distribution of experimental points above and below the regression line would be a sign of no lack-of-fit.

In general, the evidence points to an experiment with high-quality data, as the confidence interval virtually overlaps with the regressed line. The predictive interval is also in a relatively narrow band. If the experimental data exhibited more variation, both confidence and predictive intervals would be wider, thus reducing confidence in the model.

Figure 4. VAM Solution DIERS Round Robin in the VSP2, self-heating rate vs. temperature. Confidence and predictive intervals

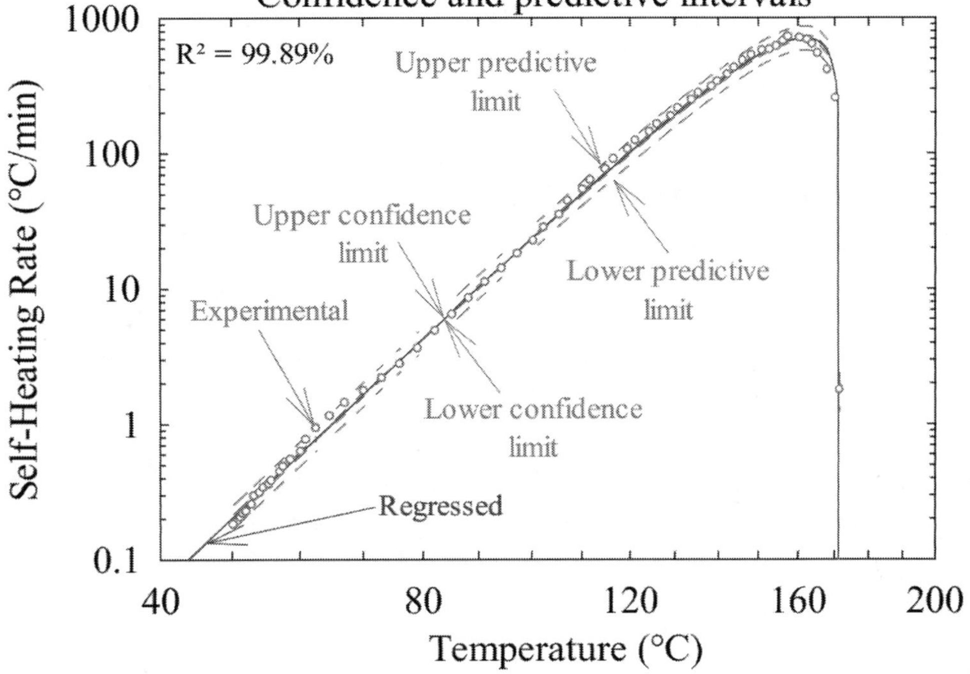

2.4 Residual analysis

A residual is the difference between an experimental point and its respective regressed value. Mathematically, it is defined as $y_i - \hat{y}_i$, where y_i is a given experimental point of ln(d*T*/d*t*) and $\hat{y}_i$ is its respective regressed value. Residual analysis is a measure of the quality of the model. Large residuals indicate great differences between regressed and experimental values. The larger the residuals are, the lower the coefficient of determination R^2. The size of the residuals, however, is hardly the only indicator of model quality. How they are spread is of paramount importance.

Figures 5 through 8 indicate how well the residuals are distributed. These kinds of graphs usually appear in statistical software when performing regression. They are presented here as Microsoft Excel 2010 graphs.

Figure 5 is the normal probability plot of the residual. It is highly desirable to have normally distributed residuals. Had all points been over the red line, the residuals would have been in a perfectly normal distribution. The null hypothesis for normality test states that the data points are normally distributed. Three normality tests were applied in Figure 5: Anderson-Darling (D'Agostino, 1986), Ryan-Joiner (Ryan and Joiner, 1976) and Kolmogorov-Smirnov (Lilliefors, 1969). These are the same options available in Minitab 17 statistical software and they have been adapted to Microsoft Excel.

For a significance level of 0.05, the residuals are normally distributed with a confidence level of 95% when the calculated P-value is larger than 0.05. As can be seen in Figure 5, the Anderson-Darling and Kolmogorov-Smirnov tests support the normality of the residuals, while Ryan-Joiner does not. This is common in statistics in borderline situations, depending on the test parameters. The most common normality test is Anderson-Darling, so from this perspective the residuals can be considered normally distributed (P-value = 0.0639 > 0.05).

When performing linear regression, the sum of residuals, and therefore their mean, should be zero. This is confirmed by the Mean box in Figure 5.

Figure 5. Normal Probability Plot of Residuals

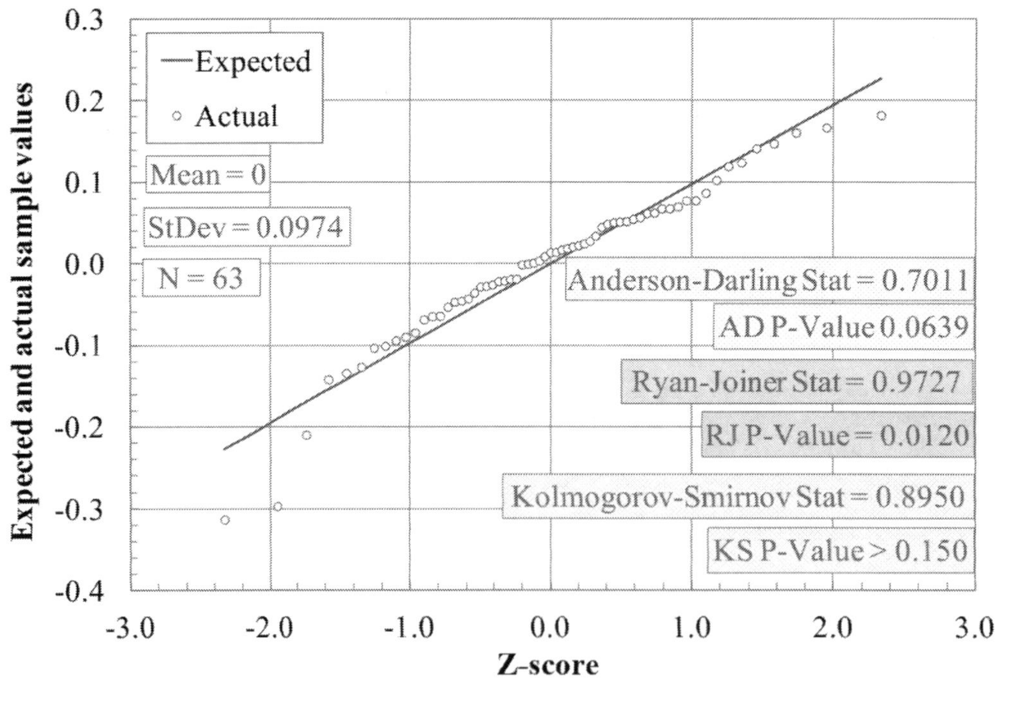

Figure 6 plots lagging residuals versus residuals. In a column of residuals, if a residual is in row i, the lagging residual will be in row i-1 of the same column. In the most desirable situation, the lagging residuals would be randomly distributed, with no visible trend. As Figure 6 indicates, there is a trend, meaning that the residuals might be autocorrelated. A statistical test called Durbin-Watson verifies autocorrelation of residuals. A Durbin-Watson statistic is calculated first. Then, this statistic is compared to tabulated values. In this particular case, the calculated value of 0.608 leads to the conclusion that residual autocorrelation is statistically significant and positive, as graphically seen in Figure 6.

Total randomness is also sought when looking at the residuals versus fits and/or observation order. As Figures 7 and 8 illustrate, it is not possible to claim that the residuals are in a random order. In Figures 7 and 8, the residuals meander around and through the zero line, even though the differences between experimental and regressed values are small, as seen in Figure 4.

This lack of randomness is typical in thermal kinetics regression and usually it cannot be removed, in part because temperature values are in sequence. It is necessary to realize that data as seen in Figures 7 and 8 lead to the statistical test for lack-of-fit to issue P-values below 0.05, thus confirming the absence of randomness in the distribution of the residuals.

Figure 6. Lagging residuals vs. residuals

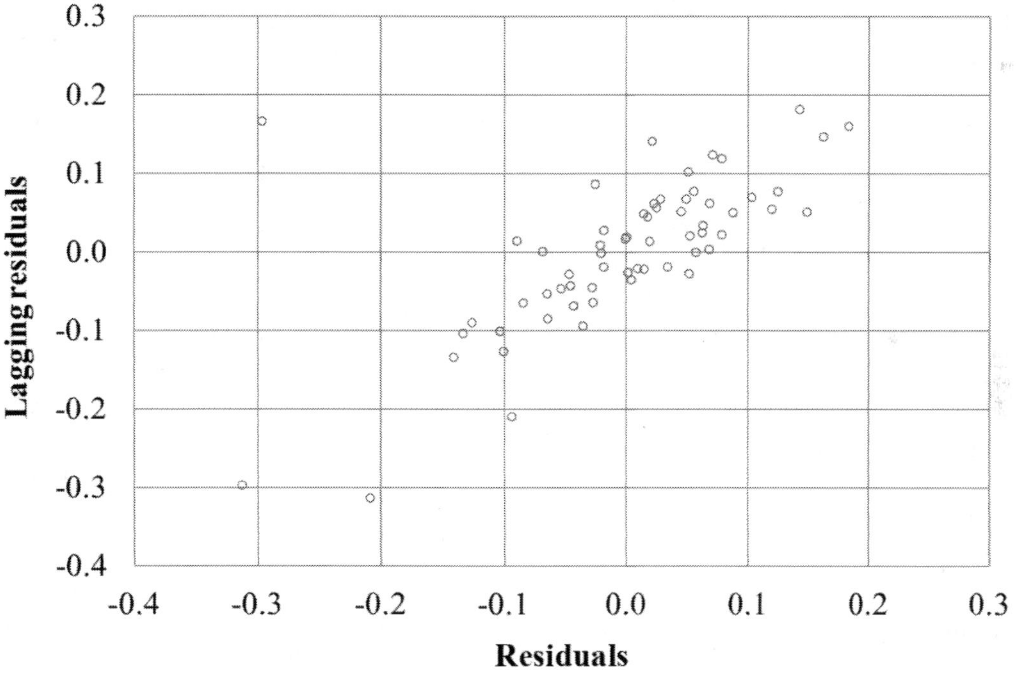

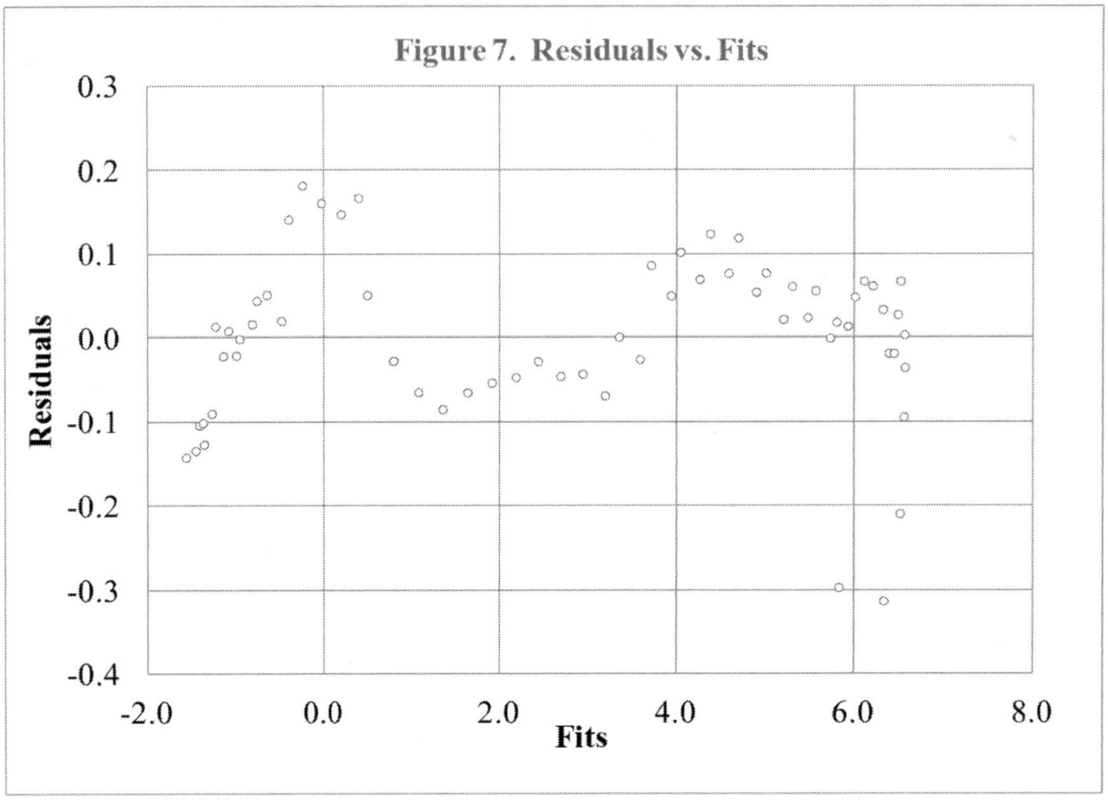

Figure 7. Residuals vs. Fits
Residuals
Fits
0.3
0.2
0.1
0.0
-0.1
-0.2
-0.3
-0.4
-2.0
0.0
2.0
4.0
6.0
8.0

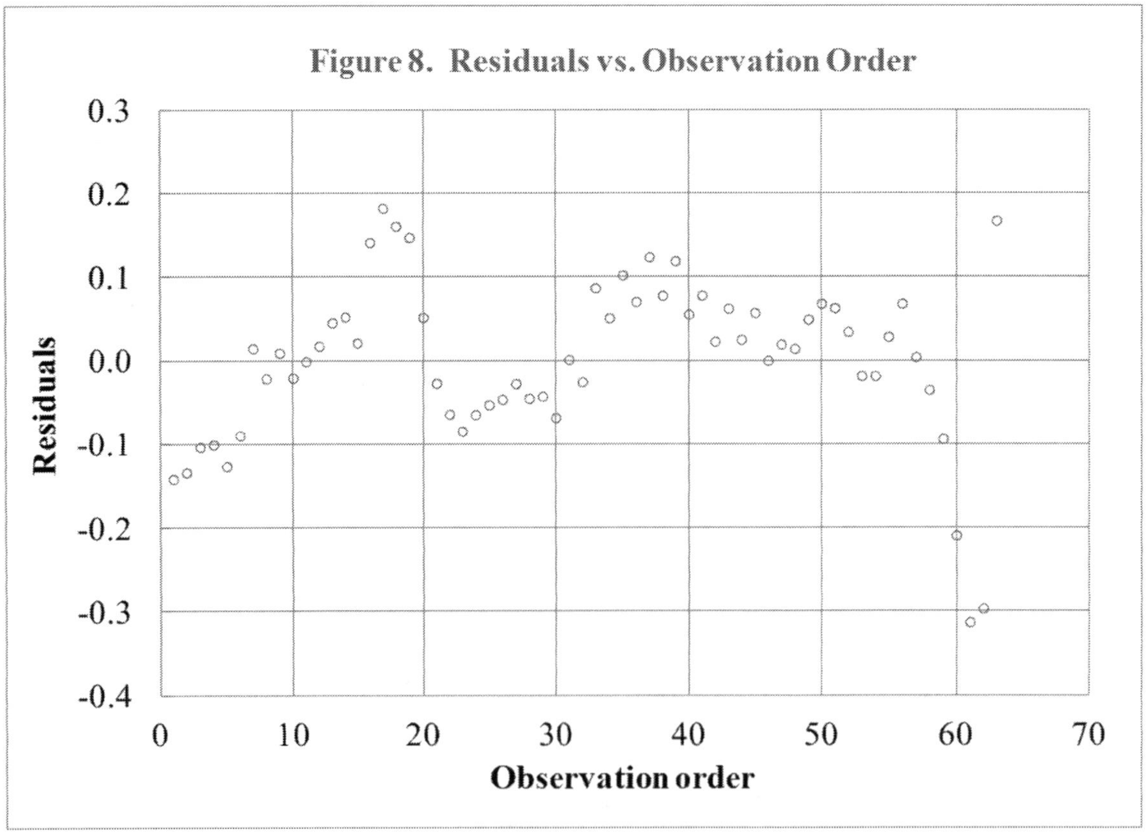

Figure 8. Residuals vs. Observation Order
Residuals
Observation order
0.3
0.2
0.1
0.0
-0.1
-0.2
-0.3
-0.4
0
10
20
30
40
50
60
70

With the issues seen in Figures 6, 7 and 8, it is reassuring to know that the residuals are normally distributed. Figure 9, obtained with Minitab 17, complements the information in Figure 7. The data would fit very nicely in a normal distribution, indicated by the red curve in Figure 9, if it were not for two data points that appear at residual values of -0.3. These two points are the second and third from the last, as the self-heating rate goes down in Figure 4.

There is an explanation for this result. Typically, high-quality calorimeters keep the contents of a test cell in adiabatic mode by providing heat to the containment vessel, so that the temperatures inside and outside the test cell remain about the same during a runaway reaction. A perfect tracking of temperature by the external heater or heaters remains in effect up to a certain temperature, which varies from calorimeter to calorimeter and with the type of heater. Above the threshold, the system is not fully adiabatic. While regression compensated for this deficiency because it is based on the behavior of the entire data set, the experimental data fell behind slightly in self-heating rate towards the more-demanding end of the runaway reaction.

Note that $y_i - \hat{y}_i$ = -0.3 represents $y_i - \hat{y}_i = \ln\left(\frac{dT}{dt}\right)_{\text{expr}} - \ln\left(\frac{dT}{dt}\right)_{regr} = \ln\frac{(dT/dt)_{\text{expr}}}{(dT/dt)_{regr}} = -0.3$,

where *expr* means experimental and *regr* stands for regression. If the logarithm is inverted, $\frac{(dT/dt)_{\text{expr}}}{(dT/dt)_{regr}} = \exp(-0.3) = 0.74$. Hence, the second and third experimental points from the last are short by about 26% compared to the regressed data. It is difficult to visualize this finding on a logarithmic scale, but it is quite clear in Figure 4 that those points are below the regression line.

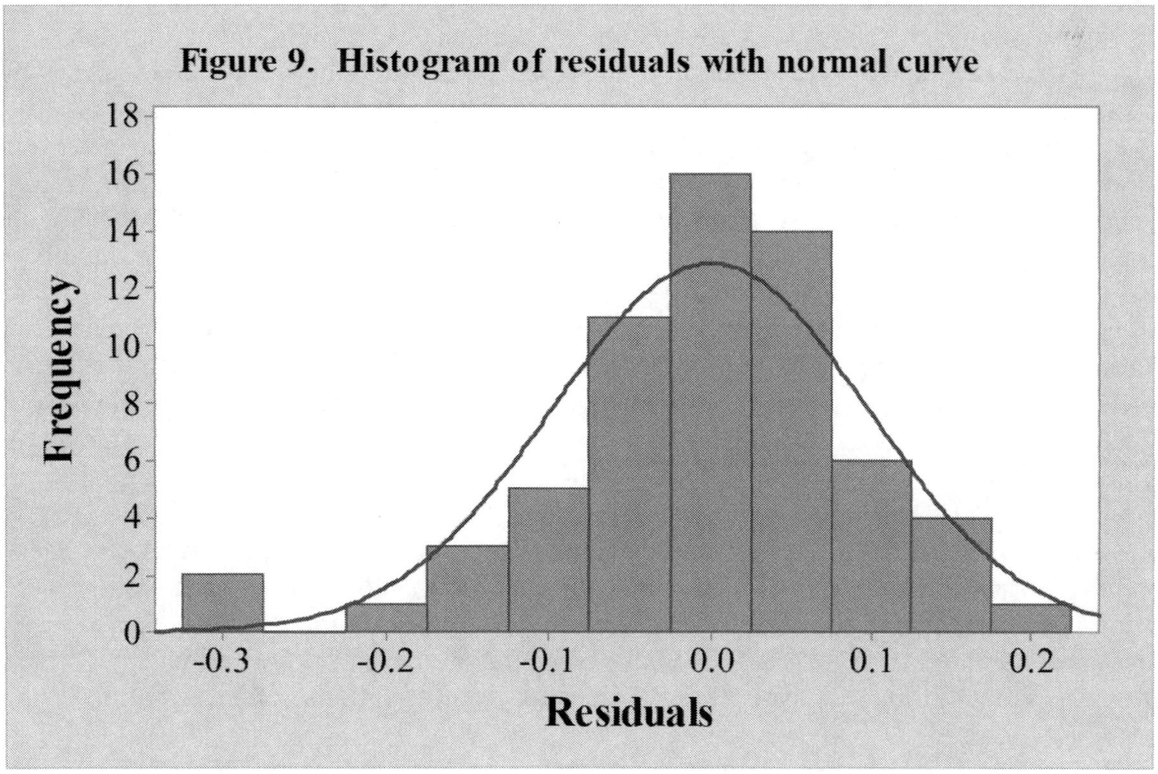

2.5 Reparametrization

In this particular application, reparametrization is a technique used to decrease correlation between the pre-exponential factor A and the activation energy term E/R, and, to a lesser degree, between the pre-exponential factor A and the reaction order n.

The starting point is Eq. [12] stated below and with the previously defined terms:

$$\frac{dT}{dt} = \exp\left(\ln A - \frac{E}{RT}\right)\left(T_f - T\right)^n \qquad [12]$$

With reparametrization through a reference temperature, Eq. [12] is transformed into Eq. [13].

$$\frac{dT}{dt} = \exp\left[\ln A^* - \frac{E}{R}\left(\frac{1}{T} - \frac{1}{T_{ref}}\right)\right]\left(T_f - T\right)^n \qquad [13]$$

where

$$\ln A = \ln A^* + \frac{E}{RT_{ref}} \qquad [14]$$

The method requires the identification of a reference temperature T_{ref}, which minimizes a sensitivity function r^2 that is the sum of squared correlations of each pair among the three coefficients, $\ln A$, E/R and n. The generation of the sensitivity function is explored in detail by Schwaab and Pinto (2007) and applied by Kumpinsky (2008) specifically for the case of runaway reactions. It involves the use of statistical functions and matrix inversions. The solution can be obtained by trial and error upon varying T_{ref} manually until r^2 is minimized. This is good for a spreadsheet like Microsoft Excel without the use of Solve. Otherwise, with programmable language, the iterations can be automated for minimal operator intervention.

The sensitivity function can be minimized for each pair of coefficients, or for an overall compromise, which minimizes the combination of the three pairs together. The latter is the preferred way to handle this problem. It is important to clarify that reparametrization may not eliminate correlations at all. For kinetics of runaway reactions, the benefit may be a reduction, not elimination, in correlations between predictors.

Table 4 shows correlation coefficients and P-values before and after reparametrization. Each 3x3 set is a symmetric matrix. It can be seen that reparametrization indeed has benefits by reducing, for the most part, the correlation between $\ln A$ and E/R. Table 4 is divided as follows: 4.a and 4.c show the correlation coefficients between the predictors before and after reparametrization; 4.b and 4.d display the P-values for the correlations before and after reparametrization.

Table 4. Correlation coefficients and P-values before and after reparametrization

4.a. Correlation coefficients before reparametrization			
	ℓn*A*	*E*/*R*	*n*
ℓn*A*	---	0.976646	0.446245
E/*R*	0.976646	---	0.612863
n	0.446245	0.612863	---
4.b. P-values for correlation coefficients before reparametrization			
	ℓn*A*	*E*/*R*	*n*
ℓn*A*	---	0.000000	0.000250
E/*R*	0.000000	---	0.000000
n	0.000250	0.000000	---
4.c. Correlation coefficients after reparametrization			
	ℓn*A*	*E*/*R*	*n*
ℓn*A*	---	0.381576	-0.421379
E/*R*	0.381576	---	0.612863
n	-0.421379	0.612863	---
4.d. P-values for correlation coefficients after reparametrization			
	ℓn*A*	*E*/*R*	*n*
ℓn*A*	---	0.002045	0.000589
E/*R*	0.002045	---	0.000000
n	0.000589	0.000000	---

For the pair (E/R,ln$A$), reparametrization with r^2 minimization reduced the correlation coefficient from 0.977 to 0.382, which is a great improvement, but the correlation is still significant, as the P-value increased from zero to only 0.002045. Hence, the null hypothesis (no correlation) still has to be rejected. For the pair (n,ln$A$), reparametrization of this type had virtually no effect. The correlation coefficient changed from 0.446 to -0.421, i.e., it went from positive to negative with about the same intensity, while increasing the P-value from 0.000250 to only 0.000589, which still leads to the rejection of the null hypothesis. The pair (E/R,n) was not significantly affected by reparametrization.

Figure 10, generated with SigmaPlot 13, shows the sensitivity function r^2 as it varies with the reference temperature. It reaches a minimum of 0.699 at T_{ref} = 200.5°C and rises asymptotically to 1.53 as T_{ref} increases boundlessly. There are two other curves in Figure 10, i.e., the P-value for the pairs (E/R,ln$A$) and (n,ln$A$) as T_{ref} varies. As can be seen, both curves reach a peak P-value of 1 at different reference temperatures. When P = 1, the pair is totally uncorrelated. However, those peaks in P-value are far from the minimum in r^2. When the peaks in P-value are reached for either pair, the overall level of correlation, represented by r^2, is higher than the minimum. What is observed in Figure 10 for this particular example is that the sum of squares of correlation terms (r^2) is the lowest at a reference temperature at which all pairs of parameters are correlated to a high degree.

Figure 10. Correlation sensitivity function r^2 and P-values

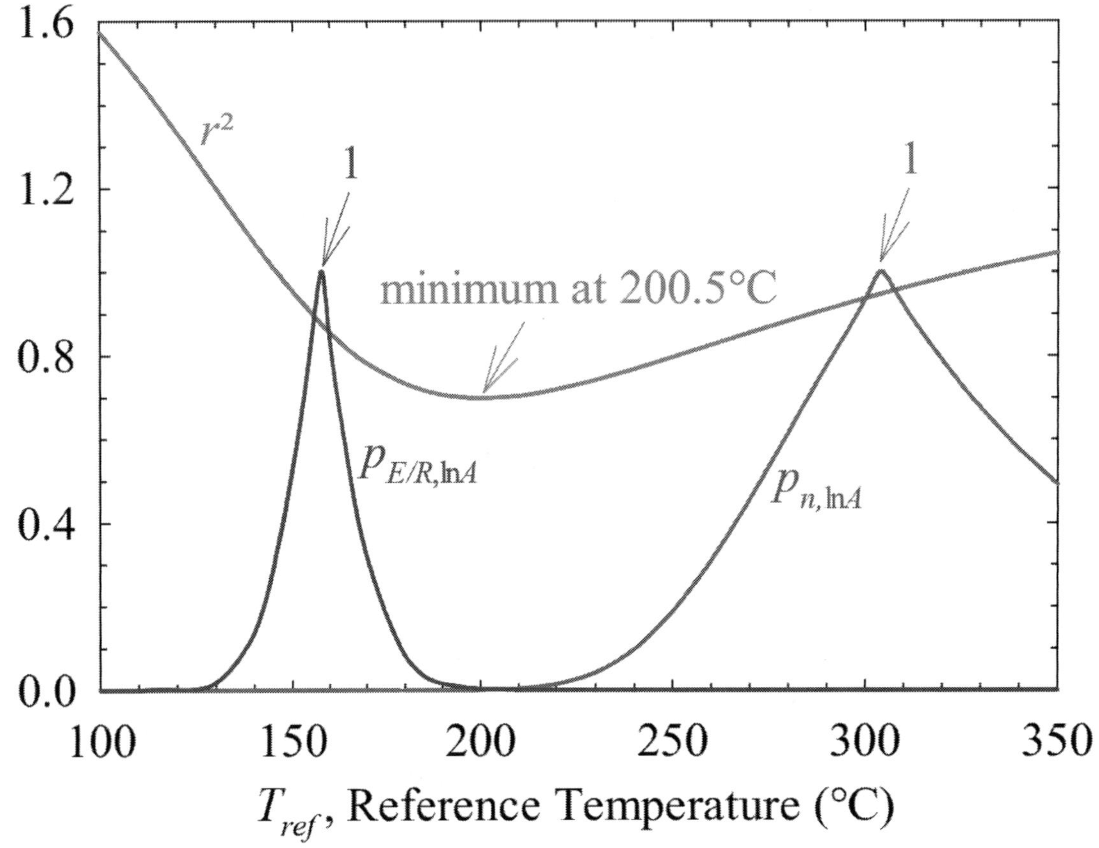

Given the findings in Figure 10, is it still worthwhile to perform reparametrization? Figure 11, also created with SigmaPlot 13, shows the 95% joint confidence region for the pair (E/R,ln$A$). This is a two-dimensional view of the confidence interval of each parameter. This means that, if the adiabatic experiment were to be repeated multiple times, there would be a 95% chance that the (E/R,ln$A$) pair of data would fall within the elliptical region. It can be seen that the major axis of the ellipse, with reparametrization, has a lower slope compared to the ellipse with no reparametrization. A decrease in slope in the joint confidence region signals a decrease in interdependence (correlation) between those coefficients. They would be totally uncorrelated with a horizontal major axis. This analysis shows that reparametrization is still useful, even in a case like this.

Figure 11. Joint 95% confidence region for ln$A$ and E/R

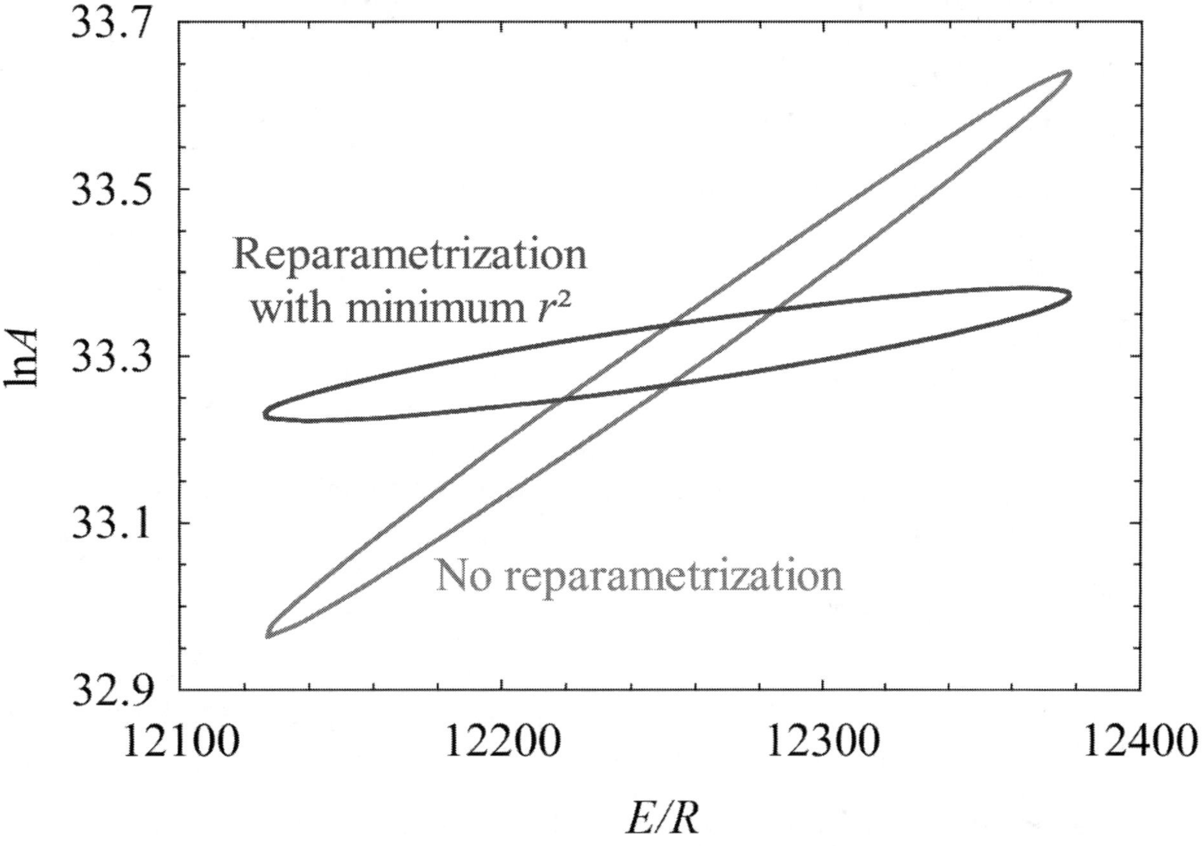

3. Conclusions

- Statistical tools were used to evaluate and improve the quality of the kinetic equation of vinyl acetate runaway reaction. The tools are the following:
  - Collinearity
  - Variance inflation factors
  - Confidence and predictive intervals
  - Residual analysis
  - Reparametrization
- The purpose of this study was to obtain more reliable kinetic parameters for hazards evaluations.
- These calculations can be done almost entirely in Microsoft Excel without using Solve, Goal Seek or Visual Basic. No programming is necessary.
- There is always collinearity between the predictors in this kind of regression, because they all depend on temperature. However, it is beneficial to reduce collinearity to improve model quality.
- More is not better. The best model is the one with the least number of parameters that can well represent the data.
- The model becomes inflated with too many predictors. The parameters may lose their meaning in an inflated model.

- Consider choosing only one among $1/T$, $(1/T)^{1/3}$ and $(1/T)^{2/3}$ when performing regression of runaway reaction experimental data. These predictors are highly collinear.
- T^d is significantly collinear with each of $1/T$, $(1/T)^{1/3}$ and $(1/T)^{2/3}$. Thorough consideration must be given to justify the inclusion of T^d in a runaway reaction kinetic model.
- Some characteristics of the quality of the experimental data can be assessed with the determination of confidence and predictive intervals at a given confidence level, typically 95%. Narrow intervals are qualities inherent to a good experiment.
- The quality of experimental data can also be evaluated through residual analysis. Normal distribution of residuals and their complete randomness throughout the tabulated data are traits of a good model. Normality, autocorrelation and lack-of-fit of residuals can be tested by means of statistical tools.
- Reparametrization is useful to decrease – not necessarily eliminate – correlations between certain parameters.
- Consider including statistical analysis when evaluating the kinetics of runaway reactions. Statistical tools enhance the understanding of the limitations of the regressed model and offer ways to improve it.

4. References

D'Agostino, R. B. (1986). *Tests for the Normal Distribution*. In D'Agostino, R.B.; Stephens, M.A. Goodness-of-Fit Techniques. Marcel Dekker: New York, **1986**.

Fisher, H. G.; Forrest, H. S.; Grossel, S. S.; Huff, J. E; Muller, A. R.; Noronha, J. A., Shaw, D. A., Tilley, B. J. *Emergency Relief System Design Using DIERS Technology: The Design Institute for Emergency Relief Systems (DIERS) Project Manual*. Wiley/AIChE: New York, **1992**; Appendix VI-A: Experimental ERS Sizing - Some do and Do Not Recommendations, pp 386-393.

Geladi, P. and Kowalski, B. Partial Least-Squares Regression: A Tutorial. *Analytica Chimica Acta*, **1986**, 185, 1-17.

Kumpinsky, E. Lumped Kinetics of Self-Heating Runaway Reaction: A Statistical Treatment. *Industrial Engineering Chemistry Research*, **2008**, 47, 1750.

Leung, J. C. The Omega Method for Discharge Rate Evaluation. Paper No. 46e, *Update on Relief Design*, AIChE Meeting, Denver, CO, unpublished. August 16, **1994**.

Leung, J. C. The Omega Method for Discharge Rate Evaluation. *International Symposium on Runaway Reactions and Pressure Relief Design*. Fisher, H. G.; Melhem, G. A., Editors. 1995 Boston, MA. August 2-4, **1995**; pp 367-393.

Lilliefors, H. W. On the Kolmogorov-Smirnov Test for Normality with Mean and Variance Unknown. *Journal of American Statistics Association*, **1969**, 62, 399.

Melhem, G. A. *Methyl, Ethyl and Butyl Acrylate Polymerization Kinetics*. ioMosaic

Corporation, **2004**.

Melhem, G. A. *SuperChems Expert. Software for pressure relief and flare systems design, chemical reactivity assessment and management, consequence analysis, and quantitative risk analysis*. Salem, NH, version 7.26, **2014**.

Montgomery, D. C.; Peck, E. A.; Vining, G. G. *Introduction to Linear Regression Analysis*. 5th. Ed. Wiley: New York, **2012**.

Ryan Jr., T. A.; Joiner, B. L. *Normal Probability Plots and Tests for Normality*, Technical Report, **1976**, Statistics Department, The Pennsylvania State University. Available from Minitab Inc.

Schwaab, M; Pinto, J. C. Optimum Reference Temperature for Reparametrization of the Arrhenius Equation. Part 1: Problems Involving One Kinetic Constant. *Chemical Engineering Science*, **2007**, 62, 2750.

Townsend, D. I.; Tou, J. C. Thermal Hazard Evaluation by an Accelerating Rate Calorimeter. *Thermochimica Acta*, **1980**, 37, 1.

11TH GLOBAL CONGRESS ON PROCESS SAFETY

Emergency relief system design for reactive systems using direct scale-up methods

Dr. Surendra K. Singh
ioKinetic, LLC.
95 Stiles Road
Salem, NH 03079
singh@iokinetic.com

Michelle R. Murphy
ioKinetic, LLC.
95 Stiles Road
Salem, NH 03079
murphy@iokinetic.com

Prepared for Presentation at
American Institute of Chemical Engineers
2015 Spring Meeting
11th Global Congress on Process Safety
Austin, Texas
April 27-29, 2015

UNPUBLISHED

Emergency relief system design for reactive systems using direct scale-up methods

Dr. Surendra K. Singh
ioKinetic, LLC.
95 Stiles Road
Salem, NH 03079
singh@iokinetic.com

Michelle R. Murphy
ioKinetic, LLC.
95 Stiles Road
Salem, NH 03079
murphy@iokinetic.com

Keywords: Reactive systems, adiabatic calorimetry, runaway reactions, vent sizing, emergency relief systems design, direct scale-up methods, APTAC, dicumyl peroxide

Abstract

Pressure relief systems design for reactive chemicals requires knowledge of reaction rates, vent material characteristics, heat of reaction, and relief rates. To do it accurately, the reactive material characteristics are acquired by calorimetric testing and a reaction kinetic model is developed. Alternative simpler methods have been presented by The Design Institute for Emergency Relief Systems (DIERS). These methods involve acquisition of calorimetry data using a low thermal inertia instrument that can be directly scaled-up. In this study, an Automatic Pressure Tracking Adiabatic Calorimeter (APTAC™) was used to collect chemical reactivity data for the dicumyl peroxide and toluene system. DIERS analytical and direct scale-up methods were used to determine the emergency relief venting requirements for a storage tank under upset conditions.

1. Introduction

DIERS[1] mainly established two practices for pressure relief systems design:

1. Detailed dynamic computer simulation
2. Simplified (analytical and direct scale-up) methods

The detailed methods described in the DIERS Project Manual[1] and by Melhem[2], advocate a fundamental approach to pressure relief design, especially for a reactive system. The reaction chemistry is qualified using small-scale experiments. A kinetic model of the system is developed, including an estimation of the vapor-liquid equilibrium of the reactants and products. Simulations are completed of the full scale system coupling fluid dynamics to design the emergency relief system. The method is highly accurate, can handle complex systems, and is valuable for sensitivity analysis (i.e., iterative design and what-if analysis). The direct evaluation of the impact of temperature, pressure, composition, fill level, solvent boiling point, reduced charge, etc. can be simply and quickly completed. The detailed method also provides the necessary flow data for vent containment design (if required), structural support, etc. It requires expert skill for establishing kinetic information from the rate of chemical energy release and thermophysical properties of the system. This is followed by computer dynamic simulation.

The simplified methods are mostly applicable to non-reactive systems but often are applied to reactive systems because they are simple, and easy to use. The scale-up data is only valid for the system composition and thermal inertia represented by the small-scale experiments; the results are often conservative (especially for gassy systems), and the presence of long inlet lines and the impact of downstream equipment are not considered. Sensitivity analysis requires additional experiments. The analytical methods described in Appendix VI-A of the DIERS Project Manual, include FAI's (Fauske & Associates, LLC.) nomograph/analytical method, and Leung's analytical methods[3]. The DIERS Project Manual also presents a direct scale-up method based on relief area to vessel charge scaling.

The present paper illustrates how the DIERS methodologies are used to estimate the required relief area for a reactor under upset conditions. These methods include Leung's analytical and the direct scale-up methods presented in the DIERS Project Manual. The system under study is 50% dicumyl peroxide in toluene. The analysis is based on experimental data presented by Murphy [2].

The analytical method for pressure relief system (PRS) evaluation estimates the relief area requirement from adiabatic calorimetery tests. In this method, sizing is completed for both vapor and gassy systems. The sizing is based on selecting the larger of the relief area required. Each method and the accompanying equations are presented here.

1.1. PRS sizing for vapor only flow regime

Leung's Analytical Method-I[3,4], also described in Appendix VI-A of the DIERS Project Manual, is valid for a vapor only flow regime (volatile/tempered reaction), and the required relief area is estimated using Equation 1:

$$A = \frac{m_o(\bar{q})}{G\left[\sqrt{\left[V/m_o\right)\left[T\left(\frac{dP_v}{dT}\right)\right]\right]} + \sqrt{(C_p\,\Delta T)}\right]^2} \qquad \text{[Eq.1]}$$

Where m_o is reactor content mass (kg), $\bar{q}$ is reaction energy release rate (J/s), G is mass flux (kg/m^2-s), V is volume of vessel (m^3), and T is temperature at operating pressure (K), C_p is specific heat (J/kg/K), dP_v/dT is change in pressure per unit temperature (bar/K).
Melhem[5] also reported an approach for the estimation of relief area for a vapor system (see Equation 2):

$$\frac{A}{M} = \frac{q_s}{(0.9)C_d\,\Delta P\,\sqrt{T_s\,C_{p,l}}} \qquad \text{[Eq.2]}$$

Where A is relief area in inch^2, M is the vessel mass in kilograms, ΔP is the vessel overpressure over the set point in Pascal, T_S is the relief set temperature in Kelvins, q_s is the reaction energy release rate in J/kg/s, and $C_{p,l}$ is the liquid heat capacity in J/kg/K and C_d is discharge coefficient.

1.2. PRS sizing for gassy systems

For a gassy system, the temperature can continue to increase after relief due to negligible tempering effects of non-condensable gas generation. The required relief area can be estimated by equating the rate of volume generation due to reaction and the rate of volume discharge from the relief device as described by Melhem[5]:

$$\frac{C_d\, G\, A_h}{\rho_o} = \left(\frac{M_o}{M_{cell}}\right)\left(\frac{V_{cell}}{P_o}\right)\left[\frac{dP}{dt}\right]_{max} \qquad \text{[Eq.3]}$$

Where G is a two-phase mass flux to be computed using the Omega method[4,6] for a non-flashing system ($\omega = \alpha_o$). The pressure rise rate data is derived from a suitable open system calorimeter. P_o is the maximum allowable pressure accumulation in the vessel. M_o is the initial mass charged (kg) to the vessel, V_{cell} and M_{cell} represent the void volume (m^3) in the test cell and the mass (kg) in the test cell. The relief area is calculated by:

$$A_h = \rho_o\left(\frac{1}{G\, C_d}\right)\left(\frac{M_o}{M_{cell}}\right)\left(\frac{V_{cell}}{P_o}\right)\left[\frac{dP}{dt}\right]_{max} \qquad \text{[Eq.4]}$$

The two-phase flow rate (G, kg/m^2-s) can be estimated using equation 5, where m is the mass of the vented mixture (kg), A is the relief area of vent (m^2) and t is the time taken to vent (sec)

$$G = \frac{(m)}{A \times t} \qquad \text{[Eq.5]}$$

DIERS proposes a simple two-parameter equation of state with two adjustable parameters, a and b, that are calculated from either an isentropic or isenthalpic equilibrium flash calculation described in Chapter II of the DIERS Project Manual.

$$\frac{\rho_o}{\rho} - 1 = a\left(\frac{P_o}{P} - 1\right) + b\left(\frac{P_o}{P} - 1\right)^2 \qquad \text{[Eq.6]}$$

Leung[4,6] simplifies this equation by setting b = 0 and a = ω leading to a generalized form for mass flux for nozzle flow using the ω pressure/volume equation:

$$G^* = \frac{\sqrt{-2[\omega \ln \eta + (\omega - 1)(1-\eta)]}}{1 + \omega\left[\frac{(1-\eta)}{\eta}\right]} \qquad \text{[Eq.7]}$$

Where $G^* = \frac{G}{\sqrt{P_o \rho_o}}$, G is the two phase mass flux (kg/m^2.s), $\eta = P/P_o$, exit pressure P, stagnation pressure P_o (bar), density of mixture at stagnation pressure ρ_o (kg/m^3), and Omega constant ω. This equation is evaluated for different ratios of P/P_o to find the point at which the two-phase mass flux (G) is maximum. P ranges from the stagnation pressure condition to whatever back pressure is imposed on the system. The value of ω ranges from 0 to 1 for non-flashing flow.

1.3. Direct scale-up method

Low thermal inertia tests can be used for direct sizing of emergency relief devices by conducting vented tests (top or bottom) as described by Fauske[7, 8]. The method is simple but is valid only as long as the experimental scale does not affect the two-phase vapor-liquid onset/disengagement behavior. Designing an appropriate test is crucial. Using the scaling equation (Eq.8), the relief area of a scaled vessel with a top mounted pressure relief vent can be estimated assuming a homogenous vent mixture:

$$A_S = A_M \left(\frac{\Delta t_e}{\Delta t_a}\right)\left(\frac{m_S}{m_M}\right) \qquad \text{[Eq.8]}$$

Where A_s is the scaled area (m^2), A_M is the vented area of measurement (m^2), Δt_e is the emptying time (s), Δt_a is the time to raise the pressure from set pressure to peak allowable pressure in a closed adiabatic system (s), m_s scaled mass (kg), and m_M measured mass (kg).

In the sections that follow, a description of the experimental approach and an example of how these methods apply to the estimation of the relief area of a system are presented. The data of the 50% dicumyl peroxide in toluene system was generated using an automatic pressure tracking adiabatic calorimeter (APTAC™). Closed and vented test data, as published by Murphy[9], were used to estimate the relief size area using analytical and direct scale-up methods. Closed test data were used to estimate the relief area at a set pressure of 58 psig using the vapor flow regime. Two vented tests at a set pressure of 58 psig were run to estimate the relief area using the gassy flow regime as well as area:charge direct scale-up method.

2. Experimental

2.1 Experimental approach

The automatic pressure tracking adiabatic calorimeter (APTAC™) generates data that can be used to size pressure relief systems by direct scale up or by analytical methods. It is based on the technology developed and promoted by DIERS. The key characteristics of the APTAC™ are listed below and described in the Operation Manual APTAC 264[10]:

1. Both adiabaticity (>400 °C/min) as well as thermal inertia of the test cell are close to full scale vessels (>1.1)
2. Pressure balancing outside the cell allows for use of very thin walled test cells
3. An exotherm detection limit of 0.002°C/min in iso-fixed mode, and <0.1°C/min at 400 °C/min in heat-wait-search mode
4. Different modes of heat operation like isothermal/iso-aging, heat-wait-search, constant heat rate, and constant temperature difference
5. Computer controlled venting (i.e., simple vent to simulate a rupture disc and controlled vent to simulate a safety relief valve)

6. Solid ceramic insulation which results in consistent and low heat loss
7. Calibration adjustment resulting in a better exotherm detection limit
8. Two injection modes: metered injection and one shot injection
9. Small sample size (25cc to 100cc) for safe and easy handling of hazardous materials
10. Smooth data (i.e., no data smoothing necessary)

2.2 APTAC test design

A total of 3 tests were conducted, one closed test and two vented tests. Both vented tests were vented at 58 psig - one vented to an open vessel and the other vented to a closed vessel. The following procedure was used for all three tests:

1. A recipe amount of 50% dicumyl peroxide in toluene was prepared and added to an APTAC steel test cell
2. The cell was connected to the calorimeter and the head space was replaced with nitrogen
3. Tests were run in the heat-wait-search heating mode from 50°C to 350°C
4. The temperature and pressure responses were recorded while the sample was heated
5. The cool down temperature and pressure data were recorded once the sample reached 40°C
6. The end weight of the test cell and residue were recorded

Test 1 was a closed adiabatic test. Test 2 was vented (at 58 psig) to an open knockout pot through 1/16" tubing (0.016" wall thickness and 25 cm long) and a 1/16" Swagelok ball valve. Test 3 was similar to Test 2, except that it was vented to a closed knockout pot (starting pressure of 14.7 psia). At the end of Tests 2 and 3, the weights of the vented liquid in the knockout vessel were recorded. In the case of Test 2 the weight of the vented liquid was also recorded after the non-condensable gas was released. The test parameters for all tests are reported in Table 1.

Table 1: Test Information

Test Parameters	*Test-1 (closed)*	*Test-2 (vented to open vessel)*	*Test-3 (vented to closed vessel)*
50% dicumyl peroxide in toluene mass (g)	35.4	35.16	35.24
Weight of test cell (g)	93	92	93
Thermal inertia	1.7	1.7	1.7
Vessel volume (ml)	133.4	128	133.4
Start temperature (°C)	50	90	50
Exotherm sensitivity (°c/min)	0.06	0.10	0.06
P_{set} (psig)	NA	58	58
Volume of vented vessel (ml)	NA	50	500

3. Results and Discussion

3.1 Test Results

Closed test results

The APTAC closed test revealed an exotherm with an onset temperature of 107°C. The calorimeter lost adiabaticity due to a self-heating rate exceeding 400°C/min. The maximum heating rate and pressure rise rate measured were 439°C/min and 1,459 psi/min, respectively. The detailed results are reported in Figures 1, 2, and 3 and Table 2.

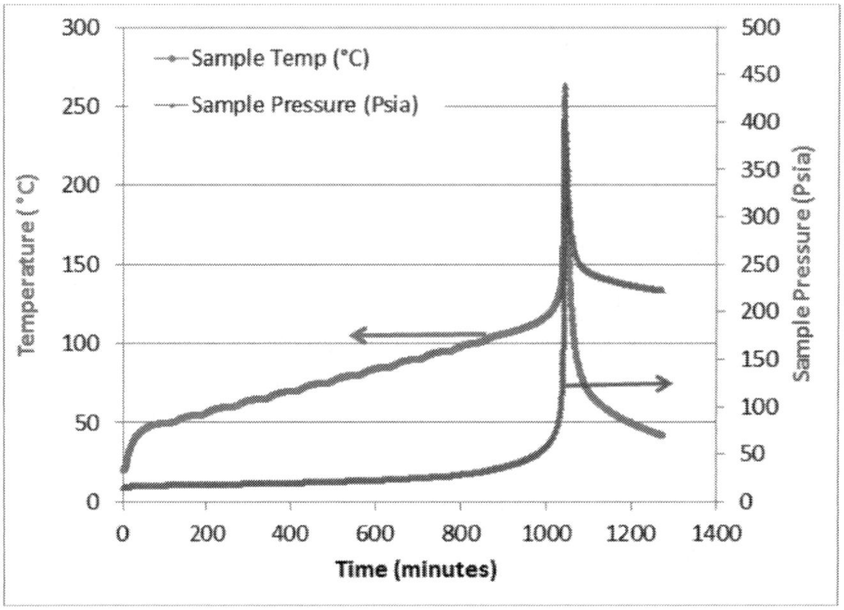

Figure 1: Temperature and pressure histories of 50% dicumyl peroxide in toluene

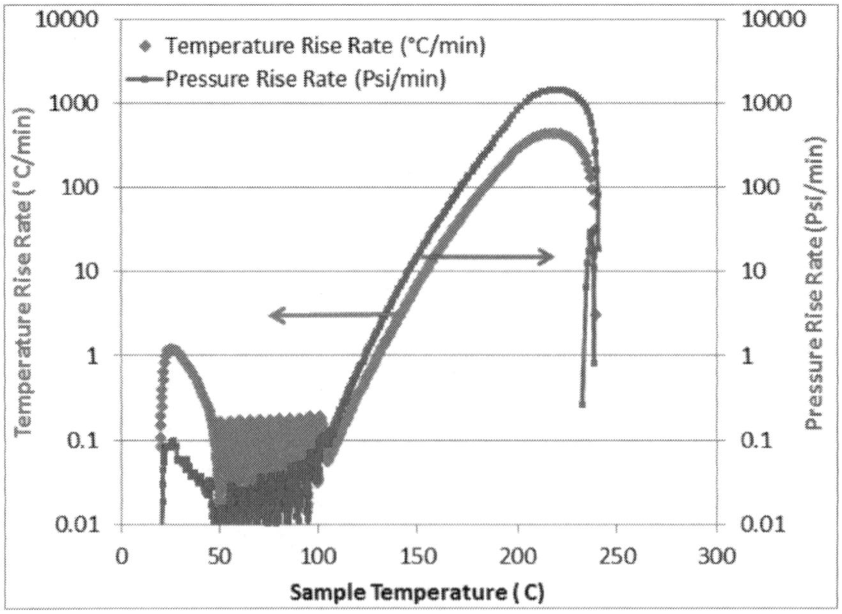

Figure 2: Temperature rise rate and pressure rise rate vs. temperature plots of 50% dicumyl peroxide in toluene

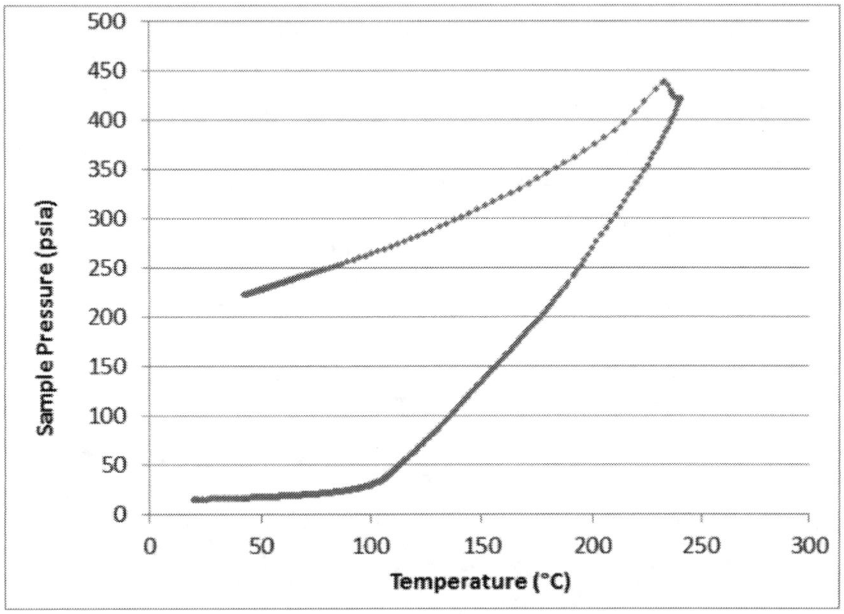

Figure 3: Pressure vs. temperature plot of 50% dicumyl peroxide in toluene

Vented Test Results

Two vented tests were performed as described in Section 3.1. For Test 2, the pressure was vented to an open knockout vessel (open vessel). For Test 3, the pressure was vented to a closed knockout vessel (closed vessel) with starting ambient pressure. Test result details are summarized in Table 2.

In Test 2, with venting to an open knockout vessel, the following behavior was seen:

1. Decomposition of dicumyl peroxide increases temperature and generates pressure in the test cell
2. Vent opens at 58 psig, relieving to a knock out pot that is open to atmosphere, pressure drops to ambient and temperature drops to 125.5°C due to the increase in volume
3. Temperature continues to rise and pressure rebounds up to 41.8 psig at the peak of the reaction rate
4. Pressure in the test cell starts increasing as soon as the gas generation rate is more than the venting rate, causing choked flow
5. Once the reaction completes, the pressure equalizes across the entire volume

In Test 3, with venting to a closed knockout vessel, the following behavior was seen:

1. Decomposition of dicumyl peroxide increases temperature and generates pressure in the test cell
2. Vent opens at 58psig, relieving to a closed knock out pot, pressure drops to 9 psig due to expansion of volume, no drop in temperature is noticed

3. Temperature and pressure of the test cell keep rising and pressure inside the test cell rebounds up to 62 psig at the peak of reaction the rate
4. Once the reaction completes, the pressure equalizes across the entire volume at 48 psig after the completion of reaction.

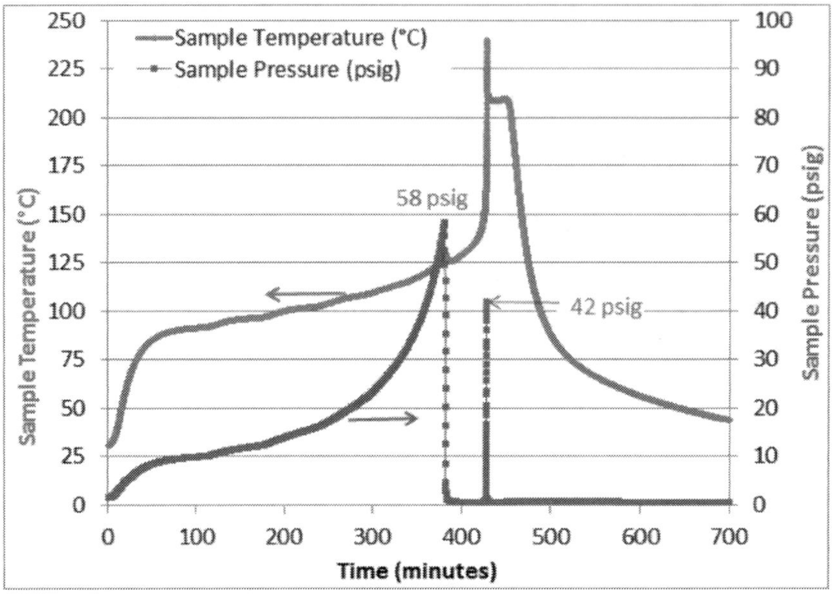

Figure 4: Temperature and pressure histories for a vented test to an open knockout vessel (Test 2)

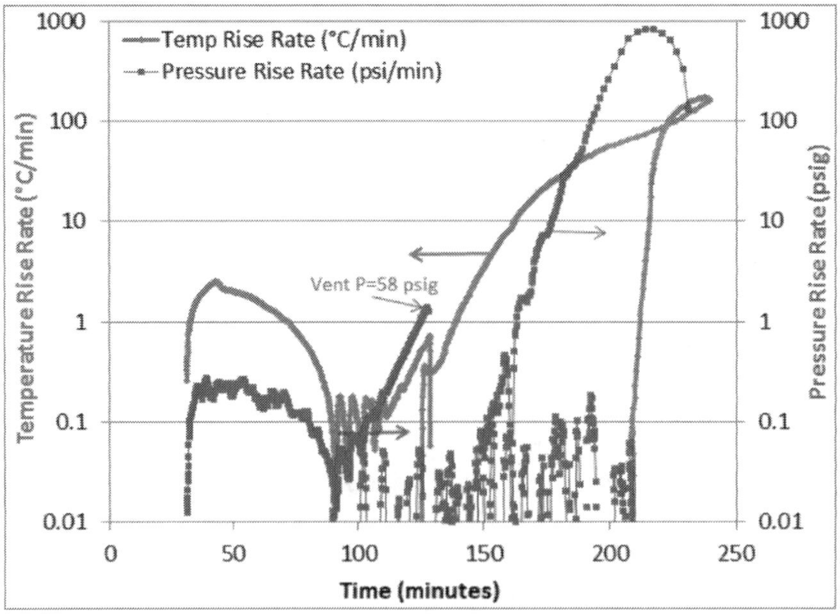

Figure 5: Temperature and pressure histories for a vented test to an open knockout vessel (Test 2)

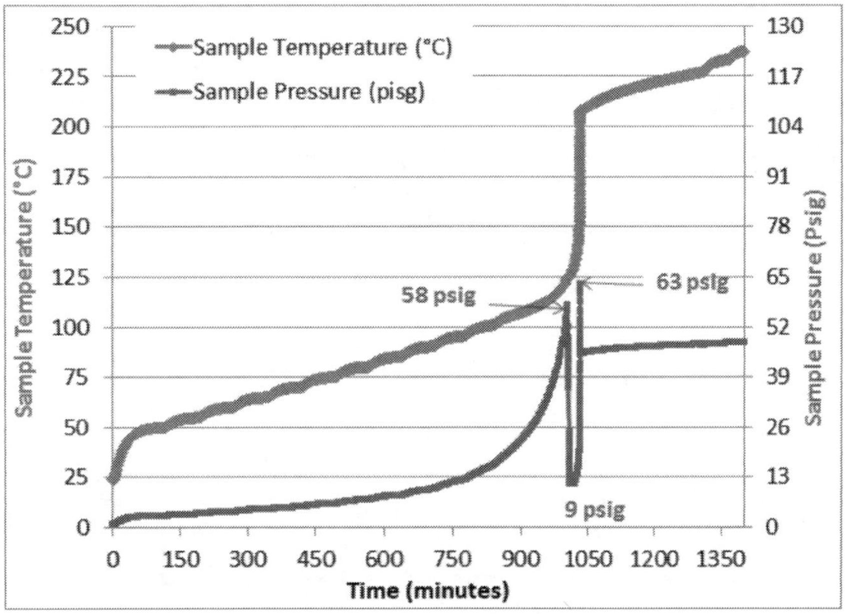

Figure 6: Temperature and pressure vs. time for a vented test to a closed knockout vessel (Test 3)

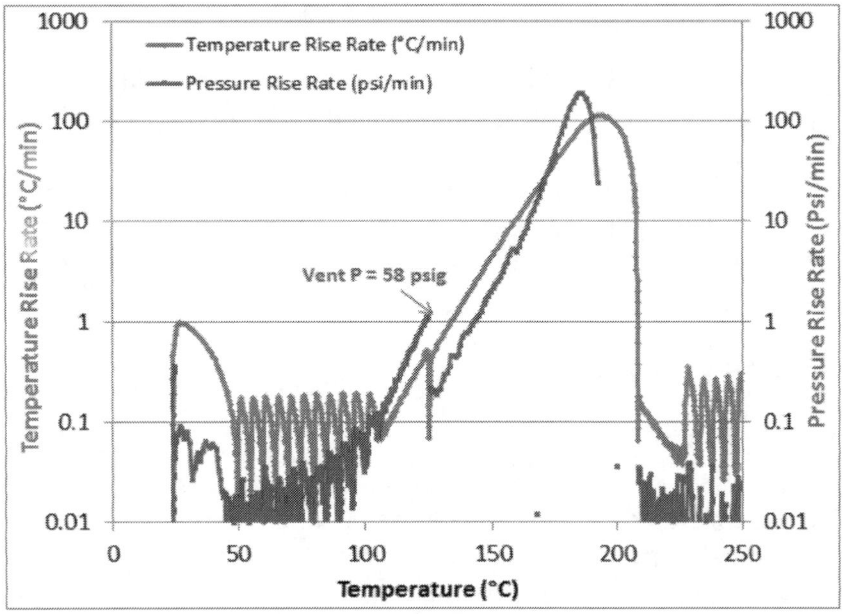

Figure 7: Temperature and pressure vs. time for a vented test to a closed knockout vessel (Test 3)

Table 2: Test results

Test Parameters	*Test-1 (Closed)*	*Test-2 (Vented to open vessel)*	*Test-3 (Vented to closed vessel)*
50% dicumyl peroxide in toluene mass (g)	35.40	35.16	35.24
System Φ factor	1.73	1.70	1.74
Onset (°C)	107.0	107.7	106.7
dT/dt_{max} (°C/min)	439.0	171.3	112.9
dP/dt_{max} (psi/min)	1459	836	190
Temperature at dP/dt_{max} (°C)	222	237	193
ΔT (°C)	>131.7	100.0	100.4
ΔT_{ad} (°C)	>227.8	170.0	174
ΔH_r (J/g dicumyl peroxide)	>771	NA	NA
Activation energy (cal/gmol)	7721	NA	NA
Pre-exponential factor (s^{-1})	3.74×10^{16}	NA	NA
P_{set} (psig)	NA	58	58
dP/dt at P_{set} (psi/min)	NA	2.53	1.12
P_{min} after vent open (psig)	NA	0.5	9.2
P_{max} after vent open (psig)	NA	41.8	63.2
dP/dt_{max} after vent open (psi/min)	NA	837	191
Weight of liquid just after venting (g)	NA	0.4	NA
Weight of total vented liquid (g)	NA	15.2	NA
Total weight loss (%)	8.2	60.4	79

As shown in the plots and Figure 4, venting to an open vessel resulted in a higher temperature at 427 min. (at peak pressure rise rate). This was due to a loss of solvent. Ultimately this resulted in a higher pressure rise rate than venting to a closed vessel. In closed vessel venting, the rate of solvent evaporation was slow due to larger swelling and a higher heat sink, resulting in a lower peak decomposition temperature. It was found that 79% of the sample was vented in the case of venting to a closed vessel in comparison to only 60% in the case of venting to an open vessel.

3.3 Application to Full Scale Vessel

A 400 gallon vessel containing 400 kg of 50% Dicumyl peroxide in toluene solution has an MAWP of 58 psig (MAAP of 63.8 psig) and is equipped with a top mounted rupture disc. The rupture disc opens directly to ambient pressure with a set pressure (P_{set}) of 58 psig. The vent area for the rupture disc was calculated based on both the open and closed knockout vessel tests to compare relief size requirements. The calculations were completed using the DIERS simplified analytical methods and area to charge scaling method.

3.4 Rupture disc size estimation

DIERS analytical method

Rupture disc sizes were estimated using both relief regimes of vapor (Eq. 2) and gassy (Eq.4). Time, temperature and pressure data generated by an APTAC closed test were corrected for thermal inertia as described by Townsend and Tou[11]. The two-phase mass flux and relief area were estimated using Equations 5 and 7, respectively. Venting was assumed to be two-phase non-flashing.

The required relief area was estimated considering a vapor regime from the calorimetry data generated by Test 1. The required vent area, vent diameter, and area/mass were 3.813 in^2, 2.2 inches, and 0.0055 in^2/kg respectively. Table 3 provides the vessel parameters used in the calculations.

The required relief area was also estimated considering a gassy regime from the calorimetry data generated by both Tests 2 and 3. The estimated required relief areas were 6.12 inches based on venting to an open vessel and 3.0 inches based on venting to a closed vessel. Table 3 provides the vessel parameters used in and the results of the calculations.

Venting to an open vessel rendered a larger relief size requirement than venting to a closed vessel. The maximum pressure rise in the open vessel was significantly lower than that in the closed vessel. Venting to an open vessel causes tempering of the system due to the loss of solvent vapor, resulting in a higher concentration of dicumyl peroxide and a lower heat sink in the vessel. An increased decomposition reaction results along with higher pressure and temperature rise rates. This shows the importance of designing the test to closely match the actual full-scale design, including venting to atmosphere, an open, or a closed vessel.

Table 3: Vessel Parameters and rupture disc size calculations

Description	***Test 2 (vented to open vessel)***	***Test 3 (Vented to closed vessel)***
Test Type	Vented at 58 psig	Vented at 58 psig
Mass in vessel, Kg	400	400
MAWP, barg	4.0	4.0
MAAP, barg, 10% additional	4.4	4.4
Test sample mass, g	35.16	35.24
Void volume for test cell, m^3	1.28×10^{-4}	0.98×10^{-4}
Measured max. pressure rise rate, bar/s	0.961	0.219
Liquid density at vent, kg/m^3	264.2	264.2
Discharge coefficient assumed	0.5	0.5
Two-phase mass flux, $kg/m^2/s$	6414	6414
P_{set} (psig)	58	58
Measured temperature at Pset, K	401.3	397.3
Measured dT/dt at Pset, K/s	0.011	0.008
Required rupture disc diameter, gassy system, in	6.12	3.00

Direct area to charge scale-up method

The relief areas were also estimated using the area to charge scale-up method. A set pressure of 58 psig was used along with 10% maximum allowable over pressure. The data from the two vented tests were applied, using Eq.8. The results are summarized in Table 3. The rupture disc sizes estimated for venting to an open vessel and venting to a closed vessel were 1.3 inches and 1.78 inches, respectively. Venting to an open vessel resulted in a shorter time to vent which rendered a smaller rupture disc size.

Table 4: Rupture disc size calculation using area:charge scaling equation

Description	*Test 2 (Vented to open vessel)*	*Test 3(Vented to closed vessel)*
Pset (psig)	58	58
Vent area of test, m^2	4.72×10^{-7}	4.72×10^{-7}
Emptying time, min	2.8	5.45
Closed system adiabatic pressure rise time from set pressure to peak allowable pressure , min	4.52	4.52
Mass in scaled vessel, Kg	400	400
Mass of measured, Kg	0.035	0.035
Required relief area, in^2	4.1	10
Required relief diameter, in	1.3	1.78

3.5 Discussion

Figure 8 provides a comparison of the results of the various calculation methods along with experimental results. First, a set of calculations were completed using the gassy flow regime method, explained in Section 1.2, to define the vent area/mass ratio as a function of percent overpressure. This is shown as blue diamonds in Figure 8.

The relief areas were estimated using the area to charge scale-up method for both open and closed discharge vessels, Tests 2 and 3, respectively. The resulting required vent area/mass ratios were 0.008 in^2/kg and 0.025 in^2/kg respectively. These are shown as the red box and green diamond in Figure 8.

As a comparison with an actual vent, the actual experimental vent area to charge mass from Test 2 is shown as the purple X in Figure 8. This value was calculated from the experimental design to be 0.021 in^2/kg. This experimental design was shown to be an adequate vent area/mass ratio for the system under study. A review of Figure 4 shows the vent opening at 58 psig and the maximum pressure reached after opening of 42 psig.

The area to mass ratio estimates for the gassy regime were found to be higher than those calculated using the direct scale-up method and those measured as adequate in experimental vented tests.

Table 5 provides a summary of all estimates completed above.

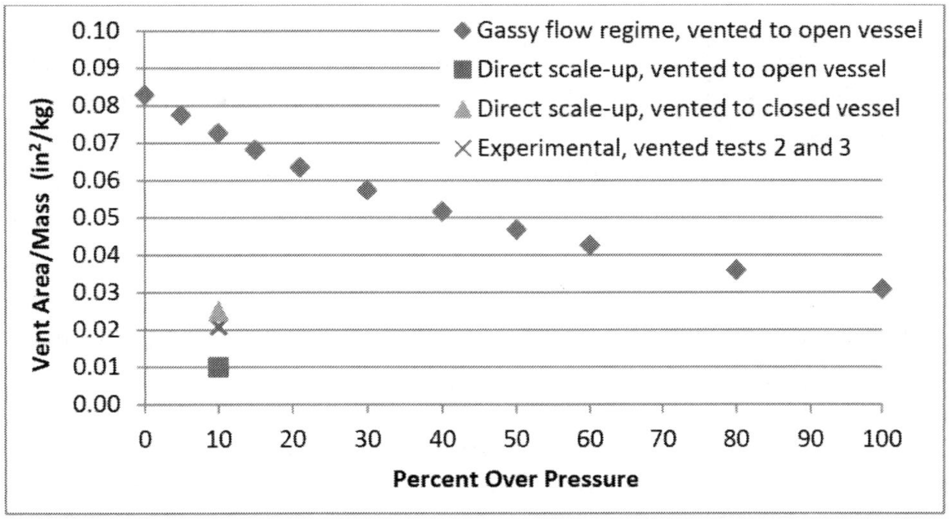

Figure 8: Comparison of estimated vent area to charge mass ratios

Table 5: Summary of rupture disc size

Description	***Relief Size***		***Relief area/mass***
	Area (in^2)	Diameter (in)	(in^2/kg)
Vapor Regime	3.81	2.2	0.006
Gassy Regime			
Vented to open vessel	29.5	6.12	0.074
Vented to closed vessel	7.1	3.00	0.018
Area:Charge Scaling (direct scale-up)			
Vented to open vessel	4.1	1.30	0.010
Vented to closed vessel	8.6	3.31	0.025

4. Conclusion

The rupture disc size estimations using the analytical method for the thermal runaway reaction of 50% dicumyl peroxide in toluene are significantly higher if the system is considered gassy rather than vapor.

The temperature and pressure profile of the test vented to an open vessel indicate that the system is tempered. This is followed by evidence of gas generation as the temperature continues to rise during venting.

In the case of venting to a closed vessel, the test results indicate that the system temperature kept rising even after the drop in pressure from the opening of the vent. This suggests a non-tempered system. Contrary to the case in the open system, where the drop in temperature and pressure indicates tempering.

The direct scale-up method rendered a smaller rupture disc size when venting to an open vessel versus venting to a closed vessel.

The analytical method using the gassy system resulted in the largest required vent size. It would be the recommended rupture disk vent size because it is the most conservative.

5. References

[1] The Design Institute for Emergency Relief System (DIERS) Project Manual, American Institute of Chemical Engineers, New York, NY 1992.

[2] G.A Melhem, H.G. Fisher and D.A. Shaw, An advanced method for the estimation of reaction kinetics, scale up, and pressure relief design, Process safety progress, vol. 14, no. 1, January (1995),1-21

[3] J.C. Leung and H.K. Fauske, Runaway system characterization and vent sizing based on DIERS Methodology, Plant Operation Progress, Vol.6, No.2, April 1987, pg. 77-83

[4] J.C. Leung and M.A. Grolmes. The discharge of two-phase flashing flow in a horizontal duct, AIChE Journal, 33(3), (1987), 524-527.

[5] G.A. Melhem, Personal communication, 2006.

[6] J.C. Leung, Easily size relief devices and piping for two-phase flow, Chemical Engineering Progress, December (1996), 28-50.

[7] H.K. Fauske, M.A. Grolmes, and R.E. Henry, Emergency relief systems- sizing and scale-up, Plant/Operation Progress, Vol.2, No.1, 1983), 27-30.

[8] H.K. Fauske, Scale-up for safety relief of runaway reactions, Plant/Operation Progress, Vol.3, No.1 (1984), 7-11.

[9] M.R. Murphy Model based scale-up of dicumyl peroxide decomposition. DIERS User Group meeting May 9, 2012, Concord, MA.

[10] Operation Manual APTAC 264 , NETZSCH Instruments North America, LLC., Burlington, MA, USA.

[11] D.I. Townsend and J.C. Tou, Thermal hazard evaluation by an accelerating rate calorimeter, Thermochemica Acta, 37 (1980), 1-30.

11TH GLOBAL CONGRESS ON PROCESS SAFETY

Engineering Safe Pressure Relief for Existing Flare Systems

Jay Riha
Contek Solutions, LLC
6221 Chapel Hill Blvd, Suite 300
Plano, TX 75093
jr@contekllc.com

Steve Streblow
Contek Solutions, LLC
6221 Chapel Hill Blvd, Suite 300
Plano, TX 75093
sstreblow@contekllc.com

Prepared for Presentation at
American Institute of Chemical Engineers
2015 Spring Meeting
11th Global Congress on Process Safety
Austin, Texas
April 27-29, 2015

UNPUBLISHED

Engineering Safe Pressure Relief for Existing Flare Systems

Jay Riha
Contek Solutions, LLC
6221 Chapel Hill Blvd, Suite 300
Plano, TX 75093
jr@conteklc.com

Steve Streblow
Contek Solutions, LLC
6221 Chapel Hill Blvd, Suite 300
Plano, TX 75093
sstreblow@conteklc.com

Keywords: Flare, relief, existing

Abstract

This paper will discuss the primary objectives of relief system analysis, some industry trends and disasters that led to the emphasis on relief systems and some common deficiencies in flare and relief systems.

1. Introduction

This paper presents a practical approach for evaluating an existing flare and relief system. We will present the most common elements that are found to be "working" and "not working well". We will discuss the common bottlenecks in a relief system. This will help the evaluation team to focus their efforts on the items that are most likely to need attention. The result is a set of recommendations that determines the minimum modifications necessary to reduce the risk of uncontrolled releases to an acceptable level.

2. The Primary Objectives of Flare/Relief System Analysis

- Ensure the safe design and operation of relief devices and effluent handling equipment during an overpressure scenario that may affect multiple pieces of equipment relieving through multiple relief devices or otherwise challenging the design of the flare system in some way
- Ensure the backpressures on pressure relief devices does not adversely affect their capacities to function correctly
- Ensure the condition of the piping and network equipment fits within their stated design limits

Designing a safe and cost-effective pressure relief system is a constant concern for engineers working to design safe facilities. Existing flare systems are an integral part of a plant's overall safety system and have been built and modified, sometimes over the course of many years and through several plant expansions. Often these expansions have been done in a piecemeal manner as new vessels and pressure valves are added over the course of several modifications. These systems are very expensive and time-consuming to replace or retrofit and require long lead times to design and build. We have seen lead times in excess of 18 months for a new flare and pressure protection system for a new plant. Retrofitting often leads to plant shutdowns for installation or maintenance. We will be discussing an approach to performing a study on an existing flare system in order to suggest modifications to assure that the risk of uncontrolled releases and harm to personnel and the environment is minimized.

3. High Profile Disasters

The evaluation and proper design of flare and relief systems has been discussed in industry standards for many years. With the advent of formalized Process Safety Management in the 1990's and several high-publicity incidents, the emphasis on safe relief systems has increased.

Some of the highest profile disasters were directly attributed to failures of the pressure relief system design:

1976: Seveso, Italy Disaster. Due to the release of dioxins into the atmosphere, 3,000 pets and farm animals died and 70,000 animals were slaughtered to prevent dioxins from entering the food chain. In addition, 193 people in the affected areas suffered from chloracne. This was the highest known exposure to these dioxins in residential populations (over 100 parts per million), which gave rise to numerous scientific studies and standardized industrial safety regulations.

1979: Three Mile Island Accident. Partial nuclear meltdown. Mechanical failures in the non-nuclear secondary system, followed by a stuck-open pilot-operated relief valve in the primary system, allowed large amounts of reactor coolant to escape. Plant operators initially failed to recognize the loss of coolant, resulting in a partial meltdown. The reactor was brought under control but not before up to 13 million curies of radioactive gases were released into the atmosphere

1984: The Bhopal disaster in India is one of the largest industrial disasters on record.
A runaway reaction in a tank containing poisonous methyl isocyanate caused the pressure relief system to vent large amounts to the atmosphere. Estimates of its death toll range from 4,000 to 20,000. The disaster caused the region's human and animal populations severe health problems to the present.

2005: Texas City Refinery Explosion. An explosion occurred at a BP refinery in Texas City, Texas. Over 100 were injured, and 15 were confirmed dead. Several level indicators failed, leading to overfilling of a knock out drum, and light hydrocarbons concentrated at ground level throughout the area. A truck running nearby set off the explosion.

4. Why the Emphasis on Relief Systems?

As companies and facilities have grown in size and public awareness of environmental hazards has increased, there has been a need to guarantee the safe venting of harmful gases and liquids in industrial operations.

Company safety audits of their relief systems and OSHA's National Emphasis Program for flare systems have also put pressure on many companies to re-evaluate their existing flare systems. This requires a formal flare study which involves many variables that the engineer must evaluate. These variables include:

- the properties of the fluids that are being processed,
- operating parameters,
- relieving pressures and temperatures,
- piping and vessel characteristics,
- loss of containment scenarios,
- safety systems and
- operating practices.

Each of these different variables can have a tremendous impact on the effectiveness of a flare and relief system for a facility.

5. Common Deficiencies in Relief Systems

A major concern in addressing an existing flare system is the cost and time associated with modifying or replacing it. This cost must be weighed against the costs in terms of potential loss of life, environmental damages, physical plant damages, loss of production and negative publicity. There are many options to modify deficient flare and relief systems short of replacement. These are detailed in Tables 1 & 2 below.

Some of the most common deficiencies and most effective "fixes" involve the following:

5.1 Not developing and modeling over-pressure cases to understand worst case, multiple relief scenarios.

- Multiple relief valve fire cases need to be developed, usually involving equipment protected by PSV's within a 50 foot proximity or fire circle
- Model all valves relieving simultaneously
- This approach provides insight to the highest potential pressures and flow rates
  - Most relief systems are not sized using this method in initial design
  - Relief valves are usually sized for a non-fire, over pressure situation
- We often find that knock out drums were not sized for this scenario which results in potentially undersized knockout drums and/or liquid flow to the flare
- Consider modeling all liquids in a fire case separately from the vessels with vapor to understand the maximum liquid flows
- Knockout drums and flare stacks are undersized in 60% of cases

5.2 Under-sizing inlet and outlet piping

- Maximum flow in tail pipes can occur when PSV's relieve independently due to lower back pressures
- Most commonly need outlet piping "one size up" to handle flow and back pressure buildup in worst case scenarios.
  - For example: for a 2 inch outlet nozzle, size the tail pipe at 3 inches
  - Where sub headers combine increase the size by one step.... i.e. 4 inches to 6 inches
- This is the most common problem we see and occurs in over 50% of cases we have analyzed
- This contributes to excessive back pressure in 30-40 plus percent of cases
- We find that sub headers are undersized in 60% of cases
- Noise and velocity issues occur in over 45% of cases
- Choked flow occurs in over 15% of cases

5.3 Not taking extreme pressure and temperatures into account when specifying piping and design specifications for relief system vessels.

- Extreme pressures and temperatures during relieving conditions can cause phase changes that can exceed vessel capacities... i.e. 2 PSV's releasing vapors at high pressure can transform to a liquid resulting in choked flows in the sub headers
- 40% or more of the PSVs have issues with outlet temperatures higher than the maximum specified piping design temperature or vessel design temperature
- Process nozzle connections on knockout drums are often undersized

5.4 **Not maintaining complete and up to date records for relief valve sizing calculations and piping specifications**.

- Need "Fire Case" relieving conditions for PSV's.... Typically the worst case for relieving
- These are not commonly provided during plant design due to lack of plot plan definition when the process is designed
- Design limits for piping and vessels are typically not provided
- Not having this information involves additional time and expense in relief study

Table 1. Significant Issues with Specific Relief System Components Summarized Across Multiple Flare Studies

Component Design Issues from US On-shore Gas Plants	Plant 1	Plant 2	Plant 3	Plant 4	Plant 5	Percentage of Occurrences in Total Number of PSV's
Number of PSV's	27	10	11	14	28	Total of 90 PSV's
PSV Inlet/Outlet pipes needed to increase in size		8	11		28	**53%**
PSV relieving exceeded data sheet back pressure	25		11	6		**48%**
Relief flow exceeded .7 Mach number	22		11	9		**48%**
Relief flow exceeded maximum piping design temperature	27		2	7		**41%**
Relief flow ub header exceeds maximum design temperature	25	4				**33%**
Relief flow exceeded PSV typical maximum back pressure	10		11	6		**31%**
Relief flow exceeded max piping design pressure	15			7		**25%**
Slug/Choked flow present in piping	4			9		**15%**
Excessive liquid to knock out drum		4			4	**9%**

Table 2. Significant Issues with Major Relief System Components Summarized Across Multiple Flare Studies

System Design Issues from US On-shore Gas Plants	Plant 1	Plant 2	Plant 3	Plant 4	Plant 5	Percentage of Occurrences in Total Number of Plants
Sub headers needed to increase in size	Y		Y		Y	**60%**
Flare stack undersized	Y	Y	Y			**60%**
Knock out drum undersized		Y	Y		Y	**60%**

6. Conclusion

The primary reason for a flare study is to ensure the safe operation of a pressure relief system in a worst case scenario. As plants grow and evolve we need to re-evaluate the effectiveness of existing flare systems. There is a growing interest from OSHA and industry in ensuring the safety of relief systems. We believe the most effective approach for ensuring operational safety is to model the system under worst case scenarios whenever significant changes or additions are made to a plant. Finally we discussed some of the most common findings we have encountered in recent flare and relief system evaluations. The most common of these are undersized outlet piping and sub-headers, and relief system pressures exceeding PSV back pressure limits. This resulted in undersized relief system piping, flare stacks and knock out drums in three out of five plants evaluated.

AUTHOR INDEX